BIOFILMS II

WILEY SERIES IN
ECOLOGICAL AND APPLIED MICROBIOLOGY

EDITED BY
Ralph Mitchell
Division of Applied Sciences
Harvard University

ADVISORY BOARD
Ilan Chet
Faculty of Agriculture
Hebrew University of Jerusalem

Madilyn Fletcher
Belle W. Baruch Institute for Marine
  Biology and Coastal Research
University of South Carolina

Peter Hirsch
Institut für Algemeine
Mikrobiologie Universität Kiel

David L. Kirchman
College of Marine Studies
University of Delaware

Kevin Marshall
School of Microbiology
University of New South Wales

James T. Staley
Department of Microbiology
University of Washington

David White
Institute for Applied Microbiology
University of Tennessee

Lily Y. Young
Center for Agricultural Molecular Biology
Cook College, Rutgers University

RECENT TITLES

THERMOPHILES: General, Molecular,
and Applied Microbiology
Thomas D. Brock, Editor, 1986

INNOVATIVE APPROACHES TO PLANT
DISEASE CONTROL
Ilan Chet, Editor, 1987

PHAGE ECOLOGY
Sagar M. Goyal, Charles P. Gerba, and
Gabriel Bitton, Editors, 1987

BIOLOGY OF ANAEROBIC
MICROORGANISMS
Alexander J.B. Zehnder, Editor, 1988

THE RHIZOSPHERE
J.M. Lynch, Editor, 1990

BIOFILMS
William G. Characklis and Kevin C.
Marshall, Editors, 1990

ENVIRONMENTAL MICROBIOLOGY
Ralph Mitchell, Editor, 1992

BIOTECHNOLOGY IN PLANT DISEASE
CONTROL
Ilan Chet, Editor, 1993

ANTARCTIC MICROBIOLOGY
E. Imre Friedmann, Editor, 1993

EFFECTS OF ACID RAIN ON FOREST
PROCESSES
Douglas L. Godbold and Aloys
Hütterman, Editors, 1994

MICROBIAL TRANSFORMATION AND
DEGRADATION OF TOXIC ORGANIC
CHEMICALS
Lily Y. Young and Carl E. Cerniglia,
Editors, 1995

BACTERIAL ADHESION: Molecular and
Ecological Diversity
Madilyn Fletcher, Editor, 1996

EXTREMOPHILES: Microbial Life in
Extreme Environments
Koki Horikoshi and W.D. Grant,
Editors, 1998

WASTEWATER MICROBIOLOGY
2nd Edition
Gabriel Bitton, 1999

MICROBIA ECOLOGY OF THE OCEANS
David L. Kirchman, Editor, 2000

BIOFILMS II: Process Analysis and
Applications
James D. Bryers, Editor, 2000

BIOFILMS II
PROCESS ANALYSIS AND APPLICATIONS

EDITED BY

JAMES D. BRYERS

The Department of Chemical Engineering
The Center for Biomaterials
The University of Connecticut Health Center
Farmington, Connecticut

A JOHN WILEY & SONS, INC., PUBLICATION

New York • Chichester • Weinheim • Brisbane • Singapore • Toronto

This book is printed on acid-free paper. ∞

Copyright © 2000 by Wiley-Liss, Inc. All rights reserved.

Published simultaneously in Canada.

No part of this publication may be reproduced, stored in a retrieval system or transmitted in any form or by any means, electronic, mechanical, photocopying, recording, scanning or otherwise, except as permitted under Sections 107 or 108 of the 1976 United States Copyright Act, without either the prior written permission of the Publisher, or authorization through payment of the appropriate per-copy fee to the Copyright Clearance Center, 222 Rosewood Drive, Danvers, MA 01923, (978) 750-8400, fax (978) 750-4744. Requests to the Publisher for permission should be addressed to the Permissions Department, John Wiley & Sons, Inc., 605 Third Avenue, New York, NY 10158-0012, (212) 850-6011, fax (212) 850-6008, E-Mail: PERMREQ@WILEY.COM.

For ordering and customer service, call 1-800-CALL-WILEY.

Library of Congress Cataloging-in-Publication Data:

Biofilms II: analysis, process, and applications / edited by James D. Bryers.
    p. cm.—(Wiley series in ecological and applied microbiology)
 Includes index.
 ISBN 0-471-29656-2 (cloth : acid-free paper)
  1. Biofilms. 2. Biofilms—Industrial applications. I. Bryers, James D. II. Series.
 QR100.8.B55 B582 2000
 660.6'2—dc21   99-055613

Printed in the United States of America.

10 9 8 7 6 5 4 3 2 1

Create, teach, wonder, inquire, stimulate, provoke, amuse, excite, inspire, intrigue, and love

Thanks to my sons, Morgan and Andrew,
for all the above and so much more

CONTENTS

Preface	ix
Acknowledgments	xi
Contributors	xiii

PART 1. FUNDAMENTAL ASPECTS OF BIOFILM SYSTEMS — 1

1 Biofilms: An Introduction — 3
  James D. Bryers
2 Process Engineering — 13
  James D. Bryers
3 Biofilm Formation and Persistence — 45
  James D. Bryers and M. Fletcher
4 Molecular Ecology of Biofilms — 89
  Søren Molin, Alex T. Nielsen, Bjarke B. Christensen, Jens Bo Andersen, Tine R. Licht, Tim Tolker-Nielsen, Claus Sternberg, Martin C. Hansen, Cayo Ramos, and Michael Givskov

PART 2. BENEFICIAL ASPECTS OF BIOFILM SYSTEMS — 121

5 Biofilms in Porous Media — 123
  Edward J. Bouwer, Huub H. M. Rijnaarts, Al B. Cunningham, and Robin Gerlach
6 Innovative Biofilm Treatment Technologies for Water and Wastewater Treatment — 159
  Valentina Lazarova and Jacques Manem
7 Biofilms Applied to Hazardous Waste Treatment — 207
  Bruce E. Rittmann, Alex Otto Schwarz, and Pablo Baldomero Saez

PART 3. DETRIMENTAL ASPECTS OF BIOFILM SYSTEMS — 235

8 Biofouling of Engineered Materials and Systems — 237
  Gill G. Geesey and James D. Bryers
9 Biocorrosion — 281
  Gill G. Geesey, Iwona Beech, Philip J. Bremer, Barbara J. Webster, and D. Bret Wells

10	Biofilms—Impact on Hygiene in Food Industries	327
	Gun Wirtanen, Maria Saarela, and Tiina Mattila-Sandholm	
11	Biofilm Control by Antimicrobial Agents	373
	Philip S. Stewart, Gordon A. McFeters, and Ching-Tsan Huang	
	Index	407

PREFACE

Biofilm engineering is a new technical discipline that has emerged in response to the tremendous opportunities and significant costs resulting from pervasive microbial activity at interfaces. *Biofilms* (Characklis and Marshall, 1990) was the first text for an interdisciplinary research area beginning to focus on understanding and modulating the combination of biological and chemical reactions, as well as the transport and interfacial processes, affecting microbial accumulation and activity at surfaces. In 1990, the scientific community was just beginning to suspect that biofilm processes were, from a systems viewpoint, far more complex and fascinating than processes occurring in other microbial systems.

Interfacial properties and processes affect virtually every aspect of engineering and biotechnology. In *Biofilms*, Characklis and Marshall focused on the following: (1) protocols and methods to analyze biological phenomena at surfaces and interfaces, (2) the relationship between interfacial structure and biological function at interfaces, and (3) the advancement of process engineering models for fundamental biological/chemical interactions at surfaces. *Process analysis*—a systematic examination that combines essential thermodynamic and transport phenomena with process kinetics and stoichiometry to describe complex, interconnected processes—was used by Characklis and Marshall to unite the diverse fundamental disciplines necessary to understand biofilm phenomena.

Engineers and scientists analyze problems on a continuum of scales ranging from microscale to macroscale. Since the 1980s, engineering analysis of biofilm processes has generally been accomplished at the *mesoscale* ($10^{-3} \rightarrow 10$ meters). Studies at cellular or molecular dimensions (i.e., the *microscale*, $\leq 10^{-3}$ meters) have generally been the domain of biologists and chemists. Biofilms in environmental systems (e.g., an alpine stream) and industrial systems (e.g., a recirculating cooling tower system) are *macroscale* phenomena (>10 meters). In *Biofilms*, Characklis and Marshall integrated the microscale and mesoscale phenomena and introduced methods for scaling these processes to the macroscale. They related the properties and functional characteristics of the biofilm at the mesoscale (e.g., heat exchange tube) to the cell and interface properties at the microscale (e.g., the biofilm), the contacts and connectivity of the cells with the interface, extracellular components, and the surrounding environment. Critical factors in the design and operation of biofilm reactors at the macroscale were also discussed.

One barrier in the past to further development of biofilm science and technology was an inability to integrate the diverse disciplines required to understand microbial processes at surfaces, especially in large industrial and environmental systems. The interdisciplinary nature of interfacial microbial processes precluded mounting a coherent attack until microbiologists, chemists, and process engineers began to jointly share their expertise to address these complex systems. *Biofilms* went a long way in facilitating the merger of these two disciplines. Members of the biofilm community owe a tremendous debt of gratitude to the vision that Kevin Marshall and Bill Characklis have always displayed.

THE NEED FOR A SEQUEL TO *BIOFILMS*

Biofilm processes profoundly impact industrial productivity and competitiveness. Because of their pervasive effects on water quality, power generation, energy efficiency, human and animal health, and product quality, the detrimental effects of biofilms are observed in many, if not most, industries. While in some circumstances biofilms can be extremely costly to industry and harmful to health, they can also be highly beneficial in other applications. Indeed, the opportunities presented by biofilm systems for process innovation, analytical diagnosis, instrumentation development, biotechnological advances, and environmental benefits are tremendous. As a consequence of the combined interdisciplinary research efforts in biofilm science and engineering, major advances in biofilm diagnostics, ecology dynamics, genetic processes, and transport phenomena have emerged in the past decade sufficient to warrant a sequel to *Biofilms*.

As in *Biofilms*, *Biofilms II* combines the fundamental principles of mathematical models, conservation of mass and energy, and hydrodynamics with the basics of biochemistry and molecular microbiology to explain the rapid advances in biofilm science and technology. *Biofilms II* has distilled into two intensive chapters (chapters 2 and 3) the fundamentals of process analysis: stoichiometry, kinetics, reactor design, and transport phenomena (mass, momentum, and energy transport). These chapters not only reiterate the fundamentals of the "process analysis" or "systems engineering" approach, but they also apply this approach to accommodate recent discoveries in biofilm processes such as plasmid gene retention and transfer, cell:cell signaling control of adhesion and gene transfer, and biofilm structural heterogeneity.

Biofilms II presumes either that the audience is familiar with the basics of general microbiology and concepts of bacterial growth and replication, or that readers will be able to glean this information from the original *Biofilms* text. *Biofilms II* does provide an entirely new treatment of biofilm ecology (Chapter 4) that highlights the significant advances in genetics and molecular biology, as applied to biofilm ecosystems. As in the first tome, *Biofilms II* provides a number chapters that detail real systems where biofilms are either used intentionally for a beneficial application (chapters 5, 6, and 7) or where undesired biofilms cause severe damage (chapters 8, 9, 10, and 11).

Biofilms II is intended as a stand-alone extension of *Biofilms* and as such its audience will be senior-level undergraduate students, graduate students, academic and industrial scientists, and engineers. Obviously, *Biofilms II* is intended for academic and industrial scientists and engineers specifically engaged in biofilm research and education. However, this book is also intended for any scientist or engineer, engaged in any area of biotechnological research, that wishes to increase his or her understanding of the impact of complex microbial systems.

James D. Bryers

ACKNOWLEDGMENTS

Mine has been an easy task of assembling the intellectual quality of major researchers in the biofilm community. Advances in biofilm research that include topics ranging from oligonucleotide probe diagnosis of biofilm populations and their specific activities to biofilms in the food sector, from cell:cell signaling control of plasmid transfer and concomitant biofilm initiation to biofouling and biocorrosion have been discussed. Once again, the reader will develop a sense of the breadth and depth of the advances in biofilm research. There are a select group of researchers around the world that have made significant major contributions to the understanding of biofilm processes. My only regret is that I could not include in this single book every bright star in the biofilm universe. But I can extend my gratitude to the following persons whose research creativity has had a major influence on my research, this text, and understanding biofilms. They are Bill Characklis, Kevin Marshall, Henk Busscher, Madilyn Fletcher, Bjorn Christensen, Poul Harremoes, Willi Gujer, Oskar Wanner, Mark van Loosdrecht, Peter Wilderer, Toyo Tanaka, Jean-Claude and Marie Clare Block, Allan Hamilton, Geoffrey Hamer, Peter Wilderer, Keith Cooksey, Henny van der Mei, Gill Geesey, Harry Bungay, Mark van Loosdrecht, Bruce Rittmann, Irving Dunn, Julian Wimpenny, David Davies, and (the original biofilm engineers) Michael Trulear, Michael Nimmons, and Nicholas Zelver.

Pragmatically, someone has to take the chaos of 11 manuscripts and make a bouillabaisse from the ingredients. My thanks to the editors at John Wiley & Sons who made this book happen including Robert M. Harington, Luna Han, and Kristin Cooke Fasano.

For every fresh idea. For every calm word. For every wide-eyed stare of wonder. When no one else questioned the status quo, there was this one voice that kept penetrating the dogma. "Why is this so? What causes that?" For that sense of community, obligation, persistence, intelligence, and caring . . . what can I say but thank you, Alma.

In the 1970s, the concept that bacteria in a biofilm were "operating" at conditions that were different than the bulk fluid environment was blasphemy; the concept that biofilm formation was the "net" result of several physical, chemical, and biological processes, was startling. Concepts are established and, as science progresses, established concepts must give way to new findings. The one-dimensional slab model of biofilm, consisting of equally active bacteria uniformly distributed throughout the biofilm, has given way to a "toadstool" caricature of a biofilm, where the bulk fluid can flow through a perforated biofilm structure. In addition, research now suggests that bacteria in the Y2K are capable of communicating and controlling this new perforated structure. In the future, the paradigm of heterogeneous bacterial biofilms will surely be replaced, perhaps with the detailed nucleotide sequence information for a biofilm-determining gene. What will be the next generation's new gospel? Where will the study of complex biofilm ecosystems go next? Bill Characklis once said, "Any study of the biofilm phenomena that does not simultaneously consider chemistry, microbiology, surface science, transport phenomena and molecular biology is . . . incomplete." No matter what the new chapters of the biofilm saga are, the message in *Biofilms* and reiterated here in *Biofilms II* is that what is always needed is a true interdisciplinary scrutiny of biofilms.

In the words of Ralph Waldo Emerson, "We live amid surfaces and the true art is to skate well on them." Well, what are you waiting for?

James D. Bryers

CONTRIBUTORS

Jens Bo Andersen Department of Microbiology, Technical University of Denmark, DK2800 Lyngby, Denmark

Iwona Beech School of Pharmacy, Biomedical and Physical Sciences, Portsmouth University, Portsmouth, P01 2DT, United Kingdom

Edward J. Bouwer Department of Environmental Engineering and Geography, Johns Hopkins University, Baltimore, MD 21218

Philip J. Bremer Crop and Food Research, University of Otago, Dunedin, New Zealand

Bjarke B. Christensen Department of Microbiology, Technical University of Denmark, DK2800 Lyngby, Denmark

Al B. Cunningham Center for Biofilm Engineering, Montana State University, Bozeman, MT 59715

Gill G. Geesey Department of Microbiology, Montana State University, Bozeman, MT 59717

Robin Gerlach Center for Biofilm Engineering, Montana State University, Bozeman, MT 59715

Michael Givskov Department of Microbiology, Technical University of Denmark, DK2800 Lyngby, Denmark

Martin C. Hansen Department of Microbiology, Technical University of Denmark, DK2800 Lyngby, Denmark

Ching-Tsan Huang Department of Agricultural Chemistry, Natioanl Taiwan University, Taipei, Taiwan 10674, ROC

Valentina Lazarova Lyonnaise des Eaux, CIRSEE, 38 rue du président Wilson, 78230 Le Pecq, France

Tine R. Licht Department of Microbiology, Technical University of Denmark, DK2800 Lyngby, Denmark

Gordon A. McFeters Center for Biofilm Engineering and Department of Microbiology, Montana State University, Bozeman, MT 59717-3980

JACQUES MANEM   Societe Lyonnaise des Eaux-Dumez, 38 rue du Président Wilson, Le Pecq, 78230, France

TIINA MATTILLA-SANDHOLM   Biotechnology and Food Research, VTT (Finnish Federal Research Institute), P.O. Box 1501, Espoo 02044 VTT, Finland

SØREN MOLIN   Department of Microbiology, Technical University of Denmark, DK-2800 Lyngby, Denmark

ALEX T. NIELSEN   Department of Microbiology, Technical University of Denmark, DK2800 Lyngby, Denmark

CAYO RAMOS   Department of Microbiology, Technical University of Denmark, DK2800 Lyngby, Denmark

HUUB H. M. RIJNAARTS   Department of Environmental Biotechnology, TNO Institute of Environmental Sciences, Energy Research and Process Innovation, P.O. Box 342, 7300 AH Apeldoorn, The Netherlands

BRUCE RITTMANN   Department of Civil Engineering, Northwestern University, Evanston, IL 60208-3109

MARIA SAARELA   VTT Biotechnology and Food Research, Tietotie 2, Espoo, Finland

PABLO BALDOMERO SAEZ   Catholic University of Chile, Santiago, Chile

ALEX OTTO SCHWARZ   Catholic University of Chile, Santiago, Chile

CLAUS STERNBERG   Department of Microbiology, Technical University of Denmark, DK2800 Lyngby, Denmark

PHILIP S. STEWART   Department of Chemical Engineering, Montana State University, Bozeman, MT 59717

TIM TOLKER-NIELSEN   Department of Microbiology, Technical University of Denmark, DK2800 Lyngby, Denmark

BARBARA J. WEBSTER   Industrial Research Ltd., Lower Hutt, New Zealand

D. BRET WELLS   Industrial Research Ltd., Auckland, New Zealand

GUN WIRTANEN   VTT Biotechnology and Food Research, Tietotie 2, Espoo, Finland

PART 1

FUNDAMENTAL ASPECTS OF BIOFILM SYSTEMS

BIOFILMS: AN INTRODUCTION

JAMES D. BRYERS

The University of Connecticut Health Center, Farmington, Connecticut

---

"Begin at the beginning," the King said gravely, "and go on 'till you come to the end; then stop."

—Lewis G. Carroll, *Through the Looking Glass*

1.1 A ROSE BY ANY OTHER NAME

In 1943, Zobell observed microbial cells "attached in layers" to bottle walls and that added glass rods increased the biological activity of batch suspended cultures. Atkinson and coworkers (1964, 1967) coined the term *microbial or biological film* to represent the gelatinous layer of cells and their adherent by-products on bioreactor vessel walls. Topiwala and Hamer (1971) and Howell et al. (1972) referred to mucilaginous layers of bacterial cells and their exopolymers as "wall growth." Characklis (1973a,b) provided an extensive two-part literature review on the basic fundamentals and practical implications of "microbial slimes." Atkinson (1964) and Atkinson and co-workers (1967) and Harremoës (1977) applied heterogeneous catalyst mathematics to describe simultaneous mass transport and biological reaction within "microbial films." A consensus of the leaders in biofilm research in 1984 termed a *biofilm* as a collection of microorganisms, predominantly bacteria, enmeshed within a three-dimension (3-D) gelatinous matrix of extracellular polymers secreted by the microorganisms (Marshall, 1984). In the first edition of *Biofilms*, Characklis and Marshall (1990) require one entire page to define a biofilm. To summarize, a biofilm is "a surface accumulation, which is not necessarily uniform in time or space, that comprises cells immobilized at a substratum and frequently embedded in an organic polymer matrix of microbial origin."

In an aqueous environment, a support, termed a *substratum*, is immediately biased by dissolved organic molecules and macromolecules that adsorb rapidly from the liquid phase. Bacterial cells present in the fluid contact the substratum by a variety of transport mechanisms.

Biofilms II: Process Analysis and Applications, Edited by James D. Bryers.
ISBN 0-471-29656-2   Copyright © 2000 Wiley-Liss, Inc.

Just prior to or upon arriving at a substratum, bacterial cells alter certain gene expression patterns; enzymes and pathways are altered (induced or repressed) to create a phenotypically adherent cell. Once at the substratum, the cells can adsorb either reversibly or irreversibly. Provided the cells remain at the surface sufficient time, they will secrete extracellular polymers that attach the cells tenaciously to the substratum. Attached cells metabolize prevailing energy and carbon sources (either dissolved within the surrounding fluid, adsorbed to the substratum surface, or existing as a constituent of the substratum itself), reduce terminal electron acceptors, grow, replicate, and produce insoluble extracellular polysaccharides, thus accumulating an initial viable biofilm community. Inert particles, bacterial cells of the same or different species, and higher life forms (e.g., algae, amoeba, protozoa) continue to be recruited from the fluid and incorporated into the biofilm community.

As the biofilm bacterial communities mature, the adherent populations oxidize electron donors and reduce existing terminal electron acceptors in an order of decreasing redox potential. As the biofilm community becomes denser and thicker, the penetration depth of one electron acceptor overlaps that of another; thus in a natural ecosystem, a succession of different microbial populations can be established. Biofilm layers near oxygen-saturated water harbor aerobic heterotrophic or autotrophic activity, reducing the available oxygen. As the distance from the aerobic interface increases and oxygen is depleted, a sequence of terminal electron acceptors are utilized by specific microbial populations: facultative denitrifiers using nitrate and nitrite; anaerobic sulfate reducers using sulfate; fermentative microbes partially reducing organic carbon compounds; and finally, anaerobic methane producers. Depending on the balance between mass transfer rates and microbial reaction rates, biofilms can develop stratified ecosystems, on a microscopic scale, similar to those observed in lake or marine sediments. In engineered systems, highly active aerobic activity within a biofilm can completely deplete oxygen concentrations over distances of 10–20 μm, creating anaerobic layers capable of supporting obligate, anaerobic sulfate-reducing bacteria. Consequently, it should come as no surprise to discover anaerobic sulfate reduction and hydrogen sulfide corrosion of the metallic surfaces in heat exchange systems that utilize highly aerated cooling water to condense steam.

As a result of hydrodynamic forces and stresses exerted by replication, there can be a continual erosion of cells and extracellular material from the biofilm back to the bulk fluid. A more random stochastic process known as *sloughing* can occur where either large sections or the entire biofilm become displaced from the substratum and enter the liquid.

1.2 THE IMPACT OF BIOFILM FORMATION

Biofilms are as versatile as they are ubiquitous. Intentional use of biofilms can serve many benefits as in the water and waste treatment industry, in bioremediation applications, and in industrial biotechnology.

1.2.1 Beneficial Biofilms

Benefits afforded by biofilms in a continuous reactor situation arise chiefly because the cell population is immobilized and thus the "residence time" of the cells in the reactor is independent of the fluid phase residence time. In continuous suspended culture bioreactors (i.e., chemostats), the mean residence time of the system cannot be less than the generation time of the bacterial species, otherwise cells do not have sufficient time to replicate within the reactor and are eventually "diluted" from the system.

Immobilized and biofilm-bound cells remain in a continuous reactor system independent of the fluid phase, thus the mass loading of limiting substrate (or influent pollutant in the case of a wastewater treatment reactor) can be increased well beyond the growth rate limit imposed on suspended cultures. Consequently, immobilized-cell or biofilm reactors can provide (*a*) added volumetric reactivity, (*b*) more stable operating performance, (*c*) an inherent ease in biomass:fluid separation, and (*d*) the prospect of staging different bioconversion processes in sequential reactors. Due to these inherent advantages, the use of biofilm reactors is not confined only to bacterial cells, but also comprises plant and animal cell applications.

Bacterial biofilm reactors are employed either in commodities production or in wastewater treatment applications. Biofilm reactors have been reportedly used to produce acetic acid (Park and Toda, 1992), L-lysine (Velizarov et al., 1992), gluconic acid (Moresi et al., 1991), kojic acid (Kwak and Rhee, 1992a, b), ethanol (Tzeng et al., 1991; Demuyakor and Ohta, 1992), and in the epoxidation of propene (Kovalenko and Sokolovskii, 1992). Such biofilm reactors are operated either as packed- or fluidized-bed reactor systems, with cells either attached to an inert support particle or artificially immobilized within a gel matrix.

One major application that relies on a microbial culture's ability to form biofilms is wastewater treatment. Biofilm reactor configurations, applied in both pilot- and full-scale wastewater treatment, include packed bed "trickling filters," high rate plastic media filters, rotating biological contactors, fluidized-bed biofilm reactors, and membrane immobilized cell reactors. Examples of biofilm reactors employed for wastewater treatment recently reported in the literature include polychlorinated hydrocarbon degradation (Earthapure and Vogel, 1991), toluene degradation (Arcangeli and Arvin, 1992), denitrification (Coelhoso et al., 1992; Lemoine et al., 1992), cadmium removal (Scott and Karanjker, 1992), anaerobic butyrate degradation (Zellner and Gevcke, 1991), nitrification (Sumino et al., 1992), glyphosphate degradation (Hallas et al., 1992), anaerobic propionate degradation (Heppner et al., 1992), phenolic wastewaters (Anselmo and Novais, 1991), uranium removal (Macaskie et al., 1992), and anaerobic carbon removal (Gonzalez et al., 1992; Fennel et al., 1992).

1.2.2 Detrimental Biofilms

Unintentionally formed biofilms can create such detriments as biofouling of heat exchange systems and marine structures; microbial induced corrosion of metal surfaces or the deterioration of dental surfaces; contamination of household products, food preparations and pharmaceuticals; and the infection of short- and long-term indwelling biomedical implants and devices. Such detriments can range in severity from being a mere nuisance to being life threatening.

Uncontrolled biofilm formation within natural, engineered, and biomedical systems can create numerous detriments as detailed by Marshall, 1984 and Bryers, 1991. Detriments arise by way of a biofilm's influence on the transport of mass, momentum, and energy. Detriments attributed directly to bacterial biofilms include (*a*) material deterioration and corrosion (refer to Chapter 9), (*b*) increase in both frictional and heat transfer resistances (refer to Chapter 10), (*c*) attachment to and infection of biomedical implant devices (Reid et al., 1991; Quirynen et al., 1991; Pratt-Terpstra et al., 1991; Brokke et al., 1991; Busscher et al., 1991; Harkes et al., 1991; Chang and Merritt, 1992), and (*d*) operational problems that plague lab-scale and full-scale bioreactors (Bryers, 1991).

1.3 BIOFILMS 1960s–1990s VERSUS BIOFILM-Y2K

In 1965, the impression of a biofilm structure was that seen in Figure 1.1: a slab or layer that resembled the geometry of its underlying substratum. Biofilms were considered uniform in structure although permeable to the transport of solute molecules. Bacterial populations were assumed uniformly distributed throughout the biofilm, despite observations of a real biofilm structure (Fig. 1.2). Beginning in the 1960s and continuing to today, most researchers chose to model biofilm systems as if they were uniform in biomass density, cell populations, and physical properties (Figs. 1.3).

Although many researchers in the 1970s and 1980s recognized that biofilms were structurally irregular (Fig. 1.4; Atkinson et al., 1967), that the gelatinous viscoelasticity of a biofilm led to increases in fluid frictional resistance (Picologlou et al., 1980), that biofilm

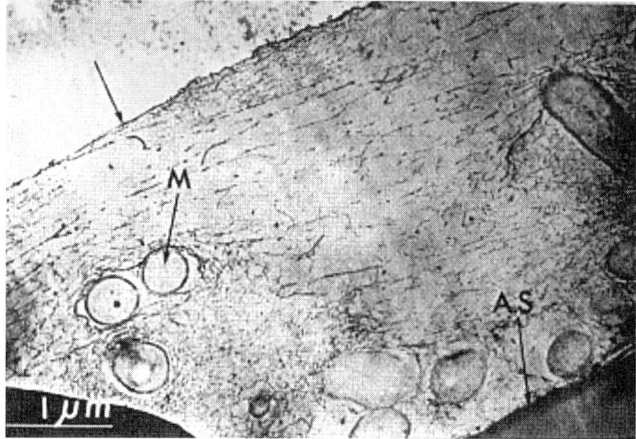

Figure 1.1 Transmission electronic micrograph of a mixed culture biofilm (Jones et al., 1969).

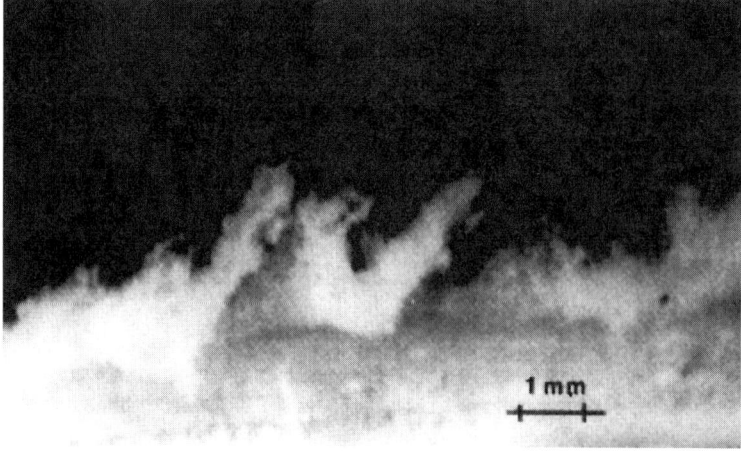

Figure 1.2 Biofilm growing on surface of a gas permeable membrane (Characklis and Wilderer, 1989).

1.3 BIOFILMS 1960s–1990s VERSUS BIOFILM-Y2K

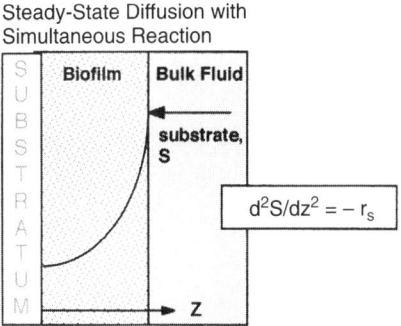

Figure 1.3 Classic mathematical concept of a steady state uniformly reactive bacterial biofilm consuming a single growth-limiting substrate, S.

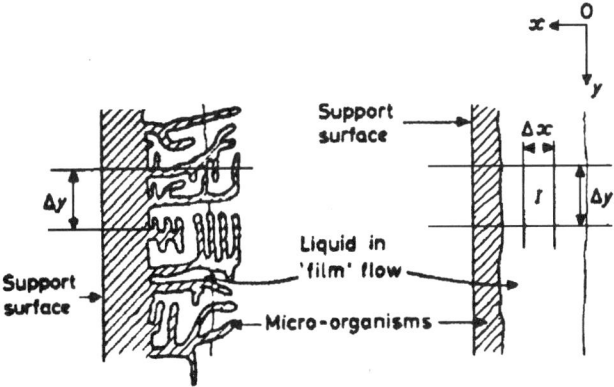

Figure 1.4 Concepts of biofilm reality and simulation in 1967 (Atkinson et al., 1967).

filamentous structure was set by hydrodynamics that subsequently determined the biofilm's influence on frictional resistances (Bryers, 1980), and that biofilm detachment and sloughing were complex spatially local processes that could not be modeled as a uniform "surface erosion" (Howell and Atkinson, 1976), many of these very topics are now resurfacing as the "hot" areas of biofilm research of the late 1990s, receiving significant attention and publicity.

What has transpired in the last 10 years is an immense explosion in the analytical methods and diagnostic tools available to now quantify, at the micro- and now molecular-scale, many subtle biofilm processes. The advent of noninvasive microscopy (single photon confocal and two-photon excitation-laser scanning microscopy) in combination with a plethora of cytological stains, enzyme active fluorescent stains, oligonucleotide genetic probes, and immunofluorescent probes allows optical dissection of an intact active biofilm. Now we have the capacity to "film" a 3-D biofilm as it develops, depict covective flow through a "real" biofilm structure (Fig. 1.5), or map local specific populations or locate expression of local phenotypic activity (refer to Chapter 8). Ion-specific microprobes (tip sizes < 2 μm) and fluorescent stains responsive to pH, redox levels, or specific ions

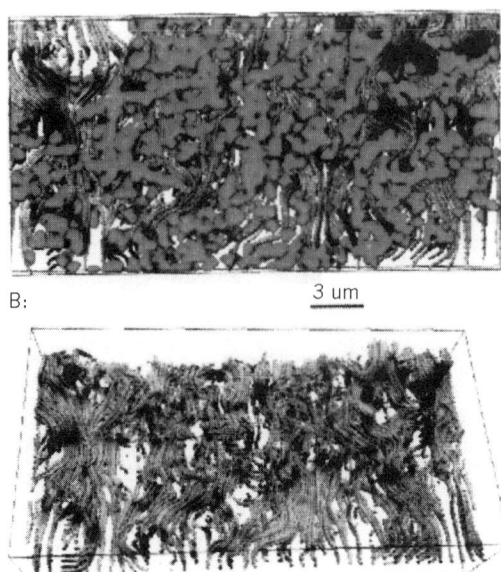

Figure 1.5 Propagational analysis of flow through a 3-D biofilm microstructure as recorded by confocal laser scanning microscopy. Flow is from front of picture to back (Singleton et al., 1997). Figure also appears in Color Figure section.

now allow for minimally disruptive estimations of local ion gradients with a biofilm. Noninvasive surface analytical tools (attenuated total reflectance Fourier transform infrared spectroscopy (ATR FT-IR) and surface enhanced Raman spectroscopy) coupled with destructive surface chemistry diagnostics such as x-ray photoelectron spectroscopy (XPS), time of flight secondary ion mass spectroscopy (ToF SIMS), atomic force microscopy (AFM), now provide dynamic analysis of the molecular interactions between cells and substratum, at the point of adhesion. Ten years ago, the biofilm community could mathematically describe local, instantaneous populations or substrate concentrations in a one-dimensional slab version of a biofilm (Fig. 1.3). Now, the ability to mathematically describe and simulate biofilm processes in three and four dimensions (space and time) (refer to Chapter 3) has increased concomitantly with the speed of computation of numerical solutions and applications of novel locally discrete modeling methods (e.g., cellular automata).

1.4 GOALS OF THIS BOOK

> Though this be madness, yet there is a method in't.
> —W. Shakespeare, *Hamlet*, Act II, Sc. 2.

The goal of *Biofilms II* is to provide new concepts, experimental methods, and mathematical techniques for the analysis and control of biofilms. *Biofilms I* provided a detailed framework for the understanding of complex biofilm processes (i.e., process analysis) that combines the fundamentals of transport phenomena, reaction kinetics, and reactor design with those of microbiology and chemistry. *Biofilms I* provided a means to approach any biofilm system, to understand biofilm impacts on system performance, and provided scientists and engineers from diverse backgrounds a common venue and language for further interdisciplinary research.

Biofilms II has distilled the fundamentals of process analysis—stoichiometry, kinetics, reactor design, transport phenomena (mass, momentum, and energy transport)—provided originally in six chapters into two intensive chapters. Chapters 2 and 3 not only reiterate the fundamentals of the process analysis or systems engineering approach, but they also apply this approach to accommodate recent discoveries in biofilm processes, such as plasmid gene retention and transfer, cell:cell signaling control of adhesion or gene transfer, and biofilm structural heterogeneity.

Biofilms I provided in two chapters a much-needed synopsis of general microbiology and biofilm biological processes, always placed in the context of the systems approach, from which engineers could appreciate the depth of biological complexity posed by biofilms. *Biofilms II* presumes the audience is either familiar with the schematic cross-section of a generic bacteria and general concepts of bacterial growth and replication or will be able to garner this information from the first edition. *Biofilms II* does provide an entirely new treatment of biofilm ecology (Chapter 4) that embraces the significant advances in genetic and molecular analysis applied to microbial ecosystems. As in the first edition, *Biofilms II* provides several chapters that detail real systems where biofilms are used either intentionally for a beneficial application (Chapters 5–7) or where unintentional biofilms create severe detriments (Chapters 9–11).

REFERENCES

Anselmo, A. M., and J. M. Novais. 1991. Biological treatment of phenolic wastes—Comparison between free and immobilized cell systems. *Biotech. Lett.* 14: 239–244.

Arcangeli, J. P., and E. Arvin. 1992. Toluene biodegradation and biofilm growth in an aerobic fixed-film reactor. *Appl. Microbiol. Biotech.* 37: 510–517.

Atkinson, B. 1964. *Biochemical Reactors.* London: Pion Press.

Atkinson, B., E. L. Swilley, A. W. Busch, and D. A. Williams. 1967. *Trans. Inst. Chem. Eng.* 45: T257–T264.

Beech, I. B., C. C. Gaylarde, J. J. Smith, and G. G. Geesey. 1991. Extracellular polysaccharides from *Desulfovibrio desulfuricans* and *Pseudomomas fluorescens* in the presence of mild and stainless steel. *Appl. Environ. Microbiol.* 35: 65–71.

Benbouzid-Rollet, N. D., M. Conte, J. Guezennec, and D. Prieur. 1991. Monitoring of a *Vibrio natriegnes* and *Desulfovibrio vulgarius* marine aerobic biofilm on a stainless steel surface in a laboratory tubular flow system. *J. Appl. Bacteriol.* 71: 244–251.

Boopathy, R., and L. Daniels. 1991. Effect of pH on anaerobic mild steel corrosion by methanogenic bacteria. *Appl. Environ. Microbiol.* 57: 2104–2108.

Bremer, P. J., and G. G. Geesey. 1991. Laboratory-based model of microbiologically induced corrosion of copper. *Appl. Environ. Microbiol.* 57: 1956–1962.

Brokke, P., J. Dankert, J. Carballo, and J. Feijen. 1991. Adherence of coagulase-negative staphylococci onto polyethylene catheters *In vitro* and *In vivo:* A study on the influence of various plasma proteins. *J. Biomat. Appl.* 5: 204–226.

Bryant, R. D., W. Jansen, J. Boivin, and E. J. Laishley. 1991. Effect of hydrogenase and mixed sulfate-reducing bacterial populations on the corrosion of steel. *Appl. Environ. Microbiol.* 57: 2804–2809.

Bryers, J. D. 1991. Understanding and controlling detrimental bioreactor biofilms. *Trends in Biotechnol.* 9: 422–426.

Bryers, J. D. 1980. Dynamics of early biofilm formation in a turbulent flow system, Ph.D. Dissertation, Rice University, Houston, TX.

Busscher, H. J., H. C. van der Mei, and J. M. Schakenraad. 1991. Analogies in the two dimensional spatial arrangement of adsorbed proteins and adhering bacteria: Bovine serum albumen and *Streptococcus sanguis. J. Biomat. Sci. Polym. Ed.* 3: 85–94.

Chang, C. C., and K. Merritt. 1992. Microbial adherence on poly(methyl methacrylate) surfaces. *J. Biomed. Mater. Res.* 26: 197–207.

Characklis, W. G., and K. C. Marshall. 1990. *Biofilms.* New York: Wiley-Interscience.

Characklis, W. G., and P. A. Wilderer. 1989. *Structure and Function of Biofilms.* New York: J. Wiley & Sons.

Characklis, W. G. 1973a. Attached microbial growths: I. Attachment and growth. *Water Research* 7: 1113–1127.

Characklis, W. G. 1973b. Attached microbial growths: II. Frictional losses due to microbial slimes. *Water Research* 7: 1249–1258.

Coelhoso, I., R. Bonaventura, and A. Rodrigues. 1992. Biofilm reactors: An experimental and modeling study of wastewater denitrification in fluidized bed reactors of activated carbon particles. *Biotech. Bioeng.* 40: 625–633.

Deckena, S., and K. H. Blotevogel. 1992. Fe_0 Oxidation in the presence of methanogenic and sulfate-reducing bacteria and its possible role in anaerobic corrosion. *Biofouling* 5: 287–294.

Demuyakor, B., and Y. Ohta. 1992. Promotive action of ceramics on yeast ethanol production and its relationship to pH, glycerol and alcohol dehydrogenase activity. *Appl. Microbiol. Biotech.* 36: 717–721.

Deshmukh, M. B., I. Akhtar, R. B. Srivastava, and A. A. Karande. 1992. Marine aerobic and anaerobic bacteria inducing corrosion of 304 stainless steel. *Biofouling* 6: 13–32.

Dowling, N. J. E., J. Guezennec, J. Bullen, B. Little, and D. C. White. 1992. Effect of photosynthetic biofilms on the open circuit potential of stainless steel. *Biofouling* 5: 315–322.

Earthapure, B. Z., and T. M. Vogel. 1991. Complete degradation of polychlorinated hydrocarbons by a 2-stage biofilm reactor. *Appl. Environ. Microbiol.* 57: 3418–3422.

Fennel, D. E., S. E. Underhill, and W. J. Jewell. 1992. Methanogenic attached film reactor development and biofilm characteristics. *Biotech. Bioeng.* 40: 1218–1232.

Franklin, M. J., and D. C. White. 1991. Biocorrosion. *Current Opinion Biotechnol.* 2: 450–456.

Gonzalez, G., F. Ramirez, and O. Monroy. 1992. Development of biofilms in anaerobic reactors. *Biotech. Lett.* 14: 149–154.

Hallas, L. E., W. J. Adams, and M. A. Heitkamp. 1992. Glyphosate degradation by immobilized bacteria—Field studies with industrial wastewater effluent. *Appl. Environ. Microbiol.* 58: 1215–1219.

Harkes, G., J. Feijen, and J. Dankert. 1991. Adhesion of *Escherichia coli* on to a series of poly(methacrylates) differing in charge and hydrophobicity. *Biomaterials* 12: 853–860.

Harremoës, P. 1977. Half order reactions in biofilm and filter kinetics. *Vatten* 2: 122–143.

Heppner, B., G. Zellner, and H. Dickmann. 1992. Start-up and operation of a propionate-degrading fluidized-bed reactor. *Appl. Environ. Microbiol.* 36: 810–816.

Howell, J. A., C. T. Chi, and U. Pawlowsky. 1972. Effect of wall growth on scale-up problems and dynamic operating characteristics of the biological reactor. *Biotech. Bioeng.* 14: 253–265.

Howell, J. A., and B. Atkinson. 1976. Sloughing of microbial film in trickling filters. *Water Research* 10: 307–315.

Jones, H. C., I. L. Roth, and W. M. Sanders. 1969. Electron microscopic examination of a slime layer. *J. Bacteriol.* 99: 316–322.

REFERENCES

Kovalenko, G. A., and V. D. Sokolovskii. 1992. Epoxidation of propene by microbial cells immobilized on inorganic supports. *Biotech. Bioeng.* 39: 522–528.

Kwak, M. Y., and J. S. Rhee. 1992a. Controlled mycelial growth for kojic acid production using ca-alginate immobilized fungal cells. *Appl. Microbiol. Biotech.* 36: 578–583.

Kwak, M. Y., and J. S. Rhee. 1992b. Cultivation characteristics of immobilized *Aspergillus oryzae* for kojic acid production. *Biotech. Bioeng.* 39: 903–906.

Lemoine, D., T. Jouenne, and G. A. Junter. 1992. Biological denitrification of water in a 2-chambered immobilized cell bioreactor. *Appl. Microbiol. Biotech.* 36: 257–264.

Marshall, K. C. (Ed.). 1984. *Microbial Adhesion and Aggregation.* Berlin, Germany: Springer-Verlag.

Macaskie, L. E., P. J. Clark, J. D. Gilbert, and M. R. Tolley. 1992. The effect of aging on the accumulation of uranium by a biofilm bioreactor, and promotion of uranium deposition in stored biofilms. *Biotech. Lett.* 14: 525–530.

Moresi, M., E. Parente, and A. Mazzatura. 1991. Effect of dissolved oxygen concentration on repeated production of gluconic acid by immobilized mycelia of *Aspergillus niger. Appl. Microbiol. Biotech.* 36: 320–323.

Park, Y. S., and K. Toda. 1992. Multi-stage biofilm reactor for acetic acid production at high concentration. *Biotech. Lett.* 14: 609–612.

Picologlou, B. F., N. Zelver, and W. G. Characklis. 1980. Biofilm growth and hydraulic performance. *J. Hydr. Div., ASCE* 160(HY5): 733–741.

Pratt-Terpstra, I. H., J. Mulder, A. H. Weerkamp, J. Feijen, and H. J. Busscher. 1991. Secretory IgA adsorption and oral streptococcal adhesion to human enamel and artificial solid substrata with various surface free energies. *J. Biomater. Sci. Polym. Ed.* 2: 239–253.

Quirynen, M., M. Marechal, D. Van Steenberghe, H. J. Busscher, and H. C. van der Mei. 1991. The bacterial colonization of intra-oral hard surfaces *In vivo:* Influence of surface free energy and surface roughness. *Biofouling* 4: 187–198.

Reid, G., H. S. Beg, C. A. K. Preston, and L. A. Hawthorn. 1991. Effect of bacterial, urine, and substratum surface tension properties on bacterial adhesion to biomaterials. *Biofouling* 4: 171–176.

Scott, J. A., and Karanjkar, A. M. 1992. Repeated cadmium biosorption by regenerated *Enterobacter aerogenes* biofilm attached to activated carbon. *Biotech. Lett.* 14: 737–740.

Singleton, S., R. Treloar, P. Warren, G. K. Watson, R. Hodgson, and C. Allison. 1997. Methods for microscopic characterization of oral biofilms. *Advances in Dental Research* 11(1): 133–149.

Sumino, T., H. Nakamura, N. Mori, Y. Kawaguchi, and M. Tada. 1992. Immobilization of nitrifying bacteria in porous pellets of urethane gel for removal of ammonium nitrogen from wastewater. *Appl. Microbiol. Biotech.* 36: 556–560.

Topiwala, H. H., and G. Hamer. 1971. Effect of wall growth in steady state continuous culture. *Biotech. Bioeng.* 8: 919–922.

Tzeng, J. W., L. S. Fan, Y. R. Gan, and T. T. Hu. 1991. Ethanol production using immobilized cells in a multistage fluidized bed bioreactor. *Biotech. Bioeng.* 38: 1253–1258.

Velizarov, S. G., E. I. Rainina, A. P. Sinitsyn, S. D. Varfolomeyev, V. I. Lozinsky, and A. L. Zubov. 1992. Production of L-lysine by free and PVA-cryogel immobilized cornebacterium-glutamicium cells. *Biotech. Lett.* 14: 291–296.

Zellner, G., M. Gevcke, E. C. Demacario, and H. Dickmann. 1991. Population dynamics of biofilm development during start-up of a butyrate-degrading fluidized-bed reactor. *Appl. Environ. Microbiol.* 36: 404–409.

Zobell, C. E. 1943. The effect of solid surfaces upon bacterial activity. *J. Bacteriol.* 46: 39–56.

PROCESS ENGINEERING

JAMES D. BRYERS
The University of Connecticut Health Center, Farmington, Connecticut

2.1 PROCESS ANALYSIS

Process analysis refers to the application of systematic methods to recognize, define, and clarify problems and to develop methodologies for their solution. Biofilm formation and persistence in both natural and engineered systems are governed by a collage of complex physical, chemical, and biological processes; each process is dependent on a unique set of system parameters. Process analysis applied to biofilm formation provides an integrated approach that incorporates microbial physiology, reaction engineering, and transport phenomena to understand, control, and exploit biofilm processes. Application of process analysis allows one to (*a*) interpret the operation of an existing biofilm system, (*b*) design new biofilm reactor systems, and (*c*) understand the complexities of natural biofilm systems. It is increasingly apparent that any research into biofilm processes that does not comprise microbiology, chemistry, and fundamental process engineering aspects is incomplete.

There are three main applications of biofilm process analysis with which we are ultimately concerned: (*a*) operation of existing biofilm reactors or other technical equipment, (*b*) design of new or modified biofilm reactors and technical equipment, and (*c*) analysis of biofilm processes. In the area of operations, both control and optimization of performance stand out as two of the main functions that require sophisticated analysis of biofilm processes. For example, computers must be instructed so that relations describing individual steps of biofilm processes can be combined into an overall biofilm control system; basic parameters in those relations (e.g., kinetic and stoichiometric coefficients) must be evaluated and qualitative observations (e.g., biofilm physical properties) must be quantified. For these and many related reasons, effective control and optimization of biofilm systems rest on sound process analysis.

The second task, biofilm reactor design, is more difficult. Actual biofilm process plant data are frequently not available beforehand and the engineer must rely on intuition.

Biofilms II: Process Analysis and Applications, Edited by James D. Bryers.
ISBN 0-471-29656-2 Copyright © 2000 Wiley-Liss, Inc.

However, when modifying existing equipment or designing new equipment similar to those already built, the engineer can draw more heavily on experience. As a consequence, construction of theoretical or semitheoretical mathematical models of the biofilm process frequently is a necessary prelude.

The final task, analyzing natural biofilm processes, has become extremely important. We frequently wish to exploit or control biofilm processes and, at the same time, minimize the human impact of them. Biofilm processes in open systems frequently possess a greater challenge to analyze than industrial or technological biofilm processes because of their oscillatory nature and the large number of variables that significantly affect them.

Here, we present the concepts of *process analysis* and the *rate concept* approach to mathematically describing complex reaction systems such as the formation and persistence of a biofilm. The basis of process analysis lies in the fundamental conservation equations of mass, energy, and momentum. Application of process analysis and the rate concept provides a systematic protocol with which to either interpret an existing reaction system or to design and operate a new system. First, the components of the material balance are defined paying particular attention to the role of *stoichiometry* and *kinetics*. Next, basic ideal reactor concepts are introduced within the process analysis framework. Finally, design, operation, and analysis of the most common laboratory biofilm study systems is provided.

Process analysis is simply a method of mathematically describing a complex phenomena as the net result of a number of individual fundamental processes. The rate concept more specifically assumes that changes in the state of a system with time can be systematically treated as the summation of the rates of the individual processes acting on the system. As implied, some degree of modeling of the biological system is posed, predictions of the model compared to observation, and the model verified or refuted. Thus, modeling must be, by nature, an evolutionary process in order to provide insight about a system.

Modeling is basically the scientific process of forming and testing a hypothesis. A model need not be mathematical nor need it exactly replicate reality. A road map is a model of a section of the Earth that allows one to successfully navigate from point A to B without representing every house or structure that actually exists. Models need not (and can not) simulate all details of a system to be of use. Models are simply our inherent tendency to explain unknown phenomena based on perceived principles. A good portion of basic science is predicated on proving or refuting such hypotheses. Mathematical models do provide a rigid structure with which to formulate concepts regarding a complex system. Mathematical models provide a means of verifying the goodness of a model by way of many statistical criteria. Important variables and system parameters can be easily identified and, through the use of dimensionless groups of such variables, mathematical models can expedite experimentation. And should the mathematical model prove successful in correlating observable data from a system at one set of conditions, then one has the ability to extrapolate, predicting the responses of the system under a variety of different conditions.

2.1.1 The Continuum Approach

Three basic types of models are: (1) transport or continuum models, (2) population models, and (3) empirical, statistical, or cellular automata models. Residence time distributions and synchronous growth are examples of population models whereas statistical regression of data sets is an example of empirical models. Random walk, cell motility, and two-dimensional (2-D) cell adhesion patterns are examples of cell automata models (for more details see Chapter 3). Transport models, the type considered here, essentially account for

2.1 PROCESS ANALYSIS

the changes of either mass, energy, or momentum in a continuum or defined control volume. Greater detail about transport phenomena is given in Chapter 3.

Consider the control volume illustrated in Figure 2.1. The control volume can exchange mass, energy, and momentum with its surroundings as indicated by the arrows. The conservation equation for mass equates the rate of accumulation of component C within the control volume to the net rate of C input into the control volume plus the summation of all process rates producing or consuming C mass in the volume. A verbal form of the conservation equation for the mass of a component C can be expressed as follows:

Net Rate of Accumulation of C in Control Volume $=$ $+$ (rate of transport of C into volume across area S_A) $-$ (rate of transport of C out of volume across area S_A) $+ \Sigma$ (all transformation processes generating or consuming C within volume V)

Mathematically, the general conservation equation for component C in the control volume can be written simplistically as,

$$V\frac{dC}{dt} = \overline{C} + \overline{\overline{R_c}} V \qquad (2.1)$$

where the left-hand side of the equation is the net rate of C accumulation (M/t) in the volume V, C is the net input rate of C into the volume (M/t), and $\overline{\overline{R_c}}$ is the sum of all rate processes consuming or producing the component C(M/t). This latter term combines two basic concepts: reaction stoichiometry and kinetics.

2.1.2 System Rates Versus Process Rates

In Equation (2.1), the net accumulation rate, VdC/dt, and the net rate of component C transport into and out of the volume, $\overline{C}$, are both termed *rates of change* or *system rates* in that they represent the observed change in the measured component, C. System rates are *extensive* quantities and are by definition system-specific and can not be correlated to fundamental parameters (e.g., temperature, velocity) describing the system. Unfortunately, all too often many try.

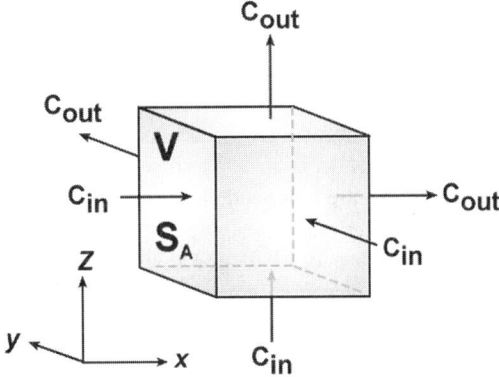

Figure 2.1 Control volume in 3-D Cartesian coordinate system.

Transport rates, the second term in Eq. (2.1), comprise *bulk transport* (movement of components due to the flow of the bulk liquid), *interfacial or interphase transport* (transport across an interface between two phases), and *intraphase transport* (transport within one phase due to a gradient—e.g., diffusion). One common criterion regarding a transport rate is that changes in a component's concentration (in the case of mass transport) are not due to a molecular change in the component.

Conversely in Eq. (2.1), the third term $\overline{\overline{R}}_c$ is a process rate. Process rates describe transformations that occur due to either chemical, biochemical, or biological reactions that either produce or consume the component. Process rates are fundamental intensive quantities in that they can be correlated to system parameters such as temperature, pressure, concentration, and velocity. Process rates and their correlation to such fundamental parameters are independent of the system in which they occur.

2.2 STOICHIOMETRY OF BIOLOGICAL PROCESSES

Stoichiometry defines the relationship that one chemical species has to another in a reaction. Kinetics involves the rate at which the reaction proceeds and the dependence of that rate upon the concentration of the chemical species and environmental factors such as temperature and pH. Stoichiometry and kinetics are so fundamental to the fields of chemistry and biology that a simple recounting of the titles of the books and journal articles written on the subject would consume this entire chapter. As a result, this chapter is limited to a general discussion of stoichiometry as it relates to single and multiple reaction systems and how it can be applied to reactions that are important to researchers and practitioners in the field of biofilm processes.

Four key concepts are developed in this section. The first is that the fundamental approach to stoichiometry described by Aris (1969) for chemical reactions can be easily extended to biological systems. Irvine (1980) and Irvine et al. (1980) established this approach earlier but without the modifications included herein (Irvine and Bryers, 1986). The second is that the molar-based approach necessary in so many chemical reaction engineering applications, must be replaced by a mass-based approach because of the uncertain chemical composition of both the reactants and products in biological products. Herbert (1976) developed this notion rather nicely but restricted his analysis to pure cultures and did not use the more formal approach of Aris (1969). The third concept is that electron equivalents or chemical oxygen demand (COD) should be used as the basis of measurement for all reactants and products. Implementing the concepts of system energetics described by Roels (1980), application of his principles reduces each reaction in a multiple reaction system to a simple mass balance with manageable stoichiometric coefficients, while allowing the user to obtain a macroscopic understanding of energetics. The fourth and final concept is that structure must be added to biological systems if necessary design and operational decisions are to be made in the field of biofilm processes.

Procedures established here are independent of the reactor types and configurations selected. That is, concepts developed here can be used to analyze both fixed film and suspended growth systems. The rates of formation that are represented simply represent one term in a material balance that is used to model a physical system.

Stoichiometry for a given set of components comprising a reaction establish the thermodynamics of the system. For example, consider the chemical oxidation of glucose,

$$C_6H_{12}O_6 + 6O_2 \rightarrow 6CO_2 + 6H_2O \qquad \mathbf{r} \qquad (2.2)$$

2.2 STOICHIOMETRY OF BIOLOGICAL PROCESSES

which states that 1 mole of glucose reacts with 6 moles of oxygen to form 6 moles of carbon dioxide and water. Mathematically, a reaction stoichiometry for N number of reaction components A can be expressed as an equation,

$$\sum_{i=1}^{N} v_i A_i = 0 \qquad (2.3)$$

where v_i is the stoichiometric coefficient of the i-th component and A_i is component i. The convention is that stoichiometric coefficients, v_i, for reactants have negative values whereas products have positive coefficients. Thus, in Eq. (2.2), the coefficients for glucose, oxygen, carbon dioxide, and water are -1, -6, $+6$, and $+6$, respectively. The term **r** in Eq. (2.2) is the rate of reaction and has units (moles of reaction/L^3-time).

2.2.1 General Development for a Single Reaction

A reaction can be written in the general form shown of an equation (Eqs. 2.3 or 2.4, following) and can be readily adapted to a wide variety of biological applications in terms of both mechanism development and kinetic implementation,

$$v_1 A_1 + v_2 A_2 + \ldots + v_{n-1} A_{n-1} + \ldots + v_n A_n = 0 \qquad \mathbf{r} \qquad (2.4)$$

where v_i = stoichiometric coefficient for the i-th component (moles), A_i = the i-th component, n = total number of components (reactants plus products), **r** = rate of reaction (moles/L^3 time), by convention we set $v_i < 0$ for reactants, $v_i > 0$ for products, $v_i = 0$ for conservative components.

The value for mechanism development is discussed later under multiple reaction systems. As far as kinetic implementation is concerned, all mass balances, whether they be for suspended growth or fixed film systems, require knowledge of the rate of formation of the ith component, r_i. If the notation defined by Eq. (2.4) is used, the *rate of appearance* of component i, on a molar basis, r_i, is defined in Eq. (2.5),

$$r_i = v_i \mathbf{r} \qquad (2.5)$$

where r_i has the units (moles-i/L^3-t). Because of the established convention, r_i takes on negative values for components that are consumed by the reaction and positive values for components that are produced.

The molar relationship given in Eq. (2.5) can be generalized to fit a mass relationship, as is shown in Eq. 2.6.

$$\omega_1 A_1 + \omega_2 A_2 + \ldots + \omega_n A_n = 0 \qquad \mathbf{r'} \qquad (2.6)$$

where $\omega_i = v_i MW_i$ and $\mathbf{r'}$ = the rate of reaction (mass/L^3-t). Thus, the rate of *appearance* of component i, is r_i (mass-i/L^3-t) = $\omega_i \mathbf{r'}$.

Equation (2.6) can be further simplified into a *unit-mass based relationship*, once a decision is made as to the specific component to be used ($A_i \rightarrow A_n$) as the normalizing factor. Typically, the rate-limiting reactant is used to normalize an equation. For example, assuming reactant A_1 is selected as the normalizing component, then Equation (2.6) in a unit mass-based relationship can be written as,

$$Y_1 A_1 + Y_2 A_2 + \ldots + Y_n A_n = 0 \qquad \mathbf{r^\cdot} \qquad (2.7)$$

where $Y_i = \upsilon_i MW_i / |\upsilon_1 MW_1|$ has the units

$$\left[\frac{\text{mass of component } A_i \text{ consumed or formed}}{\text{mass of component } A_1 \text{ consumed}} \right]$$

Typically, the rate-limiting reactant is selected as the normalizing component unless convention dictates otherwise.

The rate of formation of the i-th component in Eq. (2.7) is given by

$$\mathbf{r_i^{\cdot}} = Y_i \mathbf{r^{\cdot}} \tag{2.8}$$

where the sign depends on whether A_i is a reactant or product and the units of $\mathbf{r_i^{\cdot}}$ are given by

$$\left[\frac{\text{mass of component } A_i \text{ consumed or formed}}{\text{mass of component } A_1 \text{ consumed} \cdot L^3 \cdot \text{time}} \right].$$

As can be seen from this analysis, the rate of reaction, $\mathbf{r_i^{\cdot}}$, is tied directly to the component identified as the normalizing component. While the functional form of $\mathbf{r^{\cdot}}_i$ does not depend on the selection of A_1, the numerical value of the reaction rate constant does.

2.2.1.1 Example.
Busch (1971) summarized earlier studies involving glucose utilization by mixed cultures in dilute batch systems and found the overall reaction reported in Eq. (2.9) to be representative.

Molar basis

$$24\ C_6H_{12}O_6 + 59 O_2 + 17 NH_3 \rightarrow 17 C_5H_7NO_2 + 59 CO_2 + 110 H_2O \tag{2.9}$$
(glucose) \hspace{4cm} (cells)

This equation actually represents the net result of many intracellular and extracellular reactions but is presented here in overall form so that principles developed previously can be described simply. The molecular formula for cells is obviously approximate and would, of course, depend on the organism in question and the overall physiological state of the organisms involved (see Section 2.2.2.1).

Equation (2.2) is on a molar basis. On a mass of glucose basis, this equation can be rewritten as follows:

Glucose-mass basis

$$1\ C_6H_{12}O_6 + 0.44\ O_2 + 0.07 NH_3 \rightarrow 0.44\ C_5H_7NO_2 + 0.60 CO_2 + 0.46 H_2O \tag{2.10}$$

All stoichiometric coefficients have been expressed in terms of the grams of reactant (e.g., oxygen) or product (e.g., cells) consumed or produced for each gram of glucose utilized. As written, the total mass of substrates consumed on the left-hand side of the reaction (1.51) must, within round-off error, equal the total mass produced on the right-hand (1.51) side, and each of the stoichiometric coefficients can be thought of as being a yield coefficient. For example, 0.44 g of biomass would be produced for each gram of glucose consumed.

Equation (2.10) can also be rewritten as a carbon atom balance which considers only carbon-containing compounds; here all stoichiometric coefficients are reduced to the basis of a unit of glucose carbon mass consumed.

2.2 STOICHIOMETRY OF BIOLOGICAL PROCESSES

Unit Mass Glucose basis

$$1 \text{ G} \rightarrow 0.59 \text{ Cells} + 0.41 \text{ CO}_2 \qquad (2.11)$$

or in equation form,

$$0.59 \text{ Cells} + 0.41 \text{ CO}_2 - 1\text{G} = 0$$

where G = glucose-carbon (mass), Cells = cell-carbon (mass), CO_2 = carbon dioxide-carbon (mass), (−1) = stoichiometric coefficient for G (dimensionless), (+0.59) = stoichiometric coefficient for cells (cell mass formed per unit mass G consumed), (+0.41) = stoichiometric coefficient for CO_2, (mass CO_2 carbon formed per unit mass G consumed), and **r·** = rate of reaction (mass/L^3 time). Note that the sum of the stoichiometric coefficients in a unit-based mass stoichiometry must equal zero when the reaction is written in the form described by Eqs. (2.10) or (2.11). In Eq. (2.11), the mass of carbon consumed is equal to the mass of carbon produced. Maximum yield on this basis would be 0.59 g of cell carbon produced for each gram of glucose carbon consumed.

2.2.2 Estimating Stoichiometry for Biological Reactions

Verbally, the equation for microbial growth can be written as:

Carbon source + Energy source + Terminal electron acceptor + Nutrients $\underrightarrow{\text{microorganisms}}$ Microorganisms + End-products

It would be desirable to be able to write a quantitative equation of the same form, for any specific situation, no matter what the carbon source, energy source, or hydrogen acceptor (we exclude photosynthetic reactions, because they greatly complicate the issue and are only rarely used in biochemical operations). Fortunately, McCarty (1972) has devised a framework, based on the concept of oxidation-reduction half reactions, so that the single reaction stoichiometry can be derived.

As depicted in Figure 2.2, all nonphotosynthetic microbially mediated reactions consist of two components, one for synthesis and one for energy. Carbon in the synthesis component

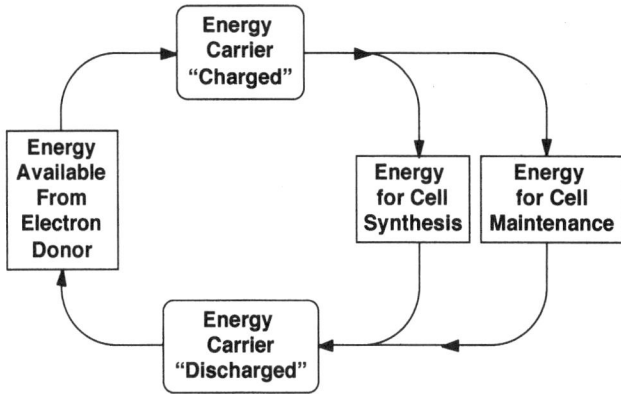

Figure 2.2 Theoretical distribution of microbial chemical energy.

is converted to cell material, whereas the carbon in the energy stream is eventually oxidized to carbon dioxide. Microbial reactions are oxidation-reduction reactions that involve the transfer of electrons from a donor to an acceptor.

McCarty (1972) has developed a framework to establish the stoichiometry of a single reaction that is the overall equivalent for numerous microbial bioconversions. The framework is based on the summation of three types of redox half reactions, one for cell material (R_c), one for the electron donor (R_d), and one for the electron acceptor (R_a). In the McCarty framework, all half reaction equations are written in terms of a single unit mole of electrons. Thus, the R_d half reaction states the amount of electron donor converted to a unit mole of electron equivalent. Similarly, R_c is the half reaction for the amount of cellular biomass required to generate one mole of electron equivalent, and R_a is the half reaction for the amount of electron acceptor required to form one mole of electron equivalent.

The overall stoichiometric equation (R) is just the sum of the individual half reactions,

$$R = R_d - f_e R_a - f_s R_c \qquad (2.12)$$

The term f_e is the fraction of the electron donor that is shuttled into the energy pathway and f_s represents the portion used for synthesis. In order for Eq. (2.12) to balance properly,

$$f_e + f_s = 1.0 \qquad (2.13)$$

McCarty and co-workers have fortunately catalogued numerous half reactions based on variations in electron acceptor and electron donor, with variations in the cellular biomass half reaction being dependent upon the nitrogen source (ammonia vs. nitrate). A small collection of these half-reactions are provided in Table 2.1 with more extensive collections available in Grady and Lim (1980) and McCarty (1972). To complete any biological reaction stoichiometry using this approach, the empirical formulas for cell biomass, electron donor, and terminal electron acceptor must be stipulated.

2.2.2.1 Cellular Biomass.
One of the oldest and most widely accepted empirical formulas for cell mass is $C_5H_7NO_2$ (Porges et al., 1956). Other formulas consisting of the same elements have also been used, but they all result in about the same electron equivalents, when based on a balanced stoichiometric equation representing the complete oxidation of the cell carbon.

While the elemental composition may remain relatively constant, the chemical composition of microbial cells at different culture "ages" and environmental conditions is not necessarily a constant. The chemical composition of microorganisms can be considerably affected by the growth rate and the nature of the growth-limiting nutrient (Herbert, 1958).

2.2.2.2 Electron Acceptors.
Composition of the electron acceptor depends on the environment in which the microbes are growing and the microbial culture itself. The environment and the culture is *aerobic* if the electron acceptor, oxygen, is delivered in sufficient quantities and rate. If the bacterial inocula are facultative anaerobes and prevailing conditions are anaerobic, then any ambient NO_3^{-2} can serve as an electron acceptor. If nitrate and oxygen are not available, oxidized forms of sulfur (sulfate, sulfite, and thiosulfate) can serve as terminal electron acceptors for sulfate-reducing bacterial species. Finally, should the surroundings be devoid of oxygen, nitrate, and sulfate, organic molecules can serve as

2.2 STOICHIOMETRY OF BIOLOGICAL PROCESSES

TABLE 2.1. Examples of Biological Oxidation Half Reactions

Reactions for Bacterial Cell Synthesis (R_c)

Ammonia as nitrogen source:
$$1/20\ C_5H_7NO_2 + 9/20\ H_2O = 1/5\ CO_2 + 1/20\ HCO_3^- \ 1/20\ NH_4^+ + e^-$$
Nitrate as nitrogen source:
$$1/28\ C_5H_7NO_2 + 11\ H_2O = NO_3^- + 5/27\ CO_2 + 29/28\ H^+ + e-$$

Reactions for Electron Donors (R_d)

ORGANIC DONORS (HETEROTROPHIC REACTIONS)

Domestic wastewater:
$$1/50\ C_{10}H_{19}O_3N + 9/25 H_2O = 9/50 CO_2 + 1/50 NH_4^+ + 1/50 HCO_3^- + H^+ + e^-$$
Carbohydrates (cellulose, starch, sugars):
$$1/4\ CH_2O + 1/4\ H_2O = 1/4\ CO_2 + H^+ + e^-$$
Acetate:
$$1/8\ CH_3COO^- + 3/8 H_2O = 1/8\ H_2O + 1/8 HCO_3^- + H^+ + e^-$$
Ethanol:
$$1/12\ CH_3CH_2OH + 1/4 H_2O = 1/6 CO_2 + H^+ + e^-$$

INORGANIC DONORS (AUTOTROPHIC REACTIONS):

$$Fe^{++} = Fe^{+++} + e^-$$
$$1/16\ H_2S + 1/16 HS^- + 1/2 H_2O = 1/8\ SO_4^{-2} + 19/16\ H^+ + e^-$$

Reactions for Electron Acceptors (R_a)

Oxygen:
$$1/2\ H_2O = 1/4\ O_2 + H^+ + e-$$
Nitrate:
$$1/10\ N_2 + 3/2\ H_2O = 1/5\ NO_3^- + 6/5\ H^+ + e^-$$
Sulfate:
$$1/16\ H_2S + 1/16\ HS^- + 1/2\ H_2O = 1/8\ SO_4^{-2} + 19/16 H^+ + e^-$$
Carbon dioxide (methane fermentation):
$$1/8\ CH_4 + 1/4\ H_2O = 1/8\ CO_2 + H^+ + e^-$$

Source: (Adapted from Grady and Lim, 1980 and McCarty, 1972).

both electron donor and acceptor in anaerobic fermentation. For example, in lactic acid fermentation, pyruvic acid is the acceptor. During anaerobic methane production, carbon dioxide can be considered the terminal electron acceptor.

2.2.2.3 Electron Donors.
Under controlled closed systems (i.e., laboratory or experimental situations), the exact formula of the electron donor is most often known. Should acetate be selected for the carbon and energy source, its empirical formula CH_3COOH would be used. If a synthetic medium is used, then the empirical formula and half reaction for each are written separately, then combined into a composite R_d on a proportional basis.

In natural open systems or environmental wastewater treatment systems, the chemical composition of the electron donor is neither simple and is most often unknown. One could analyze the wastewater frequently for C, H, O, and N, to develop an empirical formula. A half reaction could then be derived for that particular formula, as per McCarty (1972). This approach is not practical because the analytical methods are tedious and not trivial, plus the electron donor mixture may fluctuate in composition on a daily basis. However, if the COD (proportional to electron equivalents), organic nitrogen, organic carbon, and volatile solids content of the waste were determined, they could be used to generate the foregoing half reaction, R_d.

2.2.2.4 Estimation of the Energy Distribution.
Once the electron donor and acceptor are selected, along with the chemical composition of cell, then the fraction of energy used for synthesis, f_s, must be determined before Eq. (2.12) can be evaluated. f_s is related to the overall yield of the culture, which must be experimentally measured. Units of f_s are "equivalents of cells formed per equivalent of electron donor consumed," whereas the units of Y can be reported in any number of realistic or arcane combination of units. Preferably, the observed Y should be written in or converted into the units of "mass of cells formed per mass of substrate equivalent."

2.2.3 Multiple Reactions

Literally thousands of reactions take place within a biological system. Clearly only a limited number of these can be considered in any reasonable reaction scheme. Those reactions selected must somehow be important in research or in the design and operation of a full-scale biological process. A reasonable test for importance might be that omission of one of the reactions would result in such loss of information that the research findings would have little practical use.

Procedures described in subsequent paragraphs can be easily adapted to develop any set of reactions that are felt necessary for proper system understanding. The procedures are general and are independent of the type of systems involved, whether they be suspended growth or biofilm. The conservation laws used to model the entire system simply include information from the reaction scheme developed below. That is, the net rate of formation of any component is equal to the sum of the incremental rates of consumption or production of that component in each of the reactions. If the net rate of formation is negative, that component is being consumed in the entire system; if positive, the component is being produced.

The multiple reaction scheme for molar quantities that is equivalent to Eq. (2.4) is presented in Eq. (2.14).

$$\begin{aligned}
v_{1,1}A_1 + v_{1,2}A_2 + \ldots v_{1,n-1}A_{n-1} + v_{1,n}A_n &= 0 \quad \mathbf{r_1} \\
v_{2,1}A_1 + v_{2,2}A_2 + \ldots v_{2,n-1}A_{n-1} + v_{2,n}A_n &= 0 \quad \mathbf{r_2} \\
\cdot \quad \cdot \quad \quad \cdot \quad \quad \cdot \quad \quad \cdot & \\
\cdot \quad \cdot \quad \quad \cdot \quad \quad \cdot \quad \quad \cdot & \\
v_{m,1}A_1 + v_{m,2}A_2 + \ldots v_{m,n-1}A_{n-1} + v_{m,n}A_n &= 0 \quad \mathbf{r_m}
\end{aligned}$$

(2.14)

where $v_{i,j}$ = stoichiometric coefficient for the j-th component in the i-th reaction; negative for reactants, positive for products (moles), $\mathbf{r_i}$ = rate of the i-th reaction (moles/L^3-time), m = number of reactions, and n = number of components.

The rate of formation of the i-th component, r_i, in the j-th reaction is given by Eq. (2.15)

$$\bar{r}_i = v_{i,j}\mathbf{r_i} \qquad (2.15)$$

2.2 STOICHIOMETRY OF BIOLOGICAL PROCESSES

Each of these reactions can be transformed to a *mass basis* using the procedures discussed earlier. In the single reaction development, the *molar-based* reaction was transformed to the unit mass-based reaction, converting the stoichiometric coefficients to yield coefficients by identifying one of the components as the *normalizing stoichiometric coefficient*. Conceptually, the selection of the component is arbitrary and left to the discretion of the specific researcher and system under consideration. However, be aware that certain selections of the unit base stoichiometric coefficient result in stoichiometric coefficients that have more conventional connotations.

In the multiple reaction system, each reaction should be treated separately. The particular component (a reactant) selected to have the stoichiometric coefficient of (-1) should be determined for each reaction so that each resulting normalized stoichiometric coefficient has a physical meaning compatible with the user's thinking. Equation (2.16) represents a unit mass-based reaction scheme for Eq. (2.14),

$$
\begin{aligned}
(-1)A_1 + Y_{1,2}A_2 + \ldots + Y_{1,n-1}A_{n-1} + Y_{1,n}A_n &= 0 \quad \mathbf{r_1^*} \\
Y_{2,1}A_1 + Y_{2,2}A_2 + \ldots + Y_{2,n-1}A_{n-1} + Y_{2,n}A_n &= 0 \quad \mathbf{r_2^*} \\
\cdot \qquad \cdot \qquad \cdot \qquad \cdot \qquad \cdot & \\
\cdot \qquad \cdot \qquad \cdot \qquad \cdot \qquad \cdot & \\
Y_{m,1}A_1 + Y_{m,2}A_2 + \ldots + Y_{m,n-1}A_{n-1} + Y_{m,n}A_n &= 0 \quad \mathbf{r_m^*}
\end{aligned}
\tag{2.16}
$$

where $Y_{i,j} = \upsilon_{ij} MW_j / |\upsilon_{i,\alpha}| MW_\alpha$, $Y_{i,j}$ = yield of component A_j in reaction i (mass of component A_j consumed or formed in the i-th reaction per unit mass of component A_α consumed in the i-th reaction), α = component A_α, that component selected to be the normalizing stoichiometric coefficient resulting in a value of (-1) in the i-th reaction.

The rate of formation of component A_j in the i-th reaction $(r_j)_i$, is given by Eq. (2.17).

$$(\overline{r_j})_i = (Y_{i,j})\mathbf{r_i^*} \tag{2.17}$$

The units of $(r_j)_i$ are the mass component A_j formed in the i-th reaction/L^3 time. As a result, the net amount of component A_j consumed or produced by all the reactions in the system, (r_j), is given by Eq. (2.18).

$$(r_j) = \sum_{i=1}^{m}(r_j)_i = \sum_{i=1}^{m}\sum_{j=1}^{n}(Y_{i,j}\mathbf{r_i^*}) \tag{2.18}$$

That is, the net rate of formation of any component is equal to the sum of the incremental rates of consumption or production of that component in each of the reactions.

As a matter of practical importance, each reaction must be isolated as much as possible so that each of the stoichiometric and kinetic coefficients can be intrinsically determined. Unfortunately, complete definition of all coefficients in all reactions is not easy and often forces users to reduce the reaction set to a more "manageable" number of reactions. By doing this, important information may be lost from any analysis undertaken. In such cases, qualitative information should be obtained from the expanded reaction set using estimates for the "unknown" coefficients.

2.3 KINETICS OF BIOLOGICAL TRANSFORMATIONS

The other portion of the process rate, $\overline{\overline{R_c}}$, is the reaction rate expression itself. While stoichiometry relates to the molar ratios of components in a transformation, kinetics refers to how fast the transformation occurs and what independent parameters influence that rate. If a reaction rate is *homogeneous*, it is said to occur uniformly throughout the reaction volume V and is expressed as the number of moles (or mass) reacted per unit reaction volume per time. *Heterogeneous* reactions occur at the interface between two phases and do not occur uniformly throughout the reaction volume; heterogeneous reaction rates are more commonly expressed on a *per unit surface area* or *per unit weight* of reacting surface. Homogeneous reactions assume no concentrations gradients within the reaction volume, whereas heterogeneous reactions tend to create concentration gradients. Thus, heterogeneous reactions could require partial differential equations to describe the changes in reactant or product concentration with time and space, whereas homogeneous reaction systems are typically described by either ordinary differential equations or algebraic equations.

2.3.1 Homogeneous Reaction Kinetics

Rate expressions are mathematical functions that relate the reaction rate to basic fundamental parameters, for example,

$$\mathbf{r} = f(\text{temperature, concentration, etc.}) \tag{2.19}$$

Typically, chemical reactions are described assuming this overall function can be written as the product of a number of separate functions,

$$\mathbf{r} = f_1(\text{temperature}) f_2(A_1, A_2, \ldots A_i) \tag{2.20}$$

For example, in chemical (vs. biological) reaction kinetics, the most common approach is to assume a "power-law" expression where,

$$\mathbf{r} = k\,(A_1)^w (A_2)^x \ldots (A_i)^z \tag{2.21}$$

and k, the specific reaction rate constant, is a function of temperature only. The exponents in Eq. (2.21) are referred to as the *orders* of the components in the reaction. For example, in the reaction, aA + bB + cC → dD + eE, experimental data indicate the rate expression is,

$$\mathbf{r} = k\,C^1 D^{0.5} E^{-2} \tag{2.22}$$

Thus, the reaction rate is said to be first order in component C, half order in component D, and a -2 order in component E. Tacitly, the reaction rate is also zero order in components A and B. Note that reaction orders are empirical parameters that must be determined by experiment and can not be presumed from stoichiometry. Thus, one can never assume that orders of a component are equal to its stoichiometric coefficient.

For biological reactions, recall by stoichiometry, the rates of reaction of the energy source, electron donor (r_{ed}), cell or biomass growth, (r_x) and terminal electron acceptor (r_{tea}) are related as:

2.3 KINETICS OF BIOLOGICAL TRANSFORMATIONS

$$\frac{-r_{tea}}{1 - Y_{X/S}} = \frac{r_X}{Y_{X/S}} = -r_{ed} \qquad (2.23)$$

Consequently, one usually finds in the literature rate expressions mostly for cell or biomass growth, r_X since all other rates are related by stoichiometry. Biological rate expressions for r_X are written typically as a function of the growth-limiting substrate (or substrates) and are usually not power law expressions as discussed previously. Examples of the more traditional biological growth rate dependencies on substrate are:

Monod Kinetics: $r_x = \mu X = \mu_{max} SX/(K_s + S)$ \qquad (2.24)

Tessier Kinetics: $r_x = \mu X = \mu_{max} X(1 - e^{-S/K_s})$ \qquad (2.25)

Dual Substrate Kinetics: $r_x = \mu X = \mu_{max} S_1 S_2 X/(K_{S_1} + S_1)(K_{S_2} + S_2)$ \qquad (2.26)

2.3.2 Ideal Reactors

2.3.2.1 Batch Reactors. Many biochemical processes involve batch growth of microorganisms. After seeding or inoculating a liquid medium of appropriate composition with living cells, nothing (except possibly oxygen) is added or removed from the culture (except possibly CO_2) as growth proceeds. The batch reactor is generally well mixed such that spatial gradients in component concentration or fluid density do not exist. In a batch process, nutrients are depleted, products accumulate, and all components dynamically change with time.

Consider the schematic of a batch reactor as depicted in Figure 2.3. A material balance describing a batch reactor is based on the following assumptions:

- The reaction volume, V, is constant.
- At time equal zero, all components are placed into the system in their initial concentrations and at their required stoichiometric ratios.
- There is no exchange of any component across the reaction boundary.
- The homogeneous reaction, **r**, for the reaction $v_a A \rightarrow$ Products, occurs uniformly throughout the reaction volume.

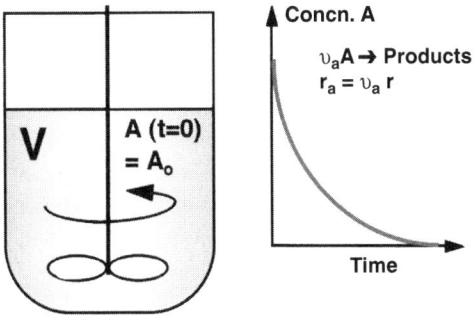

Figure 2.3 An ideal batch reactor.

A material balance for reactant A in this system would be, according to Eq. (2.1),

Net Rate of A Accumulation in Volume	=	Net Rate of A Input to Volume	+	Rate of A Conversion within Reaction Volume
$V dA/dt$	=	0	+	$v_a \mathbf{r} V$	(2.27)

or, simply

$$dA/dt = v_a \mathbf{r} \tag{2.28}$$

which states that component A in the batch system changes with time due to only the chemical reaction occurring in the system.

Batch reactor systems are frequently used to study bacterial growth and substrate utilization kinetics due mainly to their simplicity. Unfortunately, biofilm unintentionally formed upon the surfaces can create a number of operational complications in a batch reactor (e.g., strain contamination, biased plasmid retention in culture, operational errors in pH and dissolved oxygen control, errors in kinetic parameter estimation). Regrettably, literature is replete with examples of bacterial adhesion studies or biofilm formation studies that were carried out in batch systems. Batch systems for biofilm studies have the complication that all parameters in the bulk fluid dynamically change with time, along with those in the biofilm phase. Consequently, all biofilm parameters would vary during formation, under bulk fluid conditions that were also changing in time. To alleviate the dynamic conditions prevailing in a batch reactor, two forms of steady state continuous reactors were derived which differ from each other based on their relative degree of mixing: a continuously fed well-mixed reactor and a plug flow reactor.

2.3.2.2 Continuously Fed Well-Mixed Reactor.

This ideal reactor concept has been inappropriately designated in most reactor design texts as a *continuously stirred tank reactor* and thus has come to be known as a CSTR. This nomenclature is confusing, because a batch reactor is also a well-mixed, gradient-less reactor. The real distinction between a batch reactor and a CSTR is that the latter continually receives an influent of reactants and discharges, in its effluent, products, and unreacted reactants.

In an ideal CSTR, like a batch system, agitation is assumed to be sufficiently vigorous that mixing is complete. If a drop of dye were placed into an ideal CSTR, it would disperse instantaneously. No real reactor can exhibit perfect mixing but, in practice, the behavior of a real reactor can be made to closely approach that of an ideal CSTR. The usual trait of CSTR behavior is simply that the composition of the liquid stream leaving the reactor is the same as that of a sample withdrawn anywhere from the reactor.

Considering Figure 2.4, let A_{in} be the concentration (amount per unit volume) of some substance, biotic or abiotic, in the reactor influent while A is the concentration of that substance in the reactor and its effluent. Then, application of the principle of conservation of mass to the CSTR yields the following:

Net Rate of A Accumulation in Volume, V	=	Net Rate of A Input to Volume, V	+	Rate of A Conversion within Reaction Volume, V

2.3 KINETICS OF BIOLOGICAL TRANSFORMATIONS

or, mathematically,

$$VdA/dt = F(A_{in} - A) + \upsilon_a \mathbf{r} V \quad (2.29)$$

where A = component A concentration in system and leaving in reactor (M/L^3), A_{in} = component A influent concentration (M/L^3), F = influent and effluent volumetric flow rate (L^3/t), $\mathbf{r}$ = rate of production per unit volume (M/L^3-t), V = liquid volume in the vessel (L^3). The units of each group in Eq. (2.29) are mass per unit time (M/L^3).

At steady state, $(dA/dt = 0)$ and Eq. (2.29) reduces to

$$F(A_{in} - A) = -\upsilon_a \mathbf{r} V \quad (2.30)$$

such that the difference in the net rate of A input to the reactor is equal to the rate of chemical transformation of A in the reactor. The composition of the feed stream undergoes a discontinuous change when the stream enters the CSTR (i.e., composition changes from A_{in} to A). Thus the reaction rate, $\mathbf{r}$, in Eq. (2.30) is happening at a concentration of A, not A_{in}. In CSTR literature, the ratio of V/F is defined as τ_{CSTR}, the mean hydraulic residence of the reactor. Simply, τ_{CSTR} can be considered the time required to process one reactor volume of fluid.

A *chemostat* is a CSTR in which a microbial culture is grown continuously on one or more limiting substrates. Since a CSTR, or chemostat, operates at steady state it is the most common means for attaining and sustaining a bacterial culture at a constant cell concentration and known net growth rate.

A chemostat is essentially the same design as the CSTR in Figure 2.4; it consists of a vessel with an overflow that maintains a constant reaction volume. Sterile medium is continuously delivered into this vessel and immediately mixed into the culture suspension, thus displacing an equal volume of the stirred culture through the overflow. The culture medium is constituted so that one essential nutrient is the growth-limiting substrate (i.e., the organisms are nutrient-limited). In a chemostat experiment, it is usual practice to limit the carbon supply, but it is possible to limit N, Mg, P, S, or any other essential nutrient. Using the CSTR and a growth limiting substrate, the bacterial concentration (and extent of bacterial growth) may be controlled. Even more important, by controlling the rate of liquid addition, F, the growth rate of bacterial growth may be regulated.

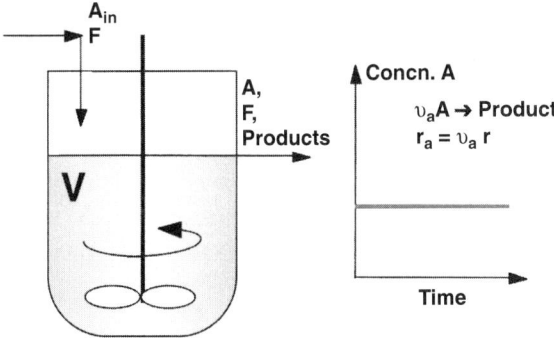

Figure 2.4 Schematic of an ideal continuously fed, well-mixed constant volume reactor.

Repeating the material balance approach above, a separate balance on cell mass in the chemostat yields

$$VdX/dt = F(X_{in} - X) + \mu XV \qquad (2.31)$$

where X = concentration of cells in the reactor (M_x/L^3), X_{in} = inlet cell concentration (which is zero in a sterile feed) (M_x/L^3), and μ is the microbial growth rate (t^{-1}). At steady state, $dX/dt = 0$, then,

$$FX = \mu XV \qquad (2.32)$$

which states the rate of cell mass leaving the chemostat is equal the rate of cell mass production. Rearranging,

$$F/V = \mu \qquad (2.33)$$

or

$$D = \mu \qquad (2.34)$$

where D = the chemostat "dilution rate," which is equal to $1/\tau_{CSTR}$. A chemostat, operated at one dilution rate, soon establishes steady state conditions. There is no change in bacterial numbers because bacterial growth exactly balances the bacteria lost by dilution. There is a very gradual decline in steady state bacterial concentration and a very gradual increase in steady state nutrient concentration as the dilution rate is increased from one experiment to another until, finally, dilution rate exceeds the highest possible growth rate, μ_{max} (Figure 2.5). At this point, D_c, the rate at which bacteria are diluted out, exceeds the rate at which they divide, resulting in a rapid decline in bacterial numbers and "wash-out" of the cell suspension, terminating operation.

Using a CSTR to study biofilm formation dynamics or substrate removal rates is an attractive reactor option because it offers the prospects of steady state bulk-fluid conditions. But be aware; this benefit is only realized under very special conditions. Biofilm accumulation has a significant effect on the performance of monoculture chemostats because de-

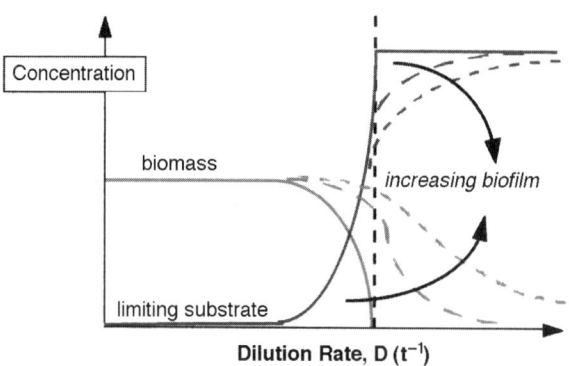

Figure 2.5 Biomass and limiting substrate concentrations as a function of chemostat dilution rate, D. Solid lines indicate theoretical behavior without biofilm; dashed lines indicate deviations from ideality caused by presence of a biofilm.

2.3 KINETICS OF BIOLOGICAL TRANSFORMATIONS

tachment of cells from the biofilm serves as a continuous source of bacteria (i.e., a continuous inoculation) for the liquid phase. The result of biofilm formation within a chemostat is a significant deviation (Figure 2.5) from the foregoing mathematical model describing chemostat behavior, especially at high dilution rates (Topiwala and Hamer, 1971; Howell et al., 1972; Molin and Nilsson, 1983). Biofilms in mixed culture chemostats can also impact population dynamics in a chemostat (Bryers, 1984, 1986). In essence, during surface inoculation and biofilm formation, the reactor theoretically violates the basic assumptions of CSTR conditions.

As per Bryers (1986), material balances describing adherent biofilm bacterial concentration in a continuously fed well-mixed reactor can be written as follows:

$$\begin{array}{c} \text{Net Accumulation} \\ \text{of Biofilm Biomass} \end{array} = \begin{array}{c} \text{Rate of} \\ \text{Biomass} \\ \text{Deposition} \end{array} + \begin{array}{c} \text{Rate of} \\ \text{Biofilm} \\ \text{Growth} \end{array} - \begin{array}{c} \text{Rate of} \\ \text{Biofilm} \\ \text{Detachment} \end{array}$$

$$AdB/dt = (R_{dep} + R_{bg} - R_{det})A \qquad (2.35)$$

where B = biofilm mass (M_x/L^2) and A = reactor surface area, (L^2).

A modified version of the suspended biomass balance, Eq. (2.31), that now takes into account biofilm detachment can be written as follows:

$$VdX/dt = -FX + \mu XV + R_{det}A \qquad (2.36)$$

Typically, in most biofilm studies, suspended growth of bacterial cells is a nuisance and is experimentally eliminated by operating the reactor at a residence time far less than the culture growth rate ($\tau = V/F <<< \mu$), such that any suspended cell in the system leaves before dividing. Thus, any cells present in the effluent must originate by way of the detachment of biofilm mass. Thus, at $D >> \mu_{max}$, Eq. (2.36) reduces to

$$VdX/dt = -FX + R_{det}A \qquad (2.37)$$

For Eq. (2.37) to attain a steady state, $dX/dt = 0$, requires

$$R_{det}A = FX \qquad (2.38)$$

and X in the effluent to be constant with time. This will only occur when both groups in Eq. (2.38) are constant. For $R_{det}A$ to become a constant requires that the biofilm mass reach a constant in time, i.e., dB/dt = zero, such that

$$R_{dep} + R_{bg} = R_{det} \qquad (2.39)$$

With R_{dep} being insignificant compared to the biofilm growth rate R_{bg}, Eq. (2.39) becomes, approximately

$$R_{bg} \approx R_{det} \qquad (2.40)$$

Thus, at a steady state biofilm amount, any biofilm mass generated is balanced by the detachment of this mass.

Figure 2.6 illustrates the effluent data for suspended biomass and limiting substrate (glucose) leaving a well-mixed continuously fed (sterile influent) rotating annular reactor, as a function of time. Also displayed is the biofilm amount (i.e., biofilm cell carbon and polymer carbon) with time. The system only attains true steady state behavior when the biofilm reaches a constant thickness, indicating a balance between biofilm growth and detachment (Eq. 2.40). Since all biomass detached from the surface must enter the reactor effluent, the suspended biomass concentration becomes constant only when biofilm reaches a steady state.

2.3.2.3 Plug Flow Reactor.
Completely opposite to the CSTR in system mixing, the other type of "continuously fed" ideal reactor is the plug flow reactor (PFR). In a PFR, culture medium and/or organisms are continuously fed to the reactor, where again microbial growth and other processes occur. However, the liquid phase within an ideal PFR is not uniformly mixed throughout the reactor; thus elements of liquid move progressively through the fermenter without mixing in the direction of flow (Figure 2.7). Conceptually think of the small "plugs" of reacting fluid being well mixed within themselves but not exchanging mass with the plug fore or aft; thus the plugs move through the PFR much like cars on a roller-coaster.

Let A be the concentration of some substance in the liquid. Assuming no radial gradients in the reactor, application of the principle of conservation of mass to a system of infinitesimal length dz moving with a velocity, $v (= F/a_x)$ through the reactor yields the partial differential equation:

$$\partial A/\partial t \quad + \quad v_z(r)\, \partial A/\partial z \quad = \quad r_A = \upsilon_A \mathbf{r} \qquad (2.41)$$

rate of + net rate of transport by = rate of production
accumulation   bulk fluid convection   of A

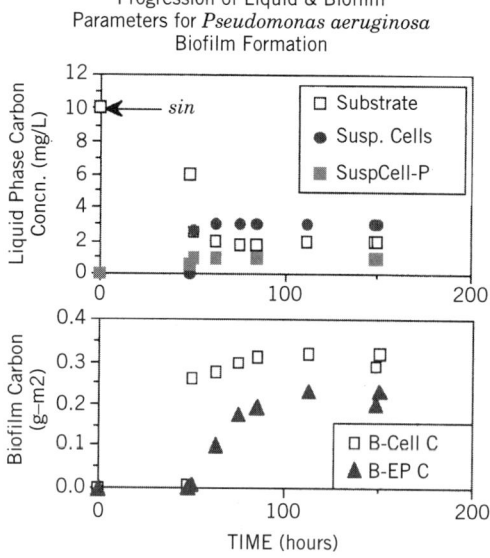

Figure 2.6 Effluent concentrations of substrate and suspended biomass relative to biofilm formation in a steady state CSTR (Trulear, 1983). Susp. Cells (·) = suspended cell carbon concn., SuspCell-P (■) = suspended cell exopolymer carbon concn., B-Cell C (□) = biofilm cellular carbon concn., and B-EPC (▲) = biofilm exopolymer carbon concn.

2.3 KINETICS OF BIOLOGICAL TRANSFORMATIONS

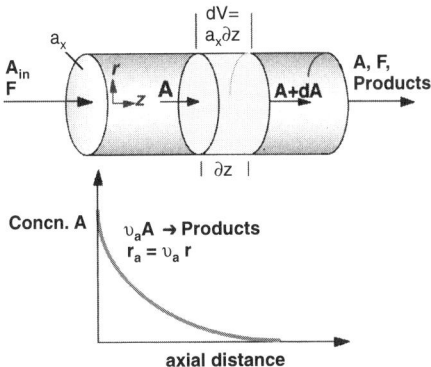

Figure 2.7 Schematic of ideal plug flow reactor.

where z = axial distance from the inlet, (L)

$v_z(r)$ = fluid velocity in the z direction as a function of r, (L/t)

This equation also assumes that diffusion in the z-direction is negligible to the convective transport term.

Assuming steady state operation, Eq. (2.41) becomes

$$v \partial A / \partial z = r_A \tag{2.42}$$

which is

$$F/a_x \partial A / \partial z = r_A \tag{2.43}$$

which becomes

$$F \partial A / \partial V = r_A \tag{2.44}$$

Separating and integrating Eq. (2.44),

$$\frac{1}{F} \int dV = \int \frac{dA}{r_A} \tag{2.45}$$

Integrating Eq. (2.45), over the entire volume of the reactor yields

$$\frac{V}{F} = \int \frac{dA}{r_A} \tag{2.46}$$

which is also τ_{pfr}, the residence time of the plug flow reactor. The actual value of τ_{pfr} depends on the limits of the integration and the dependency of r_A on A.

The most important feature of the PFR from the microbial physiology standpoint is the progressive change in environmental conditions seen by an organism traversing the reactor. This is in marked contrast to the constant conditions seen by an organism traversing an ideal CSTR. Since many fouling biofilms affect industrial systems that are of a circular cross-section tube geometry, ideal PFR behavior is an obvious first choice at mathematically describing the effect of biofilms on the bulk fluid concentration profiles.

However, just as with the CSTR, a biofilm growing at the outer radius of a tubular reactor (refer Figure 2.8) seriously compromises the validity of PFR behavior, as defined by Eq. (2.46). As shown in Figure 2.8, initially starting with a uniform biofilm in the reactor, a continuous supply of nutrients would create the following nonidealities:

- As nutrients are consumed by the biofilm, axial and perhaps radial gradients in substrate concentration will arise.
- Regardless of the kinetic expression, biofilm-bound cells near the entrance of the tube will "see" a higher bulk concentration than those further along the tube. The biofilm will experience a time- and space-dependent bulk condition in limiting substrate. As time progresses, an axial gradient in biofilm will develop along the length of tube (as is seen in many biofouling cases).

Rather than PFR behavior (Eq. 2.46), Eq. (2.41) would be more appropriate except it too is incorrect. The r_A reaction rate in both Eqs. (2.41) and (2.46) is a homogeneous reaction that pertains only to the bulk fluid and thus does not account for the biofilm, which occurs at the boundary of the tube. Using the approach of Kirkpatrick et al. (1980), rewriting Eq. (2.41) to describe a biofilm developing in a tube of circular cross-section (Figure 2.9) results in

Bulk Fluid Phase

$$\underset{\substack{\text{Radial Diffusion of } A \\ \text{in the positive radial } r \text{ direction}}}{} - \underset{\substack{\text{Convective Transport} \\ \text{of } A \text{ in axial } z\text{-direction}}}{} = 0$$

$$\frac{1}{r}\frac{\partial}{\partial r}\left[D_{\text{liq}} r \frac{\partial A}{\partial r}\right] - v_z \frac{\partial A}{\partial z} = 0 \qquad (2.47)$$

Biofilm Phase

$$\underset{\substack{\text{Radial Diffusion of } A \\ \text{in Biofilm}}}{} = \underset{\substack{\text{Biological Reaction Rate(s)} \\ \text{in Biofilm}}}{}$$

$$-D_{\text{eff}}\left(\frac{1}{r}\frac{\partial}{\partial r}\left(r\frac{\partial A}{\partial r}\right)\right) = r_A(r,z) \qquad (2.48)$$

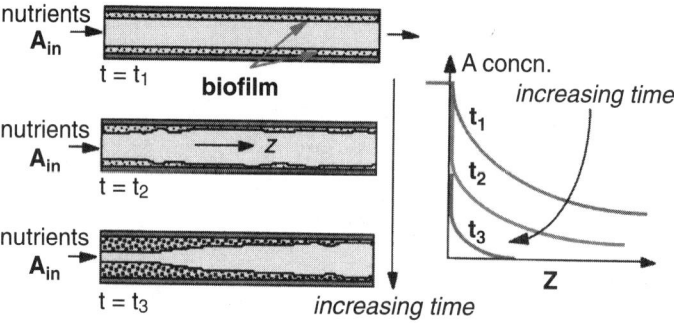

Figure 2.8 Effects of biofilm accumulation on dissolved nutrients in a tubular reactor.

2.3 KINETICS OF BIOLOGICAL TRANSFORMATIONS

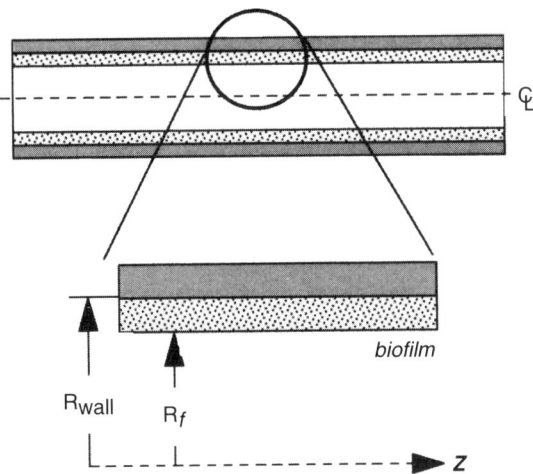

Figure 2.9 Unsteady state development of a biofilm at the wall of a circular tube.

where D_{liq} is the diffusivity of A in the bulk liquid [L²/t], D_{eff} is the effective diffusivity of A in the biofilm, and r_A is the consumption rate of A by the biofilm. To solve the foregoing problem, one needs the following boundary conditions:

Initial Condition: @ $t \leq 0$, all $z = 0$, and all r, A = some known A concentration

Boundary Conditions: @ $t > 0$
- @ $r = 0$ all z, $\partial^2 A/\partial r^2 = 0$ sets symmetry at the centerline for fluid concentrations
- @ $r = R_{wall}$, $\partial^2 A/\partial r^2 = 0$ states the tube wall is impermeable to A
- @ $r = R_f$ (film:fluid interface) $N_A(R_{f-},z) = N_A(R_{f+},z)$ states that the flux of A leaving the fluid at Rf is equal to the flux of A entering the biofilm
- @ $r = R_f$, $A(R_{f-},z) = A(R_{f+},z)$ states that the concentrations on either side of the interface must be equal.

Note in this system the biofilm is growing such that R_f is varying with time, which requires an additional "boundary" condition for the moving biofilm. One can estimate from a material balance between two time points the amount of A consumed by the biofilm then by stoichiometry determine the amount of biomass generated which is averaged over the biofilm to produce an increase in biofilm thickness and thus a decrease in R_f. Thus,

$$2Y_{X/S} \int_t^{t+\Delta t} \left(rD_{eff} \frac{\partial A}{\partial r} \right)_{R_{f-}} dt = \rho(R_f^{2(t+\Delta t)} - R_f^{2(t)}) \qquad (2.49)$$

A further complication to this treatment of biofilm formation in a circular tube (Kirkpatrick et al., 1980) occurs when the biofilm begins to directly influence the fluid velocity flow field (Lewandowski et al., 1994), thus changing the functional form of the velocity profile, $v_z(r)$.

Faced with solving Eqs. (2.47) and (2.48) simultaneously, is it no wonder that one would prefer to use Eq. (2.46) to simulate biofilm presence in a tubular system?

2.3.3 Heterogeneous Reactions

Reiterating, homogeneous reactions occur uniformly through the reaction volume whereas heterogeneous reactions occur at the interface between two phases. Heterogeneous reactions complicate the mathematical analysis of reaction systems since, for a heterogeneous reaction to occur, reactants must move through one phase to reach the reaction site and may need to migrate through the other phase to react. The reverse is true for reaction products. For reactants and products to move, mass transport must take place, which implies a driving force for the flux of material: either bulk fluid motion (convection) or a concentration gradient (molecular diffusion). Mass transport of reactants to a reaction site can be slower than the surface-based reaction thus rendering the overall depletion or reactant or product formation mass transport rate limited.

Consider Figure 2.10a, where an extremely thin layer of cells attached a substratum starts to metabolize the growth limiting substrate, S. In that instant, S molecules adjacent to the surface are consumed, rendering the S concentration at the interface lower than S further away from the wall. Thus the surface reaction creates a gradient in S in the negative z direction ($-dS/dz$) which is proportional to a mass flux of S being transported to the surface. At some point the system reaches a balance between the rate of S consumption and the supply or flux of S to the surface. Depending on the degree of mixing in the fluid adjacent to the substratum, the bulk fluid could be well mixed (and the S gradient in Figure 2.10a would remain at that shown for time = 0) or poorly mixed, leading to a gradient and a system that is *externally mass transport limited*. In this latter case, the surface reaction must "wait" for a molecule of S to arrive in order to react. To a large extent, external mass transport limitations of a reaction can be avoided by increasing the fluid mixing and the bulk substrate concentration.

In Figure 2.10b, a thick biofilm has developed in a very well-mixed fluid environment. Bulk fluid is well mixed such that external mass transfer limitations can be ignored. How-

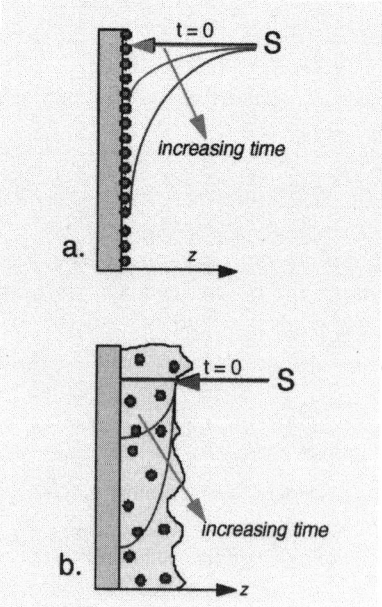

Figure 2.10 *a*. External mass transfer and *b*. internal mass transfer.

2.3 KINETICS OF BIOLOGICAL TRANSFORMATIONS

ever, due to either the effective diffusivity of the substrate S being lower in the biofilm versus the bulk fluid, or an extremely rapid substrate reaction rate, or a very dense thick biofilm, the mass flux of S is unable to maintain a constant S in z within the biofilm (gradient at $t = 0$). Thus, due to *internal mass transport limitations*, cells in the biofilm "see" a decrease in available S and that concentration depends on where the cells are in the biofilm. Consequently, not all cells in a biofilm "see" the same conditions and those conditions can be dramatically different than those of the bulk fluid. It is hard to comprehend that biofilms as thin as 10 μm can create anaerobic conditions at their base despite being in a highly aerated system. Unlike external mass transfer limitations, very little can be done to a system to reduce the effects of internal mass transport limitations (short of disrupting the biofilm).

Historically, it has been easier to measure *per reactor area* observed transformation rates as seen in the bulk liquid (e.g., electron donor removal rate, electron acceptor uptake rate, total biofilm mass accumulation) than to measure growth, replication, and death of cells *directly* in a biofilm. The three ideal reactor types and their correspondent equations (Eqs. (2.28), (2.30), and (2.46)) are all based on homogeneous kinetics for the reaction rates. To account for biofilm formation, local uptake of substrates in the biofilm, or multiple species dynamics in the biofilm would require complex multiple variable, partial differential equations (or discrete:continuous cell automata) for both the biofilm and fluid phases. Thus, it is compelling to attempt to modify the ideal reactor equations used for homogeneous reactions to account for heterogeneous biofilm reactions. This "pseudo-homogeneous" compromise is only successful in certain limiting cases, for example, estimating the overall consumption of limiting substrate by a biofilm reaction. To simulate or interpret more complex processes, where such a compromise would provide insufficient information, more complex mathematics must be used.

The steady state models (Harremoës, 1977, 1978) are useful to estimate the observed flux of growth rate limiting substrate (M_S/L^2 t) into a biofilm of fixed thickness, density, and reactivity, taking into account both external and internal molecular diffusion coupled with a simultaneous biological reaction rate. Based on identical models for one-dimensional heterogeneous catalysis, these models assume a constant biofilm concentration B (tacitly implying a spatially uniform reactivity) and diffusion path (biofilm thickness) which allows one to predict (*a*) the concentration profile of limiting substrate (and by stoichiometry all other nutrients) with biofilm depth and (*b*) the maximum substrate uptake or flux to the biofilm.

The ratio of the reaction rate observed under practical system conditions (possibly controlled by mass transfer effects) to the true intrinsic reaction rate evaluated at system conditions assuming no mass transfer effects, is defined as the *effectiveness factor* for the biofilm in question.

Models by Harremoës (1977, 1978) and Vos et al. (1990) of immobilized cell or biofilm kinetic studies, define the effectiveness factor as a function of a dimensionless group of pertinent observable system parameters. Such models define a reacting geometry (one-dimensional slab, cylinder, or sphere) with uniform distribution of the biological catalyst throughout the reacting volume. For a one-dimensional biofilm of thickness L in the z direction (perpendicular to the substratum) and uniform areal biomass concentration, B (M/L^2), metabolizing a growth limiting substrate S, according to a biological reaction, r_g, the second order differential equation describing the concentration of S in the biofilm as a function of position is Eq. (2.50),

$$D_{eff}(d^2S(z)/dz^2) = r_g \qquad (2.50)$$

where D_{eff} = the effective diffusivity of S in the biofilm. Solution to Eq. (2.5) for S as a function of the spatial coordinate z, depends upon specific system boundary conditions and the particular reaction rate dependency on local substrate concentration (i.e., first or second order, or saturation kinetics). For the case of saturation kinetics, one can define the following dimensionless variables,

$$k = S_B/K_S \quad S^* = S/S_B \quad z^* = z/L \tag{2.51}$$

thus rewriting Eq. (2.50) as,

$$d^2S^*/d(z^*)^2 = \theta^2 S^*/(1 + kS^*) \tag{2.52}$$

$$\text{where} \quad \theta = \text{Thiele modulus} = \{A_S \mu_{max} BL^2/Y_i K_S D_{eff}\}^{0.5} \tag{2.53}$$

and A_S is the surface area to volume ratio, μ_{max} is the maximum growth constant for the microorganism, B is the areal concentration of microorganisms, and K_S is the saturation constant in the growth rate expression. Note, the Thiele modulus conceptually is the ratio of the maximum reaction rate ($\mu_{max} BA/YK_s$) possible to the maximum mass transfer rate (L^2/D_{eff}).

Boundary conditions for the dimensionless Eq. (2.52) are correspondingly

$$S^* = 1 \quad \text{at} \quad z^* = 1$$
$$dS^*/dz^* = 0 \quad \text{at} \quad z^* = 0$$

Once the specific substrate profile is determined, an *effectiveness factor* can be derived according to the following definition

$$\eta = \frac{\text{rate observed with mass transfer}}{\text{true reaction rate without mass transfer effects}}$$

$$\eta = \frac{r_{obs}}{r_{true}} = [-D_{eff} dS/dz]_{|z=L} \bigg/ \left\{ \frac{-A_S \mu_{max} BL^2 S_{bulk}}{K_S + S_{bulk}} \right\} \tag{2.54}$$

or in dimensionless terms

$$\eta = [(1 + k)/q^2] \, dS^*/dz^*|_{z^*} \leq 1 \tag{2.55}$$

The most convenient means of depicting the influence of operating parameters on the effectiveness factor was presented by Pitcher (1975) using a modified Thiele modulus defined as

$$\theta_P = [\theta \, k/(1 + k)] \, [2k - 2\ln(1 + k)]^{0.5} \tag{2.56}$$

Classically, in heterogeneous enzyme catalysis or in steady state biofilm models, this derivation holds if the concentration of surface-bound catalyst (i.e., enzyme or biofilm bacteria) and the diffusion path (i.e., immobilized enzyme support or biofilm thickness) are assumed constant in time. Also, the reaction rate constant μ_{max} is assumed to pertain to a single substrate conversion process.

Modifications have been made in this approach to estimate the unsteady state flux of multiple growth rate-limiting substrates in several biological reactions in series. Tanaka and Dunn (1982) derived four unsteady state substrate balances (ammonia, nitrite, nitrate, and oxygen) for autotrophic nitrification. This system of equations assumes a constant biofilm thickness and that the concentration of the two bacterial species involved is constant in both time and space in order to solve the set of coupled partial differential equations. A similar, steady state model of denitrification of nitrate to nitrite then to nitrogen was presented in Boaventura and Rodrigues (1988) and Droste and Kennedy (1986). Here two separate effectiveness factors and corresponding Thiele moduli are defined for nitrate and nitrite conversion but no distinction between bacterial species is made; in essence, a uniform biofilm density is assumed constant. Bryers (1993) evaluated an effectiveness factor for the case of an unsteady state biofilm with a multiple bacterial species biofilm.

With estimates of the appropriate diffusion coefficients, a Thiele modulus and its corresponding effectiveness factor can be evaluated. Using a dimensionless parameter approach, a single generic relationship can be derived between η and θ for a particular geometry and type of reaction rate expression. Thus, the numerical value of θ can determined from reaction system values, the value of the effectiveness factor η determined from the generic relationship, and then used to correct the ideal reaction rate without mass transfer, that is:

$$r_g = \eta \, \mu \, (\text{at } S=S_{bulk}) = \eta \, \mu^* S_{bulk}/K_S + S_{bulk}) \tag{2.57}$$

Thus, using the effectiveness factor to compensate for mass transfer effects allows one to employ a pseudo-homogeneous treatment of the biofilm where any one of the ideal reactor equations can be used with the adjusted growth rate in Eq. (2.57).

2.4 PRACTICAL OPERATION OF A BIOFILM STUDY SYSTEM

2.4.1 Biofilm Laboratory Reactors

Table 2.2 illustrates the basic design geometry and design advantages and disadvantages for the most commonly used biofilm study reactor systems: the rotating annular reactor, the parallel plate flow cell, a Robbins device (either rectangular or circular cross-section), the radial flow cell, and a laboratory waste-gas biofilm reactor.

In theory, any of this biofilm reactors and their environmental support systems can be operated to simulate any one of the three ideal reactor concepts (i.e., CSTR, plug flow, and batch) presented previously. Following are details as to how these individual reactors can be operated to mimic these ideal reactor behaviors.

2.4.2 Creating CSTR Behavior

CSTR ideality presumes complete mixing of the reaction volume, with no concentration gradients, along with a continuous input of limiting reactant and an effluent flow to affect a constant volume. Obviously, a rotating annular reactor (Fig. 2.11) can easily accommodate these restrictions while providing sufficient surface area for the biofilm. Residence time in the rotating annular reactor can be set at any value above or below the maximum growth rate of the pure or mixed culture, simply by adjusting the input nutrient volumetric flow rate. However, often in biofilm studies, it is expedient to minimize the effects of

TABLE 2.2. Summary of Common Biofilm Laboratory Reactor Geometries

Reactor	Advantages and Disadvantages
Rotating Annular Reactor Comprises two concentric cylinders, one inside another where the inner cylinder is free to rotate and the outer cylinder is stationary (Figure 2.11). Removable sample slides can be fitted flush to the inner surface of the outer cylinder.	*Advantages:* Provides high surface area to volume ratio. Ample sample area. Defined flow dynamics in most of the reactor. Allows estimation of biofilm frictional resistance by directly measuring changes in torque. *Disadvantages:* High rotation speeds can create Taylor voticies and poor mixing. Design not amenable to small construction; thus due to volume, requires high influent flow rates to minimize residence times. Not suitable to bacterial adhesion studies. Ill-defined mixing of fluid at tops and bottoms of reactor.
Parallel Plate Flow Cells Comprises two parallel flat plates creating a thin slit flow field (Figure 2.12). Plate closest to microscope must be transparent while the target substratum can be of any material.	*Advantages:* Bacterial cell suspensions can be delivered to flow cells to inoculate the surface. Once inoculated the bacterial cell challenge can continue or it can be replaced by sterile nutrient feed. Dimensions intended for microscopic image analysis. Allows direct enumeration of bacterial cells adhering to the substratum. *Disadvantages:* To authentically create well-defined flow field, the gap width between plates must be thin and the cross-sectional area a certain aspect ratio. Also, microscopic focal lengths require the gap width be small. This restricts fluid flow to very low laminar, if not creeping flow, conditions. This design does not provide for direct sampling of the biofilm.
Robbins Flow Devices Comprises a continuous tube of either circular or rectangular cross section where removable sample plugs are made to fit flush with the inner surface of the fluid flow field (Figure 2.13).	*Advantages:* Can be simply fabricated of polymer materials, and can be sealed permanently thus obviating periodic and tedious opening of the device. Provides as many removable samples as desired, provided length of reactor is no problem. Flow channel can be larger than parallel-plate flow cells, so turbulent flow is possible. *Disadvantages:* Once sealed, difficult to clean. To increase the number of sample plugs, one must increase length but longer reactors increase the prospects of axial gradients in substrate and biofilm.

(continued)

suspended cell metabolism. Thus, input nutrient flow rate can be set to establish a residence time ($\tau = V/F$) in the reactor that is much less than the maximum generation time of the culture. Consequently, any suspended biomass found in the reactor effluent can be assumed

2.4 PRACTICAL OPERATION OF A BIOFILM STUDY SYSTEM   39

TABLE 2.2. *(Continued)*

Reactor	Advantages and Disadvantages
Radial Flow Devices Consists of two identical sized discs separated by a thin gap (Figure 2.14). Bacterial suspensions are introduced at the center of the bottom disc, impinge on the upper disc, and flow radially outward to a weir collection system at the periphery. Top plate is transparent for microscopic surveillance.	*Advantages:* Allows direct observation of bacterial adhesion. Allows the observation of adhesion as a function of variable shear stress, which decreases with distance from the center. Different substrata can be used. *Disadvantages:* No direct sampling of substrata. Best used for bacterial adhesion at laminar flow conditions. Flow fields not defined if biofilm extensive or flow in turbulent region. Construction difficult.
Waste Gas Bioreactor Consists of a completely enclosed gas-tight chamber where the bottom plate contains removable sample plugs (Figure 2.15). A biofilm is cultivated on the bottom plate. Wastegas is introduced either counter- or co-current with a liquid nutrient-containing influent stream. The liquid stream is supplied at a flow rate to just barely wet the biofilm surface. Gastight septa are incorporated into the top to allow microelectrode sampling of liquid and biofilm variables (e.g., pH, DO).	*Advantages:* Simulates a small section of a complex gas phase biofilter. Allows periodic removable sampling of biofilm and direct access for microsensor determination of dissolved components in fluid and biofilm. *Disadvantages:* Large, cumbersome. Volumetric flow rate of gas high. Hydrodynamics ill defined. Axial and depth gradients in all parameters created.

to originate by way of detachment from the biofilm. This operational technique also serves to simplify the mathematics of material balances on suspended biomass by eliminating the term for planktonic growth.

Although their geometry connotes plug flow behavior, flow cell reactors (Figs. 2.12, 2.13, and 2.15) can also be operated in a CSTR mode by utilizing a fluid recycle or recirculation loop as illustrated in Figure 2.16. Residence time is set by the ratio of the total system volume to the inlet flow rate. By recycling a portion or all of the reactor contents, the axial fluid phase concentration gradients, characteristic of plug flow behavior, can be eliminated. Referring to Figure 2.16, to operate a flow cell reactor in the CSTR mode, one must ensure that the recycle flow rate, F_r, is much greater than F, the nutrient feed flow rate. As with all laboratory reactor systems, the degree of ideal mixing must be verified for the operating conditions in question (Levenspiel, 1972).

2.4.3 Creating Plug Flow Behavior

Many biofilm systems of industrial and ecological importance occur in plug flow systems, such as biofouling of medical catheters, corrosion of pipes and sewers, or the ecology of a natural surface- or ground-water flow. Plug flow behavior in biofilm systems is characterized by longitudinal (and perhaps, radial) concentration profiles of nutrients in the liquid phase. Even if the surface of the reactor system were uniformly inoculated with cells, those attached colonies nearer the entrance of the reactor would be exposed to the highest concentration of limiting nutrient which, after time, would establish larger amounts of biofilm at the flow cell entrance. This would eventually lead to longitudinal gradients, not only in soluble nutrients, but also in attached cell concentrations. Similar spatial gradients are common to either packed bed or trickling filter reactor systems operated on a once-through basis.

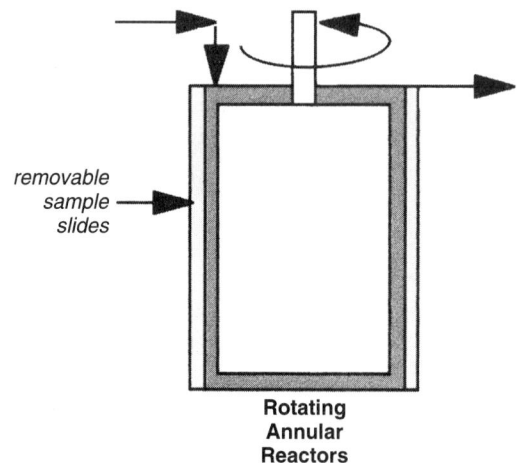

Figure 2.11 Rotating annular reactor.

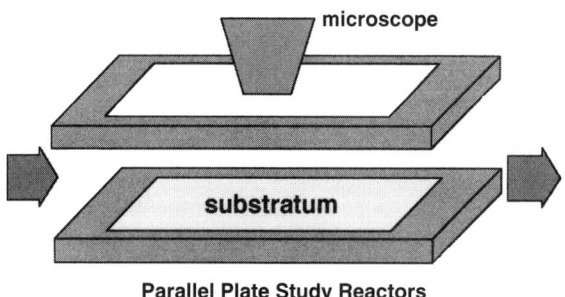

Figure 2.12 Schematic of a parallel plate flow cell.

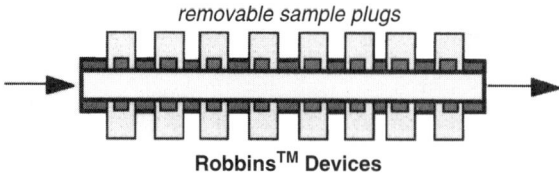

Figure 2.13 Tubular reactor with removable sample plugs.

2.4 PRACTICAL OPERATION OF A BIOFILM STUDY SYSTEM

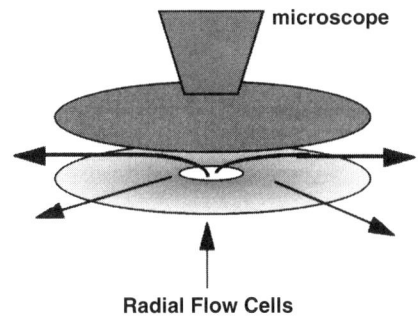

Figure 2.14 Radial disc flow cell.

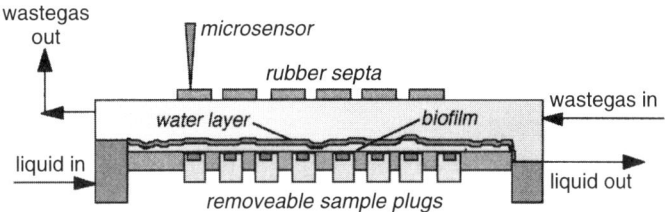

Figure 2.15 Laboratory-scale biogas biofilm reactor.

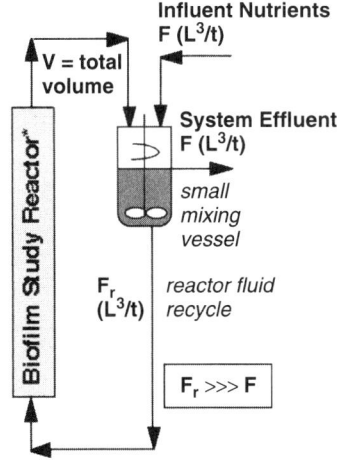

Figure 2.16 Flow cell reactor operated as a CSTR using fluid recycle.

Circular flow cells and Robbins devices of circular cross-section are obviously geometrically identical to steam condenser heat exchange tubes, water distribution lines, and wastewater sewers. The hydrodynamics of flow in a circular geometry are well defined (Schlichting, 1968). Rectangular flow cells and capillaries, while providing the accessibil-

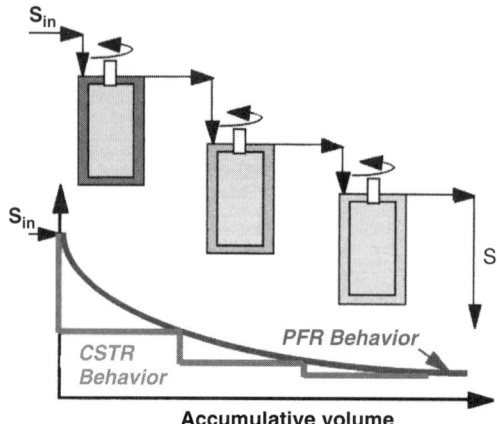

Figure 2.17 Simulated plug flow reactor behavior using multiple CSTR's in Series.

ity for microscopic observation and surface diagnostics, also have similar hydrodynamics as circular tubes (provided certain geometric criteria are met) and have been used with great success. Operation of flow cells in a plug flow mode is achieved quite simply by passing the nutrient solution (or a side stream from an operating system in the field) down the flow cell; although, as indicated previously, the system soon departs true PFR behavior as the biofilm develops an axial gradient.

Plug flow behavior can best be established with a rotating annular reactor, or more accurately, a number of rotating annular reactor units operated in series (Fig. 2.17). Theoretically, as the number of rotating annual reactor units increases, the overall system performance approaches that of plug flow behavior.

REFERENCES

Aris, R. 1969. *Elementary Chemical Reactor Analysis*. Englewood Cliffs, NJ: Prentice-Hall.

Boaventura, R. A., and A. E. Rodrigues. 1988. Consecutive reactions in fluidized bed biological reactors: Modeling and experimental study of wastewater denitrification. *Chem. Engrg. Sci.* 43: 2715–2728.

Bryers, J. D. 1984. Biofilm formation and chemostat dynamics: Pure and mixed culture considerations. *Biotechnol. Bioengrg.* 26: 948–958.

Bryers, J. D. 1986. Stability analysis of a binary culture chemostat experiencing biofilm formation. *Bioprocess Engrg.* 1: 3–11.

Bryers, J. D. 1993. Evaluation of the effectiveness factor for a multiple species biofilm. *Biofouling* 6: 363–380.

Busch, A. W. 1971. *Aerobic Biological Treatment of Wastewaters, Principles and Practice*. Houston, TX: Gulf Publishing.

Droste, R. L., and K. J. Kennedy. 1986. Sequential substrate utilization and effectiveness factors for fixed biofilms. *Biotechnol. Bioengrg.* 28: 1713–1720.

Grady, C. P. L., and H. C. Lim. 1980. *Biological Wastewater Treatment*. New York: Marcel Dekker, pp. 269–304.

REFERENCES

Harremoës, P. 1977. Half-order reactions in biofilm and filter kinetics. *Vatten* 33: 122–143.

Harremoës, P. 1978. Biofilm kinetics. In *Water Pollution Microbiology*, Vol. 2, R. Mitchell, ed. New York: J. Wiley, pp. 82–109.

Herbert, D. 1958. Some principles of continuous culture. In *Recent Progress in Microbiology*, G. Tunevall, ed. Oxford, UK: Blackwell Scientific Publishers, pp. 381–396.

Herbert, D. 1976. Stoichiometric aspects of microbial growth. In *Proceedings of the 6th International Symposium on Continuous Culture Applications and New Fields*, A. C. R. Dean et al., eds. Chicester: Ellis Horwood, pp. 1 30.

Howell, J. A., C. T. Chi, and U. Pawlowky. 1972. Effect of wall growth on scale-up problems and dynamic operating characteristics of the biological reactor. *Biotechnol. Bioengrg.* 24: 253–265.

Irvine, R. L. 1980. Activated Sludge: Stoichiometry, Kinetics and Mass Balances. ACS Audio Course, cat. no. Z-57. Washington, D.C.: American Chemical Society.

Irvine, R. L., and J. D. Bryers. 1986. Stoichiometry and kinetics of waste treatment. In *Comprehensive Biotechnology*, M. Moo-Young, ed. New York: Pergamon Press, pp. 757–772.

Irvine, R. L., J. E. Alleman, G. Miller, and R. W. Dennis. 1980. Stoichiometry and kinetics of biological waste treatment. *J. Water Pollut. Control Fed.* 52: 1997–2006.

Kirkpatrick, J. P., L. V. McIntire, and W. G. Characklis. 1980. Mass and heat transfer in a circular tube with biofouling. *Wat. Research* 14: 117–127.

Levenspiel, O. 1972. *Chemical Reaction Engineering*, 2nd ed. New York: J. Wiley, 143.

Lewandowski, Z., P. Stoodley, S. Altobelli, and E. Fukishima. 1994. Hydrodynamics and kinetics in biofilm systems—Recent advances and new problems. *Wat. Sci. Tech.* 29: 223–229.

McCarty, P. L. 1972. Energetics of organic matter degradation. In *Water Pollution Microbiology*, R. Mitchell, ed. New York: Wiley-Interscience, pp. 91–118.

Molin, G., and I. Nilsson. 1983. Effect of different environmental parameters on biofilm build-up of *Pseudomonas putida* ATCC-11172 in a chemostat. *Eur. J. Appl. Microbiol. and Biotechnol.* 18: 114–119.

Pitcher, W. H. 1975. Engineering of immobilized enzyme systems. *Catalyst Reviews—Science and Engineering* 12: 37–69.

Porges, N., L. Jasewicz, and S. R. Hoover. 1956. Principles of biological oxidation. In *Biological Treatment of Sewage and Industrial Wastes*, J. McCabe and W. W. Eckenfelder, eds. New York: Rheinhold Publishing Corp., pp. 35–48.

Roels, J. A. 1980. Application of macroscopic principles to microbial metabolisms: A bioengineering report. *Biotechnol. Bioeng.* 22: 2457–2514.

Schlichting, H. 1968. *Boundary Layer Theory*, 6th ed. New York: McGraw-Hill.

Tanaka, H., and I. J. Dunn. 1982. Kinetics of biofilm nitrification. *Biotechnol. Bioengrg.* 24: 669–689.

Topiwala, H. H., and G. Hamer. 1971. Effects of wall growth in steady-state continuous culture. *Biotechnol. Bioengrg.* 13: 919–922.

Trulear, M. G. 1983. Cellular Reproduction and Extracellular Polymer Formation in the Development of Biofilms, Doctoral Dissertation, Montana State University, Bozeman, MT.

Vos, H. J., P. J. Heederik, J. J. M. Potters, and K. C. Luyben. 1990. Effectiveness factors for spherical biofilm catalysts. *Bioprocess Engineering* 5: 63–72.

BIOFILM FORMATION AND PERSISTENCE

JAMES D. BRYERS
The University of Connecticut Health Center, Farmington, Connecticut

3.1 PROCESSES GOVERNING BIOFILM FORMATION AND PERFORMANCE

Figure 3.1 illustrates, for an arbitrary analytical measure, the typical accumulation of biofilm at a surface as a function of time. Initially, the substratum is conditioned and cells attach reversibly, then irreversibly. Next, attached cells grow, reproduce, and secrete insoluble extracellular polysaccharidic material. As the biofilm matures, biofilm detachment and growth processes come into balance, such that the total amount of biomass on the surface remains approximately constant in time. Ten years ago, when the first edition of *Biofilms* was published, it was revolutionary to recognize that biofilms were the net result of multiple physical, chemical, and biological processes that operate in each phase (Fig. 3.2). Further, the concepts that bacteria at a surface or deep within a biofilm were carrying out physiological processes at local conditions that were not equal to those elsewhere in the biofilm and were not equal to conditions in the bulk fluid was totally new.

At the time, processes governing biofilm formation and persistence included:

1. Biasing or preconditioning of the substratum either by macromolecules present in the bulk liquid or intentionally coated on to the substratum.
2. Transport of planktonic cells from the bulk liquid to the substratum.
3. Adsorption of cells at the substratum for a finite time followed by,
4. Desorption (release) of reversibly adsorbed cells.
5. Irreversible adsorption of bacterial cells at a surface.
6. Transport of substrates to and within the biofilm.

Biofilms II: Process Analysis and Applications, Edited by James D. Bryers.
ISBN 0-471-29656-2   Copyright © 2000 Wiley-Liss, Inc.

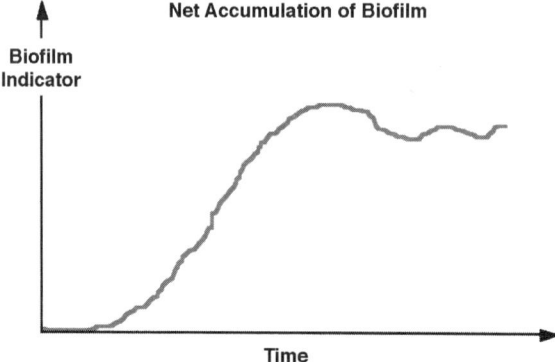

Figure 3.1 Net accumulation of biofilm.

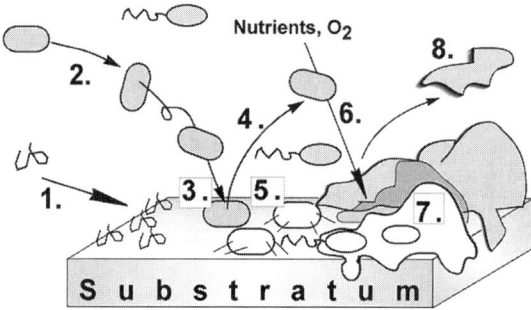

Figure 3.2 Processes governing biofilm formation (definitions provided in text).

7. Substrate metabolism by the biofilm-bound cells and transport of products out of the biofilm. These processes are accompanied by cellular growth, replication, and extracellular polymer production.
8. Biofilm removal (detachment or sloughing).

Research in the past 10 years has expanded our understanding of the molecular and genetic parameters that control many of these macroscopic processes. Biofilms are no longer considered uniform biological structures in time or space, and processes that control this heterogeneity have been characterized and are being mathematically described. This chapter focuses on past concepts, recent advances, and future directions in research on processes that govern biofilm formation.

3.2 TRANSPORT OF MICROBIAL CELLS TO THE SUBSTRATUM

When a clean surface is immersed in natural water containing dispersed microorganisms, nutrients, and organic macromolecules, transport of these components to the substratum can control the initial rate of adherent cell or biofilm accumulation. In very dilute dispersions of microbial cells and nutrients, such as open ocean waters or distilled water storage

3.2 TRANSPORT OF MICROBIAL CELLS TO THE SUBSTRATUM

tanks, transport of microbial cells to the substratum may be the rate-controlling step in biofilm accumulation for long periods of time. So, mass transport of cells and/or nutrients is critical in a rate analysis of biofilm accumulation.

Mass transport processes are influenced strongly by the mixing in the bulk fluid, which is generally related to the fluid flow regime. Laminar and turbulent flow are the two distinct fluid flow regimes that influence mass transport. Transport of molecules and small particles (< 10 μm) in quiescent or laminar flow is controlled by either sedimentation, motility, or molecular diffusion. In turbulent flow, both convective and diffusive transport prevail.

3.2.1 Quiescent Conditions

Under quiescent conditions, transport of bacteria from a bulk fluid phase to a surface is by either gravitational forces (i.e., sedimentation), Brownian diffusion, or by motility for those organisms capable of motility. Sedimentation rates are small for bacteria because of their size and specific gravity (approximately 1.05–1.10). For example, the terminal settling velocity for a bacterial "particle" can be calculated from Eq. (3.1) as

$$V_t = g(\rho_P - \rho_L) d_P^2 / 18 \mu \tag{3.1}$$

where V_t = terminal settling velocity (cm/sec), g = gravitational acceleration constant (cm/sec^2), ρ_P, ρ_L = density of particle and liquid, respectively (gm/cm^3), d_P = particle diameter (cm), μ = fluid viscosity (gm/cm-sec).

Because of their size diameters (1.0 μm) and specific gravity (1.05), sedimentation rates are extremely slow even in quiescent conditions.

Microorganisms of a size 1–4 μm^3 are limited in Brownian motion and, hence, have a small Brownian diffusivity (see Section 3.2.1.1). Therefore, motility may be a more important transport process in quiescent systems. Many microbes are capable of motility through their own internal energy independent of fluid forces. Motility is frequently related to some form of taxis (i.e., cell motility induced by external stimuli) in response to a concentration gradient.

3.2.1.1 Motility.
In an unbounded fluid medium, flagellated, motile bacteria move in a manner resembling a three-dimensional random walk. That is, they swim in nearly a straight line for about a second (running) and then tumble in place, then begin a run in another direction (Berg and Brown, 1974). If there is not a chemical or external stimulus (chemotaxis), then the angle of deviation between one run and the subsequent run is totally random. Angles are influenced in the case of positive chemotaxis so that, on the average, cells move toward the source of the chemical attractant; the opposite is true for negative chemotaxis.

Jang and Yen (1985) determined the Brownian diffusivity of different microorganisms to be approximately 5.0×10^{-6} mm^2 s^{-1} as compared to the calculated non-Brownian diffusivity (due to motility) of approximately 0.4×10^{-3} to 5.6×10^{-3} mm^2 s^{-1}. The following equation was used to calculate non-Brownian diffusion:

$$D_c = v_r d_r / 3 (1 - \cos \alpha) \tag{3.2}$$

where D_c = diffusivity (L^2 t^{-1}), v_r = velocity of motility (L t^{-1}), d_r = free length of random run (L), α = angle of turn (−).

If chemotaxis is not occurring, α is 90°. Thus, ignoring motility as a contributing factor in the process of transport to the substratum, transport rate could be underestimated by as much as 20 to 50 times. The non-Brownian diffusivity can be calculated from the velocity and the mean free path length. Vaituzis and Doetsch (1969) measured speeds up to 55.8 m/s for *Pseudomonas aeruginosa* with "track photography." Their results suggest a mean free path length of random run in the range of 50 to 85 μm. These values yield a non-Brownian diffusivity (Eq. (3.2)) of 10^{-3} mm^2/s for *P. aeruginosa*.

In their work studying the hydrodynamic effects of nearby surfaces on the swimming of the motile bacterium, *Escherichia coli.*, Frymier et al. (1995) observed that swimming cells remained near the solid surface (glass) for long periods, swimming parallel to the solid plane; in circles of ~ 50 μm in diameter. They postulated that the cells remained near the surface because they were reversibly adhered to the surface in a secondary free energy minimum, as described by the Derjaguin, Landau, Verwey, and Overbeek (DLVO) theory. However, experiments by Vigeant and Ford (1997), while documenting the attraction between the glass and the bacterial cells, did not follow expected trends predicted by the DLVO theory. The DLVO theory predicts significant differences in both amount of time spent swimming near the surface and the tendency to approach a surface as a function of solution ionic strength but neither trend was observed.

3.2.2 Laminar Flow

For laminar flow, the mechanism for mass transport of cells in the liquid phase is molecular diffusion as described by Fick's first law of law of diffusion, Eq. (3.3).

$$N_{Ax} = -D_{AB}(dC_A/dx) \qquad (3.3)$$

Fick's law states that diffusive flux of solute A in solvent B in the x-direction is proportional to the concentration gradient in that direction. Fick's law is not only used to describe diffusion of soluble components but can also be applied for large molecules or small particles, such as microbial cells, diffusing in water. In the case of particles, the Brownian (or non-Brownian) diffusion coefficient is used in Fick's law. If the cells are motile, their transport rate is increased significantly and can be estimated using Eq. (3.2).

3.2.3 Turbulent Flow

Within a turbulent flow regime, larger particles suspended within the fluid are transported to the solid surface primarily by fluid dynamic forces. Particle flux to the surface increases with increasing particle concentration. However, particle flux is also strongly dependent on the physical properties of the particle (e.g., size, shape, density) and is influenced by many other forces near the attachment surface.

Larger particles develop a sluggishness with respect to the surrounding fluid. As the particle approaches the wetted surface, eddy transport diminishes and the viscous sublayer exerts a greater influence. For soluble matter and small particles, diffusion can adequately describe transport in the viscous layer (Schlichting, 1968). For larger particles, other mechanisms must be considered to explain experimental observations. Transport of microbial cells (0.5–1.0 μm effective diameter) from the bulk fluid to the wetted surface can be influenced by several mechanisms, including the following:

- Diffusion (Brownian and non-Brownian)
- Gravity

3.2 TRANSPORT OF MICROBIAL CELLS TO THE SUBSTRATUM

- Thermophoresis
- Fluid dynamic forces
- Inertia
- Lift
- Drag
- Drainage
- Turbulent bursts

Eddy diffusion may be instrumental in dispersing particles in the turbulent core region, thus maintaining a relatively uniform concentration in that region. Particles in turbulent flow are transported to within short distances of the surface by eddy diffusion and are propelled into the viscous (or laminar) sublayer under their acquired momentum. Turbulent eddies supply the initial impetus and viscous drag slows down the particle as it penetrates the sublayer (Friedlander and Johnstone, 1957; Beal, 1970). For microbial cells, the inertial forces are very small because of their small diameter and density (in relation to water).

If the particle is traveling along the wall but faster than the fluid in the region of the wall, the lift force directs the particle toward the wall (Rouhiainen and Stachiewicz, 1970). This would normally be the case if particle density is greater than fluid density and the particle is settling toward the wall. Frictional drag forces can be significant, especially in the viscous sublayer region. The drag force slows down the particle as it approaches the surface and is proportional to the difference between particle velocity and fluid velocity.

If the mass density of the particle differs substantially from the fluid density, the gravity force may be significant. For microbial cells in turbulent flow, the gravity force is generally negligible. Thermophoresis is only relevant when particles are being transported in a temperature gradient (Lister, 1979). If the surface is hot and the bulk fluid is cold (e.g., a power plant condenser), the thermophoretic force will repel the particle from the surface. Brownian diffusion contributes little to the transport of microbial cells (> 1.0 μm diameter) in turbulent flow. For starved cells, which are on the order of 0.2–0.4 μm, Brownian diffusion will be more important. Fluid drainage forces can also be significant in aqueous systems (Lister, 1979). The drainage force supplies the resistance encountered by the particle near the wall due to the pressure required in the draining fluid film between the two approaching surfaces. This force is quite large for a microbial cell as it approaches the wall. Recent evidence indicates that "turbulent bursts" of fluid, resembling microscopic tornadoes, originate in the turbulent core and penetrate all the way to the wall (Cleaver and Yates, 1975, 1976). Particles in the bulk fluid can be transported in this manner to the wall by these turbulent bursts. Aside from the lift force, this is the only fluid dynamic force directing the bacterial cell to the wall. Thus, these turbulent bursts appear to be quite important in the transport of cells in turbulent flow. For a Reynolds number = 30,000 in a circular tube, the resulting bursts would have the following characteristics: burst diameter 1.1 mm, average axial distance between bursts 5.0 mm, and mean time between bursts 0.006 sec.

3.2.4 Surface Topography Effects

One factor contributing to transport and potentially to physicochemical effects on attachment is the influence of surface topographical features. Historically, it is assumed that bacteria preferentially stick to rougher surfaces for three reasons: (1) a higher surface area available for attachment, (2) protection from shear forces, and (3) chemical changes that cause preferential physicochemical interactions (Fig. 3.3). For example, work with bacterial suspensions has shown that a rough surface (matt steel) had 1.4 times more microorganisms attached than a smooth surface (electropolished steel) (Pedersen, 1990). Adhesion rate constants of *P. aeruginosa* to electropolished 316-L stainless steel plates were 100 times lower than those to 120-grit hand-polished surfaces (Vanhaecke, et al., 1990). Other work with stainless steel has shown that bacteria were preferentially associated with the grain boundaries (Gillis, 1993); although one should realize that grain boundaries exhibit not only a change in topography, but also a change in chemistry (Lumsden and Stocker, 1981). Bacterial cells tend to be associated with the chemically heterogeneous weld structures in stainless steel, but the issue is complicated by the presence of stagnant water at the weld crown and roots (Walsh et al., 1993). Cells deposit in pipe thread grooves, at the cusp of pipe and fittings, and in milling crevices hypothetically because they are protected from shear arising from flow (Duddridge et al., 1982). Perturbances in fluid flow create zones of negative pressure and thus eddy currents immediately downstream of the outcropping. Numerous studies have considered a wide range of shear stresses and found cells transported to the surface attached reversibly and were washed off again at rates increasing with the surface shear stress. There was a critical surface shear stress of 6–8 Nm^2 above which the extent of attachment dropped off sharply (Duddridge et al., 1982). Unfortunately, prior to 1999, most studies investigating the effects of surface topography on bacterial cell adhesion were unable to independently vary topography without also changing surface chemistry.

In a benchmark paper, the effect of substratum topography on bacterial surface colonization was studied by Scheuerman et al. (1999) using a chemically homogeneous silicon coupon as substratum. Their goals were (1) to discover how bacterial adsorption is affected by defined topographical features on a surface for which the surface chemistry is held essentially constant; (2) to determine the relative importance of bulk water fluid dynamics and bacterial motility on bacterial attachment; and (3) to assess the ability of strict colloid

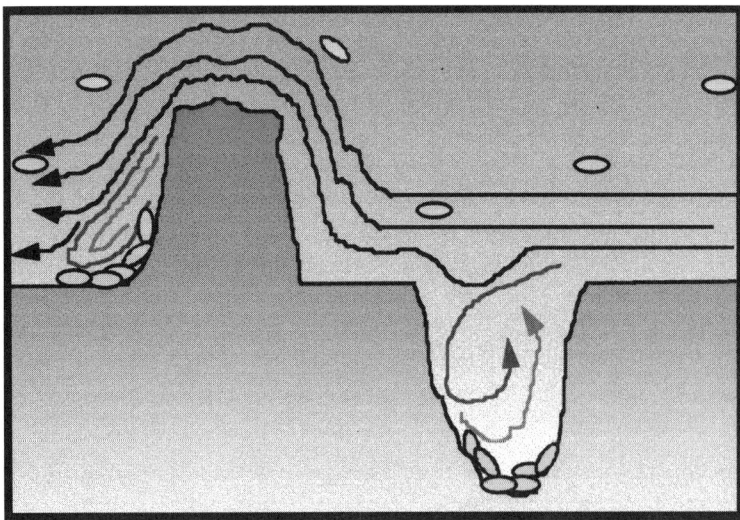

Figure 3.3 Prevalent concept regarding the effect of substratum roughness on bacterial adhesion. Pits and appendages were considered to provide bacteria "havens" of reduced fluid shear, thus promoting adhesion within the depths of such crevasses or behind obstructions. Work by Sheuerman et al. (1999) may dispel this widely held paradigm.

theory to explain bacterial attachment to surfaces. Their experiments used two motile bacterial species, one nonmotile strain of one species, and an inert colloidal particle. Colonization of an etched silicon coupon in a flow cell was observed through a window in the flow cell. "Grooves" 10 μm deep and 10, 20, 30, and 40 μm wide were etched on the coupon perpendicular to the direction of flow. Flow (Re = 5.5) of a bacterial suspension (10^8 cells/mL) was directed through a parallel plate flow chamber inverted on a confocal microscope.

Quantitative image analysis was used to document adsorption patterns and calculate rates of adhesion. Images were collected for each of three strains of bacteria: *P. aeruginosa*, and motile and nonmotile strains of *P. fluorescens*. A higher velocity experiment (Re = 16.6) and an abiotic control using hydrophilic, negatively charged microspheres were also performed. Using a colloidal deposition expression, the initial rates of bacterial attachment were compared.

P. aeruginosa attached at a higher rate than *P. fluorescens* mot+ which attached at a higher rate than *P. fluorescens* mot−. For all bacteria, the rate of adhesion was independent of groove size and was greatest on the downstream edges of the grooves. There was a significant effect of the presence of the grooves on the rates of attachment of the cells, with preferential attachment seen on the downstream edges. The rates of attachment followed the general trend of being highest on the downstream edge and lowest at the flat, control sections of the coupon. This effect was less pronounced for the higher shear experiment, but the trend was still present.

While the presence of grooves had a pronounced effect on bacterial attachment, there was no significant difference in attachment due to groove widths. This is somewhat surprising because models predict the hydrodynamics were substantially different. Hydrodynamic models predicted no disruption of the streamlines in the vicinity of a 10 μm wide groove but for the 40 μm groove, there is expected marked perturbation of flow, including corner eddies. In all cases, there was a compression of the streamlines at the edges, indicating that there was increased flow velocity and pressure, but this effect was expected to be more pronounced for the wider grooves.

In the Scheuerman experiments, only motile bacteria could be found regularly in bottoms of the grooves at numbers comparable to those on the control surfaces. Nonmotile organisms and colloidal beads could not be found in the grooves, suggesting that the presence of organisms in these troughs is a nonselective function of motility. This finding was somewhat surprising, as hydrodynamic models suggest that there would be eddies in the corners of the larger grooves that should have resulted in localized hydrodynamic entrainment of the cells and particles. Diffusion is a critical component of cell–surface interaction and can be influenced by cell motility, which can act to increase the effective diffusivity of the cells to the surface by up to four orders of magnitude (Mueller et al., 1992; Mueller, 1990). Piette and Idziak (1991) have shown that *P. fluorescens* flagellated cells attached in greater numbers than deflagellated cells. Because even nonflagellated cells attach, this effect has been attributed entirely to the added ability of motile organisms to reach the surface. The importance of motility in bacterial transport to the surface has also been reported by Camper et al. (1993).

The bead experiments yielded results substantially different from those of the bacteria, including the nonmotile strain. There was no edge effect noted with the beads and they were not found in the bottoms of the grooves. Although the colloidal particle experiments were limited in number, apparently inert particle behavior can not predict bacterial attachment. Every effort was made to match size, charge, and hydrophobicity of the beads with the cells, but it is possible that undetermined differences may be partially responsible for the lack of correlation. For example, it is likely that the surface characteristics of the cell are sufficiently complicated so that a simple measure of hydrophobicity is inadequate for predicting attachment behavior. This finding is critically important because recent studies on the fate of bacterial transport in porous media have elected to use inert colloidal particles to mimic bacterial cells.

The existing theories regarding the attachment of cells to a substratum, DLVO, and thermodynamics (see Section 3.3.2.1), assume that the transport of the cells to the substratum is not limiting. Care was taken to ensure that the surface chemistry was uniform and the cells within each experiment were at the same growth conditions. However, the cells showed preferential attachment to the downstream edges of the grooves where the boundary layer was thinner. Also, only the motile organisms adhered in the bottoms of the grooves whereas the nonmotile cells and beads did not. This difference cannot be explained as a physicochemical effect because there were no physicochemical differences between the tops and bottoms of the grooves. For motile bacteria, the initial attachment rate at the control position between grooves on the surface was similar to the bottom of the grooves. When the cell concentration was held constant and the flow velocity increased, the rate of attachment increased, with a decrease in the importance of position relative to the grooves. The widths of the grooves displayed no effect, even though the hydrodynamic model predicted differences in flow relative to the widths. Inert beads did not behave like the nonmotile bacteria, except that both beads and nonmotile cells showed little coverage at the bottoms of the grooves. The beads did not preferentially attach at the edges of the grooves. Results suggest that the transport of the cells to the surface, especially transport due to motility, dominates physicochemical effects in predicting cell attachment in flowing systems.

3.3 MOLECULAR ASPECTS OF BACTERIAL ADHESION

Nonspecific adhesion interactions are defined as interactions between a cell and surface or a cell and another cell that do not involve molecular structures on the cell surface–receptors–binding in a lock-and-key fashion, to a complementary ligand molecule on an surface. Nonspecific interactions are thus not biochemically specific, but they do act to increase or decrease the overall strength of the interaction. The three relevant types of nonspecific forces for cell–cell and cell–surface adhesion (Fig. 3.4), electrostatic forces, steric stabilization, and van der Waals forces, are discussed in Section 3.2.

Specific adhesion refers to the involvement of receptor:ligand bonds in cell adhesion (Fig. 3.5). In many cases, it is believed that permanent adhesion would not occur without these interactions, and thus the expression of receptors on a cell and/or the modulation of receptor affinity or receptor number with time controls the types of surfaces with which a cell will interact.

3.3.1 Substrata Preconditioning

3.3.1.1 Transport of Molecules. Transport of molecules and small particles (< 0.01–0.1 μm) in quiescent or laminar flow is described satisfactorily in terms of molecular diffusion by Fick's Law. In turbulent flow, the diffusion equation must be modified to include turbulent eddy transport (an eddy is a current or bundle of fluid moving contrary

3.3 MOLECULAR ASPECTS OF BACTERIAL ADHESION

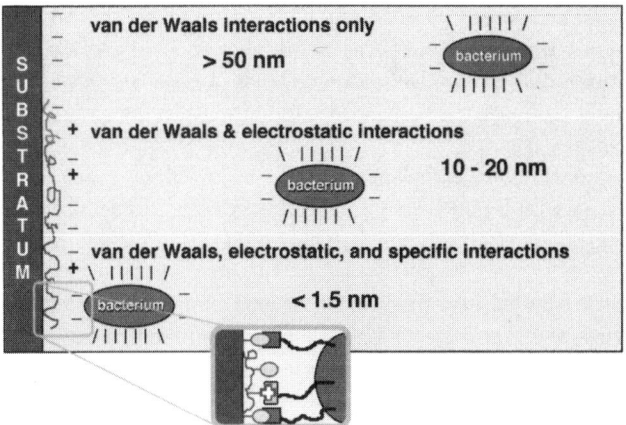

Figure 3.4 Summary of processes contributing to the nonspecific adhesion of bacterial cells to a substratum.

to the main current). Compared to larger particles such as bacterial cells, the transport of molecules and small particles is quite rapid. Consequently, adsorption of an organic conditioning film is frequently reported to occur "instantaneously."

3.3.1.2 Adsorption of the Conditioning Film. Adsorption of an organic film is an interfacial transfer process (i.e., the molecule is transferred from the bulk liquid compartment to the substratum compartment) and occurs within minutes of exposure causing changes in the properties of the wetted surface. Little and Zsolnay (1985) measured as much as 0.8 mg m^{-2} organic matter on stainless steel after 15 min exposure in seawater. Bryers (1980) observed 15 mg m^{-2} of organic material within minutes on glass in a laboratory system. The conditioning film on platinum in sea water may reach a thickness of 0.03–0.08 μm in 10 h (Loeb and Neihof, 1975).

Adsorption of an organic conditioning film is very rapid as compared to the other biofilm processes. Investigators have shown that materials with diverse surface properties (e.g., wettability, surface tension, electrophoretic mobility) are rapidly conditioned by adsorbing organic molecules when exposed to natural waters with low organic concentrations. In open natural ecosystems, these organic molecules appear to be polysaccharides or glycoproteins. In the case of biomedical implants, preconditioning molecules are predominantly blood plasma or extracellular matrix proteins or glycolipids. The layer is not static as evidenced by the results of Brash and Samak (1978) indicating significant turnover in molecular (proteinaceous) films on polyethylene. Protein molecules in the bulk fluid (laminar flow) were continuously exchanging with adsorbed proteins. Little and Zsolnay (1985) indicate that some of the adsorbed molecules on metal surfaces in sea water desorb or disappear with exposure time even though total accumulated material increases or remains unchanged. With increasing molecular weight, polymers adsorb more strongly due to multiple binding sites and they may displace molecules of lower molecular weight (Cohen-Stuart et al., 1980).

The conditioning film may be dynamic, that is, turnover of molecules may occur. Since polymers may have as many as 10^{+5} units in their chain, they may have that many bonds to the substratum. The probability of breaking all its segments on the substratum at one time is small, since a polymer may have between 30–60% of its segments in contact with the substratum (Barnett et al., 1981; Botham and Theis, 1970; Cafe and Robb, 1982). Thus, once approximately 10^{+4} segments are bonded to the surface, the polymer is unlikely to desorb. Consequently, the rate at which segments adsorb may control the period for molecular desorption and, thus, turnover at the substratum.

The conditioning film is generally observed or presumed to be uniform in both composition and coverage. There is little conclusive evidence that the spatial distribution of the conditioning film is uniform, so that a "patchy" distribution is possible. The film may be heterogeneously distributed over the substratum and may not cover the entire surface, especially when viewed at the scale of a microorganism or compared to the size of the appendages and polymers, which first interact with the substratum.

3.3.1.3 Composition of the Conditioning Film. Different substrata may accumulate conditioning films of different composition because of their differing surface properties such as potential, charge, and critical surface tension. Geesey and co-authors elaborate on this topic in greater detail in Chapter 9. Until recently, it was believed that the specific substratum had little influence on the composition of the organic conditioning film (Little and Zsolnay, 1985). This may actually be true in the case of open fresh and marine ecosystems (Baier, 1975; Baier and Weiss, 1975; Marshall, 1979; Loeb and Neihof, 1975; Zisman, 1964). However, most of the analytical methods employed in these earlier works were unable to detect differences in conditioning layer amounts, uniformity, and composition at the scale of the adhering bacteria.

Wiencek and Fletcher (1995) report self-assembled monolayers (SAMs) constructed from alkanethiols to produce a range of substrata that exposed different functional groups, that is, methyl and hydroxyl groups and a series of mixtures of the two. Percentages of hydroxyl groups in the SAMs and substratum wettability were measured by X-ray photoelectron spectroscopy and contact angles of water and hexadecane, respectively. SAMs exhibited various substratum compositions and wettabilities, ranging from hydrophilic, hydroxyl-terminated monolayers to hydrophobic, methyl-terminated monolayers. The kinetics of attachment of an estuarine bacterium to these surfaces in a laminar flow chamber were measured over periods of 120 min. The initial rate of net adhesion, the number of cells attached after 120 min, the percentage of attached cells that adsorbed or desorbed between successive measurements, and the residence times of attached cells were quantified by phase-contrast microscopy and digital image processing. The greatest numbers of attached cells occurred on hydrophobic surfaces. The initial rates of adhesion and the mean numbers

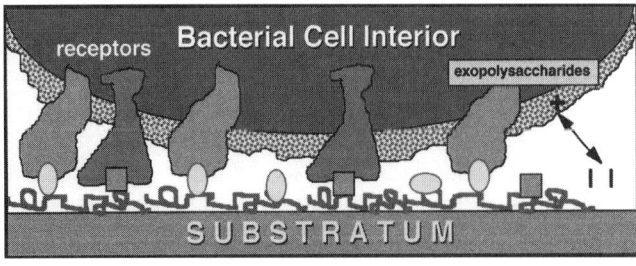

Figure 3.5 Schematic of specific receptor:ligand mediated bacterial cell adhesion.

of cells that attached after 120 min increased with the methyl content of the SAM and the contact angle of water. The percentage of cells that desorbed between successive measurements (ca. 2 min) decreased with increasing substratum hydrophobicity. With all surfaces, 60–80% of the cells that desorbed during the 120-min exposure period had residence times of less than 10 min, suggesting that establishment of firm adhesion occurred quickly on all of the test surfaces.

3.3.1.4 Influence of Conditioning Film on Substratum Properties.

Alterations in substratum properties as a result of conditioning film include (*a*) decreases in hydrophobicity (Baier, 1975); (*b*) both positively and negatively charged surfaces acquire a net negative charge (Loeb and Neihof, 1975); and (*c*) zeta potentials, contact potentials, and critical surface tensions are increased or decreased (Baier, 1975), depending on the initial surface energy. Adsorption of a conditioning film decreases the surface energy of clean, high energy surfaces (70 dynes cm^{-1}) but has little effect on low energy surfaces (20 dynes cm^{-1}) (Baier, 1975). One would expect that surfaces of initially differing energies, even after conditioning with similar adsorbed layers of protein, would continue to influence bacterial adsorption. This appears to be the case. Baier (1975) has shown that apparent strength of adhesion, as measured by spread areas of mammalian cells, is related to the substratum surface energy prior to the conditioning process. Thus, siliconized surfaces promoted desorption of adsorbed cells, even after protein conditioning of those surfaces. Baier (1980) concludes that the configurations of even similarly composed molecules within the conditioning films must be influenced by the initial surface state of the substratum.

Surface modification by organic macromolecules usually does change the surface charge. Loeb and Neihof (1975) showed the convergence of surface charge on various types (by means of microelectrophoresis experiments) of particles when exposed to natural seawater.

3.3.2 Nonspecific Adhesion Processes

Adhesion is an ubiquitous aspect of microbial life in most natural and engineered systems. Bacterial adhesion is often studied from a biological viewpoint, that is, based on the assumption that adhesion is brought about by specific molecules, appendages, or sites at the cell surface, called adhesins or receptors. Alternatively, general models for the description of adhesion can use a physiochemical viewpoint, for which literature provides two approaches. The first one is based on the Gibbs energy involved in the destruction and creation of interfaces (Absolom et al., 1983; Busscher et al., 1984) while the second approach is based on the DLVO theory for colloidal stability (Rutter and Vincent, 1984; vanLoosdrecht et al., 1989).

Based on observations, initial bacterial adhesion has been divided into two separate stages, namely reversible and irreversible adhesion. *Reversible adhesion* refers to that association of a bacteria to a surface where the bacterial cell continues to exhibit a two-dimensional Brownian motion and can be removed from the surface by relatively weak forces including the bacterium's own mobility. *Irreversibly adherent* bacteria no longer exhibit Brownian motion and cannot be removed by moderate shear forces.

Treating bacterial adhesion as a physicochemical process is complicated by the nature of bacterial cells, which are not "ideal" particles. They have no simple geometry, definitive boundary, or uniform exterior molecular composition. Internal chemical reactions can lead to changes in molecular composition both in the interior and at the cell surface, with molecules and ions constantly crossing the bacterium/water interface. Although altered, these chemical processes also continue after adhesion. Therefore, the adhered cells are rarely in complete chemical equilibrium with their environment. So, while many have tried to model bacterial adhesion processes with colloidal theory, be aware that interpretations must be regarded with caution.

3.3.2.1 Thermodynamic Approach. To thermodynamically describe bacterial adhesion, the major assumption is that the interfaces between solid/liquid (SL) and bacterium/liquid (BL) are replaced by a solid/bacterium (SB) interface (Absolom et al., 1983; Busscher et al., 1984). The change in the interfacial excess Gibbs free energy upon adhesion ($\Delta_{adh}G$, expressed in J/ m^2), is described in Eq. (3.4):

$$\Delta_{adh}G = G_{SB} - G_{SL} - G_{BL} \tag{3.4}$$

If $\Delta_{adh}G$ is negative, adhesion is thermodynamically favored and will proceed spontaneously. If the molecular composition of the interface, the pressure, and the temperature do not change, Eq. (3.4) may be written as a balance of interfacial tensions (γ, expressed in J/ m^2):

$$\Delta_{adh}G = \gamma_{SB} - \gamma_{SL} - \gamma_{BL} \tag{3.5}$$

Equations (3.4) and (3.5) only apply if both interacting surfaces make direct contact.

By measuring the contact angle (θ) of a drop of liquid on a solid surface, it is possible to obtain information about the interfacial tensions by applying Young's equation (subscript V stands for vapor).

$$\gamma_{SV} = \gamma_{SL} + \gamma_{LV} \cos\theta \tag{3.6}$$

Thus, for a given liquid, $\cos\theta$ depends on the difference, $\gamma_{SV} - \gamma_{SL}$. The smaller γ_{SL} is (i.e., the more the surface properties of the solid and liquid are alike), the higher $\cos\theta$. The contact angle of water on a surface is therefore large ($\cos\theta$ small) when the surface contains many nonpolar (hydrophobic) groups.

To determine the solid/liquid and solid/vapor interfacial Gibbs energies, a second relation in addition to Eq. (3.6) is needed. Two approaches have been proposed, namely the "equation of state" approach (Absolom et al., 1983; Neumann et al., 1974) and the "geometric mean" approach (Busscher et al., 1984). Both procedures have the same underlying principle based on a model proposed by Fowkes (1964), and both therefore lead to comparable results (vanLoosdrecht et al., 1987; vanLoosdrecht, 1988). In the "equation state" approach only one contact angle measurement is required for the calculation of the solid/liquid and solid/vapor interfacial Gibbs energy. For the "geometric mean" approach, it is necessary to measure contact angles of a range of liquids. This allows one to also obtain the solid/vacuum interfacial Gibbs energy (an intrinsic material property). Because of the assumptions that have to be made to calculate interfacial Gibbs energies from contact angle measurements, most researchers consider the contact angle itself as a measure for the hydrophobicity.

3.3.2.2 Colloidal Approach. Bacteria can also be considered as living colloidal particles. At the pH of most natural waters, bacteria exhibit a net negative surface charge. The DLVO theory assumes that the total long range interaction (> 1 nm) between two colloidal

3.3 MOLECULAR ASPECTS OF BACTERIAL ADHESION

particles is a summation of Van der Waals and Coulomb interactions (Rutter and Vincent, 1984). In contrast to the thermodynamic approach, the colloidal approach describes the interaction between a particle and a surface as a function of the separation distance (for separation distances > 1 nm).

Extended DLVO theory assumes that there are three major types of nonspecific interactions important for cell adhesion processes: electrostatic forces, steric stabilization, and van der Waals or electrodynamic forces (Bongrand and Bell, 1984; Israelachvili and McGuiggan, 1988). Although all are present at once, each is dominant at a different cell–cell or cell–substrate separation distance.

Electrostatic forces, which may be attractive or repulsive, result when two charged surfaces are brought together. The surface of a cell consists not only of an approximately 70 Å thick lipid bilayer containing receptors and other embedded molecules but also, in some cases, a extracellular capsular coating. Surface receptor binding sites can appear within this coat.

For cell:cell adhesion, the bringing together of two negatively charged surfaces results in an overall repulsive electrostatic force between the cells. For cell–surface adhesion, the electrostatic forces may be positive or negative, depending on the composition of the surface. For example, a positively charged surface will attract most cells, so nonspecific cell–surface adhesion may result regardless of the availability of receptors and complementary ligands.

Figure 3.6 shows three variations of the total interaction Gibbs energy (G_{tot}) as a function of separation (H) between two likewise charged particles, for different ionic strengths. At low ionic strength (Case a), $G_{tot}(H)$ has a positive maximum that constitutes a barrier for adhesion in the primary minimum. The maximum in $G_{tot}(H)$ is suppressed by increasing ionic strength, due to a reduction of G_E. At certain intermediate values of the ionic strength (Case b), the maximum is so low that a fraction of the particles may contain sufficient thermal energy to pass the barrier (i.e., slow adhesion takes place). At higher ionic strength (Case c), when the maximum in $G_{tot}(H)$ is suppressed to a value = 0, all particles can reach the primary minimum. This results in a strong, irreversible binding.

At a somewhat larger separation another, more shallow, minimum in $G_{tot}(H)$ exists: the so-called secondary minimum. It is most pronounced at intermediate ionic strengths and is deeper for systems having a larger Hamaker constant and for relatively large particles, like

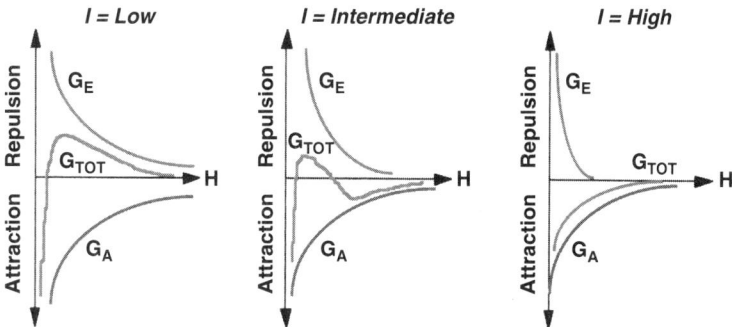

Figure 3.6 Effect of fluid ionic strength on Gibb's nonspecific adhesion interaction energy for a substratum and cell particle of the same net surface charge.

microbial cells. Usually the secondary minimum does not reach large negative values so that particles are reversibly captured in this minimum. Needless to say that opposite charges on the interacting surfaces yield negative values for G_E and hence G_{tot} at all separations, leads to primary minimum adhesion.

At short separation, say $H < 1$ nm, short range interactions (e.g., hydrogen bonding, ion pair formation, etc.) are effective. They determine the strength of adhesion in the primary minimum. The DLVO theory is only able to predict whether primary minimum adhesion occurs, but the depth of this minimum cannot be predicted because short range interactions are not incorporated in this theory.

The capsular layer, around a bacterial cell or covering a substratum coated in an existing biofilm, consists of polymers in a hydrated environment. As two polymer coats approach each other, the layers overlap and some of the water of hydration is pushed out. A repulsive force termed *steric stabilization* results because of the osmotic tendency of water to return and because of the steric compression of the polymer chains. This force dominates at small separation distances and likely acts to prevent significant interpenetration of the slime layers on two adhering cells.

3.3.3 Receptor–Ligand Mediated Specific Adhesion

3.3.3.1 Observations of Bacterial Specific Adhesion. Although it may appear in the following section that bacterial specific adhesion is dominated by protein:protein or lectin interactions, this is only because the major source of definitive research comes from the infection:pathogenesis microbiology sector or literature on biomedical device-based infections. This is not to say that bacteria in open water or in heat exchanger tubes do not employ specific adhesion mechanisms; they most likely do, but in these cases the ligands may be complex carbohydrates, humic acids, or proteins. Unfortunately, less quantitative information about these latter receptor:ligand pairs exists in comparison to that found in the infection literature.

In intact organisms, most extracellular matrices (ECM) are covered by epithelial or endothelial cells and therefore are not available for bacterial binding and colonization. However, any type of trauma (e.g., injury, surgery, biomedical implant placement) that damages the host tissue may expose the ECM and allow colonization by bacteria. Accordingly, many microorganisms that cause opportunistic infections have been shown to express microbial surface components recognizing adhesion matrix molecules (termed MSCRAMMs) by Höök and co-workers (Patti and Höök, 1994; Patti et al., 1994). A bacterium can simultaneously express several MSCRAMMs that recognize a variety of matrix proteins. Furthermore, some microorganisms, such as enteropathogenic *Yersinia* (Schulze-Koops et al., 1993) and *Porphyromonas gingivalis*, appear to express an adhesin that can bind multiple host ligands (Lantz et al., 1991a,b). Ligand-binding sites in MSCRAMMs appear to be defined by relatively short contiguous stretches of amino acid sequences (motifs). Because a similar motif can be found in several different species of bacteria, it appears as though these functional motifs are subjected to interspecies transfer. However, the ligand-binding sites in only a few MSCRAMMs have been defined so far, and therefore, generalizations are risky.

First, to be classified as an MSCRAMM, the molecule of interest must be bound to the microbial cell surface. Second, the microbial component must recognize a macromolecular ligand that can be found within the extracellular matrix. These ligands include components such as collagen and laminin that are found exclusively in the ECM, whereas other molecules we have defined as ligands (e.g., fibronectin, fibrinogen, and vitronectin) are part-time ECM molecules and also occur in soluble forms in body fluids such as blood plasma. Other potential MSCRAMM ligands, such as heparin sulfate proteoglycans, occur both in ECM forms and as intercalated cell membrane proteoglycans. Third, the MSCRAMM's interaction with the extracellular matrix component should be of high affinity and exhibit a high degree of specificity; that is, unrelated molecules should not be able to significantly interfere with the interaction between the MSCRAMM and its ECM ligand. Thus, adhesins of the lectin type that recognize carbohydrate determinants present on many different classes of molecules should not be classified as MSCRAMMs even though they may bind to ECM components. One could argue that adhesins recognizing glycosaminoglycans are both MSCRAMMs and lectins. Höök and co-workers have chosen to classify these adhesins as MSCRAMMs. Although numerous bacteria have been shown to bind a variety of ECM components (see Tables 1 and 2 in Patti et al., 1994), the molecules involved in these interactions have in many cases not been identified nor characterized at a molecular level. Before a microbial component can be classified as an MSCRAMM, its interaction with the ECM ligand should be characterized in sufficient detail to show that it fulfills the foregoing criteria.

A single MSCRAMM can bind several ECM ligands. For example, the plasmid-encoded outer membrane protein *YadA*, which appears to be a collagen-binding MSCRAMM on enteropathogenic *Yersiniae* (Emödy et al., 1989), can also bind laminin and a form of fibronectin (Thole et al., 1992). A fibrinogen-binding MSCRAMM present on *Porphyromonas gingivalis* seems to also recognize fibronectin (Lantz et al., 1991a, 1991b). In addition, a microorganism can express several MSCRAMMs that recognize the same matrix molecule. For example, *Staphylococcus aureus* appears to express several fibrinogen-binding proteins (Boden and Flock, 1989; McDevitt et al., 1994), and *Streptococcus dysgalactiae* (Lindgren et al., 1992, 1993) and *S. aureus* (Jonsson et al., 1991) each have at least two genes encoding fibronectin-binding MSCRAMMs. This type of variation in the interactions between MSCRAMMs and their matrix ligands resembles the interactions between the eukaryotic integrins and matrix molecules, where one integrin can bind several different ligands.

The nonspecific adhesion forces described in Section 3.3.2 provide only a weak attractive force, of the order of 10^3 dyne/cm^2 (10^{-5} dyne/μm^2) for typical cell–cell separation distances. In order to strengthen adhesive interactions and provide specificity, cell surface receptors must play a role. One can examine the strength of a receptor:ligand bond from both an equilibrium and kinetic standpoint. From the equilibrium perspective, Bell (1978) estimates the strength of a single receptor:ligand bond from the relation $f_c = \Delta G/r_o$, where f_c is the force required to break the bond, ΔG is the free energy of bond formation, and r_o is the range of the bond potential energy minimum. For $r_o = 10$ Å and $\Delta G = 13$ kcal/mole (corresponding to an equilibrium dissociation constant $K_D = 10^{-9}$ M), then $f_c = 9 \times 10^{-6}$ dyne/bond. A covalent bond with $\Delta G \sim 70$ kcal/mole and $r_o \sim 1$ Å, would require $f_c \sim 4 \times 10^{-4}$ dyne/bond. Noncovalent receptor:ligand bonds with $K_D < 10^{-9}$ M (i.e., higher affinity bonds) would fall somewhere in between, with a logarithmic dependence of f_c on $(K_D)^{-1}$.

A related approach to analyzing the strength of a receptor:ligand bond is by the kinetic approach. This approach was introduced by Bell (1978) and is based on the kinetic theory of isotropic materials (Zhurkov, 1965). Considering the forward and reverse rate constants,

k_f and k_r, for receptor:ligand association and dissociation, Bell proposed that the dissociation rate constant is increased by a physical stress, as stated in Eq. (3.7),

$$k_r = k_{r,0}\exp\left\{\frac{\gamma f}{K_b T}\right\} \quad (3.7)$$

where $k_{r,0}$ is the unstressed dissociation rate constant, f is the applied force stressing a bond, and γ is a parameter loosely defined as the bond interaction range and likely of the order of rT is absolute temperature and K_b is Boltzmann's constant. Bell used Eq. (3.7) to determine the force needed to detach a cell initially attached via multiple receptor:ligand bonds. He obtained an expression for the approximate total detachment force divided by the initial number of bonds, or the *adhesion strength per bond*, F_{bond}:

$$F_{bond} \approx 0.7\left(\frac{k_B T}{\lambda}\right)\ln\left\{\frac{n_s}{K_D}\right\} \quad (3.8)$$

where n_s is the surface ligand density and K_D is the surface equilibrium dissociation constant. For $\gamma = 10$ Å, $n_s = 10^{11}$ ligand number/cm^2, and $K_D = 10^5$ number/cm^2 (corresponding to a solution value of 10^{-9} M; using an effective volume with height of 200 Å), Eqs. (3.7–3.8) yields $F_{bond} = 4 \times 10^{-6}$ dyne/bond. Note that the kinetic approach thus gives a similar but lower estimate for the bond strength than does the equilibrium approach, because it does not require all bonds to break simultaneously. An advantage to the kinetic approach is that it permits dynamic modeling. It is typically assumed that k_r is unaffected by stress, but some analyses have suggested how it could vary with strain (see Dembo et al., 1988).

Now compare estimates of specific bond interactions to previous estimates of nonspecific interactions. We stated earlier the estimate that a force per unit area of about 10^{-5} dyne/μm^2 is sufficient to detach a cell held by only *nonspecific* forces from another cell. This is equivalent to a single high-affinity receptor/ligand bond per μm^2 of cell/cell contact area. Given that the cell surface receptor number density is usually about 10–100/μm^2, receptor/ligand bonds can be expected to provide at least an order of magnitude stronger adhesive strength than nonspecific interactions.

3.3.3.2 Mathematical Model of Bacterial-Specific Adhesion.
Figure 3.7 is a representation of the specific adhesion of a bacteria to a ligand-covered substratum, which is a bacterial analogy (Wang and Bryers, 1997) of the benchmark model by Hammer and Lauffenburger (1987) for mammalian cell specific adhesion. The bacterial cell is modeled as a *rigid* sphere of radius R_c. The cell is covered uniformly with R_T number of receptors on its surface. Ligand density on the substratum is N_l. Under uniform shear flow, which is characterized by negligible inertia, the shear force and torque acting on the cell causes translational and rotational motions in addition to the shear contribution. Meanwhile, the shear force can also lift those cells in the reversible adhesion stage away from the substratum.

Upon close approach to the surface, a contact area for receptor:ligand binding is formed on the bacterial cell and the bond is built. The contact area is assumed to be a disk of radius a and is considered to remain constant throughout the adhesion period. Based on the geometric estimate by Cozens-Roberts et al. (1990), a is ~ 10% of the cell radius R_c. The bond stress is assumed to be constant within the contact area A_c.

3.3 MOLECULAR ASPECTS OF BACTERIAL ADHESION

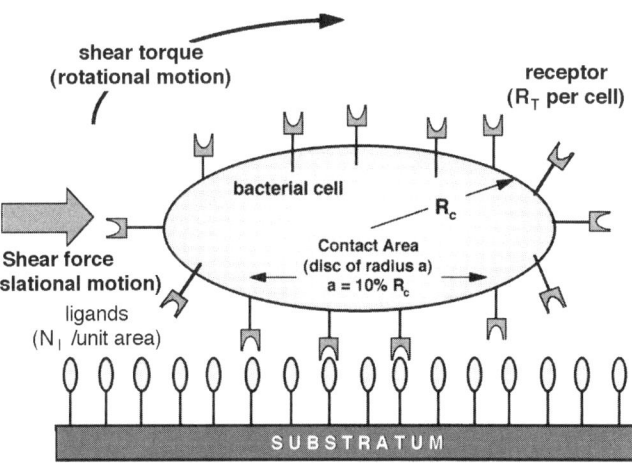

Figure 3.7 Mathematical representation of receptor:ligand mediated bacterial cell adhesion (Wang and Bryers, 1997) based on the mammalian cell model of Hammer and Lauffenburger (1987). Terms defined in accompanying text.

Unlike mammalian cells, bacteria do not spread upon surface attachment. Also, surface receptors on the bacterial cells are assumed not to move on the cell surface and the contact area does not change. Only those receptors that happen to be within the contact area participate in the specific binding and thus mediate adhesion. In addition, only one site on each receptor/ligand molecule is assumed available for binding.

The whole system is modeled as a reactor with a feed of suspended cells, X_{in}, as well as an entering growth-limiting substrate, S_{in}, and a clean ligand-coated surface. The basic assumption is that the ligand number exceeds the number of receptors available for the binding within the contact area, that is, $N_l A_c \gg R_T A_c / S_{area}$, where S_{area} stands for surface area of one cell. Therefore, the adhesion rate constant is dominated by R_T. A set of nonlinear ordinary differential equations was derived by Wang and Bryers (1997), based on mass balances of suspended cell concentration (X), attached cell concentration (B), and the substrate concentration (S).

First-order kinetics is used to express the adhesion rate (R_{adh}), detachment rate (R_{det}), and growth rate (R_g) on the substratum, with corresponding rate constants K_{adh}, k_{det}, and μ_g. As per the assumptions stated previously, the adhesion rate is dominated by the number of receptors, R_T. Therefore, the overall adhesion rate constant must be a function of receptor number on the cell, R_T. Taking R_T into account, the adhesion rate is written as:

$$R_{adh} = K_{adh} X = k_f R_T \frac{XV}{A} \qquad (3.9)$$

where $K_{adh} = k_f R_T \frac{V}{A}$ = overall adhesion rate constant (cm/min), X = number of cells suspended in bulk liquid at time t (number of cells/cm^3), k_f = specific rate constant (per cell/min − [number of receptor molecules]), R_T = receptor number (number of receptor molecules/per cell), V = reactor volume (cm^3), A = substratum area (cm^2).

Microbial growth is assumed at or near zero to assess just the contributions of adhesion and desorption.

In summary, the time rate of change in B, X, and S is given as follows:

Adherent Cells:

$$\frac{dB}{dt} = K_{adh} - k_{det}B = [k_f R_T XV/A] - k_{det}B \quad (3.10)$$

Suspended Cells:

$$dX/dt = \frac{F}{V}(X_{in} - X) - (k_f R_T X)/V + (k_{det}BA)/V \quad (3.11)$$

where B = number of cells attached to surface per unit area at time t (number of cells/ cm²), X = number of cells suspended in bulk liquid at time t (number of cells/cm³), X_{in} = inlet concentration of cells in bulk liquid (number of cells/cm³), t = time (min), F = flow rate into the reactor (cm³/min), D = dilution rate (min⁻¹) = F/V

Once the cells come in close approach to the deposition surface, the receptors and the ligands form a specific bond and thus the bacterial cells adhere to the surface. Meanwhile, the adherent cells are subject to possible detachment from the surface if the binding force between the cells and the surface is not strong enough to withstand the shearing forces of flow (Morat, 1985). The detachment rate constant is in the form of a decreasing lifetime function (Zhurkov, 1965; Bell, 1978), which is related to the characteristic bond length (γ), the force acting on each bond (F_b), and the temperature (T). When the bonds are unstressed (at equilibrium), k_{det} will take the base value of k^0_{det}:

$$k_{det} = k^0_{det} \exp\left(\frac{\gamma F_b}{K_b T}\right) \quad (3.12)$$

where K_b = Boltzmann constant = 1.38E-16 erg/mole¹ − °K¹, T = absolute temperature (°K).

F_b is calculated from the total force acting on the bonds per cell, F_t, and the density of bonds formed per cell, N_b. For this calculation, we assume that the adherent cell is in mechanical equilibrium with its surroundings. Therefore, the bonds experience the force and torque that are applied to the cell by the shear fluid (Hammer and Lauffenburger, 1987). Eq. (3.13) shown here is based on the analysis performed by Goldman et al. (1967a, 1967b) and Hammer and Lauffenburger (1987):

$$\begin{aligned} F_t &= N_b A_c F_b \\ &= 6\pi\mu R_c^2 \eta \{(1 + \frac{S_c}{R_c})^2 F_2^s (1 + \frac{9\pi^2 R_c^2}{16a^2}) + (1 + \frac{S_c}{R_c})F^s \tau^s \frac{3\pi^2 R_c^2}{4a^2} + \frac{(\pi R_c)^2}{4a^2} \tau_2^s\}^{1/2} \\ &= 6\pi\mu R_c^2 \eta \, (\text{Root}_1) \end{aligned} \quad (3.13)$$

where N_b = bond density (number of bonds · cm⁻²), A_c = contact area size (cm²/per cell), F^s = dimensionless shear force (unity), τ^s = dimensionless shear torque (unity), S_c = separation distance (cm), μ = fluid viscosity (g · cm⁻¹ · sec⁻¹), η = shear rate (sec⁻¹).

3.4 DYNAMICS OF BIOFILM REACTIVITY

Here, let

$$Root_1 = \left\{(1 + \frac{S_c}{R_c})^2 F^{s2}(1 + \frac{9\pi^2 R_c^2}{16a^2}) + (1 + \frac{S_c}{R_c})F^s \tau^s \frac{3\pi^2 R_c^2}{4a^2} + \frac{(\pi R_c)^2}{4a^2}\tau^{s2}\right\}^{1/2}$$

for simple notation. F^s and t^s are functions of separation distance only. Details of this mathematical derivation are provided in Wang (1993).

To obtain the expression for N_b, the idea of adsorption in a heterogeneous system is adapted (Marshall, 1976; Gibbons et al., 1976). In equilibrium, N_b, the bond density, is given as

$$N_b = \frac{k_f R_T N_l A_c / S_{area}}{k_f R_T A_c / S_{area} + k^0 \text{det}} \quad (3.14)$$

where N_b = bond density (number of bonds/cm^2), N_l = ligand density (number of ligand molecules/cm^2), A_c = contact area size (cm^2/per cell), S_{area} = surface area of one cell (cm^2/per cell).

Solutions to the nonlinear ordinary differential equations were obtained by Wang and Bryers (1997) using a fourth-order Runge-Kutta algorithm. Analyses on the effect of various parameters in bacteria adhesion, such as ligand density, receptor number, shear rate, and bond density, were carried out for the case of both pure and mixed culture adhesion.

3.4 DYNAMICS OF BIOFILM REACTIVITY

3.4.1 Metabolic Responses to Adhesion

3.4.1.1 Cell:Cell Signaling Control of Various Biofilm Processes

3.4.1.1.1 Gram-negative Bacteria. Davies and Geesey (1995) used reporter gene technology to observe the regulation of the alginate biosynthesis gene, *algC*, in a mucoid strain of *P. aeruginosa* in developing and mature biofilms. The plasmid pNZ63, carrying an *algC-lacZ* transcriptional fusion, was shown to not be lost by segregation in continuous culture over a period of 25 days in the absence of selection pressure. Biofilm cells under bulk phase steady state conditions demonstrated fluctuations in *algC* expression over a 16-day period, although no consistent trend was obvious. In vivo detection of *algC* up-expression in developing biofilms was carried out with a fluorogenic substrate for the plasmid-borne *lacZ* reporter gene product (β-galactosidase). Using microscopic image analysis, cells were tracked over time and analyzed for *algC* activity (via *lacZ* expression). During the initial stages of biofilm development, cells attached to a glass surface for at least 15 min exhibited up-expression of *algC*, detectable as the development of whole-cell fluorescence. However, initial cell attachment to the substratum appeared to be independent of *algC* promoter activity. Furthermore, cells not exhibiting *algC* up-expression were shown to be less capable of remaining at a glass surface under flowing conditions than were cells in which *algC* up-expression was detected.

Such studies (Davies and Geesey, 1995; Boyd et al., 1993; Boyd and Chakrabarty, 1994) have shown that alginate synthesis is up-regulated in *Pseudomonas* species when they become associated with a surface. As the alginate is synthesized, biofilm forms, resulting in the formation of cell clusters comprising cells embedded within dense alginate gel matrices with these clusters separated by open torturous channels. Recent advances in cell–cell communication in bacteria have shed light on the possible mechanism by which biofilm matrix polymer production and dissolution may be regulated. Research in gram-negative species has demonstrated that specific molecules, known as homoserine lactones (HSLs), are released by bacteria in batch cultures. These HSL molecules pass readily through the cell's membrane where they accumulate to a threshold concentration at which they are able to induce the transcription of specific genes. Due to this mode of action, these molecules are referred to as *autoinducers*. All known, small, diffusible autoinducers in gram-negative bacteria belong to the class of N-acylated homoserine lactones (Ochsner and Reiser, 1995). Two chemically and genetically distinct autoinducer-dependent regulatory circuits are found in *P. aeruginosa*. The *lasI* gene is responsible for the production of N-(3-oxododecanoyl)-L-homoserine lactone (OdDhl) (Pearson et al., 1995) and the *RHII* gene is responsible for the production of N-butylyl-L-homoserine lactone (BHL) (Latifi et al., 1995; Winson et al., 1995). In *P. aeruginosa*, quorum sensing has been shown to be involved in the regulation of a large number of exoproducts including elastase, alkaline protease, *LasA* protease, hemolysin, cyanide, pyocyanin, and rhamnolipid (Gambello et al., 1993; Latifi et al., 1995; Winson et al., 1995). Most of these exoproducts are synthesized and exported maximally as *P. aeruginosa* enters stationary phase. It is during stationary phase also, that gram-negative bacteria have been shown to develop stress response resistance that is coordinately regulated through the induction of a stationary-phase sigma factor known as *RpoS* (Hengge-Aronis, 1993). Biofilm bacteria are generally considered to show physiological similarity to stationary phase bacteria in batch cultures. Thus, it is presumed that the synthesis and export of stationary-phase autoinducer-mediated exoproducts occurs generally within biofilms. The stationary phase behavior of biofilm bacteria may be explained by the activity of accumulated HSL within cell clusters. The mechanism causing biofilm bacteria to demonstrate stationary-phase behavior is hinted at by the recent discovery that *RpoS* is produced in response to accumulation of BHL in *P. aeruginosa* cultures (Latifi et al., 1996).

The production of alginate by *P. aeruginosa* has been shown by many authors to be a stationary phase response. Furthermore, the breakdown of alginate on solid media has been shown to occur after approximately 50 h incubation. These observations indicate that HSLs may be involved in the regulation of the production and digestion of alginate in biofilms composed of *P. aeruginosa*.

By artificially manipulating the binding of homoserine lactones to their cognate receptor molecules, it might be possible to control the formation, persistence, and dispersion of microbial biofilms. Hypothetically, the addition of an analog that blocks the binding of OdDhl to its cognate receptor (*LasR*) may prevent the production of the biofilm polymer matrix as the bacteria continue to multiply. Potentially, cell aggregates formed under these conditions could be easily dispersed by the addition of simple surfactants. Further, existing biofilms could be treated with the homoserine lactone, BHL, to induce the release of enzymes (e.g., lyase) which would digest the biofilm matrix material and disperse the biofilm into the bulk medium. Thus, nontoxic treatment regimens could be used as effective means of controlling biofilms.

3.4.1.1.2 Gram-positive Bacteria. Unlike gram-negatives that employ transcription control of phenotypic expression by HSL signal molecules, in gram-positive bacteria, the diffusible molecules are small peptides that bind to membrane bound receptors.

3.4 DYNAMICS OF BIOFILM REACTIVITY

Two distinct pathogenic mechanisms, adhesion to polymer surfaces and subsequent accumulation of sessile bacterial cells, are considered important pathogenic steps in biomaterials infections caused by *Staphylococcus epidermidis*. Hussain et al. (1997) recently generated a mutant strain M7, from *SE* RP62A which is otherwise unaffected except in its accumulation on glass or polystyrene surfaces and that it lacks a 140-kDa extracellular protein. The membrane-bound protein structure was hypothesized to be a signal receptor for very small (5–8 amino acids) peptide signal molecules, secreted by *S. epidermidis* in response to proximity to a substratum. To evaluate the role of this protein in biofilm formation, Hussain et al. harvested extracellular membrane proteins from *S. epidermidis* RP62A grown on dialysis membranes placed over chemically defined medium, purified the protein by using ion-exchange chromatography, and raised antiserum to the 140-kDa receptor.

The purified 140-kDa protein was subjected to N-terminal amino acid sequence analysis, and the following sequence was obtained:

```
      V  S         T           Q
   TQ T(A)NVS(G) QTYQDPTYV (P)(K)
      A  M         K           D
```

The nine-amino-acid unambiguous sequence was evaluated for homology with the Swiss-Prot database, and no deposited protein sequence showed a high level of homology (-67%).

The antibody recognized only a single band in a Western immunoblot of the crude extracellular extract. With a microtiter biofilm test (batch test, no fluid mixing), antiserum at a dilution of $< 1:1000$ eliminated biofilm accumulation of strain RP62A, whereas preimmune serum did not. The 140-kDa receptor was found only on extracellular membranes of bacteria that would grow as under sessile populations. Of 58 coagulase-negative clinical isolates, 32 strains were 140-kDa receptor positive and produced large amounts of biofilm. The 26 strains that were 140-kDa antigen negative did not adhere to the surface nor did they form any biofilm.

One process that may influence or instigate the adhesion process is autoinduction of plasmid conjugation. The main mechanism of plasmid transfer, conjugation, requires the intimate contact of two bacterial cells, a donor and recipient. In certain bacterial species, the initiation of conjugation is within the species own control. Conjugation between sexually differentiated bacterial cells requires physical and chemical interaction. *Enterococcus faecalis*, a nonmotile gram-positive species, needs cell interactions for sex plasmid transfer during conjugation. *E. faecalis* produces a family of peptide signaling molecules designated as sex pheromones (Clewell, 1993; Dunny et al., 1979; Dunny et al., 1995). Each pheromone triggers the conjugal transfer system of a particular plasmid such as the hemolysin plasmid pAD1, the bacteriocin plasmid pPD1, or the antibiotic tetracycline resistance plasmid pCF10 (Dunny et al., 1995; Dunny et al., 1981). When the plasmid-containing donor bacteria are in close proximity to a plasmid-free recipient, the conjugal transfer system encoded on the plasmid is activated, and a copy of the plasmid is transferred to the recipient. At least 18 plasmids that encode a pheromone response have been described (Clewell and Weaver, 1989). Several pheromones have been purified and shown to be different hydrophobic octapeptides, or in one case a heptapeptide (Galli et al., 1990). Pheromones are typically active at concentrations below 5×10^{-11} M, and as few as two molecules per donor cell may be sufficient to induce the transcription of genes on the target plasmids (Clewell et al., 1982; Weaver and Clewell, 1990). Not only is the response to pheromone very sensitive, but its specificity is also high. Pheromone cAD1 is unable to induce expression (indicated by clumping by cells) from the heterologous plasmid pPDI, even at a concentration of 1 μM, which is 10^5-fold higher than that needed to induce expression from the homologous plasmid.

The pheromone-induced surface-bound adhesins specified by plasmids pAD1 and pCF10 have been purified, and their structural genes have been cloned and sequenced (Clewell and Weaver, 1989; Galli et al., 1990). These adhesins are large, closely related proteins that may form dense, hairlike structures on the cell wall of the induced bacteria (Galli et al., 1989). The ligand for the adhesin on *E. faecalis* cells is a surface constituent present on all cells, regardless of whether they carry a plasmid. Available evidence strongly favors involvement of lipoteichoic acid, the major wall antigen of gram-positive cells (Ehrenfeld et al., 1986; Trotter et al., 1990). These and other experiments imply that the adhesion system is a heterophilic, adhesin-lipoteichoic acid binding system (Dunny, 1990).

Once transferred from donor to recipient, the plasmid directs the synthesis of a plasmid-encoded inhibitor that specifically blocks the inducing action of the cognate sex pheromone (Clewell et al., 1982). The inhibitors for *cAD1* and *cPD1* have been purified and shown to be hydrophobic octapeptides that have sequences weakly related to their corresponding pheromones (Galli et al., 1990). In one case, the inhibitor has three identical residues among seven total. The inhibitor peptide neutralizes its cognate pheromone, probably by competition, thus preventing a donor cell from responding to its own pheromone.

3.4.2 Cell Replication, Maintenance, and Endogenous Decay

3.4.2.1 *Microbial Growth and Replication.*
When bacteria are inoculated into a medium containing all requirements for growth and incubated under appropriate conditions, a tremendous increase in numbers occurs within a relatively short time. With some species the maximum population is reached within 24 hours, whereas others require a much longer period of incubation. The term growth, as commonly applied to bacteria and other microorganisms, usually refers to changes in the entire crop of cells rather than to change in the individual organism. More frequently than not, the inoculum contains thousands of organisms; thus growth denotes the increase in number or mass beyond that present in the original inoculum.

The most common process for reproduction in bacteria is that known as binary fission, or transverse fission. The terminal events of this reproductive process find the single cell dividing into two, after the development of a transverse cell wall to separate the intracellular contents. This is an asexual reproductive process. the exact morphological transitions leading up to the event of binary fission are not clearly understood.

However, if we start with a single viable bacterium in a growing culture, we can postulate the following developments: the nutrients from the medium are taken into the cell by a selective process. The enzyme systems of the bacterium then convert the chemicals (nutrients) that have been assimilated into macromolecules charactcristic of the particular organism. A doubling of DNA occurs and cell elongation may follow (this is more evident in bacilli than cocci). The contents of the cell undergo reorganization to distribute the material for two cells, which are formed by a transverse wall, or septum, that subsequently develops by an invagination of the cytoplasmic membrane.

Another means of reproduction observed in some bacteria (e.g., Actinomycetes) is the formation of a filamentous growth followed by fragmentation into small units which then develop into cells of normal size. Again, some bacteria are capable of reproducing by bud-

3.4 DYNAMICS OF BIOFILM REACTIVITY

ding. An outgrowth, or bud, develops from the parent cell and after a period of enlargement separates from the parent as a new cell.

The prevailing means of bacterial reproduction, as already indicated, is binary fission; one cell divides, producing two new cells. Thus, if we start with a single bacterium, the increased in population is by geometric progression: 1 2 4 8 16 32, and so on. The time interval required for the cell to divide, or for the population to double, is known as the mean generation time. Not all bacteria have the same generation time; for some, such as *E. coli*, it may be 15 to 20.0 min; for others it may be several hours. Similarly, the generation time is not the same for a particular bacterium under all conditions. The amount and kind of nutrient available in the medium and the specific prevailing environmental conditions influence the generation time.

Bacteria growing in biofilms are copiotrophs, that is, they require relatively high levels of energy substrate for growth (Poindexter, 1981). In very low nutrient waters (oligotrophic environments), copiotrophic bacteria tend to starve. In order to survive such conditions, these bacteria exhibit size reduction (0.2 to 0.5 μm diameter) and significantly lowered endogenous metabolism (utilization of internal organic substrates). This process has been termed *starvation survival* (Morita, 1982). Zobell (1943) postulated that the advantage to bacteria of colonizing surfaces was the concentration at the surfaces of macromolecules and other molecules that could serve as nutrient sources for the bacteria. Starved bacteria have been reported to be more adhesive than well-fed bacteria (Dawson et al., 1981) and it has been demonstrated that starved bacteria in oligotrophic conditions are able to utilize energy substrates bound at surfaces for growth and reproduction (Kjelleberg et al., 1982; Power and Marshall, 1988). In the study of Kjelleberg et al. (1982), the marine *Vibrio* DW1 employed had a generation time of 37 min in a nutrient-rich medium. It did not grow in the aqueous phase of an oligotrophic medium, but showed a mean generation time of about 57 min on a surface exposed to the oligotrophic medium.

3.4.2.2 Mathematics of Microbial Growth.

The equations presented in this section describe rates of growth for an unstructured system. Thus, cell composition remains constant throughout the experimental measurements, that is, balanced growth. Balanced growth can be realized during the exponential growth phase in a batch reactor and at steady state in an open or continuous reactor.

The requisite conditions for microbial growth include a viable inoculum, an energy source, nutrients for synthesis, and a suitable physicochemical environment. The rate of microbial growth is proportional to the biomass present, which results in the following rate equation:

$$r_x = \mu X \tag{3.15}$$

where μ = specific growth rate (t^{-1}), X = biomass concentration (M/L^3)
r_x = reaction rate ($M/L^3 \, t^1$).

This equation describes exponential or logarithmic growth and is a useful description of microbial growth as long as environmental conditions remain constant and the constitution of the biomass remains constant (balanced growth). In a batch reactor, rate of biomass accumulation in the logarithmic growth phase is described by the following balance equation:

$$dX/dt = \mu X \tag{3.16}$$

The equation can be integrated to yield

$$X = X_0 \, 2^{t/g} \text{ or } g = (\ln 2)/\mu \qquad (3.17)$$

where g = generation time or doubling of the population (t), X_0 = initial biomass concentration (the inoculum) (M/L^3).

But the specific growth rate, μ, is influenced by many environmental variables. The most important variable is the concentration of the limiting nutrient, the substrate. So, $\mu = f(S)$. Other variables of interest include temperature, pH, ionic strength, and concentration of inhibiting substances. Therefore,

$$\mu = f(S, T, \text{pH}, S_1, S_2, \ldots S_i)$$

where S_i = concentration of i-th component in growth medium (M/L^3).

The most widely used expression for describing the rate of microbial growth as a function of nutrient (substrate) concentration is that attributed to Monod (1950), which describes a rectangular hyperbola

$$\mu = \frac{\mu_{max} S}{K_S + S} \qquad (3.18)$$

where K_S = rate (saturation) coefficient (M/L^3), S = substrate concentration (M/L^3).

The expression can be considered as the combination of two expressions which describe behavior at very low and very high concentrations:

$$\mu = \mu_{max} \qquad \text{for } S >> K_S \qquad (3.19)$$

$$\mu = (\mu_{max}/K_S) S = k^* S \qquad \text{for } S << K_S \qquad (3.20)$$

At low concentrations, the rate is first order with respect to substrate concentration (Eq. 3.19). In many systems of relevance to biofilm formation, the most important data are those at low substrate concentration (oligotrophic conditions). Unfortunately, analytical limitations sometimes limit the evaluation of data in this region. At low concentrations, it is also likely that diffusion or mass transport may limit the overall rate of growth. At high concentrations, the rate is independent of concentration or zero order with respect to substrate (Eq. 3.20). The basis for the rate expression can be visualized by the way in which it relates to saturation of the organism with substrate. At low concentration, the organism still has a significant reaction potential available and an increase in substrate supply increases its growth rate. As concentration is increased further, however, the cell can no longer assimilate the substrate being provided; it becomes saturated. Because of this behavior, Eq. (3.18) is frequently referred to as the *saturation rate equation*. The same equation is used to describe enzyme kinetics (Michaelis and Menten, 1913) and adsorption phenomena. All of these processes are characterized as having active sites that become saturated at high concentration.

The similarity between the equations describing microbial growth and enzyme kinetics suggests a mechanistic relationship between the two. Monod indicates that the models are of the same form but suggested that the saturation coefficient, K_S, should be expected to be related to the apparent dissociation coefficient of the enzyme involved in the first step of the breakdown of the substrate.

3.4.2.3 Maintenance and Endogenous Decay.

Microorganisms require energy to "maintain" existing structures and processes such as motility. *Maintenance* processes may significantly influence intracellular processes during starvation survival within a biofilm. Thus, maintenance may result in reduced size of biofilm cells in substrate-depleted regions of the biofilm. Maintenance rate, therefore, reflects a diversion of substrate away from the synthesis or growth process. Consequently, maintenance decreases overall (or observed) yield of cells produced from a unit of substrate. Physiologically, maintenance assumes that a certain portion of the substrate conversion rate is directed toward maintenance and is not available for cell synthesis. Thus, mathematically, maintenance appears in a substrate material balance as an extra rate of substrate depletion along with growth. Thus, maintenance makes no direct appearance in a biomass balance equation; only the substrate balance has changed. Typically, maintenance energy rates are assumed to be first order in biomass concentration,

$$r_m = k_m X \quad (3.21)$$

where k_m = maintenance rate constant (Mass-substrate/mass-biomass per time).

Another way of rationalizing a decreased observed cellular yield is by considering *endogenous metabolism*, which refers to the degradation of cellular (i.e., endogenous) components. Endogenous metabolism may also play an important role in starvation survival behavior of organisms in a substrate depleted biofilm. Unlike maintenance, endogenous decay is strictly a depletion of intercellular constituents should the cell not have sufficient exogenous supplies. Thus, an endogenous metabolism rate term is found in the biomass material balance and essentially reflects the "decay" of biomass. The rate expression for endogenous metabolism is first order in biomass concentration:

$$r_e = k_e X \quad (3.22)$$

where k_e = endogenous rate constant (t^{-1}).

A clear distinction between endogenous metabolism and maintenance may not be possible on an operational level. However, endogenous metabolism only occurs in the absence of significant exogenous substrate. Maintenance energy, on the other hand, is a requirement of the living cell, which may be obtained from an exogenous or endogenous substrate. Mathematically speaking, however, only one approach is valid, with the maintenance energy approach (incorporation of an addition term in the substrate balance) leading to an erroneous situation as substrate concentration approaches zero.

Consider the two approaches applied to balances on a batch reactor system:

	A. Maintenance	B. Endogenous Decay
Biomass Balance:	$dX/dt = \mu X$	$dX/dt = \mu X - k_e X$
Substrate Balance:	$dS/dt = -\mu X/Y - k_m X$	$dS/dt = -\mu X/Y$

Assuming $\mu = \mu_{max} S/(K_S + S)$ then as $S \to 0$, $\mu \to 0$. In Case B, this states $dS/dt \to 0$ (no substrate depletion if there is no substrate) but in Case A, dS/dt is still a finite rate $= -k_m X$; which implies substrate consumption even in the absence of substrate. Consequently, the endogenous decay approach is most often employed in modeling.

3.4.3 Gene Retention and Transfer

The close proximity of cells within a biofilm might enhance certain processes that occur only at a comparatively low frequency in planktonic populations. Such processes might include the exchange of mobile genetic elements (Belas et al., 1986). Processes such as transposition, transduction, and conjugation require an intimate contact between the donor/phage and recipient cells. *Conjugation* requires cell-to-cell contact for the transfer of plasmid DNA from a donor cell to a recipient cell. The genes necessary for conjugation to occur are encoded by the plasmid themselves (and as we will show later, these genes may be under autoinducer control by the recipient cells themselves). Plasmid-encoded transfer functions include: the production of sex pili, mobilization of the DNA, specification of a cell-envelope function termed surface exclusion (which eliminates matings of two donors), and the formation of the mating pair, either surface-obligatory or surface-preferred.

Mobilization allows a plasmid, which does not have all the conjugative transfer genes, to be cotransferred with or assisted by a conjugative plasmid. This can occur by way of two mechanisms: (1) the conjugative plasmid is transferred first, bringing the mobilizable plasmid with it or (2) by insertion of the conjugative plasmid's *oriT* sequence into the mobilizable plasmid (Guiney and Lanka, 1989). Suspended cells must rely on random collisions with potential partners in order for genetic transfer to occur, whereas dense but discreet populations of cells coexist in biofilms. In the gut or oral cavity, where most of the commensal microorganisms are organized within thick biofilms that coat tissue mucosal surfaces, incoming organisms must be incorporated into the biofilm in order to pass on/or acquire carried genetic elements. Residency in a biofilm community should provide optimum conditions for the acquisition and dissemination of antibiotic resistance determinants. However, the study of genetic exchange within biofilm communities is still in its infancy.

It has been shown that the survival and genetic transfer abilities of microorganisms depend on several factors: the nature of the bacterial host and cloning vector, the final ecological niches of the organisms, the transmissibility of the recombinant DNA to other bacteria, and the selective advantage that the DNA can confer to the host in order to survive harsh environments (Cruz-Cruz et al., 1988; Stotzky and Babich, 1994). Since little information is available, any study on the dynamics of gene transfer to and between biofilm organisms would greatly contribute to the understanding of a number of complex ecosystems. This is especially important because bacteria isolated from many different microbial environments have been shown to possess plasmids that are conjugative (Bender and Cooksey, 1986; Diels et al., 1989; Focht et al., 1996; Fry and Day, 1990; Guiney and Lanka, 1989; Hill et al., 1992; Lilley et al., 1992; Shoemaker et al., 1992; Wickman and Atlas, 1988) or have been shown to mobilize other plasmids (Hill et al., 1992; Mergeay et al., 1987). Angles et al. (1993) found transfer of the broad host range plasmid *RP1* to be higher in biofilms than in the aqueous phase. However, in their experiments they used a recipient strain that had been cultured and optimized for plasmid transfer and they measured plasmid transfer by replicate plating, which has been shown to give artificially high numbers (Walter et al., 1991).

Both the oral cavity and the gastrointestinal track provide excellent examples of a complex, multispecies biofilm in which the spatial organization of the community provides for cross-feeding and interspecies cooperation. Biofilm-bound bacteria are remarkably difficult to treat with antimicrobials. Bacteria within biofilms are actually less susceptible to antimicrobial compounds, even though these chemical agents are lethal at sub-μg/mL quantities against the same organism grown in free suspension. Biofilm-bound bacteria may enhance the segregational stability, expression, and rate of conjugative transfer of plasmid-DNA (often containing antibiotic resistance genes) between bacteria. This is important because antibiotic-resistant bacteria, in a biofilm, may transfer these resistance genes more readily to neighboring bacteria. Gene transfer could also convert a previous avirulent commensal organism into a highly virulent pathogen. For example, in the intestinal track, the genetic promiscuity of the *enterococci* and *streptococci* spp. has led many workers to regard such organisms as reservoirs of antibiotic resistance genes that might be taken up by gram-negative bacteria (Noble, 1992). In this respect Bertram et al. (1991) demonstrated transposon Tn916 to transfer between gram-positive and gram-negative bacteria in a natural system. Their study consolidated earlier reports of conjugative transfer of pDNA between gram-positive and gram-negative bacteria (Trieu-Cuot, 1987, 1988) and the presence of the enterococcal erythromycin resistance gene *Erbf3* in enterobacterial isolates taken from patients treated with this antibiotic (Arthur et al., 1987). Gruzza et al. (1994) examined the exchange of a conjugative and nonconjugative plasmid between *Lactococcus lactis* and organisms such as *Bacteroides*, *Peptostreptococcus*, *Bifidobacterium*, and *Enterococcus faecalis*, isolated from the human fecal flora. Of these possible recipients only *E. faecalis* was shown to take up the plasmids. Earlier studies by Gruzza et al. (1993) examined the *in vivo* transfer of a self-transmissible plasmid (pIL205), nontransferable but mobilizable plasmid elements (pIL252 and prL253), and a plasmid (pMSI5B) that had been integrated into the chromosome of the donor organism *Lactobacillus lactis*. Of these, only pEL205 and pIL253 were shown to be transferred. The failure of pEL252 to transfer was attributed to its low copy number and high segregational instability. This conclusion gains support from the work of Doucet-Populaire et al. (1992) which demonstrated that plasmid pAT191, encoding kanamycin resistance, was lost from an *in vivo* mouse model due to its low copy number and could not be demonstrated to have transferred from *E. faecalis* to *E. coli*.

Some conjugative plasmids can only be transferred within a limited host range, while other "promiscuous" plasmids are capable of transfer to cells of species or genera far different from the donor. Promiscuity makes conjugation more likely the major gene transfer mechanism in nature, because transduction or transformation require closely related species for gene transfer. For example, RSF1010, a gram-negative plasmid can be transferred to gram-positive strains like *Streptomyces* and *Mycobacterium*. The plasmid RK2 has been transferred from *E. coli* to such gram-positive bacteria as *Enterococcus faecalis*, *Streptococcus lactis*, *Streptococcus agalactiae*, *Bacillus thuringiensis*, *Listeria monocytogenes*, and *Staphylococcus aureus* via conjugation (Trieu-Cuot, 1987, 1988). In addition, it has been shown that conjugative transposons also exist in nature. These elements may be more involved than R plasmids in the dissemination of drug resistance in some species of *Streptococcus* (Clewell and Gauton-Burke, 1986). One such transposon, Tn916, is capable of interspecies transfer, but only among gram-positive bacteria. The fertility potential is encoded by the transposon itself, and upon transfer, the transposon preferentially inserts into the chromosome.

3.4.3.1 Autoinduction of Gene Expression. Transfer of genes, including antibiotic resistance, within biofilms is likely to be a multifactorial process with the age, architectural composition, and environmental conditions surrounding the biofilm being as important as the properties of the individual plasmid, such as host range, copy number, and stability. The understanding of biofilm gene transfer physiology may also require defining and quantifying the link between cell:cell communication and autoinducer control of plasmid acquisition, mobilization, and expression.

Intracellular signaling mechanisms are employed by microorganisms to sense, integrate, and process information from their surroundings. *Quorum sensing* is a bacterial intercellular communication mechanism for controlling gene expression in response to population density. Quorum sensing comprises two cellular components, a small diffusable signal molecule and some type of cell activator that "senses" the signal. While gram-negative bacteria control direct transcription-level manipulation with a class of signal molecules known as homoserine lactones, gram-positive bacteria employ peptide pheromones that bind to specific cell membrane receptors, with the binding instigating subsequent community-level responses.

Conjugation between sexually differentiated bacterial cells requires physical and chemical interaction; the reader is directed to a previous section (Section 3.4.1.1) for cited examples of cell signaling induced plasmid transfer.

3.4.4 Biofilm Detachment Processes

Detachment has always been considered as an "interfacial transfer process" which transfers cells and other biofilm components from the biofilm to the bulk liquid. By the 1990s, the biofilm community considered "desorption" of microbial cells from the substratum to occur from the moment of initial cell adsorption. "Detachment" was considered material loss from the biofilm matrix as opposed to material loss from the substratum. Mathematically, it became the accepted practice to use the process of detachment to offset biofilm growth and adsorption; in order for biofilm net accumulation to reach a steady state. As a consequence, detachment was assumed to occur only at the leading edge of the biofilm:bulk liquid interface. Subsequently, most hypotheses regarding mechanisms controlling detachment were based on biofilm responses to interfacial forces, such as shear stress related erosion and abrasion. Evidence does indicate that increasing shear suddenly over that which prevailed during a biofilm's development results in an increased detachment.

However, there is also evidence at constant shear stress conditions, that the detachment rate of biofilm is independent of shear stress but highly dependent on growth. Thus, advances in our understanding of biofilm detachment have occurred in the past decade that point to physiological control of biofilm detachment processes.

3.4.4.1 Molecular Control of Biofilm Detachment. Most of the alginate biosynthetic genes of *P. aeruginosa* are clustered at 34 min on the chromosome (Boyd and Chakrabarty, 1994). Alginate lyase enzymes cleave the 4-*O*-linked glycosidic bonds between uronate residues by an eliminative mechanism to produce unsaturated sugar derivatives (Gacesa, 1987). The *algL* gene, which codes for alginate lyase, is also located within the alginate gene cluster (Boyd et al., 1993; Shinabarger et al., 1993). Alginate lyase (*algL*) of *P. aeruginosa* has optimal activity against nonacetylated polymannuronic acid (Dunne and Buckmire, 1985; Linker and Evans, 1984; Nguyen and Schiller, 1989). The role of the *P. aeruginosa* alginate lyase in alginate production is intriguing. Several other microbes, including *Bacillus circulans* and two marine *Pseudomonas* species, that possess such an enzyme can utilize alginate as a carbon source (Hansen et al., 1984; Kashiwabara et al., 1969; Nguyen and Schiller, 1989). However, it appears that *P. aeruginosa* 8821 and 8830 are unable to do so. The *algL* gene of *P. aeruginosa* is indispensable for alginate production. Dis-

3.4 DYNAMICS OF BIOFILM REACTIVITY

ruption of the *algL* gene results in a nonmucoid phenotype that can be changed to a mucoid phenotype solely by the presence in *trans* of the downstream gene *algA* (Boyd et al., 1993). *AlgL* could be involved in alginate modification as it could be important for determining the molecular size of the alginate polymer produced. A decrease in polymer length could affect the properties of the alginate, including its ability to enhance attachment of the bacteria to solid surface.

Boyd and Chakrabarty (1994) hypothesized that increased expression of the alginate lyase in *P. aeruginosa* would alter the size of the alginate synthesized and this would in turn affect the adherence properties of the bacteria. The stable mucoid strain *P. aeruginosa* 8930 harboring the vector pmMB22 or the *algL* plasmid pSK700 were used for their experiments. Isopropyl β-D-thiogalactoside (IPTG) served as inducer of alginate lyase expression from the *tac* promoter of plasmid pSK700. pMMB22 served as the vector control. The level of alginate activity of *P. aeruginosa* 8830-pMMB22 grown in the presence or absence of IPTG was low. *P. aeruginosa* 8830/pSK700 grown in the absence of IPTG had a higher level of alginate lyase activity than the vector control because of the leakiness of the *tac* promoter in *P. aeruginosa*. This approximately 10-fold increase in alginate lyase specific activity was not sufficient to alter the amount of alginate produced or to affect can detachment.

However, *P. aeruginosa* 8830-pSK700 with IPTG induction exhibited a high level of alginate lyase specific activity because of increased expression of the *algL* gene from the *tac* promoter. The amount of cell detachment from a biofilm of *P. aeruginosa* 8830-pSK700 grown in the presence of IPTG was 17-fold greater than that observed for *P. aeruginosa* 8830-pMMB22. The amount of alginate produced by *P. aeruginosa* 8830-pSK700 with IPTG was similar to that produced by *P. aeruginosa* 8821. Thus, the increase in sloughing observed for *P. aeruginosa* 8830-pSK700 with IPTG can not be attributed solely to a decrease in the amount of alginate present. However, the increase in cell detachment did correlate with the degree of depolymerization of the alginate. Alginate samples of the foregoing strains were found to be quite different from each other qualitatively when they were visualized on a 5% polyacrylamide gel. The alginate of *P. aeruginosa* 8830/pSK700 with IPTG was greatly degraded as seen by its polydisperse gel pattern. The alginate samples of *P. aeruginosa* 8830-pMMB with and without IPTG and *P. aeruginosa* 8830/pSK700 without IPTG each showed a high-molecular-weight monodispersed band with little or no alginate degradation.

Boyd and Chakrabarty (1994) further investigated the effects of lyase induction on cell detachment at various stages of biofilm growth. Either *P. aeruginosa* 8830/pMMB22 or *P. aeruginosa* 8830/pSK700 were cultivated as biofilm on membranes separating two chambers. To induce the alginate lyase of pSK700, IPTG was added to the bottom chamber, either at the time of incubation (t = 0 h) or 24 h after inoculation. Biofilm were then allowed to develop for 48 h. Addition of IPTG caused a large increase in alginate lyase specific activity, and extensive alginate degradation was observed for both IPTG-induced *P. aeruginosa* 8830/pSK700 samples. Induction of the alginate lyase in *P. aeruginosa* 8830/pSK700 at 0 h resulted in a threefold reduction in the overall amount of alginate produced. The number of detached cells increased 9- to 16-fold over the number produced by the vector control. A less pronounced decrease in alginate formation was seen when IPTG was added at 24 h versus when it was added at time = 0 h. Cell detachment increased fourfold to eightfold over that of the vector control, compared with a 9- to 16-fold increase when the alginate lyase was induced at time = 0 h.

3.5 DEVELOPMENT OF BIOFILM ARCHITECTURE AND GLOBAL REACTIVITY

3.5.1 Observations of Biofilm Heterogeneity

The structural heterogeneity, nonuniform distribution of cells and polymers, variable biofilm thickness and surface topography, and variable density and porosity, has long been recognized and quantitatively measured by many researchers (Drummond, 1993; Gjaltema et al., 1994; van Loosdrecht et al., 1995). However, there are now at least two to three generations of mechanistic mathematical models that describe well the conversion of soluble substrates by biofilms and, with reasonable insight, the subsequent growth and replication of bacterial populations (Arvin and Harremoës, 1990; Characklis and Marshall, 1989; Wanner and Gujer, 1986; Wanner and Reichert, 1996). Both features, global and local substrate reaction rates and bacterial population turnover rates, can be predicted provided the biofilm architecture is known. With rare exceptions, all these mathematical models treat the biofilm as a continuum, with uniform dimensional structure, despite experimental evidence to the contrary.

Prior to 1998, no available model could predict *á priori*, spatial biofilm heterogeneity. In 1998, Picioreanu et al. (1998a,b) developed the benchmark discrete mathematical model of the solid phase (e.g., using cellular automata, CA), combined with classical continuum methods for soluble components, that can predict the development of 3-D biofilm structure.

3.5.2 Modeling Biofilm Three-Dimensional Architecture

The geometrical structures of several biological communities have already been modeled by CA methods. For bacteria colonies, a diffusion-limited aggregation (DLA) model was used by Fujikawa (1994) and a different random-walk model was applied by Schindler and Rataj (1992) and Schindler and Rovensky (1994). All these models can produce time-evolving complex growth patterns as observed on planar agar slabs, but no explicit conversion of "nutrients" was included. Biomass accumulation rate, mycelial density, and differentiation of filamentous fungi on solid surfaces have also been generated by a probabilistic CA (Laszlo and Silman, 1993). There is also a series of CA models for proliferation of animal cells (Hawboldt et al., 1994; Lim and Davies, 1990; Zygourakis et al., 1991). A more realistic representation of colony growth was made by Ben-Jacob et al. (1994) who included in their model the explicit growth of bacteria in a substrate gradient field. Both in this latter study and in that of Tackács and Fleit (1995), who modeled the filamentous bulking of activated sludge, the nutrient field was solved by finite difference methods. The first mentioned application of CA modeling in biofilm research was in a recent study by Wimpenny and Colasanti (1997), based on ideas discussed earlier by Colasanti (1992). This CA model is, however, comparable to the DLA models because growth is only considered across the *x-y* plane of the substratum (where unoccupied space is available) but not vertically (*z* direction), away from the substratum. A quite different approach was used in the model by Barker and Grimson (1993), in which both diffusion and reaction of substrates forming biomass are represented by CA algorithms.

An inherent drawback to the aforementioned discrete models is the use of abstract, mathematical parameters (e.g., "units of resource," "random-walk distance," etc.), which are not explicitly linked to the macroscopic physical/chemical/biological parameters commonly used to describe biofilm systems (e.g., yields, concentrations, rates, fluxes of nutrients). Past discrete models produced various patterns as a response to changing parameters,

3.5 DEVELOPMENT OF BIOFILM ARCHITECTURE AND GLOBAL REACTIVITY

but these patterns evolve in abstract time. By combining differential with discrete models, Picioreanu et al. (1998a) developed the first quantitative model that can predict biofilm structure, together with correct time evolution of concentrations, fluxes, and conversion rates. Due to the "discrete spreading" algorithm for biomass, it is possible to correctly predict the spatial two-dimensional (2-D) and three-dimensional (3-D) distribution of microorganisms in relation to, for example, the substrate flux and inoculation density. A summary of the Picioreanu model is given here.

3.5.2.1 Model Derivation.
The physical *space* of the model is represented by a rectangular uniform grid. Either square elements are used as tiles to fill the 2-D space or cubic volume elements are used to fill the 3-D volume (Fig. 3.8). In the $N \times M \times L$ 3D Cartesian grid, coordinates of volume elements are given by a vector $(x, y, z) \in (0 \ldots N-1, 0 \ldots M-1, 0 \ldots L-1)$.

Two *variables* are chosen to represent the state of the system that stimulates the biofilm development in the simplest case: the soluble limiting substrate concentration (**S**) and biomass density (**C**) (both in dimensionless form). In addition, a third matrix for "solid components," called **c**, is used, where information is stored about the occupation state of each grid in the space.

Each element, $S_{x,y,z}$ of the substrate matrix, **S**, takes *real* values between 0 and 1. Biomass content of each volume element (represented by matrix **C**) also varies in the same range. Any space grid in the matrix **c** is 0 if unoccupied (meaning liquid phase), 1 if occupied with biomass, and 2 if occupied with an inert carrier (a gel in an immobilized cell system). A single bacterial cell is assumed to fill an entire volume element.

The matrices, **S**, **C**, and **c** are updated according to different rules, corresponding to the different processes that can affect their state. Substrate and biomass liquid concentrations are found by differential methods, while the biofilm development is found by discrete (CA) rules.

In the general 3-D system, the mass balance for substrate is:

$$\frac{\partial c_S}{\partial t} = D_S \left(\frac{\partial^2 c_S}{\partial x^2} + \frac{\partial^2 c_S}{\partial y^2} + \frac{\partial^2 c_S}{\partial z^2} \right) - r_S(c_S, c_X) \tag{3.23}$$

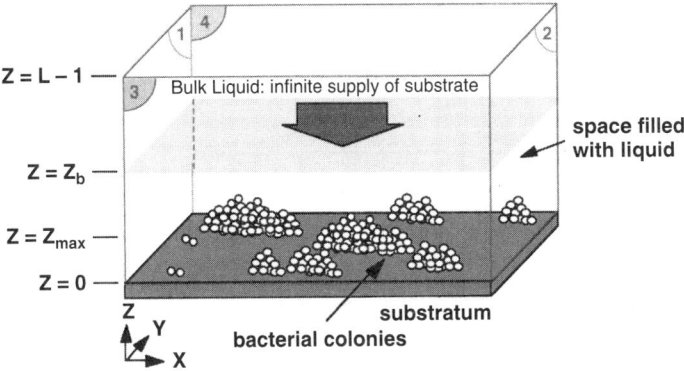

Figure 3.8 Schematic definition of a 3-D biofilm volume for Cell Automata (CA) modeling of biofilm structure development as per Picioreanu et al. (1998).

which assumes only molecular diffusion and conversion of substrate to biomass, with the boundary conditions depending on system geometry and physics. The substrate conversion rate, r_S, depends on the substrate concentration, c_S, and the biomass density in the biofilm, c_X.

In summary, variables chosen to represent the biofilm state in the simplest case are the dimensionless concentration of a soluble limiting substrate (**S**) and the biomass density in each element (**C**). In a third matrix, **c**, information is stored about the occupation of space with solid particles (i.e., bacteria, carrier, etc.). A planar substratum is placed at the bottom of the working volume where a number (n_0) of inoculum colonies is initially distributed at random. The soluble substrate (nutrient) comes from an ideal source situated at the top of the system. The model assumes that biofilm grows in a static liquid environment; that is, the substrate is transferred only by diffusion from the ideal source, through the liquid boundary layer, to the attached bacteria, where the nutrient is utilized. The spatial distribution of substrate is calculated by applying relaxation methods to the reaction-diffusion mass balance (Eq. (3.23)). The spreading of biomass is modeled by a discrete CA algorithm each time the biomass density in a grid element reaches its maximum value, c_{Xm}. The reader is directed to Picioreanu et al. (1998a,b) for details regarding the special boundary conditions required for CA models.

3.5.2.2 Quantitative Descripters of Biofilm Structure Heterogeneity.
Measures of biofilm shape must be invariant to the position, size, or orientation of the biofilm (Glasbey and Horgan, 1994). Picioreanu et al. (1998b) introduced a few selected statistical quantities to characterize the internal and external structure of simulated biofilms (Fig. 3.9). For greater detail of their derivations, the reader is directed to the original paper.

To quantitatively describe differences in biofilm structure, it is first necessary to define the *biofilm front*, also referred to in the literature as the "external perimeter" or the "border." Similar to the diffusion front (Sapoval et al., 1985; Chopard and Droz, 1990), the biofilm front is defined as follows:

- The first step is to define "the fluid," which consists of the empty sites connected by the nearest neighbor to the infinite reservoir.

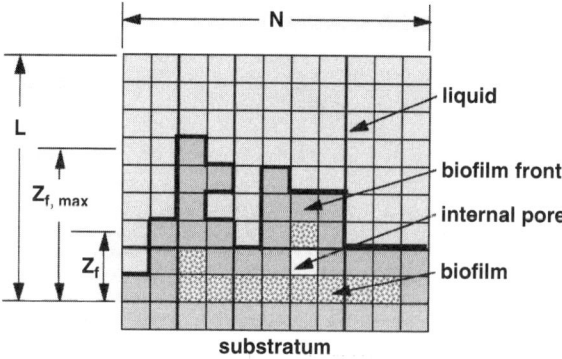

Figure 3.9 Definition of biofilm "front" parameters and mathematical descriptors of biofilm surface topography as per Picioreanu et al. (1998).

3.5 DEVELOPMENT OF BIOFILM ARCHITECTURE AND GLOBAL REACTIVITY

- The complementary cluster is the *biofilm* ("the land"), which is defined as all compartments connected to the substratum by nearest-neighbor biomass–biomass bonds. The biofilm can also encompass empty sites (pores, voids), which are finite empty clusters (a.k.a. "lakes").
- Thus, the biofilm *front* is the chain of nearest-neighbor particle–particle bonds linked in such way that each compartment of the front has at least one "fluid" (external) nearest or next-nearest neighbor (Fig. 3.9).

Biofilm surface area is an important parameter in the design and performance evaluation of biofilm reactors because surface area explicitly enters into all mass transfer calculations as the biofilm/liquid interface. Biofilm surface areas are usually calculated assuming the biofilm has the area of its carrier (substratum). However, as has been clearly documented for many natural and engineered systems, the biofilm surface shape can be very complex. Therefore, one measure of biofilm surface structure introduced by Picioreanu et al. is the ratio between the real biofilm surface area and the substratum area, called the *biofilm surface enlargement*.

To define surface enlargement, one must first define, at time t, the *average density* of biofilm front points at distance Z from the solid carrier, $\bar{c}_{f,Z}$,

$$\bar{c}_{f,Z} = \frac{1}{N} \cdot \sum_{X=0}^{N-1} c_{f,X,Z} \text{ in two dimensions (2-D)} \quad (3.24)$$

$$\bar{c}_{f,Z} = \frac{1}{M \cdot N} \cdot \sum_{X=0}^{N-1} \sum_{Y=0}^{M-1} \bar{c}_{f,X,Y,Z} \text{ in three dimensions (3-D)} \quad (3.25)$$

where $c_{f,X,Y,Z}$ is the state of the occupant matrix: 0 for a nonfront point and 1 for a front point (so that $0 \leq \bar{c}_{f,Z} \leq 1$). To compare the results, Picioreanu et al. introduced a *normalized density* of the distribution of front points:

$$E_{f,Z} = \bar{c}_{f,Z} \Big/ \sum_{Z=0}^{L-1} c_{f,Z} \quad (3.26)$$

Analogous to $c_{f,Z}$, the average substrate, $\bar{S}_Z$, and biomass, $\bar{C}_Z$, concentrations and the average space occupation, c_Z, (a measure of *surface coverage*), are also defined for the slice, Z. These variables are used later to represent front spreading.

Front length is then the number of front points (solid grid elements in Fig. 3.9). The normalized biofilm perimeter (P_f) will be obtained in two dimensions,

$$P_f = \sum_{Z=0}^{L-1} \bar{c}_{f,Z} = \left[\sum_{X=0}^{N-1} \sum_{Z=0}^{L-1} c_{f,X,Z} \right] \Big/ N \text{ in 2D} \quad (3.27)$$

whereas in the 3-D system it means the normalized biofilm surface area (A_f), called *surface enlargement*, is defined by:

$$A_f = \sum_{Z=0}^{L-1} \bar{c}_{f,Z} = \left[\sum_{X=0}^{N-1} \sum_{Y=0}^{M-1} \sum_{Z=0}^{L-1} c_{f,X,Y,Z} \right] \Big/ M \cdot N \quad (3.28)$$

The connection of the front points is only by nearest-neighbor compartments (the first option, Fig. 3.8).

The surface enlargement coefficient takes usually values greater than 1, meaning that the real surface is increased by some folded structure when compared with the bare surface. Only at short time intervals after surface inoculation could values be less than 1, which signifies the occurrence of a thin, patchy biofilm.

3.5.2.2.1 Biofilm Surface Roughness. The biofilm surface enlargement, A_f, is one measure for biofilm front complexity but it cannot take directly into account the depth of biofilm irregularities (roughness). That is why Picioreanu et al. defined another time-dependent variable to measure the biofilm surface roughness.

First, the *average biofilm front width* can be defined as the absolute deviation (σ_f) of biofilm front points ($c_{f,z}$) from the mean front position (Z_f):

$$\sigma_f = \left(\sum_{Z=0}^{L-1} |Z - \overline{Z}_f| \cdot \overline{c}_{f,z}\right) \Big/ \left(\sum_{Z=0}^{L-1} \overline{c}_{f,z}\right) \tag{3.29}$$

where the *mean front position*, Z_f, is the first moment of front-elements distribution in relation to the origin:

$$\overline{Z}_f = \left[\sum_{Z=0}^{L-1} Z \cdot \overline{c}_{f,z}\right] \Big/ \sum_{Z=0}^{L-1} \overline{c}_{f,z} \tag{3.30}$$

The absolute deviations, σ_f, depends, however, on the mean biofilm thickness. To compare the biofilm roughness obtained between simulations at different parameters or at different time intervals, a dimensionless deviation, σ, is defined as the average front width to mean biofilm thickness

$$\sigma = \frac{\sigma_f}{\overline{Z}_f} \tag{3.31}$$

Picioreanu et al. evaluated the effects of different biofilm growth conditions on the evolution of biofilm structure. The effects of both mass transport rate and biomass growth rate on biofilm architecture were captured in a dimensional group

$$G = l_z^2 \cdot \frac{\mu_m C_{Xm}}{D_S C_{So}} \tag{3.32}$$

where l_z = the average biofilm thickness, μ_m = maximum specific bacterial growth rate, C_{Xm} = maximum biomass density, C_{So} = bulk concentration of soluble grow limiting substrate, D_S = substrate diffusion coefficient. Upon inspect, one will find that Eq. (3.32) can be interpreted as,

$$G = \frac{\text{maximum biomass growth rate}}{\text{maximum substrate transport rate}}$$

Thus, G is high if either C_{so} or D_S, or both are very small which is a situation of low substrate concentrations or a low diffusivity or if the biofilm is extremely active. In this case, the system is said to be "transport rate limited." Conversely, G is low if the system is "reaction rate limited."

3.5 DEVELOPMENT OF BIOFILM ARCHITECTURE AND GLOBAL REACTIVITY

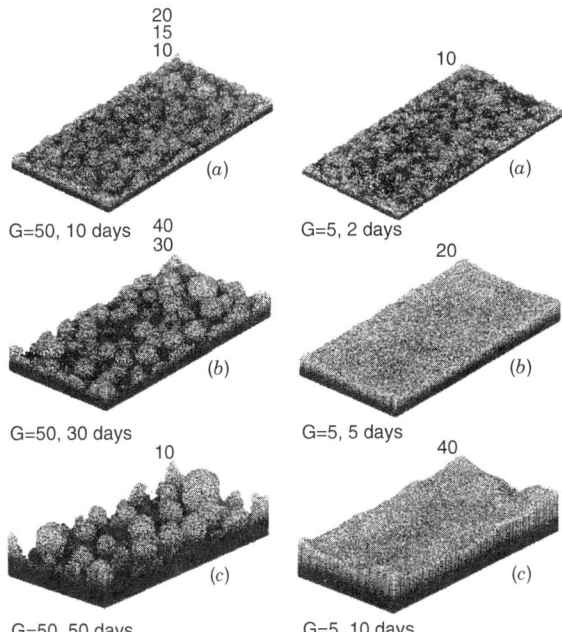

Figure 3.10 Spatial biomass distribution in a 3-D simulation for two different limiting cases of the substrate transfer rate coefficient, G. At G=50, low substrate-transfer limited case, significant biofilm structural irregularities arise over the course of 50 day simulation. At G=5, a high substrate mass transfer case, biofilm surface structure is relatively smooth. Each dot in both simulations represents a biomass cluster of 4 mm. Taken from Picioreanu et al. (1998).

Results of computer simulations of 3-D biofilm formation are shown in Figure 3.10, for two extreme values of the parameter, G; one low ($G=5$) and one high ($G=50$). In the substrate transport limited case ($G=50$), the biofilm develops a highly invaginated surface topography, with deep channels or interstices surrounded by mushroomlike biofilm cluster structures that have been observed microscopically. Alternatively, under growth rate-limited biofilm growth, a relative flat smooth biofilm topography is predicted. Estimation of the many surface topography parameters—roughness coefficient, biofilm compactness, surface enlargement, and fractal dimension as a function of the parameter G - are shown in Figure 3.11A and B. These also quantify the effect of mass transport rate on biofilm structure where every measure of surface irregularity indicates biofilms will develop highly heterogeneous structures under conditions of low substrate, slow substrate transfer, or very metabolically reactive biofilm.

3.6 SUMMARY

By the 1990s, biofilm accumulation was recognized as the net result of a number of macroscopic processes that both added and negated the overall amount of biofilm at a surface. Both experimentally and mathematically, by the 1990s, the biofilm community was able to describe local, instantaneous concentrations of both dissolved solutes (growth limiting

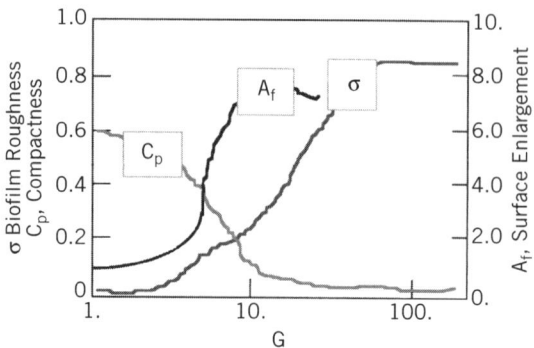

Figure 3.11 Dependence of biofilm surface structure descriptors on substrate transfer rate, G, for a 3-D biofilm simulation (Picioreanu et al., 1989). C_p represents surface compactness, σ = surface roughness coefficient, and A_f = biofilm surface enlargement parameter. All parameters are mathematically defined in the accompanying text. All indicators quantify the impression that as the substrate transfer rate decreases (increasing G), the biofilm topography becomes increasingly irregular.

substrates, essential nutrients, and metabolic by-products) and bacterial population distributions. Most of the experimental methods up to that time were destructive and either presented biofilm-averaged, lumped parameter estimations of these concentrations or labor intensive destructive methods to assess any spatial variations in these parameters.

In the past decade with the advent of noninvasive laser scanning microscopy, solute and species-specific fluorchromes, micro-ion specific chemical probes, and advances in molecular biology, the biofilm community has now begun to recognize and quantify micro- or even molecular-scale processes that control the structure and reactivity of a bacterial biofilm.

REFERENCES

Absolom, D. R., F. V. Lamberti, Z. Policova, W. Zingg, C. J. Van Oss, and A. W. Neumann. 1983. Surface thermodynamics of bacterial adhesion. *Appl. Environ. Microbiol.* 46: 90.

An, F. V., and D. B. Clewell. 1994. Characterization of the determinant (*tra*B) encoding sex pheromone shutdown by the hemolysin/bacteriocin plasmid PADI in *Enteroccocus faecalis*. *Plasmid* 31: 215–221.

Angles, M. L., K. C. Marshall, and A. E. Goodman. 1993. Plasmid transfer between marine bacteria in the aqueous phase and biofilms in reactor microcosms. *Appl. Environ. Microbiol.* 59: 843–950.

Arthur, M., A. Andremont, and P. Corvalin. 1987. Distribution of erythromycin esterease and rRNA methylase genes in members of the family *Enterobacteriaceae* highly resistant to erythromycin. *Antimicrob. Agents Chemother.* 31: 404–409.

Arvin, E., and P. Harremoës. 1990. Concepts and models for biofilm reactor performance. *Water Sci. Technol.* 22: 2177–2180.

Baier, R. E. 1975. Applied chemistry at interfaces. In *Applied Chemistry at Protein Interfaces*, Advances in Chemistry Series, 145, R. E. Baier, ed. Washington, D.C.: American Chemical Society, pp. 1–25.

Baier, R. E., and L. Weiss. 1975. Demonstration of the involvement of adsorbed proteins in cell adhesion and cell growth on solid surfaces. In *Applied Chemistry at Protein Interfaces*, Advances in

REFERENCES

Chemistry Series, 145, R. E. Baier, ed. Washington, D.C.: American Chemical Society, pp. 300–307.

Barker, G. C., and M. J. Grimson. 1993. A cellular automaton model of microbial growth. *Binary* 5: 132–137.

Barnett, K. G., T. Cosgrove, B. Vincent, D. S. Sissons, and M. Cohen-Stuart. 1981. *Macromolecules* 14: 1018–1020.

Beal, S. K. 1970. Deposition of particles in turbulent flow on channel or pipe walls. *Nucl. Sci. Engrg.* 40: 1–11.

Belas, R., M. Simon, and M. Silverman. 1986. Regulation of lateral flagellar gene transcription in *Vibrio parahaemolyticus*. *J. Bacteriol.* 167: 210–218.

Bell, G. I. 1978. Models for the specific adhesion of cells to cells. *Science* 200: 618–627.

Bender, C. L., and D. A. Cooksey. 1986. Indigenous plasmids in *Pseudomonas syringae* pv. *tomato*: Conjugative transfer and role in copper resistance. *J. Bacteriol.* 165: 535–541.

Ben-Jacob, E., O. Schochet, A. Tenenbaum, I. Cohen, A. Czirdk, and T. Vicsek. 1994. Generic modelling of competitive growth patterns in bacterial colonies. *Nature* 368: 46–49.

Berg, H. C., and D. A. Brown. 1974. Chemotaxis in *Escherichia coli* analyzed by three-dimensional tracking. *Antibiot. Chemother.* (Basel) 19: 55–78.

Bertram, J., M. Stratz, and P. Durre. 1991. Natural transfer of conjugative transposon Tn916 between gram-positive and gram negative bacteria. *J. Bacteriol.* 170: 443–448.

Boden, M. K., and J. L. Flock. 1989. Fibronectin-binding protein/clumping factor from *Staphylococcus aureus*. *Infect. Immun.* 57: 2358–2363.

Bongrand, P., and G. I. Bell. 1984. Cell-cell adhesion: Parameters and possible mechanisms. In *Cell Surface Dynamics: Concepts and Models*, A. S. Perelson, C. DeLisi, and F. W. Wiegel, eds. New York: Marcel-Dekker, pp. 459–493.

Botham, R. A., and C. Theis. 1970. *J. Poly. Sci.* (Part C) 30: 369–380.

Boyd, A., and A. M. Chakrabarty. 1994. Role of alginate lyase in cell detachment of *Pseudomonas aeruginosa*. *Appl. Environ. Microbiol.* 60: 2355–2359.

Boyd, A., M. Ghosh, T. B. May, D. Shinabarger, R. Keogh, and A. M. Chakrabarty. 1993. Sequence of the *algL* gene of *Pseudomonas aeruginosa* and purification of its alginate lyase product. *Gene* 131: 1–8.

Brash, J. L., and Q. M. Samak. 1978. Dynamics of interaction between human albumin and poly(ethylene) surface. *J. Colloid Interface Sci.* 65: 495–504.

Bryers, J. D. 1980. Dynamics of early biofilm formation in a turbulent flow system, Ph.D. Dissertation, Rice University, Houston, TX.

Busscher, H. J., A. H. Weerkamp, H. C. van der Mei, A. W. J. van Pelt, H. P. de Jong, and J. Arends. 1984. Measurement of the surface free energy of bacterial cell surface and its relevance for adhesion. *Appl. Environ. Microbiol.* 48: 980.

Cafe, M. C., and I. D. Robb. 1982. *J. Coll. Interface Sci.* 86: 411–421.

Camper, A. K., J. T. Hayes, P. J. Sturman, W. L. Jones, and A. B. Cunningham. 1993. Effects of motility and adsorption rate coefficient on transport of bacteria through porous media. *Appl. Environ. Microbiol.* 59: 3455–3462.

Characklis, W. G., and K. C. Marshall. 1989. *Biofilms*. New York: John Wiley.

Chopard, B., and M. Droz. Cellular automata approach to diffusion problems. *Springer Proc. Phys.* 46: 130–143.

Cleaver, J. W., and B. Yates. 1975. *Chem. Eng. Sci.* 30: 983–992.

Cleaver, J. W., and B. Yates. 1976. *Chem. Eng. Sci.* 31: 147–151.

Clewell, D. B. 1993. Sex pheromones and the plasmid-encoded mating response in *Enterococcus faecalis*. In *Bacterial Conjugation*, D. B. Clewell, ed. New York: Plenum Press, pp. 349–367.

Clewell, D. B., and C. Gauron-Burke. 1986. Conjugative transposons and the dissemination of antibiotic resistance in Streptococci. *Ann. Rev. of Microbiol.* 40: 635–659.

Clewell, D., and K. E. Weaver. 1989. Sex pheromones and plasmid transfer in *Enterococcus faecalis*. *Plasmid* 21: 175–184.

Clewell, D., Y. Yagi, Y. Ike, B. Brown, and F. An. 1982. Sex pheromones in *Streptococcus faecalis*: Multiple pheromone systems in strain D65, similarities of pAD1 and pAMγl, mutants of pAdl altered in conjugative properties. In *Microbiology: 1982*. Washington, DC: American Society for Microbiology, pp. 97–100.

Cohen-Stuart, M. A., J. M. H. M. Scheutjens, and G. J. Fleer. Polydispersity effects and the interpretation of polymer adsorptoin isotherms. *J. Polymer Sci.* (Part B) 18: 559.

Colasanti, R. L. 1992. Cellular automata models of microbial colonies. *Binary* 4: 191–193.

Cozens-Roberts, C., J. A. Quinn, and D. A. Lauffenburger. 1990. Receptor-mediated adhesion phenomena. Model studies with the radial-flow detachment assay. *Biophys. J.* 58: 197–125.

Cruz-Cruz, N. E., G. A. Toranzos, D. G. Ahearn, and T. C. Hazen. 1988. *In situ* survival of plasmid-bearing and plasmidless *Pseudomonas aeruginosa* in pristine tropical water. *Appl. Environ. Microbiol.* 54: 2574–2577.

Davies, D. G., and G. G. Geesey. 1995. Regulation of the alginate biosynthesis gene *algC* in *Pseudomonas aeruginosa* during biofilm development in continuous culture. *Appl. Environ. Microbiol.* 61: 860–867.

Davies, D. G., A. G. Chakrabarty, and G. G. Geesey. 1993. Exopolysaccharide production in biofilms: Substratum activation of alginate gene expression by *Pseudomonas aeruginosa*. *Appl. Environ. Microbiol.* 59: 1181–1186.

Dawson, M. P., B. D. Humphrey, and K. C. Marshall. 1981. Adhesion, a tactic in the survival strategy of a marine *Vibrio* during starvation. *Curr. Microbiol.* 6: 195–198.

Dembo, M., D. C. Torney, K. Saxman, and D. Hammer. 1988. The reaction-limited kinetics of membrane-to-surface adhesion and detachment. *Proc. Roy. Soc. London, B.* 234: 55–83.

Diels, L., A. Sadouk, and M. Mergeay. 1989. Large plasmids governing multiple resistances to heavy metals: A genetic approach. *Toxicol. Environ. Chem.* 23: 79–89.

Doucet-Populaire, F., P. Trieti-Cucot, A. Andremont, and P. Corvalin. 1992. Conjugal transfer of plasmid DNA from *Enterococcus faecalis* to *Escherichia coli* in digestive tracts of gnotobiotic mice. *Antimicrob. Agents Chemother.* 36: 502–504.

Drummond, F. E. 1993. Macromolecule transport mechanisms in living biofilm of *Pseudomonas putida*. Masters Thesis, Duke University, Durham, NC.

Duddridge, J. E., C. A. Kent, and J. F. Laws. 1982. *Biotechnol Bioengrg.* 24: 153.

Dunne, W. M., Jr., and F. L. A. Buckmire. 1985. *Appl. Environ. Microbiol.* 50: 562–567.

Dunny, G. M. 1990. Genetic functions and cell-cell interactions in the pheromone-inducible plasmid transfer system of *Enterococcus faecalis*. *Mol. Microbiol.* 4: 689–696.

Dunny, G. M., B. L. Brown, and D. B. Clewell. 1978. Induced cell aggregation and mating in *Streptococcus faecalis*: Evidence for a bacterial sex pheromone. *Proc. Natl. Acad. Sci. USA* 75: 3479–3483.

Dunny, G. M., R. A. Craig, R. L. Carron, and D. B. Clewell. 1979. Plasmid transfer in *Streptococcus faecalis*: Production of multiple sex pheromones by recipients. *Plasmid* 2: 454–465.

Dunny, G. M., C. Funk, and J. Adsit. 1981. Direct stimulation of the transfer of antibiotic resistance by sex pheromones in *Streptococcus faecalis*. *Plasmid* 6: 270–278.

Dunny, G. M., B. A. B. Leonard, and P. J. Hedberg. 1995. Pheromone-inducible conjugation in *Enterococcus faecalis*: Interbacterial and host-parasite chemical communication. *J. Bacterial.* 177: 871–876.

REFERENCES

Ehrenfeld, E. E., R. E. Kessler, and D. B. Clewell. 1986. Identification of pheromone-induced surface proteins in *Streptococcus faecalis* and evidence of a role for lipoteichoic acid in formation of mating aggregates. *J. Bacteriol.* 168: 6–12.

Emödy, L., J. Heesemann, H. Wolf-Watz, M. Skurnik, P. Kapperud, and T. Wadström. 1989. *J. Bacteriol.* 171: 6674–6679.

Focht, D. D., D. B. Searles, and S.-C. Koh. 1996. Genetic exchange in soil between introduced chlorobenzoate degraders and indigenous biphenyl degraders. *Appl. Environ. Microbiol.* 62: 3910–3913.

Fowkes, F. M. 1964. *Ind. Eng. Chem.* 56: 40.

Friedlander, S. K., and H. F. Johnstone. 1957. Deposition of suspended particles from turbulent gas streams. *Ind. Eng. Chem.* 49: 1151–1156.

Fry, J. C., and M. J. Day. 1990. Plasmid transfer and the release of genetically engineered bacteria in nature: A discussion and summary. In *Bacterial Genetics in Natural Environments*, J. C. Fry and M. J. Day, eds. London: Chapman & Hall.

Frymier, P. D., R. M. Ford, H. C. Berg, and P. Cummings. 1995. Three dimensional tracking of motile bacteria near a solid planar surface. *J. Proc. Natl. Acad. Sci. (USA)* 92: 6195–6199.

Fujikawa, H. 1994. Diversity of the growth patterns of *Bacillus subtilis* colonies on agar plates. *FEMS Microbiol. Ecol.* 13: 159–168.

Gacesa, P. 1987. Alginate-modifying enzymes/A proposed unified mechanism of action of lyases and epimerases. *FEBS Lett.* 212: 199–202.

Galli, D., R. Wirth, and G. Wanner. 1989. Identification of aggregation substances of *Enterococcus faecalis* cells after induction by sex pheromone. *Arch. Microbiol.* 151: 486–490.

Galli, D., F. Lottspeich, and R. Wirth. 1990. Sequence analysis of *Enterococcus faecalis* aggregation substance encoded by the sex pheromone plasmid pAD1. *Mol. Microbiol.* 4: 895–904.

Gambello, M. J., S. Kaye, and B. H. Iglewski. 1993. *LasR* of *Pseudomonas aeruginosa* is a transcriptional activator of the alkaline protease gene (apr) and an enhancer of exotoxin A expression. *Infect. Immun.* 61: 1180–1184.

Gibbons, R. J., E. C. Moreno, and D. M. Spinell. 1976. Model delineating the effects of a salivary pellicle on the adsorption of *Streptococcus miteor* onto hydroxyapatite. *Infect. Immun.* 14(4): 1109–1112.

Gillis, R. J. 1993. M.S. Thesis, Montana State University, Bozeman, MT.

Gjaltema, A., P. A. M. Arts, M. C. M. van Loosdrecht, J. G. Kuenen, and J. J. Heijnen. 1994. Heterogeneity of biofilm in rotating annular reactors: Occurrence, structure, and consequences. *Biotechnol. Bioeng.* 44: 194–204.

Glasbey, C. A., and G. W. Horgan. 1994. *Image Analysis for the Biological Sciences*. Chichester, UK: John Wiley.

Goldman, A. J., R. G. Cox, and H. Brenner. 1967a. Slow viscous motion of a sphere parallel to a plane wall-I. Motion through a quiescent fluid. *Chemical Eng. Sci.* 22: 637–651.

Goldman, A. J., R. G., Cox, and H. Brenner. 1967b. Slow viscous motion of a sphere parallel to a plane wall-II. Couetter flow. *Chemical Eng. Sci.* 22: 653–660.

Gruzza, M., P. Langella, Y. Duval-Iflah, and R. Ducluzeau. 1993. Gene transfer from engineered *Lactococcus lactis* strains to *Enterococcus faecalis* in the digestive tract of gnotobiotic mice. *Microbiol. Releases* 2: 121–125.

Gruzza, M., M. Fons, M. F. Ouriet, Y. Duval-Iflah, and R. Ducluzeau. 1994. Study of gene transfer *in vitro* and in the digestive tract of gnotobiotic mice from *Lactobacillus lactis* strains to various strains belonging to human intestinal flora. *Microb. Releases* 2: 183–199.

Guiney, D. G., and E. Lanka. 1989. Conjugative transfer of IncP plasmids. In *Promiscuous Plasmids of Gram-Negative Bacteria*, C. M. Thomas, ed. London: Academic Press.

Hammer, D. A., and D. A. Lauffenburger. 1987. A dynamical model for receptor-mediated cell adhesion to surfaces. *Biophys. J.* 52: 475–487.

Hansen, J. B., R. S. Doubet, and J. Ram. 1984. *Appl. Environ. Microbiol.* 47: 704–709.

Hawboldt, K. A., N. Kalogerakis, and L. A. Beheehee. 1994. A cellular automaton model for microcarrier cultures. *Biotechnol. Bioeng.* 43: 90–100.

Hengge-Aronis, R. 1993. Survival of hunger and stress: The role of *rpoS* in early stationary phase regulation in *E. coli*. *Cell* 72: 165–168.

Hill, K. E., A. J. Weightman, and J. C. Fry. 1992. Isolation and screening of plasmids from the epilithon which mobilize recombinant plasmid pD1O. *Appl. Environ. Microbiol.* 58: 1292–1300.

Huang, C.-T. 1993. Plasmid retention and gene expression in bacterial biofilm cultures, Ph.D. Dissertation, Duke University, Durham, NC.

Hussain, M., M. Herrmann, C. vonEiff, F. Perdreau-Remington, and G. Peters. 1997. A 140-Kdal extracellular protein is essential for the accumulation of *Staphylococcus epidermidis* strains on surfaces. *Infect. Immun.* 65: 519–524.

Israelachvili, J. N., and P. M. McGuiggan. 1988. Forces between surfaces. *Science* 241: 795–800.

Jang, L. K., and T. F. Yen. 1985. A theoretical model of diffusion of motile and non-motile bacteria toward solid surfaces. In *Microbes and Oil Recovery*, 1, Zajic and Donaldson, eds. International Bioresources Journal, pp. 226–246.

Jonsson, K., C. Signäs, H.-P. Muller, and M. Lindberg. 1991. Two different genes encode for fibronection-binding proteins in *Staphylococcus aureus*: The complete nucleotide sequence and characterization of the second gene. *Eur. J. Biochem.* 202: 1041–1048.

Kashiwabara, Y., H. Suzuki, and K. Nisizawa. 1969. *J. Biol. Chem.* 66: 503–512.

Kjelleberg, S., and M. Hermansson. 1984. Starvation-induced effects on bacterial surface characteristics. *Appl. Environ. Microbiol.* 48: 497–503.

Kjelleberg, S., B. A. Humphrey, and K. C. Marshall. 1983. Initial phases of starvation and activity of bacteria at surfaces. *Appl. Environ. Microbiol.* 46: 978–984.

Lantz, M. S., R. D. Allen, T. A. Vail, L. M. Switalski, and M. Höök. 1991a. Specific cell components of *Bacteriodes gingivalis* that mediate binding and degradation of human FN. *J. Bacteriol.* 173: 495–504.

Lantz, M. S., R. D. Allen, W. L. Duck, J. L. Blume, L. M. Switalski, et al. 1991b. Identification of *Porphyromonas gingivalis* components that mediate its interaction with fibronection. *J. Bacteriol.* 173: 4263–4270.

Laszlo, J., and R. W. Silman. 1993. *Biotechnol. Adv.* 11: 621–633.

Latifi, A., M. Foglino, K. Tanaka, P. Williams, and A. Lazdunski. 1996. A hierarcial quorem-sensing cascade in *Pseudomonas aeruginosa* links the transcriptional activators *LasR* and Rh1R (VsmR) to expression of the stationery phase sigma factor. *Mol. Microbiol.* 21: 1137–1146.

Latifi, A., K. M. Winson, M. Foglino, B. S. Bycroft, G. S. Stewart, A. Lazdunski, and P. Williams. 1995. Multiple homologues of *LuxR* and *LuxI* control expression of virulence determinants and secondary metabolites through quorem sensing in *Pseudomonas aeruginosa*. *Mol. Microbiol.* 17: 333–344.

Lilley, A. K., M. J. Bailey, M. J. Day, and J. C. Fry. 1992. Natural transfer plasmids from sugarbeet rhizosphere. In Sixth international symposium on microbial ecology. Barcelona, Spain.

Lim, J. H. F., and G. A. Davies. 1990. A stochastic model to simulate the growth of anchorage dependent cells on flat surfaces. *Biotechnol. Bioengrg.* 36: 547–562.

Lindgren, P.-E., P. Speziale, M. McGavin, H.-J. Monstein, and M. Höök. 1992. Cloning and expression of two different genes from *Streptococcus dysgalactiae* encoding fibronectin receptors. *J. Biol. Chem.* 267: 1924–1931.

Lindgren, P.-E., M. J. McGavin, C. Signas, B. Guss, and S. Gurusiddappa. 1993. The nucleotide sequence of two different genes coding for fibronectin binding protein from *Streptococcus dysgalactiae* and identification of active sites. *Eur. J. Biochem.* 214: 819–827.

Linker, A., and L. R. Evans. 1984. *J. Bacteriol.* 159: 958–964.

Lister, D. H. 1979. Corrosion products in power generating systems. In *Fouling of Heat Transfer Equipment*, E. F. S. Somerscales and J. G. Knudsen, eds. Washington, DC: Hemisphere, pp. 135–200.

Little, B. J., and A. Zsolnay. 1985. Chemical fingerprinting of adsorbed organic materials on metal surfaces. *J. Colloid. Int. Sci.* 104: 79–86.

Loeb, G. I., and R. A. Neihof. 1975. In *Applied Chemistry at Protein Interfaces*, Advances in Chemistry Series, 145, R. E. Baier, ed. Washington, DC: American Chemical Society, pp. 319–335.

Lumsden, J. B., and P. J. Stocker. 1981. *Metallurgica* 15: 1295.

Marshall, K. C. 1976. *Interfaces in Microbial Ecology*. Cambridge, MA: Harvard University Press.

Marshall, K. C. 1979. Growth at interfaces. In *Strategies of Microbial Life in Extreme Environments*, Dahlem Konferenzen Life Sciences Research Report 13, M. Shilo, ed. Weinheim: Verlag Chemie, pp. 281–290.

McDevitt, D., P. Francois, P. Vaudaux, and T. Foster. 1994. Molecular characterization of the clumping factor (fibrinogen receptor) of *Staphylococcus aureus*. *J. Mol. Microbiol.* 11: 237–248.

Mergeay, M., P. Lejeune, A. Sadouk, J. Gerits, and L. Fabry. 1987. Shuttle transfer (or retrotransfer) of chromosomal markers mediated by plasmid pULB112. *Mol. Gene Genet.* 209: 61–70.

Michaelis, L., and M. L. Menten. 1913. *Z. Biochemie.* 49: 333–369.

Monod, J. 1949. The growth of bacterial cultures. *Ann. Inst. Microbiol.* 3: 371–394.

Morat, H. Z. 1985. *The Inflammatory Reaction*. Amsterdam: Elsevier Science.

Mori, M., A. Isugai, Y. Sakagami, M. Fujino, C. Kitada, D. B. Clewell, and A. Suzuki. 1986. Isolation and structure of the *Streptococcus faecalis* sex pheromone inhibitor, iAD1, that is excreted by the donor strain harboring plasmid PAD1. *Agric. Biol. Chem.* 50: 531–541.

Mori, M,. H. Tanaka, Y. Sakagami, A. Isogai, M. Fujino, C. Kitada, D. B. Clewell, and A. Suzuki. 1987. Isolation and structure of the sex pheromone inhibitor, iPD1, excreted by *Streptococcus faecalis* donor strains harboring plasmid pPDI. *J. Bacteriol.* 169: 1747–1749.

Morita, R. Y. 1982. Starvation survival of heterotrophs in the marine environment. *Adv. Microb. Ecol.* 6: 171–198. K. C. Marshall, ed. New York: Plenum Press.

Mueller, R. F. 1990. M.S. Thesis, Montana State University, Bozeman, MT.

Mueller, R. F., W. G. Characklis, W. L. Jones, and J. T. Sears. 1992. Characerization of initial events in bacterial surface colonization by two *Pseudomonas* species using image analysis. *Biotechnol. Bioeng.* 39: 1161.

Nakayama, J., E. Yoshida, H. Kobayashi, A. Isogai, D. B. Clewell, and A. Suzuki. 1995. Cloning and characterization of a region of *Enterococcus faecalis* plasmid pPDI encoding pheromone inhibitor (*ipd*), pheromone sensitivity (*tra*C), and pheromone shutdown (*tra*B) genes. *J. Bacteriol.* 177: 5567–5573.

Nakayama, J., G. M. Dunny, D. B. Clewell, and A. Suzuki. 1995. Quantitative analysis for pheromone inhibitor and pheromone shutdown in *Enterococcus faecalis*. *Dev. Biol. Stand.* 85: 35–39.

Nakayama, J., R. E. Ruhfel, G. M. Dunny, A. Isorgai, and A. Suzuki. 1994. The *prgQ* gene of the *Enterococcus faecalis* tetracycline resistance plasmid, pCF10, encodes a peptide inhibitor, iCF10. *J. Bacteriol.* 176: 7405–7408.

Nakayama, J., Y. Ono, A. Isogai, D. B. Clewell, and A. Suzuki. 1995. Isolation and structrue of the sex pheromone inhibitor, iAM373, of *Enterococcus faecalis*. *Biosci. Biotechnol. Biochem.* 59: 1358–1159.

Neumann, A. W., R. J. Good, C. J. Hope, and M. Sejpal. 1974. An equation of state approach to determine surface tensions of low energy solids from contact analysis. *J. Colloid Interface Sci.* 49: 291.

Nguyen, L. K., and N. L. Schiller. 1989. *Curr. Microbiol.* 18: 323–329.

Noble, W. C., Z. Virani, and R. G. Cree. 1992. Co-transfer of vancomycin and other resistance genes from *Enterococcus faecalis* NCTC 12201 to *Staphylococcus aureus*. *FEMS Microbiol. Lett.* 72: 195–199.

Ochsner, U. A., and J. Reiser. 1995. Autoinducer-mediated regulation of rhamnolipid biosurfactant synthesis in *Pseudomonas aeruginosa*. *Proc. Natl. Academy of Sci. USA* 92: 6424–6428.

Patti, J. M., and M. Höök. 1994. Microbial adhesins recognizing extracellular matrix macromolecules. *Current Opinion in Cell Biology* 6: 752–758.

Patti, J. M., B. L. Allen, M. J. McGavin, and M. Höök. 1994. MSCRAMM-mediated adherence of microorganisms to host tissues. *Ann. Rev. Microbiol.* 48: 585–617.

Pearson, J. P., L. Pasador, B. H. Iglewski, and E. P. Greenberg. 1995. A second N-acyl homoserine lactone signal produced by *Pseudomonas aeruginosa*. *Proc. Natl. Academy of Sci., USA* 92: 1490–1494.

Pedersen, K. Biofilm development on stainless steel and PVC surfaces in drinking water. *Wat. Res.* 24: 239.

Picioreanu, C., M. C. M. van Loosdrecht, and J. J. Heijnen. 1998. Mathematical modeling of biofilm structure with a hybrid differential-discrete cellular automaton approach. *Biotechnol. Bioeng.* 58: 101–116.

Picioreanu, C., M. C. M. van Loosdrecht, and J. J. Heijnen. 1998. A new combined differential-discrete cellular automation approach for biofilm moldeling: Application for growth in gel beads. *Biotechnol. Bioeng.* 57: 718–731.

Piette, J.-P. G., and E. S. Idziak. 1991. Role of flagella in adhesion of *Pseudomonas fluorescens* to tendon slices. *Appl. Environ. Microbiol.* 57: 1635.

Poindexter, J. S. 1981. Oligotrophy: Feast or famine existence. *Adv. Microb. Ecol.* 5: 63–89. M. Alexander, ed. New York: Plenum Press.

Power, K., and K. C. Marshall. 1988. *Biofouling* 1: 163–174.

Rouhiainen, P. O., and J. W. Stachiewicz. 1970. *J. Heat Transfer* (Trans. ASME) 92: 169–177.

Rutter, P. R., and B. Vincent. 1984. Physicochemical interactions of the substratum, microorganisms, and the fluid phase. In *Microbial Adhesion and Aggregation*, K. C. Marshall, ed. Berlin: Springer Verlag, p. 21.

Sapoval, B., M. Rosso, and J.-F. Gouyet. 1985. *J. Phys. Lett. (Paris)* 46: L149.

Scheuerman, T. R., A. K. Camper, and M. A. Hamilton. 1999. Effects of substratum topography on bacterial adhesion. *J. Colloids Interfac. Sci.*, in press.

Schindler, J., and T. Rataj. 1992. Fractal geometry and growth models of a *Bacilus* colony. *Binary* 4: 66–72.

Schindler, J., and L. Rovensky. 1994. A model of intrinsic growth of a *Bacilus subtilis* colony. *Binary* 6: 105–108.

Schlichting, H. 1968. *Boundary Layer Theory*, 6th ed. New York: McGraw-Hill.

Schulze-Koops, H., H. Burkhardt, J. Heesemann, T. Kirsch, B. Swoboda, C. Bull, S. Goodman, and F. Emmrich. 1993. Plasmid-encoded outer membrane protein *YadA* mediated specific binding of enteropathogenic *Yersiniae* to various types of collagen. *Infect. Immun.* 61: 2513–2519.

Shinabarger, D., T. B. May, A. Boyd, M. Ghosh, and A. M. Chakrabarty. 1993. Nucleotide sequence and expression of the *Pseudomonas aeruginosa algF* gene controlling acetylation of alginate. *Mol. Microbiol.* 9: 1027–1035.

Shoemaker, N. B., G. Wang, and A. A. Salyers. 1992. Evidence for natural transfer of a tetracycline resistance gene between bacteria from the human colon and bacteria from the bovine rumen. *Appl. Environ. Microbiol.* 58: 1313–1320.

Stotzky, G., and H. Babich. 1994. Fate of genetically engineered microbes in natural environments. *Recomb. DNA Technol. Bull.* 7: 163–188.

Takács, I., and E. Fleit. 1995. Modelling of the micromorphology of the activated sludge floc: Low DO, low F/M bulking. *Water Sci. Technol.* 31: 235–243.

Thole, J. E. R., R. Schoningh, A. A. M. Jansson, T. Garbe, Y. E. Cornelisse, et al. 1992. Molecular and immunological analysis of a fibronectin-binding protein antigen secreted by *Mycobacterium leprae*. *Mol. Microbiol.* 6: 153–163.

Trieu-Cuot, P., M. Arthur, and P. Courvalin. 1987. Transfer of genetic information between gram positive and gram negative bacteria. In *Streptococcal Genetics*, R. Curtiss and J. J. Feretti, eds. Washington, DC: American Society of Microbiology Press, pp. 65–68.

Trieu-Cuot, P., C. Carlier, and P. Courvalin. 1988. Conjugative plasmid transfer from *Enterococcus faecalis* to *Escherichia coli*. *J. Bacteriol.* 170: 4388–4391.

Trotter, K. M., and G. M. Dunny. 1990. Mutants of *Enterococcus faecalis* deficient as recipients in mating with donors carrying pheromone-inducible plasmids. *Plasmid* 24: 57–67.

Vanhaecke, E., J.-P. Remon, M. Moors, F. Raes, D. de Rudder, and A. van Reteghem. 1990. Kinetics of *Pseudomonas aeruginosa* adhesion to 304 and 316-L stainless steel: Role of cell surface hydrophobicity. *Appl. Environ. Microbiol.* 56: 788.

van Loosdrecht, M. C. M., J. Lyklema, W. Norde, G. Schrsa, and A. J. B. Zehnder. 1987. *Appl. Environ. Microbiol.* 53: 1893.

van Loosdrecht, M. C. M. 1988. Ph.D. Thesis, Wageningen, Netherlands.

van Loosdrecht, M. C. M., H. Lyklema, W. Norde, and A. J. B. Zehnder. 1989. Bacterial adhesion: A physicochemical approach. *Microbial Ecol.* 17: 1.

van Loosdrecht, M. C. M., D. Eikelboom, A. Gjaltema, A. Mulder, L. Tijhuis, and J. J. Heijnen. 1995. Biofilm structures. *Water Sci. Technol.* 32: 35–43.

Vigeant, M. A. S., and R. M. Ford. 1997. Interactions between motile *E. coli* and glass in media with various ionic strengths as observed with a three dimensional tracking microscope. *Appl. Environ. Microbiol.* 63: 3424–3479.

Walsh, D., D. Pope, M. Danford, and T. Huff. 1993. *JOM* 45: 22.

Walter, M. V., L. A. Porteous, V. P. Fieland, R. J. Seidler, and J. L. Amstrong. 1991. Formation of transconjugants on plating media following in situ conjugation experiments. *Can. J. Microbiol.* 37: 703–707.

Wang., T.-Y. 1993. A dynamic model for receptor-mediated specific adhesion of bacteria under uniform shear flow. Masters Thesis, Duke University, Durham, NC.

Wang, G. T.-Y., and J. D. Bryers. 1997. A dynamic model for receptor-mediated specific adhesion of bacteria under uniform shear flow. *Biofouling* 11: 227–252.

Wanner, O., and W. Gujer. 1986. A multi-species biofilm model. *Biotechnol. Bioeng.* 28: 314–328.

Wanner, O., and P. Reichert. 1996. Mathematical modeling of mixed culture biofilms. *Biotechnol. Bioeng.* 49: 172–184.

Weaver, K. E., and D. B. Clewell. 1990. Regulation of the pAD1 sex pheromone response in *Enterococcus faecalis*: Effects of host strain and *tra*A, *tra*B, and C region mutants on expression of an E region pheromone-inducible *lacZ* fusion. *J. Bacteriol.* 172: 2633–2641.

Wickman, G. S., and R. M. Atlas. 1988. Plasmid fluctuations in bacterial populations from chemical stressed soil communities. *Appl. Environ. Microbiol.* 54: 2192–2196.

Wiencek, K. M., and M. Fletcher. 1995. Bacterial adhesion to hydroxyl- and methyl-terminated alkanethiol self-assembled monolayers. *J. Bacteriol.* 177: 1959–1966.

Wimpenny, J. W. T., and R. Colasanti. 1997. A unifying hypothesis for the structure of microbial biofilms based on cellular automaton models. *FEMS Microb. Ecol.* 22: 1–16.

Winson, M. K., M. Camara, A. Latifi, M. Foglino, S. R. Chhabra, M. Daykin, M. Bally, V. Chapon, G. P. C. Salmond, B. W. Bycroft, A. Lazdunski, G. S. Stewart, and P. Williams. 1995. Multiple

N-acyl-L-homoserine lactone signal molecules regulate production of virulence determinants and secondary metabolites in *Pseudomonas aeruginosa*. *Proc. Natl. Acad. Sci. USA* 92: 9427–9431.

Zhurkov, S. V., 1965. Kinetic concept of the fracture of solids. *Int. J. Fract. Mech.* 1: 311–323.

Zisman, W. A. 1964. In *Contact Angle Wettability and Adhesion*, Advances in Chemistry Series, 43, F. M. Fowkes, ed. Washington, DC: American Chemical Society, pp. 1–51.

Zobell, C. E. 1943. The effect of solid surfaces upon bacterial activity. *J. Bacteriol.* 46: 39–56.

Zygourakis, K., R. Bizios, and P. Markenscoff. 1991. Proliferation of anchorage dependent contact inhibited cells: Development of theoretical models based on cellular automata. *Biotechnol. Bioeng.* 38: 459–470.

4

MOLECULAR ECOLOGY OF BIOFILMS

SØREN MOLIN
ALEX T. NIELSEN
BJARKE B. CHRISTENSEN
JENS BO ANDERSEN
TINE R. LICHT
TIM TOLKER-NIELSEN
CLAUS STERNBERG
MARTIN C. HANSEN
CAYO RAMOS
MICHAEL GIVSKOV

Department of Microbiology, Technical University of Denmark, Lyngby, Denmark

4.1 FROM MOLECULAR MICROBIOLOGY TO MOLECULAR MICROBIAL ECOLOGY

The most important contribution from molecular biology to the understanding of the living cell has been that of the central dogma, which expresses the direction of the flow of information from the genome to the cellular functions. The universality of the genetic code and the consequential understanding of the evolutionary relationships between organisms has been a major driving force in the studies of both pro- and eukaryotes at the level of molecular genetics and the related biochemistry, and the subsequent introduction of modern molecular genetics techniques (genetic engineering methods) has allowed a vast expansion of the number of investigated organisms. In microbiology, the initial strong focus on a few bacteria—*Escherichia coli* and *Bacillus subtilis*—has been replaced by a huge diversity of model systems, due to the relative ease with which molecular biology issues can now be addressed in a broad spectrum of organisms.

The picture that has developed from the field of molecular microbiology over the last 30 years is one of contrasts: Despite the enormous diversity in the microbial world, both at the level of physiological characteristics and at the level of genome contents, there is a striking conservation of many key activities related to regulation of the principle life functions. Thus, what was a specific objective at the beginning of the molecular biology of *E. coli*—to reveal the molecular details of how this organism is able to sense and react to changing environments in an ordered and controlled fashion—is now a general goal in biology. The common features found in many different investigated organisms show that not only do bacteria share a common evolutionary origin, but they also share a range of interactive processes that control how signals from the outside world are interpreted and transformed to reactions inside the cells. As experimental biologists, we can therefore apply both techniques and biological concepts from one case to the next in our studies of new biological systems (i.e., organisms).

Molecular microbiology is a multidisciplinary activity covering topics like general microbiology (isolation of pure cell lines and characterization of physiological properties), genetics (gene mapping and mutant isolation), biochemistry (metabolic pathways, identification of signal compounds, nucleic acid analysis, etc.), cellular anatomy and morphology (specific intracellular structures, surface organels, cell shapes), and cell differentiation (stress-induced reprogramming of gene expression). The common platform for all these perspectives is the mono-species culture (normally in suspension) kept under controlled environmental conditions. Most often it is an important premise that the culture is homogenous, allowing conclusions made for a large number of cells to be drawn also for the individual.

Bacteria living in natural environments do not normally follow the foregoing description: They usually live in multispecies communities bound to surfaces under changing and uncontrollable conditions. The biofilm mode of life is the rule for bacteria, while dense populations of suspended organisms with few exceptions are rare, and typically with no or very low levels of metabolic activity. The objective of molecular ecology (at least in the present context) is to reach an understanding of the community life of bacteria from a central dogma perspective: How does information flow in a microbial community from the external environment to the community members and between the members, and how is this information converted to coordinated activities in the community? In this area of microbial ecology, the scientific rationale in molecular microbiology and the many specific and sensitive molecular tools developed for detailed functional investigations are extremely useful for analysis of defined model biofilm systems.

4.2 FOUR COMMUNITY PARAMETERS—ORGANISMS, STRUCTURE, INTERACTIONS, AND COORDINATION

Microbial biofilm communities can be viewed as random aggregates of a number of different organisms, which happen to be together in a specific place at a specific time. Alternatively, they may be seen as multicellular "organisms" with developed internal interdependencies and coordinated activities. The perspective favored here is that it is not a question of either one or the other, but instead a matter of choice of the observer. Experience from the laboratory shows that many different species of bacteria may be brought together in stable biofilm communities, even though the organisms most likely never meet under natural conditions; at the same time, such random consortia may display activities and form structures that are indicative of coordinated performance. Therefore, in the description of biofilm communities it is useful to define a set of parameters which on one hand lead to an overall phenotypic description of the system independent of its composition or location, and on the other hand may be used to define both views of microbial communities (random aggregates/multicellular organisms) (Molin and Molin, 1997). Moreover, comparative descriptions of different communities should be based on features analogous to those known from molecular microbiology, that is, the principal characteristics rather than very specific activities.

The four principal parameters forming the basis for experimental descriptions of multispecies biofilm communities are organisms (who is there?), structure (where are they?), interactions (what are they doing?), and coordination (how and why?). There is a direct line from these features of community life to the central dogma features directing studies in molecular biology. There is, however, also a clear line to the questions asked in "classical" microbial ecology from the microscopic to the global scale. Thus, molecular ecology in the present meaning is a platform connecting molecular microbiology and microbial ecology.

The four parameters are assumed to be empirically accessible using methods from microbiology, biochemistry, molecular biology, and microbial ecology. In the following sections, the methods and the methodological approaches available for the characterization of microbial communities based on these four parameters are reviewed.

4.2.1 Organisms

Natural microbial communities most often consist of complex mixtures of microbial species whose relative frequency may vary significantly. The classical approach toward an analysis of such populations is to cultivate bacteria on laboratory substrates as pure cell lines, followed by identification of the individuals. This task is often impossible due to the complexity of the community, and also because the majority of the bacteria are not culturable. However, it must be emphasized that population analysis without cultivation and identification of at least the major community constituents is problematic. It is therefore crucial that new methods for cultivation and characterization of single species be developed.

The most powerful technique at present for organism identification in environmental samples is that of rRNA identification (Pace et al., 1986). Bacterial taxonomy is changing rapidly these years due to the systematic analysis of rRNA sequences, which are almost ideal indicators of evolutionary relationships (Woese, 1987). Direct analysis of rRNA sequences present in complex samples is possible without any step of cultivation using the polymerase chain reaction (PCR) method to create libraries of probes (Giovannoni et al., 1988; Ward et al., 1990). The simplicity and speed of the PCR-based methods make detailed analyses of microbial communities possible. The obtained information relates to the complexity of the community, but also provides specific data on identification at a number of different taxonomic levels (Amann et al., 1995).

The method of rRNA hybridization is furthermore useful in the context of in situ identification of single cells (Amann et al., 1995). Specific rRNA probes labeled with fluorochromes are precise and highly sensitive tools for single-cell identification in the fluorescence microscope (Fig. 4.1), and when used in combination with the flow-cytometer they may also provide quantitative information such as the relative proportion of a certain organism (Amann et al., 1990; Møller et al., 1995). To some extent similar quantitative information can be obtained by fluorescence microscopy. The limitation of the method is that only a few species can be investigated at the time.

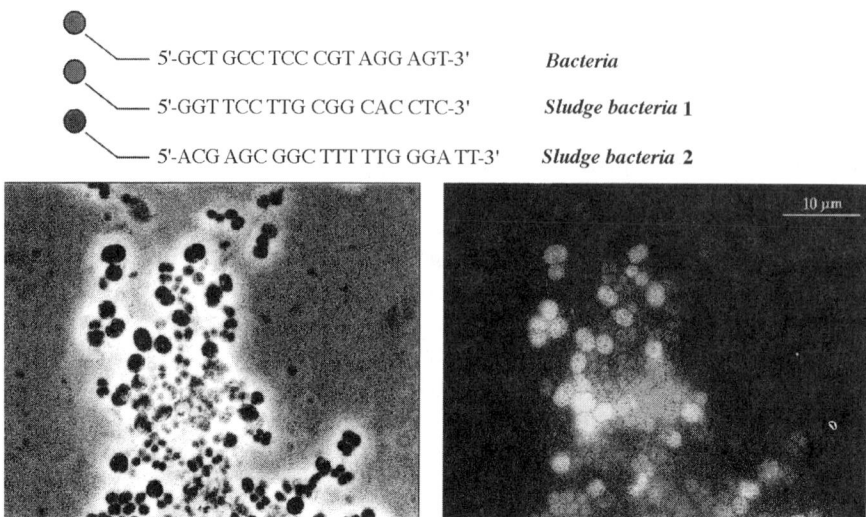

Figure 4.1 In situ rRNA hybridization for identification of organisms in an activated sludge system. In a deteriorated biological phosphorus removal reactor a novel group of bacteria have been identified, and specific rRNA probes were designed for their detection. The pictures show a representative sample from the sludge of the reactor, the left frame representing the phase contrast image from the microscope, and the right side presenting the fluorescence image. The coccus-shaped bacteria (red/yellow and blue cells) are new species identified from this reactor, whereas green cells belong to the *Bacteria*. (Reprinted from Nielsen et al., 1999) Figure also appears in Color Figure section.

Finally, molecular tagging of a distinct organism is an interesting possibility in cases where the purpose is to follow specifically one particular strain in a complex community (Chalfie et al., 1994; Jansson, 1995; Prosser, 1994). Introduction of such tags requires growth and manipulation of the organism followed by introduction of the tagged cells to the community. This scenario is often not possible in natural ecosystems, but in laboratory model systems it represents an excellent choice of specific organism analysis.

In conclusion, the organism parameter is directly accessible for analysis through a combination of traditional and modern molecular approaches. The choice of method of course depends on the complexity of the system, on the need for specific information, and on the available base of instrumentation. Community analysis has always been important in the context of microbial ecology, and there is no doubt that method improvement will progress continuously. It should be emphasized once more that there is also a strong need for improved methods of organism cultivation in order to develop a better understanding of the organisms and their different repertoires of performance in different environmental contexts.

4.2.2 Structure

Community structure in microbial ecology is often presented as a list of identified organisms present in a given environmental sample or system (Wagner et al., 1993). Here, structure is understood as the three-dimensional distribution of organisms (positioning), and the occurrence of specific compartments with functional or constructional significance (extracellular matrix, channels, pores, etc. (Massol-Deyá et al., 1995)). The scanning electron microscope has in the past offered impressive views of the structural organization of microbial biofilms (Robinson et al., 1984), but it always was a problem that the preparation methods for electron microscopy might have serious effects on the structures to be examined (and in any case they are of course destructive). Looking at microbial biofilms with the light microscope also is not ideal; the pictures are two-dimensional, and the noise from unfocused parts of the viewing field is often totally damaging to the final picture. However, the use of light microscopy offers a feature not carried by electron microscopy: simultaneous observations of organism identity (using specific hybridization probes) and of overall positioning of the cells; and the communities may be looked at directly without any seriously interfering preparations steps.

The most ideal instrument for structure analysis at the microscopic level should therefore combine the scanning possibility of the electron microscope with the repertoire of the light microscope. The scanning confocal laser microscope (SCLM) offers such an ideal analytical potential. The collection of information as discrete points and removal of information from locations that are not in focus (in the point) are coupled to the scanning of plane after plane. Finally, passing all images through image analysis leads to a detailed and well-resolved illustration in three dimensions of the observed object. The images thus obtained constitute the most obvious possibilities for discovering specific structures in the community (Fig. 4.2). The SCLM is compatible with many of the molecular probing methods involving fluorescence, and it is therefore one of the most important new methods for analysis of complex community structures (Brakenhoff et al., 1988; Caldwell et al., 1992, 1993). It should be noted, however, that the laser excitation can lead to fluorophore bleaching and to release of toxic products from the fluorophores due to the irradiation of the entire field of view at any time. These problems have been partly solved by the use of two-photon confocal microscopy (Williams et al., 1994).

One of the immediately apparent observations that is made, when studying bacterial surface communities, is the uneven distribution of cells and the existence of substructures like microcolonies, channels, extracellular polymer areas, and so forth. Such structural features

Figure 4.2 Structural organization of a composite biofilm. Confocal microscopy coupled with in situ rRNA hybridization and image analysis is a powerful method of revealing three-dimensional distributions of organisms and specific structural elements in the surface community. The image here shows that the biofilm is composed of densely populated regions as well as of void sectors most likely representing channels through which substrate and waste products may be transferred. It is seen that some cells form mounds penetrating from the substratum through the extracellular polysaccharide (EPS) matrix to the top of the biofilm. Figure also appears in Color Figure section.

may be observed in many different types of microbial communities, and after analysis of several systems it should be possible to categorize the various structural features and elements such that their presence or absence can be determined as any other system parameter.

4.2.3 Interactions

Microorganisms respond to their surrounding environment by mobilizing different parts of their genetic repertoire. Since, however, only a limited number of environmental conditions are specifically responded to by the organisms, other factors may have more general impacts on the behavior of the individuals. Thus, the overall chemicophysical conditions of the environment such as temperature, pH, ionic strength, availability of nutrients, presence of oxygen, and others may have severe effects on the performance of the organisms; these are, however, distinct from the very specific interactions that exist between particular environmental components and specific receptor systems in the cells.

In order to monitor interactions and understand their relevance to the population and the environment, it is necessary to build up a strong base of information describing the physiology of the participating organisms derived from laboratory-based experiment with pure cultures. The cellular repertoire can only be studied in precise terms from very simple experimental scenarios. However, many physiological properties in bacteria are more or less universal, which means that common tools and analytical strategies may be applied to a broad spectrum of species.

The modern molecular tools offer approaches to in situ studies of specific interactions. The presence of essential nutrients in the external environment or locally in the immediate neighborhood of a group of cells can be revealed by using molecular probes reporting on the physiological state of the cells. The ribosome content estimated by quantitative in situ rRNA hybridization reflects the growth physiology of the cells (the more ribosomes, the more physiologically active are the cells). In the absence of nutrients the concentration of ribosomes is expected to be low. Similarly, cell size and DNA content reflect the physiological state and thus the nutritional situation of the organisms (Givskov et al., 1994a; Maaløe and Kjeldgaard, 1966; Møller et al., 1995). Under extreme conditions the cells may be physiologically inactive (dormant), and hence have no interactions with the environment or with other cells in the population. (A more detailed description of useful approaches towards monitoring growth activities in complex microbial biofilms is presented in Section 3).

A number of reporter genes have been designed which—depending on their inherent features—may report on the environmental conditions and the cellular responses to these conditions. For example, constitutive light emission from bacterial luciferase requires oxygen and metabolic activity of the cell (Prosser, 1994), and consequently the "turn-off" of light indicates either lack of oxygen or lack of ATP in the cell. Fusion of reporters to genes activated by certain stress signals (starvation, draught, high osmotic pressure, etc.) can also be applied to indicate local environmental conditions, and therefore also provide information about specific interactions between surroundings and cells (Kragelund et al., 1995). The green fluorescent protein, Gfp, is a very interesting alternative reporter, which allows on-line studies of microbial activities (Chalfie et al, 1994). The great advantage of this reporter protein is its lack of substrate requirement—the protein itself fluoresces—and its compatibility with fluorescence microscopy, including the use of the confocal microscope. Many of the molecular tools supply information both about the state of the cells and about the conditions of the environment, and together this knowledge is an excellent background

4.2 FOUR COMMUNITY PARAMETERS

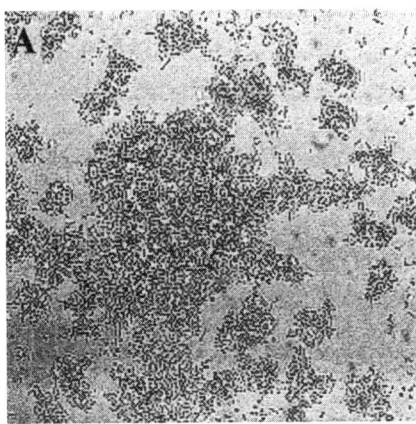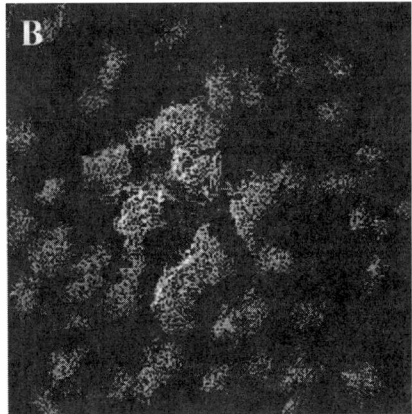

Figure 4.3 Metabolic interactions in complex biofilm communities. In a biofilm consisting of three bacterial strains, all capable of mineralizing toluene, a derivative of *P. putida* was introduced, which carries a gene fusion between the benzoate inducible TOL plasmid promoter, *Pm*, and the reporter gene, *gfp*. The community was fed benzyl alcohol as the sole carbon and energy source, and in a monoculture biofilm it was found that the specific strain of *P. putida* described showed no sign of induction of transcription from the *Pm* promoter (not shown). The presence of another community organism, *Acinetobacter spp.*, however, resulted in a significant induction of the *Pm* promoter in *P. putida* under conditions where the latter strains was colonizing the former (close proximity between the two organisms). The cells seen in *A* (phase contrast) form a dense microcolony composed of the two organisms, *P. putida* and *Acinetobacter*. The induction of green fluorescence, indicating activation of the *Pm* promoter, is coupled directly to the association between the two (*B*). The further away *P. putida* is from the *Acinetobacter* colony, the lower the Gfp signal. (Reprinted from Møller et al., 1998) Figure also appears in Color Figure section.

for reading the interactions going on in the particular location (Fig. 4.3). In addition, the level of resolution of the interactions between cells and surroundings is the single cell and its microenvironment.

4.2.4 Coordination

It is a major assumption in the present analysis of microbial communities that at least parts of the community features (structurally and functionally) are coordinated in such a way that struggling for existence is based as much on coexistence as on survival of the fittest individual. This does not rule out that individual organisms under certain conditions compete for substrate or try to eliminate each other by toxin production, and it is therefore not a question of either competition or collaboration, but rather a combination of the two.

Communication between individuals and between sectors of individuals is an essential prerequisite for coordinated activities. It is generally assumed that the language of this type of communication is chemical in nature, but in contrast to metabolic intermediates, which serve as substrates for further biochemical processes, signals for coordinated activities are targeted at cellular control loops. It is anticipated that extracellular signals emerging from some cells may influence the activity of other cells, in some cases at the level of complex phenotypic traits involving many different genes connected by joint control factors. An excellent example of this type of communication in gram-negative bacteria, is the quorum sensing systems in which a family of acylated homoserine lactones constitute the signal molecules (Dunlap and Greenberg, 1988; Greenberg et al., 1979). The signal molecules interact with dual functioning receptor proteins that transduce the signals to the final destination—promoter regions upstream of the target genes. These systems are widespread and they allow bacteria to sense and express target genes in relation to their culture density. Phenotypes regulated by these systems include traits such as Ti-plasmid conjugation in the plant pathogen *Agrobacterium tumefaciens* involved in crown gall disease, and production of antibiotics, exoenzymes, and virulence factors in pathogenic organisms such as *Erwinia carotovora* and *Pseudomonas aeruginosa* (Bainton et al, 1992; Passador et al., 1993). In fact, it has been suggested that bacteria employ these systems to coordinate their colonization and association with higher organisms. It has recently been shown that bacterial swarming (Fig. 4.4)—a highly coordinated pattern of movements of

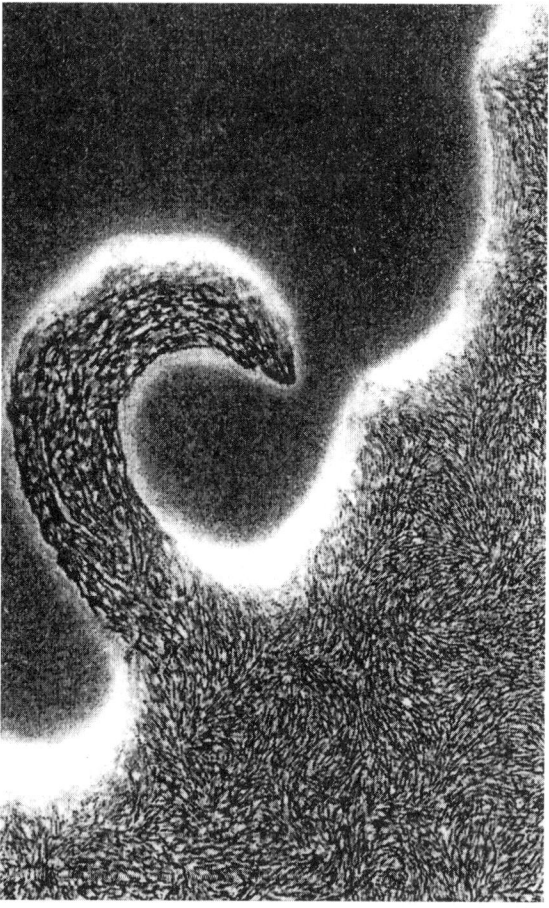

Figure 4.4 Coordinated motility of bacteria controlled by communication signals. Swarming surface associated colony of *Serratia liquefaciens*. The cells located at the brim of the colony exhibit swarming motility implying that a very large number of individuals move in a coordinated fashion as a raft of cells. Intercellular signalling between the cells is an important element in coordinating this type of behavior. (Reprinted from Eberl et al., 1996)

thousands of differentiated cells—is in part controlled by homoserine lactone derivatives (Eberl et al., 1996; Givskov et al., 1997). Of particular interest in the present context is the recent documentation by Davies et al. (1998), showing that there may be a direct correlation between the production of acylated homoserine lactones and the formation of distinct structural elements in biofilms. The presence of such signals in a complex microbial community indicates the importance of activity coordination, and it is expected that related communication systems will be discovered in the future. However, alternative explanations for the formation of biofilm structures have been proposed (Wimpenny and Colasanti, 1997).

Coordinated activities in a community may also be documented by the presence of ordered structures and activities (Shapiro and Higgins, 1989). The finding of channels and pores, through which import of nutrients and export of waste products may take place (Massol-Deyá et al., 1995), could indicate that many individuals have been involved in creating such specific structural elements within the community. In other words, they have worked together rather than against each other. This type of conclusion is indirect in the sense that observations are made that can most easily be understood if it is assumed that coordination is exerted at some level in the system. Likewise, coordinated behavior can be indicated by the systemic responses to perturbations: If one group of organisms is influenced by a perturbation from outside, and the impact is significant in other parts of the community resulting in a change of community performance, this indicates that the community performance is the result of coordinated activities.

4.3 BACTERIAL GROWTH AND STATIONARITY IN COMPLEX BIOFILMS

Determination of bacterial growth rates under a variety of substrate conditions has been a very important basis for the physiological description of single species. We must be able to bring the cells of a strain into the active growth phase before a proper analysis can be made, and any type of significant cellular performance is normally related to the active growth phase. Moreover, since growth rate measurements most often are easy to perform in liquid cultures—optical density determinations in a spectrophotometer—they have also been of great value in studies of control of macromolecular biosynthesis, specific and global gene regulation, mutant characterization, impacts of inserted genes, toxicity effects, cell composition as function of nutrient availability, and others (Maaløe and Kjeldgaard, 1966). The growth curve, with its characteristic separation into an initial lag phase (adaptation to the growth medium), the exponential growth phase (homogeneous population, same rate constant for all cellular parameters), the stationary phase (exhaustion of nutrients, production of inhibitory substances), and finally the decline in viable counts (death or dormancy) is an obvious part of almost any strain description, forming the basis for large parts of the molecular analysis of bacteria.

The major assumption behind the "trust" in the growth curve as a good basis for organism description is that all cells in a suspended culture are surrounded by the same environment and therefore they behave identically. Although this is certainly not always the case—many organisms form clumps, grow in chains, stick to the walls—and although there may be a degree of stochastic reaction pattern in cells going into and out of the exponential phase, it is by and large a reasonable assumption until the opposite is demonstrated. In surface communities, however, the situation is totally different; there is no easy way of measuring growth rates (except determining the total biomass increase; see

Fig. 4.5), the environmental conditions are not stable in all parts of the surface community, the biofilm often consists of several different species with different responses to the nutrient composition, the degree of heterogeneity is large resulting in compartmentalization of growth activities, and substrates other than those present in the external environment may be exchanged between community members or released by dead cells. Nevertheless, the importance of monitoring growth and physiological states of cells in a biofilm community is no less than in suspended cultures, when characterizing the overall phenotype of the community. It is therefore necessary to develop reliable methods for determinations of biofilm-based bacterial growth activities in situ.

4.3.1 In situ Ribosome Counting

The development of the method of in situ rRNA hybridization for identification of specific organisms in complex environments has effectively changed our possibilities for performing detailed ecological investigations. The ribosome is a phylogenetic marker of great significance (Woese, 1987), and the intracellular hybridization to either 16S or 23S rRNA is facilitated by the large number of target molecules in the cells. In the early stages of this technology development, it was realized that in situ rRNA hybridization might also yield important information about the physiological activity of the cells based on the correlation between growth rate and ribosome concentration observed previously for several different bacterial species (Schaechter et al., 1958). Thus, de Long et al. (1989) presented quantitative data based on fluorescent probes, showing that in addition to organism identification it is possible to obtain estimates of cellular growth rates. Amann et al. (1990) presented analogous information employing the flow-cytometer for quantitative fluorescence measurements, and the first growth rate estimates performed in complex biofilms were published by Poulsen et al. (1993). In these—and most subsequent—investigations, the strategy was to hybridize fixed cells in their particular environmental setting with specific

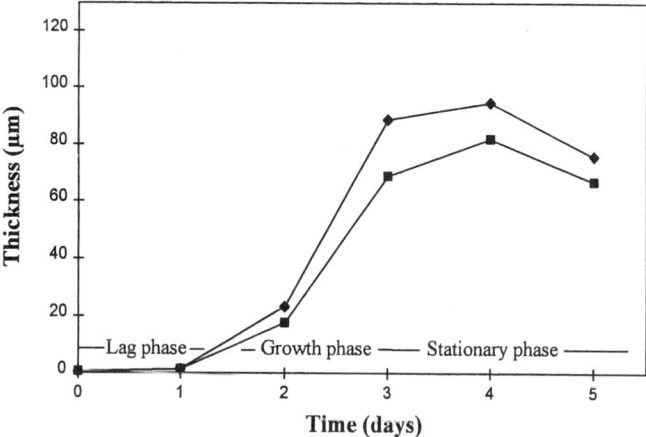

Figure 4.5 Biofilm growth curves. In a flow chamber supplied with a minimal medium substrate, colonization and growth of a bacterial biofilm on a glass surface was monitored by confocal scanning microscopy. The thickness of the biofilm was measured in the microscope, and average values across the glass slide were determined. The resulting growth curves show that also biofilm communities follow a growth cycle as we know it for suspended cultures.

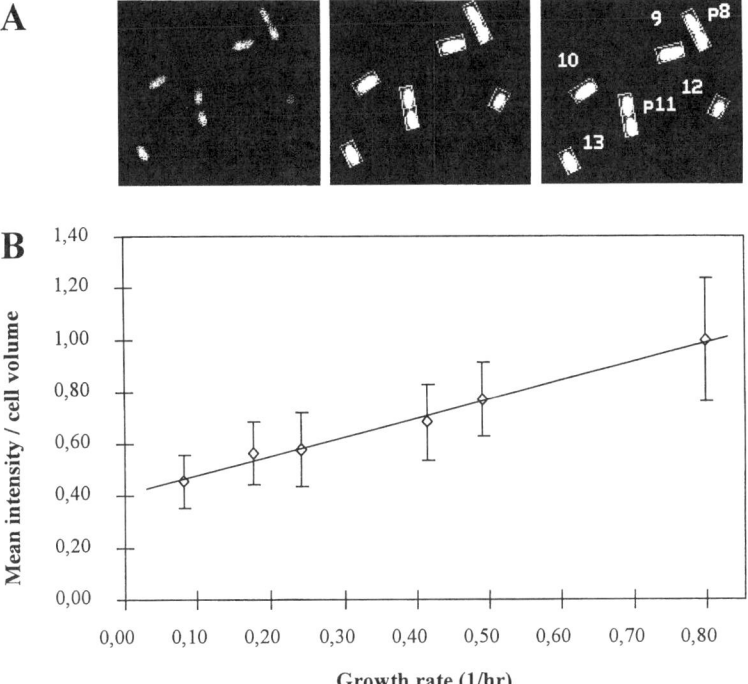

Figure 4.6 Ribosome counting as indicator of growth rate. The cellular concentration of ribosomes indicates the growth state of the cell, and through in situ rRNA hybridization quantitative estimates of the ribosome content may be obtained. Many cells have to be analyzed, and for this reason (and to avoid subjective biases in manual estimations) an image analysis software for this type of quantification has been developed—the Cellstat program—which decodes and analyzes a number of different file formats representing up to 16 million gray levels (*A*). In brief, first a morphological local background subtraction is performed, objects are detected by identifying intensity peaks, and the image is ready for analysis. The cell edge is determined as the point where the intensity is a predefined fraction of the peak intensity, for example, 20%. Next, the detected objects are accepted on basis of a number of parameters. In situ measurements of fluorescence intensities must be related to a standard curve obtained from pure cultures of the organism under investigation (*B*). The curve expresses the correlation between cellular concentrations of ribosomes (mean fluorescence intensity per cell volume) and growth rate, and it is seen in the figure that the expected correlation of increasing ribosome concentration with increasing growth rate is indeed observed.

fluorescence labeled oligonucleotide probes (15–25 nucleotides) targeted against 16S or 23S rRNA, and subsequently quantify the fluorescence intensity in the cells by image processing (Fig. 4.6). The resulting signal intensities could finally be related to a standard curve obtained for the organism under investigation, based on hybridization to cells growing under defined conditions yielding different growth rates (e.g., in chemostats). The ribosome counting technique just described for growth rate determinations is particularly interesting in microbial ecology because in situ rRNA hybridization is very often used for cellular identification. For this reason it is all the more relevant to evaluate the approach and point out the shortcomings and potential misleading results that may be the outcome of such determinations.

The first obvious problem of the method is the need to fix the cells prior to hybridization. This may be of less importance for identification purposes, but in growth measurements it is a problem that the measurements result in single points of data. In some cases, sampling from the community may be possible without seriously interfering with the system itself, allowing continuous growth monitoring of specific parts of the community, but often information about an entire community is desired, in which case this destructive method has limitations as far as the dynamics of growth is concerned. In biofilms, it is also important to relate growth characteristics of cells to their particular spatial position in the community (Fig. 4.7); in this case, it is imperative that the spatial organization of the biofilm is not severely affected by the fixation method. For laboratory setups, embedding in a semisolid matrix before fixation is often sufficient to preserve the community structure (Christensen et al., 1998), but in cases of natural communities other approaches should be considered.

The fundamental requirement for the quantification of ribosomes as growth indicator is the establishment of a standard curve. Quantification of ribosome concentrations in exponentially growing cultures with different growth rates is possible only if the organism is culturable, and if proper conditions can be defined that allow a range of different growth rates to be obtained—either in batch cultures with different carbon sources or in chemostats with limiting nutrient concentrations supplied at different flow rates. Another important issue is whether the correlation between ribosome numbers and growth rates, defined in a standard curve for suspended cells, is directly applicable to surface bound cells growing in biofilms; in other words, is growth of planktonic and sessile cells regulated by the same principal mechanism?

Among the technical problems and potential artifacts of the quantitative in situ rRNA hybridization, accessibility of the probe to its target sequence is a major one. The first barrier is the cell envelope (capsule, cell wall, membrane), which may change permeability when the general physiology of the cells change. Control experiments have been performed to ensure that the fixation condition chosen creates similar permeabilities in cells growing very differently (Poulsen et al., 1993), but in principle this problem should be dealt with when new organisms are investigated. The next barrier is the ribosome itself; some sequences are more optimal than others as targets for hybridization due to the structure of the rRNA (Amann et al., 1995), and if these structures change with the cellular physiology (as for example indicated for starved cells), the quantification by hybridization becomes problematic.

Most of the foregoing problems listed may be dealt with by carefully controlling the conditions for hybridization. However, one inherent source of misleading conclusions is impossible to eliminate: The concentration of ribosomes in a cell reflects the previous history of that cell rather than an instantaneous state of physiology, a problem particularly severe in cells going through constantly changing growth conditions. Synthesis of rRNA under stringent control implies that growing cells actively generate ribosomes, whereas stationary cells are repressed for ribosome synthesis. The stability of the ribosomes, however, results in cellular levels representing an average of growth rates, and it is easy to imagine many situations where in situ determinations of these concentrations are of no value in assessing how the actively growing cells are distributed in a community. There is, therefore, a great need for more precise methods to determine the cellular physiological states at the point of measurements, that is, methods for true rate determinations. In the following sections, a number of approaches towards this goal are reviewed, all applicable to in situ monitoring.

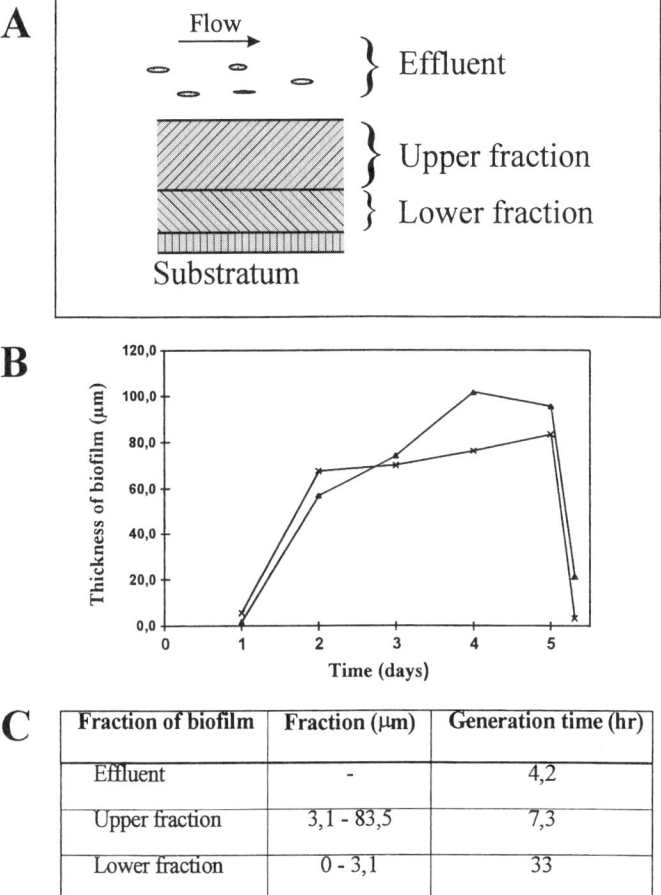

Figure 4.7 Growth activity as a function of localization in a biofilm. In a flow chamber with a flow of substrate passing over the biofilm, there is a certain degree of stratification of the community. The layer of cells in the upper part of the biofilm will be closer to the substrate flow than the cells placed near the substratum (glass surface) (*A*). Cells in the effluent will, of course, be fully embedded in the substrate and thus have complete access to the nutrients present in the substrate. Introduction of air bubbles in the flow chamber results in removal of large parts of the biofilm, and only the very low parts near the substratum remains in the chamber (*B*). It is thus possible to fractionate the biofilm into an effluent with cells in suspension, an upper part representing the cells near the substrate flow, and a lower part sticking to the surface and representing the cells furthest away from the substrate flow. The measured fluorescence intensities—when related to a standard curve as described in Figure 4.6—correspond to generation times as indicated in *C*. These values clearly show that the same organism (*P. putida* in this case) may be present in very different growth states within a small region of a biofilm.

4.3.2 Ribosomal RNA Synthesis Rates I

Cellular regulation of ribosome synthesis ensures that, in principle, no more ribosomes are synthesized than those needed for protein synthesis as defined by the energy supply provided by the environment. Ribosomes are part of the translational machinery in the cell, constituting more than 50% of the total cell weight in fast growing cells. Obviously, overproduction of these components would compromise an optimal exploitation of the resources present, and its therefore not surprising that the control system, initially characterized in *E. coli*, seems to be conserved in many different bacterial species. The primary target for the control system is the initiation of transcription of the rRNA genes. Through a coupling between the energy metabolism (with GTP as a key factor), the translation process, and the specificity of RNA polymerase directed by a GTP derivative, ppGpp, the rRNA promoters are activated in response to the environmental conditions: The better the conditions for growth, the higher rate of transcription from the rRNA promoters (Condon et al., 1995).

Recently, a method for determination of rRNA synthesis rates in bacteria was presented (Cangelosi and Brabant, 1997). It is based on the foregoing knowledge about regulation of ribosome synthesis, and it took advantage of the fact that the rRNA molecules (16S, 23S,

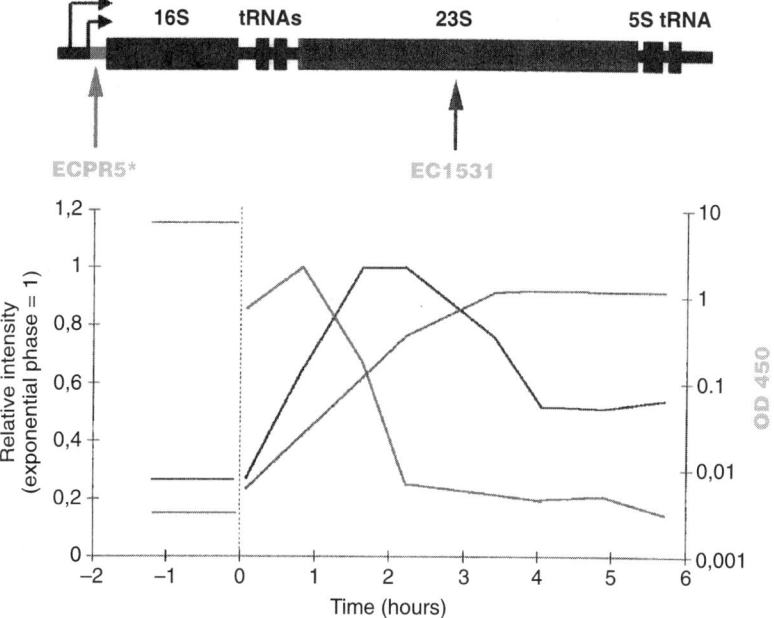

Figure 4.8 Growth activity monitored by in situ hybridization to precursor rRNA sequences. In a batch culture growth experiment the concentration of ribosomes (monitored by the use of a probe targeting the 23S rRNA) and the concentration of the precursor rRNA (monitored by the use of a probe targeting the leader sequence upstream of the 16S rRNA gene) were determined by in situ hybridization. The stability of the mature 23S rRNA and the instability of the leader sequence give rise to quite different kinetics in a simple growth experiment, where cells from an overnight stationary culture are inoculated in fresh medium at time 0. The cell density increases exponentially for a few hours after which growth gradually slows down and eventually comes to a halt (the green curve). The concentration of mature ribosomes increases slowly and decreases slowly after the end of the exponentially growth phase (blue curve). In contrast, the concentration of the precursor sequence (the red curve) immediately increases upon growth start, and decreases very rapidly when the cells begin to leave the exponential growth phase. The last curve represents the rate of synthesis of rRNA due to the instability of the target sequence, whereas the 23S rRNA content reflects a composite history of previous and present growth states, because what is measured is the accumulated levels of target sequence. (Reprinted from Licht et al., 1999)

and 5S) are transcribed together as a long precursor transcript (30S), which is posttranscriptionally processed to the final rRNA products. In this processing reaction, small precursor sequences outside the mature RNA domains are removed by nuclease activities, and these pre-RNA molecules are degraded quite rapidly. However, the lifetime of the pre-RNA molecules in growing cells is sufficient to allow their detection by quantitative hybridization. In cells that have stopped growing, the ribosome control system leads to a block of transcription of rRNA, resulting in a rapid disappearance of the pre-RNA molecules, and therefore the lack of the corresponding hybridization signals is indicative of stationarity or very slow growth.

The described method, which was originally developed for population assays in monocultures (filter assays), was subsequently transformed into an in situ technique for monitoring of single cells (Fig. 4.8) (Licht et al., 1999). With this kind of approach it is possible to combine the ribosome-counting method with direct monitoring of rRNA synthesis rates, and since in situ hybridization in biofilm communities has been developed as a standard technique, the pre-RNA hybridization method should be a fairly simple extrapolation (as long as the relevant sequence information is available). At the present time there are not many published applications of the method—in fact, there is only the foregoing based on in situ pre-RNA hybridization. In this investigation, bacterial growth (*E. coli*) in intestinal biofilms was analyzed by in situ pre-RNA hybridization. Previously, it had been found by the ribosome-counting method that enterobacteria grow quite rapidly in the large intestine in mice, and that the ribosome concentrations in various parts of the gut were nearly identical (Poulsen et al., 1995). It could be calculated that if the bacteria were growing with rates as those estimated from the ribosome counting (23S rRNA hybridization), the animal would explode due to the bacterial biomass. The direct measurements of rRNA synthesis rates measured as pre-RNA hybridization signals showed clearly that the bacteria grew in the intestinal mucus layer (approximately 10% of the population). In the intestinal contents, however, the cellular amounts of pre-RNA were surprisingly high, which subsequently was found to be caused by a growth inhibitory effect similar to that caused by addition of chloramphenicol. Thus, the high ribosome levels in the bacteria from the luminal contents reflected the previous fast growth in the mucus rather than an actual growth rate of the cells, but it also had to be concluded that the method of determining the levels of pre-RNA by hybridization could yield controversial results, which only made sense in the light of further investigations of the physiology of the bacteria (Licht et al., 1999).

4.3.3 Ribosomal RNA Synthesis Rates II

The two methods described previously for measurements of ribosome synthesis in bacterial cells are both based on in situ hybridization, and consequently imply community destruction resulting from the fixation of the cells. For dynamic studies of community physiology and ecology, in which variations in the growth status of cells, microcolonies, sectors, layers, and so forth reflect changing environmental conditions—externally or locally—there is a need for nondestructive on-line methods for determinations of physiological states of bacteria. Such methods should ideally report directly about rRNA synthesis rates without the need for any inactivating treatments of the microbial community, and there should be the same connection between the measurement and the actual status of the cells as described for the pre-RNA method. If this were achievable, it would be possible to follow over time how microbial growth activity is compartmentalized in a biofilm community, and how this distribution changes with the growth conditions and time.

The gene expression reporter, Gfp (green fluorescent protein), harbors several interesting and relevant characteristics that make it an excellent tool in molecular ecology. Most importantly, the protein carries in the polypeptide chain a fluorophore, that is excitable with blue light; thus, there is no requirement for any externally added substrate. Second, the green fluorescence emitted from the expressed protein upon excitation with light of the right wavelength is compatible with fluorescence microscopy, including scanning confocal microscopy, which makes it a relevant marker for biofilm cells. Finally, a number of variant *gfp* genes have been constructed with the following properties: better intensities (Cormack et al., 1996), changed excitation and emission wavelengths (Heim et al., 1994), reporters for both transcriptional and translational gene fusions (Ludin et al., 1996), and lately also genes expressing Gfp molecules with much reduced stability in bacteria (Andersen et al., 1998). The latter is relevant for design of reporter systems, in which it is important to monitor changing gene expression activities in complex environments.

In connection with monitoring of growth activities in biofilms, the reporter genes encoding unstable Gfp proteins have been used in fusions with an rRNA promoter. The rationale behind this strategy was to create an approach like the one described for pre-RNA hybridization, that is, direct determinations of rRNA synthesis rates as indicator of growth activity. The wildtype Gfp protein is totally stable in bacteria and therefore offers no advantage over direct counting of stable ribosomes. The degree of instability of the newly developed Gfp variants is such that continuous expression from a good promoter is easily detected in single cells by fluorescence microscopy, and inhibition of gene expression is rapidly observed as disappearance of fluorescence.

Thus, if genetic manipulation of the strain of interest is at all possible, clones comprising chromosomal insertions of rRNA promoter fusions to the gene for unstable Gfp may be obtained and used to monitor growth activity in the biofilm communities. Preliminary applications of such strains in biofilms have provided the first insights in the distribution of actively growing bacteria in simple biofilm communities, and as shown in Figure 4.9 there is a considerable heterogeneity, even in microniches of such communities, which would never have been detected with any of the other monitoring methods described previously. Through this and similar approaches it will be possible to reach a deeper understanding of the creation of microenvironments in microbial surface communities, and due to the non-destructive characteristics of analyses like this one, dynamic developments in the community can be recorded.

4.3.4 Expression of Growth Indicator mRNAs in Single Cells

"Life after log" is a term introduced by Robert Kolter (Siegele and Kolter, 1992) to represent the physiological state of bacteria when they are not growing exponentially, which they of course nearly never do outside the laboratory. The ribosome synthesis measurements described previously do provide information about the state of growing cells, and in some cases they also inform about nongrowing cells, but systematic monitoring of stationary or even stressed cells requires alternative in situ approaches. From a large pool of information it is known that even nonsporulating bacteria change the gene expression pattern as a response to stressful conditions and enter a state that phenotypically is quite different from that of exponentially growing cells (Kjelleberg et al. 1987). Like many other complex behavioral traits in bacteria the stress responses are regulated by hierarchies of regulatory loops, which to some extent are conserved across the species. In several organisms, a detailed knowledge about the change in protein synthesis pattern has

4.3 BACTERIAL GROWTH AND STATIONARITY IN COMPLEX BIOFILMS

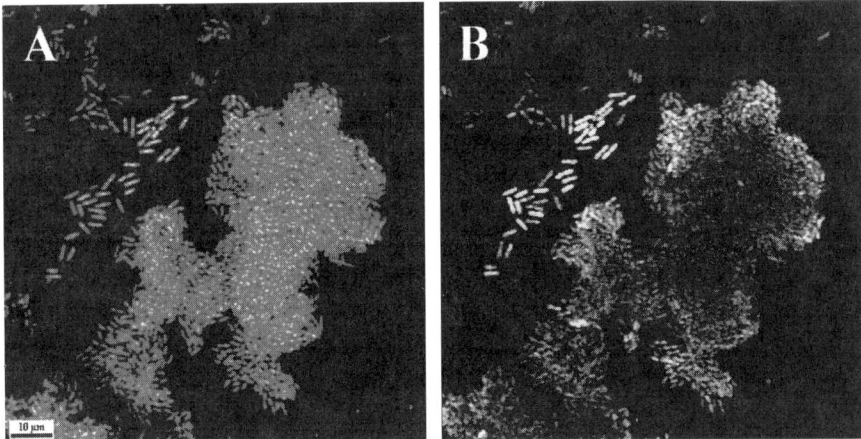

Figure 4.9 Growth activity monitored by reporter gene activity expressed from a fusion between a ribosomal promoter and the *gfp* gene. A fusion between the P1 rRNA promoter from *E. coli* (the left promoter upstream of the 16s rRNA gene shown in Fig. 4.8) and a gene encoding unstable Gfp was inserted in the chromosome of *P. putida* as part of a mini-transposon. The resulting strain was established as a biofilm in a flow chamber, and after proper fixation in situ hybridization with a fluorescent probe targeting the 16S rRNA was performed. The picture in the *A* frame shows a microcolony in the biofilm, and judging from the intensity of fluorescence it seems to be a homogenous collection of cells. The same colony was analyzed for its green fluorescence emitted from the unstable Gfp protein expressed from the gene fusion described, and it is clear from frame *B* that the cell population is fairly heterogenous with respect to green fluorescence: The cells near the surface of the colony are brighter than those in the middle of the colony, indicating that the former still express rRNA whereas the latter have stopped or reduced the rRNA expression. Since the cells on the colony surface are more likely to receive nutrients from the flow of substrate compared with the inner cells, the distribution of green fluorescence seems to more truly reflect the growth status of the cells than the ribosome counting. (Reprinted from Sternberg et al., 1999) Figure also appears in Color Figure section.

been obtained from which potential reporter genes may be developed (Givskov et al., 1994b; Nyström et al., 1990).

The approach described in Section 4.3.3 involving the Gfp reporter for monitoring rRNA synthesis rates (growth) may of course also be employed in the context of monitoring stress-induced or stationary phase-induced genes through construction of the proper gene fusions. However, here we wish to introduce an alternative molecular approach targeting the respective gene products directly. Thus, the aim is to detect in situ the mRNA of relevant genes. The major problem of mRNA detection compared with rRNA is that there are many thousand rRNA molecules in most bacterial cells, whereas mRNA concentrations range from one (or zero) to less than one hundred. Consequently, the direct in situ hybridization technique using fluorescent probes is not possible. Many attempts have been made to improve the hybridization method, but so far no real satisfactory protocol has been developed, and in particular not for complex communities in microbial biofilms.

It was therefore a breakthrough when Hodson et al. (1995) presented an alternative approach based on in situ RT-PCR, that is, intracellular reverse transcription of mRNA followed by a PCR amplification step. Modifications of this method were subsequently published, and estimates of the sensitivity indicated that with a mixture of amplification steps it is in fact possible to detect levels of mRNA down to one molecule per cell (Tolker-Nielsen et al., 1997). Additionally, it was shown that, at least to some degree, quantification of the mRNA pool in the cells is possible (Holmstrøm et al., 1998). In short, the method involves an important step of permeabilization to allow entrance of the enzymes for the nucleic acid processing, a coupled reverse transcription and PCR amplification, and finally different ways of labeling the product for detection. One important bottleneck in this procedure is the first step, because it is crucial that the cells become permeable to the necessary enzymes without losing their integrity.

In connection with identification of biofilm cells, colonies, sectors, or layers in states of growth or stationarity/stress, in situ mRNA monitoring may offer a unique combination of resolution and sensitivity. Growing cells express not only ribosomal genes but also other genes involved in the protein synthesis machinery, and the sensitivity of the RT-PCR method is much higher than what is obtainable with rRNA hybridization or Gpf reporters; therefore, more detailed mapping of activity distribution is possible if the method is applied to biofilm communities. The micrograph of a biofilm microcolony presented in Figure 4.9, showing rRNA::Gfp-fluorescent cells on the surface of the colony indicates that only the outer layer of cells is actively growing. It is likely, however, that many more cells are actually growing, but not with rates resulting in sufficient Gfp expression for its detection. In addition, in parts of a biofilm the oxygen concentration may be too low for the fluorophore of Gfp to be formed. We should thus see the in situ mRNA detection approach as a potential alternative to fine-map activity centers in a biofilm.

The gene expression activity in stationary and stressed cells is different from that in growing cells, but the fact that distinct gene products are made in the former nongrowing cells provides a basis for detection of these using the RT-PCR technique in situ. Conditions such as nutrient starvation, oxygen depletion, heat/cold shock, osmotic stress, pH stress and many more probably result in specific gene expression fingerprints, based on which relevant primers for the needed enzymatic reactions may be designed. Figure 4.10 shows such an example based on the well-studied heat shock induction of the *groEL* gene in many bacteria. The boost of expression of *groEL* mRNA upon heat treatment of the cells is clearly observed in the single cells after RT-PCR. In similar investigations, expression of other physiological marker genes can be monitored, and if sufficient DNA sequence information is available it may be possible to direct the in situ reactions from detection of a specific mRNA in specific organisms to monitoring of specific gene expression in a broad spectrum of organisms in the community.

4.3.5 Miscellaneous Cellular Reporters of Physiological States

In addition to the specific gene expression patterns, which may be used to design reporter methods for growing, stationary, or stressed cells as described previously, there are several much simpler cellular features, which at least for some bacteria may be used as supporting phenotypic expressions of specific physiological states. For example, it has been observed for many bacteria that cell size is correlated with growth rate, and that cells starved for different nutrients are smaller than their growing relatives. Thus, in situ rRNA hybridization used for identification and ribosome counting automatically also provides information about cell volumes, which is useful information in many contexts as support for conclusions concerning growth and stasis. The number of genome equivalents in a cell may also vary with the growth conditions. One is obviously the lowest number, but in fast-growing bacteria, several copies of the genome are needed to keep pace with the growth rate, and in

4.3 BACTERIAL GROWTH AND STATIONARITY IN COMPLEX BIOFILMS 107

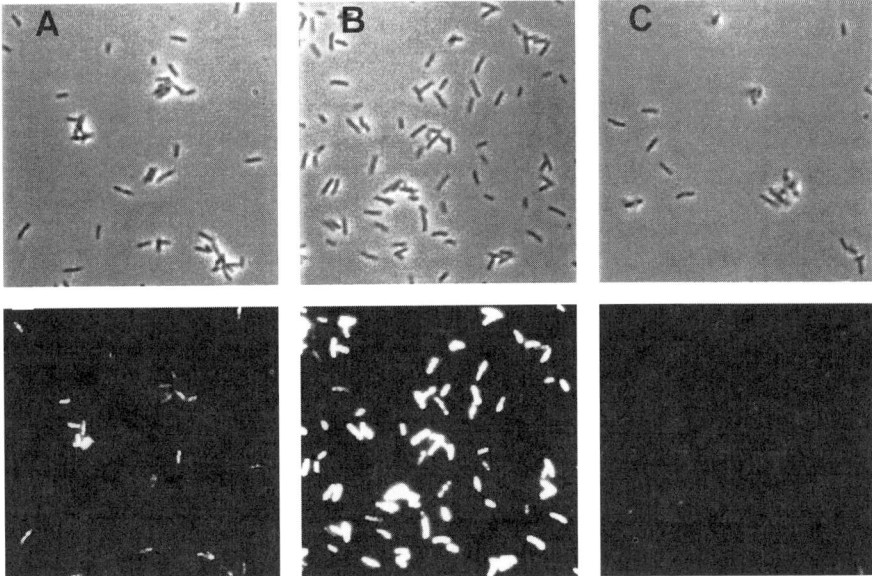

Figure 4.10 Stress induction monitored in situ by RT-PCR in single cells. Knowledge of the sequence of a gene allows design of oligonucleotide primers useful in PCR amplification of a gene or a gene sequence. When combined with the reverse transcription activity, amplification of mRNA sequences may be achieved, and if finally coupled with antibody reactions targeting antigenic substances present in the amplified DNA, an extra amplification step may be obtained if an enzymatic reaction is connected with the antibody. After permeabilization of the bacterial cells, the necessary enzymes and primers can be introduced in the cells, and results from such an in situ RT-PCR are shown here. The targeted mRNA is encoded by the heat shock inducible *groEL* gene of *S. typhimurium*, and in frame *A* the mRNA signals in normally growing cells is shown. In frame *B* the cells have been exposed to a heat shock, and in frame *C* the cells have been treated with RNase before RT-PCR (control showing that the reaction is RNA dependent; no signals are derived from the chromosomal DNA sequence (the gene itself)). All upper images are phase contrast micrographs. (Reprinted from Holmstrøm et al., 1999)

such cases monitoring of the DNA content may be a useful controlling activity to perform. Finally, the presence of a significant fraction of dividing cells indicates that the population is actively growing; however, this cellular parameter of course only provides information about a population (subpopulation), not about the individual (nondividing) cells (Møller et al., 1995).

Perhaps the most important factor in this type of estimation of bacterial physiological activities in biofilms and other complex environments is the available knowledge about the organisms and their performance phenotypes in combination with qualified guesses about the local environmental conditions. It is, for example, a good guess that the cells close to the substratum of a biofilm face less favorable nutritional conditions than those on the top being close to the external surroundings containing nutrients, and there should consequently be a gradient of growth activity among the cells from the top to the bottom of the biofilm. This theory was confirmed in an investigation of the physiological states of cells of *Pseudomonas putida* growing in a flow chamber mixed biofilm, in which it could be observed that the cells closer to the nutrient flow had more rRNA than those near the glass surface, indicating differences in growth rates (cf. Fig. 4.7).

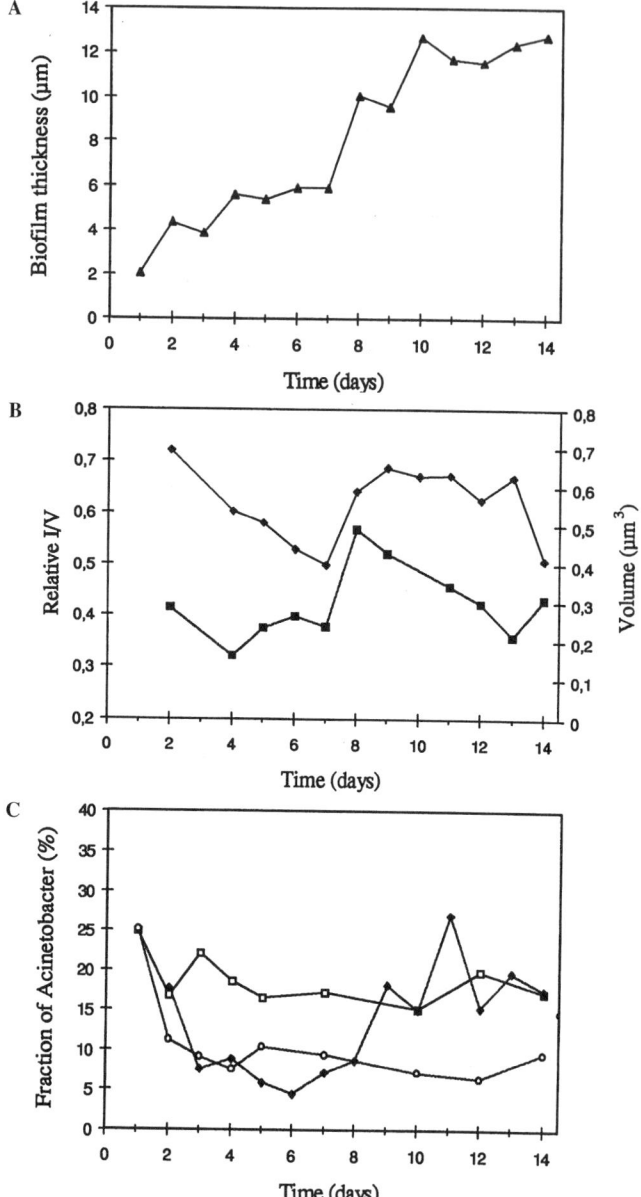

Figure 4.11 Biofilm community responses to a nutritional upshift. A mixed community of *P. putida* and a strain of *Acinetobacter* growing in a flow chamber with a poor carbon source (trichlorobenzoic acid) added to the substrate flow was shifted to a medium containing a better utilizable carbon source (benzyl alcohol) at day 7. This represents an upshift, which under conditions of growth in batch cultures, would have immediate consequences for growth and cell composition. In the biofilm community, *A* shows that the rate of increase of biomass (biofilm thickness) rapidly increases after the upshift, until a new stationary situation develops after day 10. At the cellular level (*P. putida*), cell volume increases immediately after the upshift, and similarly the ribosome concentration increases after the shift (*B*, diamonds and squares, respectively). At later times both cell volume and ribosome concentration go down as a consequence of the stationarity of the cells after day 10. In *C*, the population

4.3.6 Biofilm Community Reactions to Environmental Shifts

One of the most informative experimental strategies to reveal the responsive repertoire of an organism and the overall coordination of cellular activities is to impose sudden environmental shifts on the cells. Nutritional shifts have been fundamental to the understanding of the control of macromolecular synthesis in bacteria, temperature shifts have exposed important regulatory pathways involved in development of increased stress tolerance, and shifts to starvation conditions have been important in revealing how bacteria differentiate to more resistant life forms, enabling them to survive for very extended periods under a variety of stressful conditions.

In an analogous way, the response pattern and internal coordination scheme of a microbial community (e.g., in a biofilm) may be elucidated by introducing sudden shifts in the environmental conditions followed by careful monitoring of a range of phenotypic reactions in the community. Figure 4.11 shows how a biofilm community reacted to a sudden shift up, that is, a change from a nutrient-poor to a nutrient-rich environment. In this case, only growth physiological parameters were analyzed in detail. The total community reacted to the shift up by increasing the rate of biomass accumulation. The population profile changed, resulting in an altered postshift balance of the community members. Measurements of cell volumes showed that the cells grew larger as a consequence of the shift. The ribosome contents measured in one of the strains of the biofilm community increased immediately after the shift indicating an upshift in growth rate. At later times after the shift up the community clearly reached a state of stationarity—similar to a nutritional shift down—and the picture was reversed, showing reduction in biomass increase, cell volumes, and ribosome concentrations.

Obviously, a large number of shift conditions may be imposed on biofilm communities (at least on those controllable in the laboratory), and in addition to changing the composition of the substrate feeding the community, it may be relevant to monitor the effects of adding or removing particular organisms. One interesting perspective of such perturbation approaches is the potential for monitoring reactions in the community, which can be interpreted as coordinated responses. This requires multiparameter investigations of reproducible and controllable communities in laboratory based biofilms, but from there extrapolations to natural environments may be possible.

4.4 GENE TRANSFER IN BIOFILM COMMUNITIES

A microbial community may, under some conditions, be considered as a multicellular entity with a gene pool composed of the sum of the individual genomes present in the different species; the more diversity, the larger the gene pool. In this view, new community traits may be introduced by addition of new organisms, and such a colonization event may be a potential benefit to the other community members. However, introduction of a new species may also create a perturbation of the system, resulting in displacements and rearrangements of indigenous organisms. In both cases, the community response pattern to introduction of new organisms and their genomes is an important phenotypic reaction, from which significant

---

structure analysis shows that *Acinetobacter* constitutes a smaller fraction of the population in the poor medium (circles) than in the richer medium (squares), and a shift up induces a corresponding shift in the population structure (diamonds). The different bacteria were identified by in situ hybridization.

information about the system may be obtained. Less than entire genomes may be introduced by addition of single or few genes to specific community members by the processes of gene transfer, and their dispersal in a population may also have significant effects on the entire community.

Although gene transfer occurs naturally in three principally different ways—(1) transformation of DNA, (2) transduction by bacteriophages, and (3) conjugation of plasmids—it is mainly conjugation of plasmids that has so far attracted experimental attention in the context of microbial ecology and in particular in relation to impacts on biofilm communities. It should be emphasized in this connection that the amount of genetic information transferred by transformation and transduction is very limited compared with what may be carried on plasmids; therefore, if one objective of gene transfer studies is to investigate impacts on a community, it is more likely to be significant in case of plasmid transfer.

Introduction of new genetic information may also be seen as an element of communication within a microbial community complementary to chemical communication; the latter is responsible for localized adaptations to changing environmental conditions producing transient patterns of coordinated behavior (adjusting a subpopulation to a new situation), whereas gene transfer creates more stable population changes with the potential of altering the behavioral direction of the community. It is therefore of fundamental interest to study gene transfer in microbial communities and to assess the effects of transfer events on the population and the internal relationships between community members. In addition, gene transfer may be practically employed to increase the performance of microbial communities in cases where they perform useful activities (biological cleaning, different types of bioreactors, etc.).

4.4.1 Gene Transfer in Natural Systems

The global discussion of the potential risks of genetic engineering has stimulated a specific interest in gene transfer in natural environments. During the last decade a large number of studies concerning plasmid transfer in soils and waters have been performed, and there is now a very large body of information about a number of plasmids, bacteria, and ecosystems (Wellington and van Elsas, 1992). Since bacteria in many natural environments live on surfaces and often under biofilm conditions, there is a direct connection between the questions and problems related to gene transfer in nature and biofilm investigations, which is the focus in the present context.

There is some semiquantitative data about plasmid transfer in natural environments, but much of it is still quite descriptive. In particular, there is no, or only little, information about structure/function relationships in relation to plasmid transfer. In addition, some of these environments harbor a vast number of dormant cells whose potential role in processes such as plasmid transfer is totally unknown. If, however, conclusions should be made about plasmid transfer in soils and related environments, the most important one is that availability of nutrients is the critical factor. Colonization of particle surfaces seems to be another factor of importance. It is therefore not surprising that the plant rhizosphere represents a hot spot for conjugal transfer due to the more nutrient-rich environments created by the exudation from the roots (van Elsas and Trevors, 1990).

In water-based environments, the rates of plasmid transfer are usually much higher than those observed in soils. High levels of transfer are observed in particular in connection with biofilms forming on surfaces within the water phase, depending on the nutritional situation, and in "natural" systems such as wastewater treatment plants, where the cell densities may reach very high levels, plasmids may be dispersed with high rates (Bale et al., 1988; Fulthorpe and Wyndham, 1991; Mach and Grimes, 1982). Another specific example of plasmid transfer in a natural water-based environment is what goes on in animal guts. Here it has been found that plasmids, which normally transfer well between bacteria in suspension, do so with fairly good efficiency in the mammal gut, whereas plasmids specialized in transfer on surfaces hardly transfer in the gut (Licht, unpublished).

In conclusion, bacteria living under conditions that allow them to exploit the environment, add to the environment, and develop with the environment, also possess a potential to act as a donor or recipient for plasmid transfer; in contrast, bacteria in a state of dormancy probably neither relate to the environment nor contribute to the dispersal of new genetic information. Therefore, in order to understand how dispersal of genes in a population may participate in the adaptation and development of this community, it is necessary to first investigate populations in which a significant fraction of individuals are active community members. In the following section, such initial investigations in microbial biofilms are presented.

4.4.2 Introduction of Plasmids in Microbial Biofilm Communities

Conjugative plasmids encode their own transfer functions, and in most cases the corresponding genes constitute a large part of the entire plasmid-coding capacity. In addition to mobilizing themselves, conjugative plasmids often have the capacity to mobilize other nonconjugative plasmids carrying a transfer origin and a *mob* site (nicking site). The transfer genes of plasmids are of different types, and in the present context it is particularly relevant to realize that some plasmids transfer almost exclusively on surfaces, whereas others transfer preferentially in liquids. In biofilms, the cells are surface-bound, thus favoring the plasmids that normally transfer optimally under these conditions, and the high cell densities often found in biofilms also provide excellent conditions for plasmid transfer. Due to the frequent presence of selective markers on plasmid molecules (alternatively such markers are easily inserted in plasmids), it has been fairly simple to monitor gene transfer in a broad variety of settings by selective platings. However, the studies based on platings do not provide any information about the distribution of plasmid transfer events in communities and also do not provide any specific information about rate-limiting factors with significance for the transfer process.

The introduction of molecular tools in microbial ecology has facilitated in situ studies of gene transfer in complex microbial communities, and integrated investigations of community composition, physiological activity, structure, and dispersal of plasmids are now possible and will eventually result in a much better understanding of the impact of plasmid mobilization and provide a better background for optimizing the transfer process when desired. For the initial studies, it is important to analyze communities whose bacterial population and growth conditions are known and at least to some extent controllable. It is also important to choose plasmids that are connected to the bacteria in the community and to the environmental conditions (selection pressures). Finally, quantitative data on plasmid transfer should be combined with in situ methods for organism identification, bacterial activity, and presence or absence of plasmids, in order to map the structure/function relationships more accurately in the community.

In many studies of plasmid transfer, antibiotic resistances have been the most significant phenotypic feature of the plasmids, and transfer has been investigated under conditions of no direct selection for the presence of the plasmid. Alternatively, antibiotic addition has been used to create a selective environment, but obviously this is a very extreme selection scenario in which all bacteria but the plasmid carriers will be eliminated. In natural settings, it is likely that plasmids often contribute to the adaptation of bacteria by supply specific metabolic traits useful in particular niches. Such phenotypes give the plasmid-carrying cells a growth selective advantage without inhibiting or killing the competing organisms in the surroundings. In microbial biofilms the impact of selective pressures related to growth activity has not been clarified, and it is therefore of interest to investigate plasmid transfer in biofilms, partly in order to understand plasmid biology in such systems and partly to reach a better understanding of biofilm community adaptation and development.

Results from a toluene-degrading community, to which a donor strain harboring the TOL plasmid was added, showed that plasmids were transferred with a significant rate, and that growth rate increase does provide a conditional, selective advantage. Transfer of the TOL plasmid to a community strain of *P. putida* resulting in a 10–20% increase in growth rate resulted in a high proportion of plasmid-carrying cells, whereas a different plasmid-carrying strain of *P. putida* not related to the community could not establish despite its fast growth rate (Christensen et al., 1998). Thus, dispersal of plasmids in a biofilm community can be very successful in cases where optimal combinations of host cell background and benefits from the plasmid arise as a consequence of the transfer (Fig. 4.12). In experiments designed to monitor gene transfer, it is important to take into account the possibilities for cell growth and also to estimate the relationship between the actual conjugational transfer rate and the rate of growth of the transconjugants. What at the end of the investigation looks like a very high rate of transfer (high proportion of transconjugants) may in fact be the result of a few transconjugants proliferating with a high rate and competing plasmid-free cells out.

4.4.3 On-line Monitoring of Plasmid Transfer

The introduction of molecular tools in studies of complex microbial communities has made it possible to localize the position of specific organisms in a three-dimensional structured surface community, and also the general and specific activities (physiological states, single gene expressions) may be mapped in these systems, as described previously. The information has provided insight concerning the heterogeneity of biofilm communities and has indicated how different activities may be compartmentalized. In a similar way, it is important to map the gene transfer hot spots as an element in building up a better understanding of the community performance, the plasmid transfer physiology, and the distribution of selective domains in biofilms.

Although it should be possible to combine the methods of in situ rRNA hybridization and RT-PCR (targeting plasmid-encoded mRNA or the plasmid molecule itself), it would in fact be valuable to have a nondestructive method allowing studies of the time-dependent migration and dispersal of plasmid-carrying cells. Use of the *gfp* reporter gene is therefore an obvious alternative, and recently *zygotic induction* systems were developed for on-line monitoring of plasmid transfer in mixed communities (Christensen et al., 1996, 1998). The principle is simply to construct a donor strain harboring a plasmid with an insertion of a fusion between a regulated promoter and the *gfp* gene; in the donor strain the expression of Gfp must be repressed, whereas the fluorescent protein is expressed in the recipient after transfer of the plasmid (Fig.

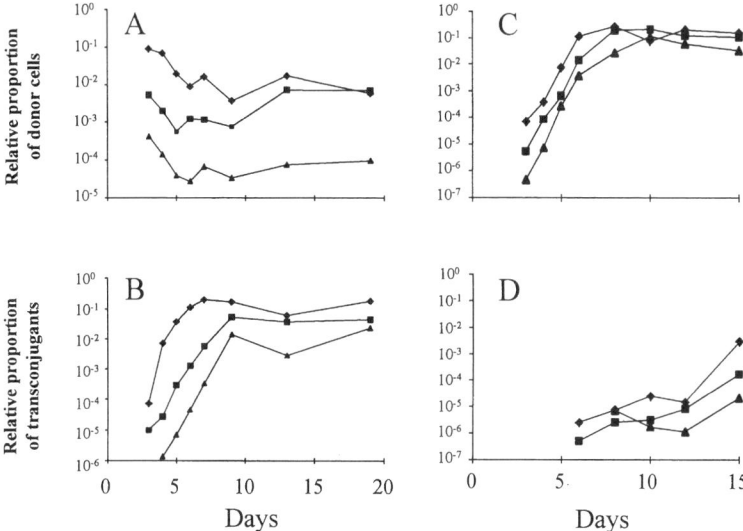

Figure 4.12 Plasmid transfer from an introduced donor strain to the indigenous biofilm population. Two different donor strains of *P. putida*—both harboring the TOL plasmid—were introduced in a flow chamber biofilm community comprising *P. putida* together with other organisms under conditions where the sole carbon source is benzyl alcohol. The TOL plasmid only transferred to the *P. putida* strain under the present conditions. In *A* and *B* the donor was a laboratory strain (KT2442), in *C* and *D* the donor was the same as the indigenous recipient, R1, derived from a toluene-degrading bioreactor. The different cells (donors, recipients, and transconjugants) could be distinguished by different antibiotic resistance markers.

The laboratory strain colonizes very poorly (*A*), but nevertheless transfers very successfully the TOL plasmid to the recipient cells in the biofilm community (*B*). The three different curves represent different donor inoculation levels. From these data the high numbers of transconjugants may be the result of either many cycles of plasmid transfer, or of a few conjugation events followed by rapid growth of the primary transconjugants.

When the donor strain is the same as the one present in the biofilm community it colonizes very well (*C*); in contrast, the number of transconjugants is very low (*D*). These results indicate that the presence of the TOL plasmid was selectively advantageous in case of the R1 strain. Once formed (or introduced) the plasmid-carrying R1 cells grew up rapidly. For the laboratory strain, the growth advantage provided by the plasmid was not sufficient for an establishment in the community. (Reprinted from Christensen et al., 1998)

4.13). Depending on the plasmid and its host range, the recipient cells may be only one species or many species in the community. The regulation of expression of the reporter gene may be through a repressor in the donor or an activator in the recipient; the former allows general studies of transfer to all compatible recipient cells in any community, the latter allows specific studies of transfer to one (constructed) strain in a controllable community.

Preliminary investigations of plasmid transfer in surface communities using the described zygotic induction technique have clearly shown that there are indeed hot spots of transfer in these. Only under conditions of complete mixing of donor and recipient cells (mixing in suspension) is it possible to obtain transfer to all recipient cells if the plasmid is fully depressed for transfer. In cases where the donor is introduced after establishment of a recipient community, on-line monitoring of transfer has shown that there are distinct zones,

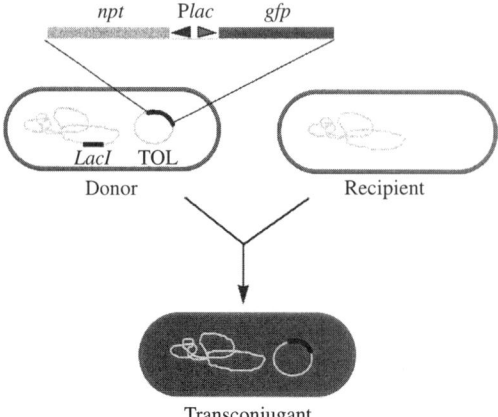

Figure 4.13 A strategy for monitoring plasmid transfer on-line in a biofilm community. The green fluorescent protein is an excellent reporter of gene expression in connection with microscopic investigations of complex communities. For studies of plasmid transfer, fusions between an inducible promoter and the *gfp* gene have been inserted in the conjugative plasmid, and in the donor strain the expression of the reporter gene is repressed by a regulatory protein, which is only synthesized in the donor. Upon transfer of the plasmid to a recipient cell, the *gfp* gene is derepressed, and thus transconjugants may be visualized in situ in a nondestructive, on-line fashion.

sectors, or sometimes unique positions where transfer occurs (Fig. 4.14). This means that overall population estimates of plasmid transfer frequencies do not provide much information about the plasmid phenotypes, but rather indicate how many sites are available for transfer to take place. In addition to this, it is very important to distinguish between conjugational events and further growth of transconjugants, as mentioned previously.

The pattern developing from these preliminary studies is that introduction of donor cells in biofilm communities leads to a very limited transfer rate, and in fact it appears as if even very close associations between plasmid-carrying cells and potential recipient cells do not necessarily result in transfer. This conclusion is based on the sharp boundaries between transconjugants and recipients observed in biofilm microcolonies and in colonies growing on agar plates (Christensen et al., 1996). As suggested by Beaudoin et al. (1998), the transconjugants may differ physiologically from the donors. It is, however, impossible at the present time to make this a general conclusion concerning all or most plasmids, but it does present one experimental route to monitor transfer patterns for other plasmids; with this kind of information it should eventually become possible to decide if the most important factors determining plasmid transfer in biofilms are plasmid associated or community related.

4.4.4 Relationship between Plasmid Transfer and Growth Activity

The physiological activity of bacteria in biofilms is determined by the availability of nutrients, and as shown previously there seems to be distinct areas of growth activity in the upper layers (Fig. 4.7), on the surface of microcolonies (Fig. 4.9), and perhaps also near water-filled channels in the biofilm matrix. A closeup of the microcolonies shows that in fact only a very thin layer of cells on the surface of the colony are actively growing as indicated by the fluorescence derived from a fusion between an rRNA promoter and the gene for unstable Gfp (Fig. 4.15). This pattern of growth is strikingly similar to the pattern of plasmid transfer observed in biofilms (under conditions where the donor cells were added subsequently to the establishment of a recipient biofilm), where essentially all transfer of the TOL plasmid took place on the surface of recipient microcolonies (Fig. 4.14). This correlation between growth activity and plasmid transfer activity suggests that there is a nutrient diffusion gradient through a microcolony in a biofilm allowing growth of only the surface cells, and that plasmid transfer only occurs between actively growing cells (Fig. 4.15).

If this is the major (or only) explanation of the heterogeneity of plasmid transfer in biofilms, it should be taken into account when planning deliberate introductions of plasmids in biofilm communities; increasing nutrient concentrations or altering the structure to make it easier for nutrients to penetrate the biofilm matrix could be ways of optimizing gene transfer. Also, predictions concerning gene transfer in a variety of different environments may become safer if knowledge about the distribution of actively growing cells is available.

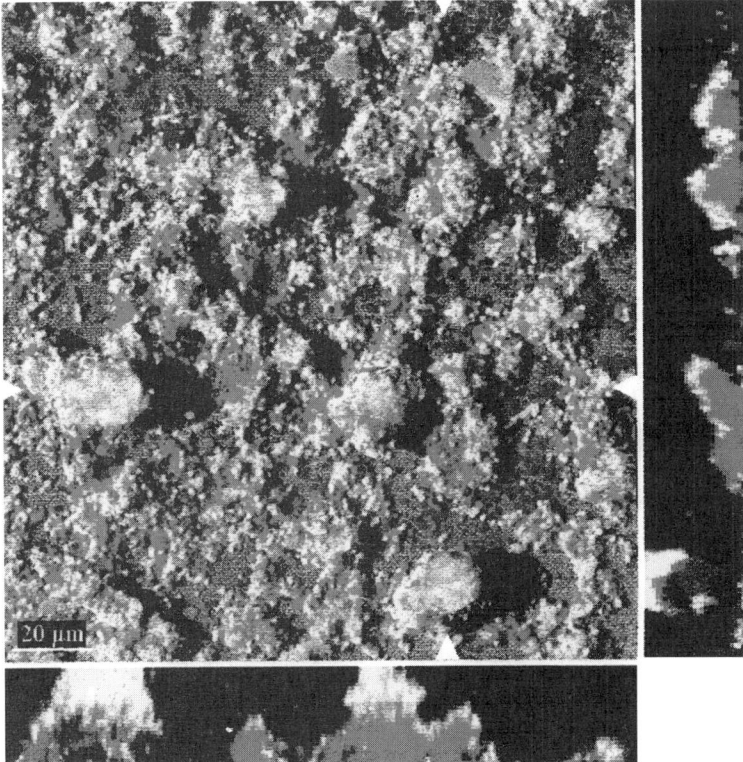

Figure 4.14 Direct monitoring of plasmid transfer in a complex biofilm community. When the strategy outlined in Figure 4.13 is combined with in situ rRNA hybridization used for strain identification, it is possible to localize specific active areas of transfer. Two organisms are identified with rRNA probes in the micrograph: *P. putida* (red cells) and *Acinetobacter* (blue cells). In addition, Gfp fluorescent cells represent transconjugants which are seen as either green or yellow cells. Z-scans (below and on the right side) show how microcolonies of *P. putida* seem to be hot spots for plasmid transfer. Note that the inner parts of the colonies do not seem to be converted to plasmid-carrying transconjugants. The TOL-plasmid, which is the one investigated in this analysis, only transfers to *P. putida*. (Reprinted from Christensen et al., 1998) Figure also appears in Color Figure section.

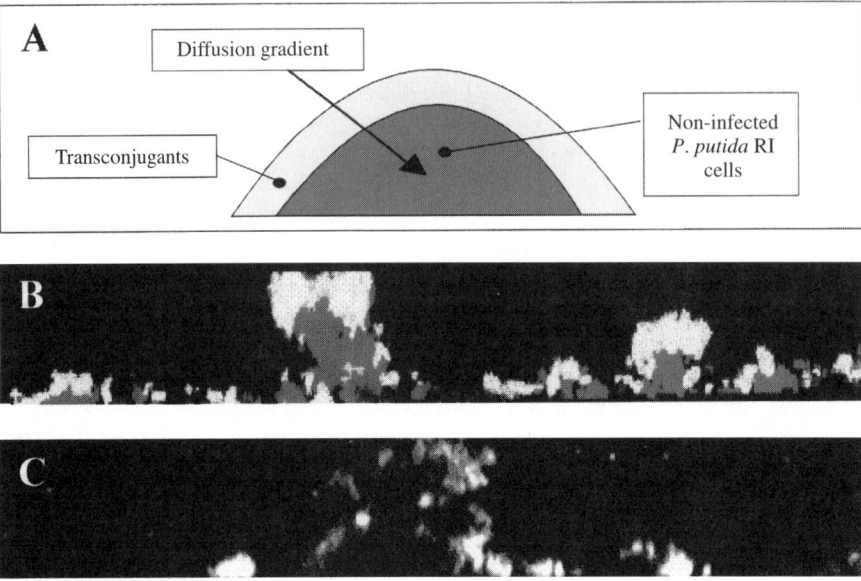

Figure 4.15 Correlation between active growth zones and plasmid conjugation zones in microcolonies in biofilms. It is likely that cells in microcolonies in biofilms face different nutritional conditions depending on their position in the colonies. *A*. A model based on nutrient diffusion gradients is suggested as an explanation for the lack of plasmid transfer to the potential recipient cells in the middle of the colony. *B*. A scan like the ones presented and explained in Figure 4.14, showing the position of transconjugant cells on the surface of a microcolony. *C*. The growth active cells are monitored through the fluorescence signal resulting from expression of an rRNA promoter/unstable Gfp fusion (cf. Fig. 4.9). There is a striking correlation between the growth active zone and the plasmid transfer active zone in these colonies. (Reprinted from Christensen et al., 1998) Figure also appears in Color Figure section.

These first investigations of gene transfer in biofilms have indicated that the structure/function relationships with respect to distribution of growth activity in the biofilm, together with the selective forces acting on the cells in the community, are perhaps more important than the conjugative properties of the plasmids in determining the rate of dispersal of plasmids through the community. It will be interesting to see if this pattern prevails in future studies of the transfer of other plasmids in other biofilm scenarios.

4.5 FINAL COMMENTS

The molecular ecology of biofilms presented here is a composite discipline comprising the objective of understanding how ecological aspects of bacterial life in complex communities like biofilms may be achieved by combining the molecular biological information about specific organisms and molecular tools developed for the same purpose. Ecological aspects refer to the social life of microorganisms including their capacities to organize themselves in specific structures and to coordinate their activities in collaborative networks. There are of course several others ways of describing biofilm communities; the present one is deliberately strongly biased toward concepts inherited from molecular microbiology.

It is sometimes argued that surface-bound cells are physiologically extremely different from the same cells in suspension, and therefore knowledge obtained from previous studies of suspended cultures is useless when analyzing biofilm communities. We should like to suggest that bacteria possess a great potential for growth and adaptation under a range of different conditions, including life in suspension and "life in slime." These different conditions require mobilization of different parts and combinations of the functional repertoire. However, the principle elements governing the exploitation of the cellular repertoire are the same no matter what the conditions are, and what is therefore very important to study are the individual windows of perception present in the organisms, and the ways these are combined at the community level to create mutually beneficial interactions that may develop the community into more optimized life in a changing environment. These windows are the receptors responding to external signals, be it metabolically useful substrates or communicative molecules. The cross-talk with the surroundings and between community members is a major target for molecular ecology, and the objective of the experimental approaches described is to achieve a listening position, so we can hear what bacteria talk about.

REFERENCES

Amann, R. I., B. J. Binder, R. J. Olson, S. W. Chrisholm, R. Devereux, and D. A. Stahl. 1990. Combination of 16S rRNA-targeted oligonucleotide probes with flow cytometry for analyzing mixed microbial populations. *Appl. Environ. Microbiol.* 56:1919–1925.

Amann, R. I., W. Ludwig, and K. H. Schleifer. 1995. Phylogenetic identification and in situ detection of individual microbial cells without cultivation. *Microbiol. Rev.* 59: 143–169.

Andersen, J. B., C. Sternberg, L. K. Poulsen, S. P. Bjørn, M. Givskov, and S. Molin. 1998. New unstable variants of green fluorescent protein for studies of transient gene expression in bacteria. *Appl. Environ. Microbiol.* 64: 2240–2246.

Bainton, N. J., B. W. Bycroft, S. R. Chabra, P. Stead, L. Gledhill, P. J. Hill, C. E. D. Rees, M. K. Winson, G. P. C. Salmond, G. S. A. B. Stewart, and P. Williams. 1992. A general role for the lux autoinducer in bacterial cell signalling: Control of antibiotic synthesis in Erwinia. *Gene* 116: 87–91.

Bale, M. J., J. C. Fry, and M. J. Day. 1988. Transfer and occurrence of large mercury resistance plasmids in river epilithon. *Appl. Microbiol. Biotechnol.* 54: 972–978.

Beaudoin, D. L., J. D. Bryers, A. B. Cunningham, and S. W. Peretti. 1998. Mobilization of broad host range plasmid from *Pseudomonas putida* to established biofilm of *Bacillus azoto formans*. I. Experiments. *Biotechnol. Bioeng.* 57: 272–279.

Brakenhoff, G. J. d. van, V. M. W. Baarslag, B. Mans, J. L. Oud, R. Zwart, and R. van Driel. 1988. Visualization and analysis techniques for three dimensional information acquired by confocal microscopy. *Scanning Microsc.* 2: 1831–1838.

Caldwell, D. E., D. R. Korber, and J. R. Lawrence. 1992. Imaging of bacterial cells by fluorescence exclusion using scanning confocal laser microscopy. *J. Microbiol. Methods* 15: 249–261.

Caldwell, D. E., D. R. Korber, and J. R. Lawrence. 1993. Analysis of biofilm formation using 2D vs 3D digital imaging. *J. Appl. Bact. Supl.* 74: 52S–66S.

Cangelosi, G. A., and W. H. Brabant. 1997. Depletion of pre-16S rRNA in starved *Escherichia coli* cells. *J. Bacteriol.* 179: 4457–4463.

Chalfie, M., Y. Tu, G. Euskirchen, W. W. Ward, and D. C. Prasher. 1994. Green fluorescent protein as a marker for gene expression. *Science* 263: 802–805.

Christensen, B. B., C. Sternberg, and S. Molin. 1996. Bacterial plasmid conjugation on semi-solid surfaces monitored with the green fluorescent protein (Gfp) from Aequorea victoria as a marker. *Gene* 173: 59–65.

Christensen, B. B., C. Sternberg, J. B. Andersen, L. Eberl, S. Møller, M. Givskov, and S. Molin. 1998. Establishment of new genetic traits in a microbial biofilm community. *Appl. Environ. Microbiol.* 64: 2247–2255.

Condon, C., C. Squires, and C. L. Squires. 1995. Control of rRNA transcription in *Escherichia coli*. *Microbiol. Rev.* 59: 623–645.

Cormack, B. P., R. H. Valdivia, and S. Falkow. 1996. FACS-optimized mutants of the green fluorescent (GFP). *Gene* 173: 33–38.

Davies, D. G., M. R. Parsek, J. P. Pearson, B. H. Iglewski, J. W. Costerton, and E. P. Greenberg. 1998. The involvement of cell-to-cell signals in the development of a bacterial biofilm. *Science* 280: 295–298.

DeLong, E. F., G. S. Wickham, and N. R. Pace. 1989. Phylogenetic strains: Ribosomal RNA-based probes for the identification of single cells. *Science* 243: 1360–1362.

Dunlap, P. V., and E. P. Greenberg. 1988. Control of *Vibrio fischeri* lux gene transcription by a cyclic AMP receptorprotein-LuxR protein regulatory circuit. *J. Bacteriol.* 170: 4040–4046.

Eberl, L., M. K. Winson, C. Sternberg, G. S. A. B. Stewart, G. Christiansen, S. R. Chabra, M. Dáykin, P. A. Williams, S. Molin, and M. Givskov. 1996. Involvement of N-acyl-L-homoserine lactone autoinducers in control of multicellular behavior of *Serratia liquefaciens*. *Mol. Microbiol.* 20: 127–136.

Fulthorpe, R. R., and R. C. Wyndham. 1991. Transfer and expression of the catabolic plasmid pBRC60 in wild bacterial recipients in a freshwater ecosystem. *Appl. Environ. Microbiol.* 57: 1546–1553.

Giovannoni, S. J., E. F. DeLong, G. J. Olsen, and N. R. Pace. 1988. Phylogenetic group-specific oligodeoxynucleotide probes for identification of single microbial cells. *J. Bacteriol.* 170: 720–726.

Givskov, M., L. Eberl, S. Møller, L. K. Poulsen, and S. Molin. 1994a. Responses to nutrient starvation in *Pseudomonas putida* strain KT2442: Analysis of general cross-protection, cell shape, and macromolecular content. *J. Bacteriol.* 176: 7–14.

Givskov, M., Eberl, L., and S. Molin. 1994b. Responses to nutrient starvation in *Pseudomonas putida* KT2442: Two-dimensional electrophoretic analysis of starvation- and stress-induced proteins. *J. Bacteriol.* 176: 4816–4824.

Givskov, M., L. Eberl, and S. Molin. 1997. control of exoenzyme production, motility and cell differentiation in *Serratia liquefaciens*. *FEMS Microbiol. Lett.* 148: 115–122.

Greenberg, E. P., J. W. Hastings, and S. Ulitzer. 1979. Induction of luciferase synthesis in Beneckea harveyi by other marine bacteria. *Arch. Microbiol.* 120: 87–91.

Heim, R., D. C. Prasher, and R. Y. Tsien. 1994. Wavelength mutations and posttranslational autoxidation of green fluorescent protein. *Proc. Natl. Acad. Sci. USA* 91: 12501–12504.

Hodson, R. E., W. A. Dustman, R. P. Garg, and M. A. Moran. 1995. In situ PCR for visualization of microscale distribution of specific genes and gene products in prokaryotic communities. *Appl. Environ. Microbiol.* 61: 4074–4082.

Holmstrøm, K., T. Tolker-Nielsen, and S. Molin. 1999. Physiological states of individual *Salmonella typhimurium* cells monitored in situ RT-PCR. *J. Bacteriol.* 181: 1733–1738.

Jansson, J. K. 1995. Tracking genetically engineered microorganisms in nature. *Curr. Opinion Biotechnol.* 6: 275–283.

Kjelleberg, S., M. Hermansson, P. Mårdén, and G. W. Jones. 1987. The transient phase between growth and nongrowth of heterotrophic bacteria, with emphasis on the marine environment. *Annu. Rev. Microbiol.* 41: 25–49.

REFERENCES

Kragelund, L., B. Christoffersen, F. J. de Bruijn, and O. Nybroe. 1995. Isolation of lux reporter gene fusions in *Pseudomonas* fluorescence DF57 inducible by nitrogen or phosphorus starvation. *FEMS Microbiol. Ecol.* 17: 95–106.

Licht, T. R., T. Tolker-Nielsen, K. Holmstrøm, K. A, Krogfelt, and S. Molin. 1999. Inhibition of *Escherichia coli* precursor-16S rRNA processing by mouse intestinal contents. *Environ. Microbiol.* 1: 23–32.

Ludin, B., R. Doll, S. Meili, S. Kaech, and A. Matus. 1996. Application of novel vectors for GFP-tagging of proteins to study microtubule-associated proteins. *Gene* 173: 107–111.

Maaløe, O., and N. O. Kjeldgaard. 1966. *Control of Macromolecular Synthesis*. New York: W. A. Benjamin.

Mach, P. A., and D. J. Grimes. 1982. R-plasmid transfer in a wastewater treatment plan. *Appl. Environ. Microbiol.* 44: 1395–1403.

Massol-Deyá, A. A., J. Whallon, R. F. Hickey, and J. M. Tiedje. 1995. Channel structures in aerobic biofilms of fixed-film reactors treating contaminated groundwater. *Appl. Environ. Microbiol.* 61: 769–777.

Molin, S., and J. Molin. 1997. CASE: Complex AdaptiveSsystems Ecology. *Adv. Microb. Ecol.* 15: 27–80.

Møller, S., C. S. Kristensen, L. K. Poulsen, J. M. Carstensen, and S. Molin. 1995. Bacterial growth on surfaces: Automated image analysis for quantification of growth rate-related parameters. *Appl. Environ. Microbiol.* 61: 741–748.

Møller et al., 1998.

Nielsen et al., 1999.

Nyström, T., N. H. Albertson, K. Flärdh, and S. Kjelleberg. 1990. Physiological and molecular adaptation to starvation and recovery from starvation by the marine *Vibrio* sp. S14. *FEMS Microbiol. Ecol.* 74: 129–140.

Pace, N. R., D. A. Stahl, D. L. Lane, and G. J. Olsen. 1986. The analysis of natural microbial populations by rRNA sequences. *Adv. Microb. Ecol.* 9: 1–55.

Passador, L., J. M. Cook, M. J. Gambello, L. Rust, and B. H. Iglewski. 1993. Expression of *Pseudomonas aeruginosa* virulence genes requires cell-to-cell communication. *Science* 260: 1127–1130.

Poulsen, L. K., G. Ballard, and D. A. Stahl. 1993. Use of rRNA fluorescence in situ hybridization for measuring the activity of single cells in young and established biofilms. *Appl. Environ. Microbiol.* 59: 1354–1360.

Poulsen, L. K., T. R. Licht, C. Rang, K. A. Krogfelt, and S. Molin. 1995. Physiological state of *Escherichia coli* BJ4 growing in the large intestines of streptomycin-treated mice. *J. Bacteriol.* 177: 5840–5845.

Prosser, J. I. 1994. Molecular marker systems for the detection of genetically modified microorganisms in the environment. *Microbiology* 140: 5–17.

Robinson, R. W., D. E. Akin, R. A. Nordstedt, M. V. Thomas, and H. C. Aldrich. 1984. Light and electron microscopy examinations of methane-producing biofilms from anaerobic fixed-bed reactors. *Appl. Environ. Microbiol.* 48: 127–136.

Schaechter, M., O. Maaløe, and N. O. Kjeldgaard. 1958. Dependency on medium and temperature of cell size and chemical composition during balanced growth of *Salmonella typhimurium*. *J. Gen. Microbiol.* 19: 592–606.

Shapiro, J. A., and N. P. Higgins. 1989. Differential activity of a transposable element in *Escherichia coli* colonies. *J. Bacteriol.* 171: 5975–5986.

Siegele, D. A., and R. Kolter. 1992. Life after log. *J. Bacteriol.* 174: 345–348.

Sternberg et al., 1999.

Tolker-Nielsen, T., K. Holmstøm, and S. Molin. 1997. Visualization of specific gene expression in individual *Salmonella typhimurium* cells by in situ PCR. *Appl. Environ. Microbiol.* 63: 4196–4203.

van Elsas, J. D., and J. T. Trevors. 1990. Plasmid transfer to indigenous bacteria in soil and rhizosphere: Problems and perspectives. In *Bacterial Genetics in Natural Environments*, J. C. Fry and M. J. Day, eds. London: Chapman and Hall, pp. 188–199.

Wagner, M., R. I. Amann, H. Lemmer, and K. H. Schleifer. 1993. Probing activated sludge with oligonucleotides specific for Proteobacteria: Inadequacy of culture-dependent methods for describing microbial community structure. *Appl. Environ. Microbiol.* 59: 1520–1525.

Ward, D. M., M. M. Bateson, R. Weller, and A. L. Ruff-Roberts. 1992. Ribosomal RNA analysis of microorganisms as they occur in nature. In *Advances in Microbial Ecology*, K. C. Marshall, ed. New York: Plenum Press, pp. 219–228.

Wellington, E. M. H., and J. D. van Elsas. 1992. *Gene Transfer Between Microorganisms in the Natural Environment*. London: Pergamon Press.

Williams, R. M., D. W. Piston, and W. W. Webb. 1994. *FASEB* 8: 804–813.

Wimpenny, J. W. T., and R. Colasanti. 1997. A unifying hypothesis for the structure of microbial biofilms based on cellular automaton models. *FEMS Microbiol. Ecol.* 22: 1–16.

Woese, C. R. 1987. Bacterial evolution. *Microbiol. Rev.* 51: 221–271.

PART 2

BENEFICIAL ASPECTS OF BIOFILM SYSTEMS

5

BIOFILMS IN POROUS MEDIA

EDWARD J. BOUWER
Department of Geography and Environmental Engineering, Johns Hopkins University, Baltimore, Maryland

HUUB H. M. RIJNAARTS
Department of Environmental Biotechnology, TNO Institute of Environmental Sciences, Energy Research and Process Innovation, Apeldoorn, The Netherlands

AL B. CUNNINGHAM
ROBIN GERLACH
Center for Biofilm Engineering, Montana State University, Bozeman, Montana

5.1 INTRODUCTION

The high specific surface area of porous media suggests that most bacteria in such systems are attached to the solid surfaces. Under low organic and nutrient concentrations, attached bacteria can outnumber suspended bacteria by several orders of magnitude and account for a major portion of all microbial activity (van Loosdrecht et al., 1990). An understanding of biofilm accumulation in porous media provides significant opportunities for improving the performance of environmental and industrial processes. Subsurface biofilms offer the potential for biotransformation of organic contaminants, thereby providing an in situ method for treating contaminated groundwater supplies (National Research Council, 1993). The rate of biotransformation is strongly influenced by porous media mass transport characteristics including media permeability and pore velocity distribution, as well as by biofilm surface roughness and other variables that affect the delivery rate of substrate and nutrients to growing cells. Industrial applications include microbially enhanced leaching of metals from ores and recovery of metals from solutions, deliberate plugging of high-permeability zones to enhance oil recovery operations, and bioreactors for water and waste treatment (Cunningham et al., 1991).

Biofilms II: Process Analysis and Applications, Edited by James D. Bryers.
ISBN 0-471-29656-2 Copyright © 2000 Wiley-Liss, Inc.

Accumulation of microorganisms on surfaces as biofilms often enhances the potential for biotransformation of organic contaminants. Biofilm growth permits greater accumulation of cells per unit volume in comparison to suspended cultures, particularly for slow-growing microorganisms, and spatial positioning may promote interaction and gene exchange between different microbial species (Ehlers and Bouwer, 1999). Attached microorganisms have an advantage over suspended cells as they are continually exposed to the source of fresh substrate and nutrients contained in either (1) the aqueous phase that flows past them or (2) sorbed to the porous material itself.

This chapter addresses some important issues governing the influence of porous media hydrodynamic properties and bacterial cell surface characteristics on the accumulation rate and spatial distribution of biofilms in porous media. These issues include relevant processes, bacteria/solid surface interactions, transport of bacteria, and biofilm accumulation. The relevance of biofilms to subsurface remediation of organic contaminants is also presented.

5.2 EXISTENCE OF BIOFILMS IN POROUS MEDIA AND CONTROLLING PROCESSES

Biofilm morphology in porous media systems can be highly variable, ranging from patchy, discontinuous colonies to thick, continuous films. For example, packed bed bioreactors for treating high-strength wastewaters or volatile organic emissions may develop biofilms several millimeters thick (Grady et al., 1999). At the other extreme, subsurface biofilms capable of biologically converting trace organic compounds generally consist of isolated microcolonies of native soil bacteria in the range of 10^5 to 10^7 cfu/g of soil (Table 5.1). The microbial distribution in the subsurface and its ramifications for modeling remains a controversial issue (Rittmann, 1993). Several studies have stressed the importance of biofilm kinetics to describe microbial growth in the subsurface (Cunningham et al., 1991; Taylor and Jaffe, 1990a; Taylor and Jaffe, 1990b). Others have questioned the existence of a "biofilm" (Baveye et al., 1992; Vandevivere and Baveye, 1992a; Vandevivere and Baveye,

TABLE 5.1. Reported Bacterial Numbers in Shallow and Deep Aquifer Materials

Location	Depth	Bacterial Numbers per Gram	Reference
Lula, OK	0–8 m	10^7–10^9	Beloin et al., 1988
Coal tar site (unidentified)	1–10 m	10^6–10^8	Madsen et al., 1991
Coal tar site, Baltimore, MD	2.7–28.5 m	10^4–10^6	Durant et al., 1995
Vejen, Denmark	4–31 m	10^7–10^9	Albrechtsen and Winding, 1992
Bordon, Ontario	<50 m	10^3–10^6	Barbaro et al., 1994
Nemaha, KS	26–86 m	10^6–10^8	Sinclair et al., 1990
Idaho	70 m (unsaturated zone)	0–10^6	Colwell, 1989
Coastal plain Sediments, MD	182 m	10^3–10^6	Chapelle et al., 1987
Depart. of Energy, Savannah River, SC	50–260 m	10^4–10^7	Hicks and Fredrickson, 1989

5.2 EXISTENCE OF BIOFILMS IN POROUS MEDIA AND CONTROLLING PROCESSES 125

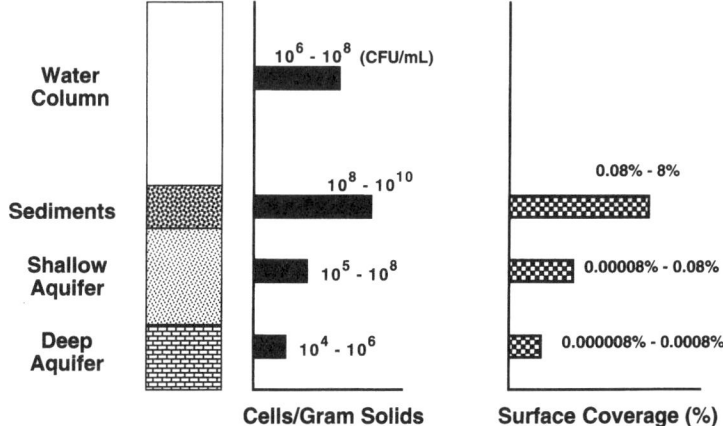

Figure 5.1 Distribution of microorganisms in water and in the subsurface environment.

1992b) because observations with soil and other natural materials have only revealed isolated bacterial aggregates.

Because soils, sediments, and aquifer materials have very high specific surface areas (range of 10 to 1000 m^2/kg, (Bear, 1964)), it is likely bacteria only colonize a small fraction of the soil surface area in the absence of significant pollution (i.e., low substrate concentration and limited nutrient availability). Assuming a dimension of 1 μm for spherical microorganisms, a single cell would occupy a projected area of about 0.8×10^{-2} m^2. For a porous media with a specific surface area of 0.1 m^2/g (100 m^2/kg) and microorganisms uniformly distributed over the solid surface, the cells at 10^{10} cells/gram of solids would occupy about 0.8×10^{-2} m^2 or 8% of the total surface area. Sediments adjacent to the water column have the highest numbers of microorganisms (10^8–10^{10} cells/gram), which correspond to 0.08 to 8% surface coverage (Fig. 5.1). The aquifer solids in shallow and deep aquifers (up to 100 feet deep) usually have low microbial numbers (10^4–10^8 cells/gram) (Table 5.1) corresponding to 0.000008% to 0.08% coverage of the surface area.

Irrespective of the amount of the attached bacteria, their net accumulation on a surface initially free of bacteria is controlled by four processes: particle transport, particle–surface interactions, microbial responses, and particle detachment. These fundamental biofilm processes have been described previously (Chapter 3) in the context of their individual contributions to net accumulation of biofilm on a substratum. The basic concept behind biofilm accumulation is that some cells will deposit on the media, utilize substrate, and grow. Biofilm is removed by biomass decay or detachment; the latter causes cells and biofilm debris to move back into suspension. The relationship between these processes and biofilm accumulation in porous media is illustrated in Figure 5.2 using in situ bioremediation as an example. A typical groundwater plume is shown in Figure 5.2a. As chemicals in the source region move with the groundwater, concentrations of the chemicals undergo physicochemical and biological changes. An engineered bioremediation system usually includes extraction and injection wells and equipment for addition and mixing of nutrients.

The accumulation and activity of biofilms varies from point to point along individual pore channels (Fig. 5.2b), and thus are considered to be microscale phenomena. The first category of microscale phenomena, transport processes, include the movement of suspended cells with the flow (termed *advective* or *convective* transport) and dispersive

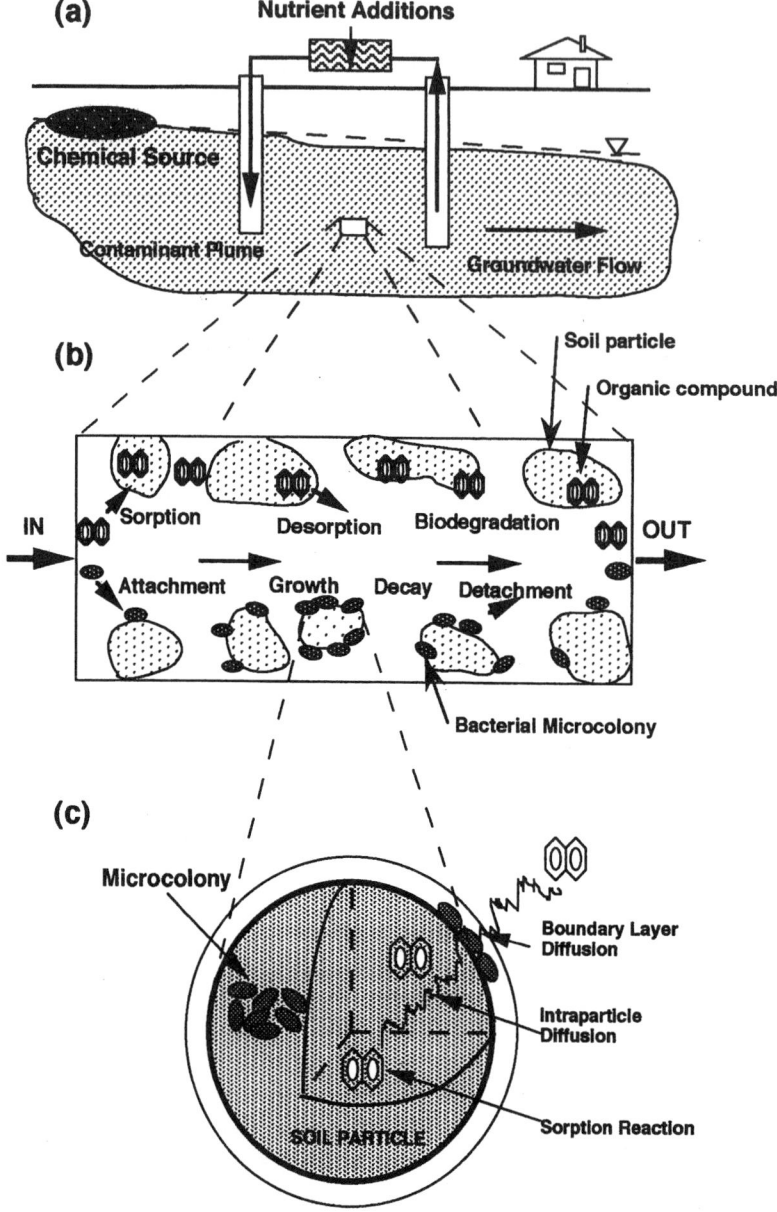

Figure 5.2 Processes controlling biofilm accumulation and contaminant behavior during in situ bioremediation. (*a*) Schematic of in situ bioremediation, (*b*) major microscale mechanisms influencing the movement and fate of bacteria and contaminants, and (*c*) idealized (model) soil particle.

processes of mixing and particle diffusion. Sedimentation of cells and the self-generated motion of motile bacteria also fall into the process of transport. These processes control the rate at which cells collide with surfaces. The phenomena that come into play during such collisions and that cause cells to be either immobilized on the surface or returned to the bulk flow constitute the second category. These particle–surface interactions depend on the chemistry of the surfaces involved and that of the aqueous phase, and influence the extent to which cells penetrate or are retained by a porous medium. Together, these processes of transport and adhesion constitute deposition. The third category, microbial responses, is reserved for phenomena that are due to the nature of bacteria as viable biological entities. Alteration of size, shape, and surface composition during growth might influence cell transport and adhesion while growth, death, and dormancy control the eventual development of the biofilm. The final category is detachment, where biomass from the biofilm is transferred to the bulk liquid. Detachment rates are generally proportional to the amount of attached biomass and the hydrodynamic shear stress (Rittmann, 1982), although other factors can be involved (Peyton and Characklis, 1993; Speitel and DiGiano, 1987).

An idealized model of a soil particle is shown in Figure 5.2c. The solid matrix is porous and sorption sites for organic contaminants are homogeneously distributed. Diffusion is the only major mass transfer mechanism within the solid, and it is assumed to occur only in the radial direction. The typical size of bacteria, 1 μm, is larger than the majority of pore diameters in soil particles. Consequently, the pore structure within the particle is assumed to be too small for penetration of bacteria. Thus bacteria tend to be located on the outer soil particle surfaces as shown in Figure 5.2c. If sufficient substrate and nutrients are available, microbial cell accumulation will develop into either patchy discontinuous colonies or colonies that have merged to form a continuous film.

5.2.1 Interactions Between Bacteria and Solid Surfaces

As just noted, the initial step in biofilm formation and growth is the deposition of planktonic cells to the substratum. Deposition is commonly viewed as sequential steps of transport and adhesion. Transport is hydrodynamically determined, whereas the adhesion step is controlled by cell–substratum interactions. This section summarizes several advances recently made in quantifying factors influencing bacterial adhesion.

Experiments with model systems and under well-defined conditions were performed to assess cell–substratum interactions in bacterial adhesion (Rijnaarts et al., 1993; Rijnaarts et al., 1995). Negatively charged Teflon and glass were used as the model substrata. Eight coryneform bacteria and four pseudomonads were selected. At pH = 7, these organisms had different negative cell surface charges (Rijnaarts et al., 1993). The hydrophobicity of solids and bacteria were determined by measuring the contact angle of drops of water on flat pieces of solid and on air-dried bacterial lawns. Glass with a contact angle (θ_w) of 12° ± 2° is hydrophilic and Teflon with θ_w of 105° ± 1° is extremely hydrophobic. Hydrophobicities of the dried bacteria varied between strongly hydrophilic ($\theta_w < 20°$) and extremely hydrophobic ($\theta_w > 100°$). Two types of experimental setups were used: (1) static batch systems containing bacterial suspension and flat pieces of Teflon or glass and (2) dynamic model porous media that consisted of water-saturated columns packed with Teflon or glass beads to which bacterial suspensions were applied. Transport of cells from bulk liquid to substratum is much more efficient in dynamic columns that in static systems: Transport is controlled by convection and diffusion under dynamic conditions but by diffusion only in static systems (Rijnaarts et al., 1993).

At an ionic strength of 0.1 M, adhesion in the batch system was irreversible and the activation energies for adhesion varied between 0 and 5 kT (k (JK^{-1}) is the Boltzmann constant; 1 kT = 4 × 10^{-21} J at 20°C) (Rijnaarts et al., 1995). A Gibbs energy barrier, located between cell and substratum and several hundreds of kT high, is created by the DLVO (Derjaguin, Landau, Verway and Overbeek) interactions (Rijnaarts et al., 1993; Rijnaarts et al., 1995). This barrier cannot be surmounted by whole cells. However, bacterial cell surface macromolecules can penetrate this energy barrier and hence reach the substratum at high ionic strength. Therefore, the interactions between the cell surface marcomolecules can penetrate this energy barrier and hence reach the substratum at high ionic strength. Therefore, the interactions between the cell surface macromolecules and the substratum, which are generally called *steric interactions*, determine adhesion at high ionic strength. The following two types of steric interactions generally occur: (1) bridging, that promotes adhesion and causes a lowering of the activation energy of adhesion, and (2) steric hindrance that inhibits adhesion and therefore increases the activation energy of adhesion.

The various mechanisms of adhesion as deduced for the different combinations of types of cell–surface coatings and solid substrata are summarized in Figure 5.3 (the capital letters given below refer to the various panels shown in this figure). Long-range electrostatic interactions as described by the DLVO theory (*A* and *B*) create strong repulsive barriers to adhesion that cannot be passed by whole cells. In addition, deep secondary minima exist for glass but not for Teflon. The secondary minima on glass result from strong Van der Waals attraction and are sufficiently deep for irreversible adhesion. Lipid or protein (nonpolysaccharide) cell surface macromolecules cause strong attractive bridging on the hydrophobic surface (Teflon) (*C*). On glass (*D*), these same cell surface macromolecules slightly inhibit adhesion and permit adhesion in the secondary DLVO minimum as demonstrated for two hydrophobic coryneforms (Rijnaarts et al., 1995). Bacteria with an amphiphilic cell surface may adhere by bridging on a hydrophobic surface (*E*) whereas strong steric hindrance prevents secondary minimum adhesion on glass (*F*). The adhesion of bacterial cells coated with anionic polysaccharides is strongly inhibited on both Teflon and glass (*G* and *H*).

At an ionic strength (*I*) of 0.1 M in a batch system, adhesion was irreversible. The bacteria/surface interactions could be quantified in terms of Gibbs activation energies for adhesion and detachment as determined from experimental adhesion and detachment rates. The adhesion efficiency α, which is the probability of a cell attaching upon reaching a sub-

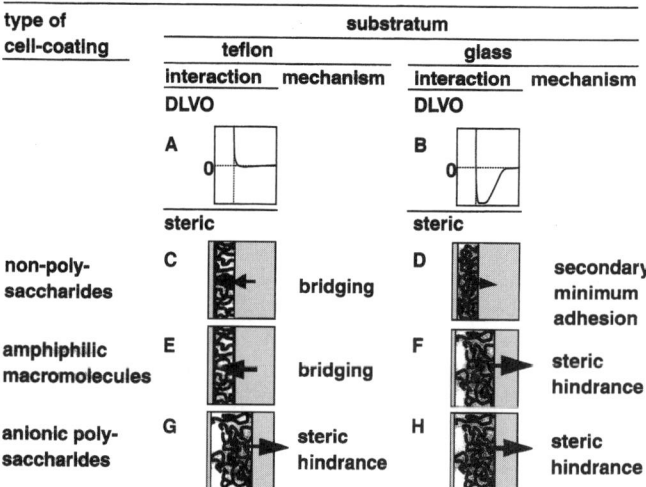

Figure 5.3 Summary of the adhesion mechanisms that occur among different bacterium/substratum combinations. The arrows indicate attraction due to bridging (arrow points to the surface, i.e., to the left) or repulsion due to steric hindrance (arrow points away from the surface, i.e., to the right).

5.2 EXISTENCE OF BIOFILMS IN POROUS MEDIA AND CONTROLLING PROCESSES

TABLE 5.2. The Level of the Activation Energy for Adhesion and the Adhesion Efficiency α for Different Adhesion Mechanisms

Adhesion Mechanism	Activation Energy for Adhesion (kT)	Adhesion Efficiency, α
Bridging	0	1
Secondary minimum adhesion	1	0.3
Steric hindrance	2.5–5	0.01–0.05

stratum, is related to the activation energy for adhesion where the activation energy is proportional to $-\ln \alpha$. Values of the activation energy for adhesion and the adhesion efficiency for the different adhesion mechanisms shown in Figure 5.3 are summarized in Table 5.2. The activation energy for detachment exceeded 5 kT for irreversible adhering bacteria. The greatest resilience against detachment was observed for hydrophobic bacteria on hydrophobic substrata. For hydrophobic/hydrophilic and hydrophilic/hydrophilic bacterium/substratum combinations, binding mechanisms not related to hydrophobicity inhibited cell detachment.

The effect of DLVO and steric interactions on adhesion as a function of the ionic strength is illustrated in Figure 5.4 (Rijnaarts et al, 1999). Steric interactions dominate at high ionic strength (0.1 M). Adhesion attains a maximum and is even 100% efficient (α = 1) in the case of bridging. Long-range electrostatic repulsion, as described by the DLVO model, starts to exert its influence when the ionic strength is reduced below a critical value I_s (Fig. 5.4) and dominates adhesion at an ionic strength of 0.0001 M. The value of I_s is determined by the distance over which the cell–surface macromolecules penetrate the electrostatic barrier. For the bacterial strains tested, the extension of cell surface macromolecules varied between 5 nm and 80 nm, which corresponds to I_s values varying between 0.1 and 0.001 M. A cell–substratum separation of 165 nm was bridged by a flagellated *Pseudomonas putida* strain. The practical consequence of these findings is that studying and controlling bacterial adhesion in porous media should include the assessment of both DLVO and steric interactions since the ionic strengths of groundwater and wastewater generally fall between 0.001 M and 0.1 M.

5.2.2 Bacterial Transport in Porous Media

The movement of bacteria through porous media is a critical factor in the fate and remediation of many polluted surface and subsurface formations. When in situ bioremediation is chosen as the remedial action, successful cleanup depends on the transport and redistribution of the endogenous strains through the entire contaminant plume. When exogenous strains with specialized metabolic features are used, their transport from the injection source to and within the contamination plume is a major factor determining the efficacy of the operation (Thomas and Ward, 1989). Alternatively, if in situ biobarriers are desired, sufficient adhesion of introduced strains in the desired locale is required (Hanna and Taylor, 1996). Bacterial transport is also important when the organisms are pathogenic or contain a modified genetic element (Trevors et al., 1990).

Bacterial transport is typically governed by aqueous phase convective movement coupled with retardation by adhesion onto surfaces and straining or trapping in interstitial pores. Adhesion is commonly thought of as the main retarding factor, while straining is

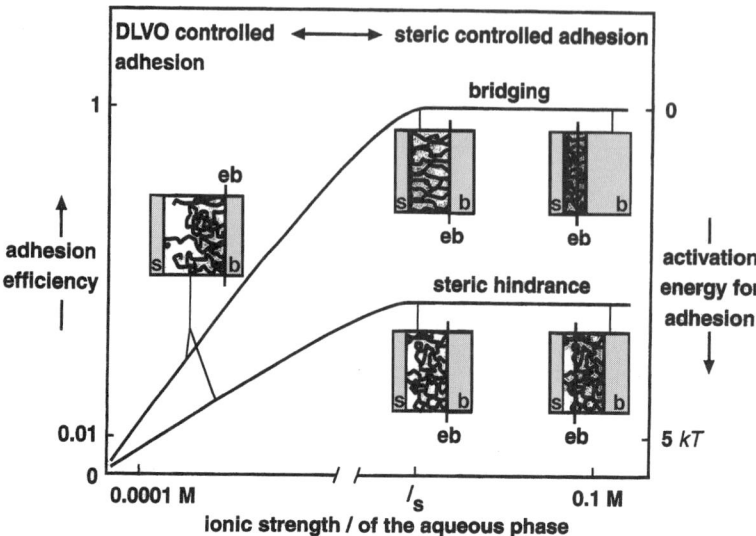

Figure 5.4 Illustration of the effect of the ionic strength on bacterial adhesion. eb = electrostatic barrier; b = bacterium; S = substratum.

important only when the diameter of the particle exceeds 5% of the mean interstitial pore size (Jenkins and Lion, 1993; McDowell-Boyer et al., 1986). Bacterial transport through porous media is influenced by several parameters including properties of the bacterial cells, solution chemistry, porous media characteristics, and interstitial fluid velocity.

Cell properties that may influence bacterial transport, retardation, and adhesion to surfaces include the presence of molecules such as proteins or polysaccharides on the cell surface (Caccavo et al., 1997; Fletcher, 1976; Fletcher and Floodgate, 1973; Rijnaarts et al., 1996b; Williams and Fletcher, 1996), the presence of pili (Busscher et al., 1990a; Busscher et al., 1990b; Busscher et al, 1992; Mueller, 1996) as well as motility and chemotaxis (Camper et al., 1993; Jenneman et al., 1985; McCaulou et al., 1995; Reynolds et al., 1989; Sharma and McInerney, 1994). Cell size and cell shape likewise have been correlated with transport through porous media (Bitton et al., 1974; Fontes et al., 1991; Gannon et al., 1991b; Weiss et al., 1995) as well as cell surface charge (Gilbert et al., 1991; Schäfer et al., 1998b; Sharma et al., 1985; van Loosdrecht et al., 1987a) and cell hydrophobicity (Caccavo et al., 1997; Fletcher and Marshall, 1982; Scholl and Harvey, 1992; van Loosdrecht et al., 1987b). In some cases the buoyant density of the bacterial cells influenced the sedimentation rates of bacteria in porous media environments (Harvey et al., 1997; Wan et al., 1995), but has also been considered negligible (Corpcioglu and Haridas, 1984). Many cell properties are influenced by the physiological state of the bacteria and can therefore be significantly different for the same bacterium, depending on environmental conditions (Grasso et al., 1996, van Loosdrecht et al., 1987a). The growth state of the bacteria and the presence of nutrients have, for instance, been shown to influence attachment (Grasso et al., 1996; Mueller, 1996; van Loosdrecht et al., 1987a). Starvation is another important physiological state of bacteria. Short-term starvation of bacteria can result in an increased tendency to attach to surfaces (Dawson et al., 1981; Kjelleberg, 1984). Long-term starvation (weeks to months) in contrast may enhance bacterial transport through porous media (Cusack et al., 1992; Gerlach et al., 1998; Lappin-Scott and Coster-

5.2 EXISTENCE OF BIOFILMS IN POROUS MEDIA AND CONTROLLING PROCESSES

ton, 1992; Lappin-Scott et al., 1988a; Lappin-Scott et al., 1988b; MacLeod et al., 1988; Sharp et al., 1999).

Solute characteristics including ionic strength, pH, temperature, concentrations of dissolved organic matter, surfactants, and nutrients have also been shown to influence bacterial transport and adhesion to surfaces. Increased ionic strength has widely been correlated with increased attachment (Fontes et al., 1991; Gannon et al., 1991a; Jewett et al., 1995; Mills et al., 1994; Scholl et al., 1990). This effect is usually attributed to the compression of the electrostatic double layer in the presence of high ion concentrations. Bitton et al. (1974), Scholl and Harvey (1992), Kinoshita et al. (1993), and Jewett et al. (1995) investigated the effect of changes in pH values on bacterial transport and attachment. However, no uniform results were found. Increased attachment with decreasing pH was found by Scholl and Harvey (1992); Kinoshita et al. (1993) observed decreased attachment with decreasing pH. However, Jewett et al. (1995) stated that changes in pH from 5.5 to 7 did not affect bacterial transport through porous media columns. Temperature has also been shown to influence bacterial attachment (McCaulou et al., 1995; Sarkar et al., 1994). Johnson and Logan (1996) investigated the influence of dissolved and sediment organic matter (DOM, SOM) on bacterial transport through porous media columns and observed an increase in travel distance in the presence of SOM. The addition of surfactants or dispersants can result in decreased attachment and therefore facilitate the transport of bacteria through porous media, however the activity or viability of the bacteria may also be altered (Goldberg et al., 1990; Gross and Logan, 1995; Jackson et al., 1994; Paul and Jeffrey, 1984; Sarkar et al., 1994).

Porous media properties that have been reported to influence bacterial transport and adhesion include pore water velocity (Gannon et al., 1991a), hydraulic conductivity (Vandevivere and Baveye, 1992b), pore size (Sharma and McInerney, 1994), the presence of Fe minerals (Johnson and Logan, 1996; Mills et al., 1994; Scholl et al., 1990), the organic matter content (Johnson and Logan, 1996), and grain and pore size distribution (Fontes et al., 1991; Sharma and McInerney, 1994; Smith et al., 1985; Wood and Ehrlich, 1978). The surface charge (Johnson and Logan, 1996; McCaulou et al., 1994; Scholl et al., 1990) and surface hydrophobicity (Absolom et al., 1983; Fletcher and Loeb, 1979; McCaulou et al., 1994; Schäfer et al., 1998a) of the porous media can also influence bacterial attachment to surfaces. Geesey and Costerton (1979), Mueller et al. (1992), and Scheuermann et al. (1998) report that the roughness of surfaces can also influence attachment. The surface roughness in dynamic systems may even become more important than the chemical nature of the surface. Trevors et al. (1990) reports that the presence of plants and plant roots can have an effect on bacterial transport through soil.

Transport of bacteria through porous media may be influenced by some combination of the foregoing parameters. Camper et al. (1993) state that bacterial adsorption rate coefficients determined under the conditions of interest are a much better predictor of microbial transport phenomena than individual characteristics of cells or the porous medium. To predict bacterial transport through porous media, one should therefore perform experiments with the aquifer material of interest and as close as possible to the expected conditions in the field. Harvey (1997) gives a good overview on how to design and standardize bacterial transport experiments.

Basic bacterial transport models have been commonly based on the convection–dispersion equation written for bacteria (Harvey and Garabedian, 1991; Jewett et al., 1995; Taylor and Jaffe, 1990c). Empirically determined collision efficiency factors have typically been used to account for the extent of bacterial attachment to surfaces. Macroscopic models of bacteria transport have, in general, ignored detailed treatment of interfacial forces between the solid surface and the bacterium. Column breakthrough experiments with well-characterized Teflon and glass collectors and bacterial strains were recently carried out to provide a better basis for bacterial transport modeling by identifying and quantifying the principal mechanisms of bacterial retention in coarse grain media (Rijnaarts et al., 1996a, 1996b).

Bacterial deposition on spherical glass and Teflon collectors was studied in vertical downflow columns (Rijnaarts et al., 1996a, 1996b). Deposition was analyzed in terms of the clean bed collision efficiency α_o (the probability of a cell attaching upon reaching a cell-free substratum), and a blocking factor B (the ratio of the area blocked by an attached cell to the geometric area of a cell). At an average interstitial fluid velocity of 200 μm s^{-1} and 0.1 M ionic strength, α_o was close to unity (0.83 ± 0.01) for the two tested bacterial strains. The cell–solid interactions were nearly completely favorable for deposition and were similar to the results of the static batch systems shown in Figure 5.3 and Table 5.2. Values of B determined from model fits to column breakthrough data increased (enhanced blocking) as cell–cell repulsion increased due to decreasing (more negative) charge of the cell and increasing hydrophilicity of the cell–surface macromolecules.

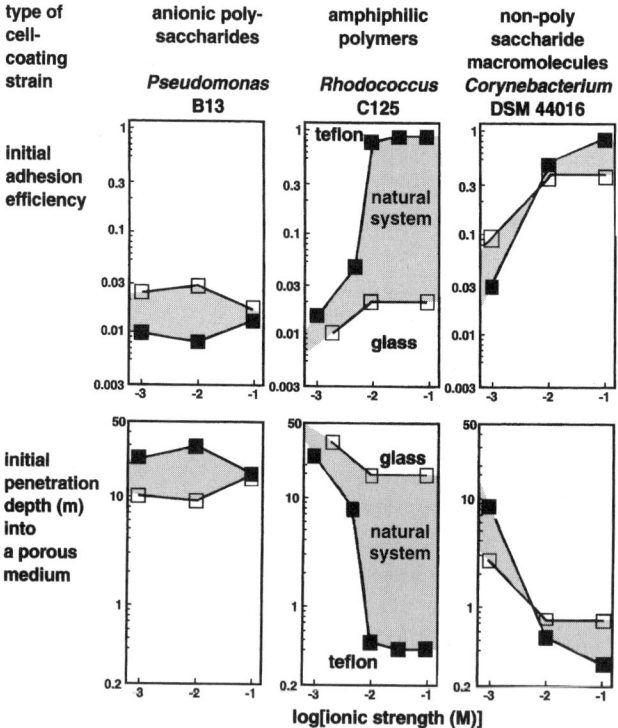

Figure 5.5 Effect of the ionic strength on adhesion and transport of bacteria in porous media. Results of the following organisms are shown: *Pseudomonas* strain B13, *Rhodococcus* strain C125, and *Corynebacterium* strain DSM 44016. Adhesion and transport for natural porous media (shaded area) probably fall between the values for glass (open symbols) and for Teflon (closed symbols). The penetration depth into the porous medium is the porous medium length required to reduce the influent cell concentration by a factor of 100 at an interstitial fluid velocity of 30 cm h^{-1}.

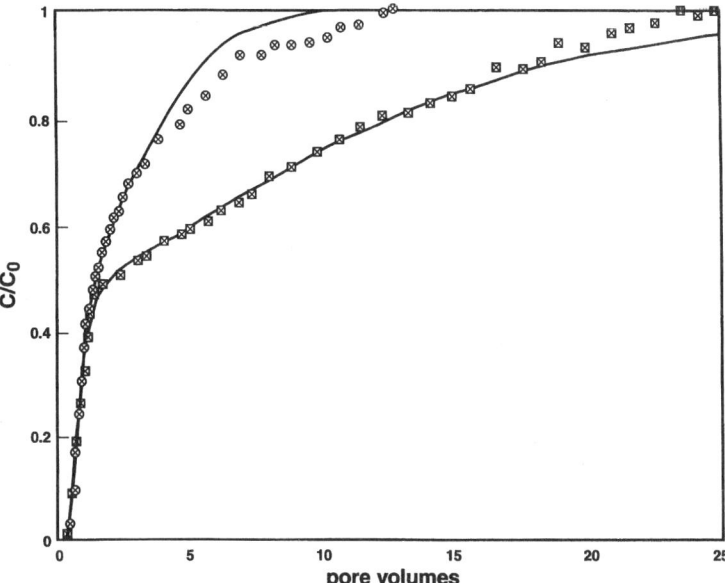

Figure 5.6 Measured data (points) and simulations obtained by incorporating α_o and B into groundwater flow models (lines) for *Bacillus sp.* breakthrough in sand columns at initial concentrations of 3×10^8 cells/cm^3 (squares) and 1.2×10^9 cells/cm^3 (circles).

Initial clean-bed collision efficiency (α_o) decreased with decreasing I, for I values smaller than the critical level I_s in a similar way as found for the static batch systems (Fig. 5.4). The level of B increased about one order of magnitude upon changing I from 0.1 to 0.001 M. The effect of cell–solid interaction and ionic strength (as inferred from the initial adhesion efficiency on bacterial transport in porous media is illustrated in Figure 5.5. Depths of bacterial penetration into a porous medium may be as high as 50 m at low ionic strength based on the laboratory conditions (collector radii of 190 μm ± 50 μm for Teflon and 225 μm = 25 μm for glass, porosity of 0.34 ± 0.02, and interstitial fluid velocity of 30 cm/hr). At high ionic strength, this penetration depth varies between about 0.25 m and 20 m depending on the type of cell–surface constituents, such as nonpolysaccharide, amphiphilic, or anionic polysaccharide macromolecules. The practical consequence of the research summarized in Figure 5.4 is that maximal control of bacterial mobility in porous media can be reached in systems for which B and α_o are high at I around 0.1 M. The high B value obtained from model fits minimizes the occurrence of pore clogging whereas the dependencies of α_o and B on I allow manipulation of deposition by varying the ionic strength. The concepts of initial adhesion (α_o) and blocking factor (B) were recently incorporated into a microbial transport model based on the widely used groundwater flow models termed MT3D and MODFLOW (Te Stroet et al., 1999). This model was able to simulate the microbial breakthrough data for sandy soil columns reported by Lindqvist and Enfield (1992) (Fig. 5.6). Consequently, this model appears to be superior to other models for describing microbial transport in natural heterogeneous subsurface systems.

In field-scale applications, however, manipulating the solute chemistry (e.g., lowering the ionic strength, changing the pH, changing the temperature of the groundwater), or manipulating aquifer characteristics to facilitate bacterial transport may pose difficult problems both from a technical as well as economical standpoint. However, manipulating the physiological properties of the bacterial inoculum to facilitate bacterial transport through porous media may be more promising. Starvation of bacteria, for example, has been proposed as a means of facilitating transport of an inoculum injected into the subsurface (Cusack et al., 1992; Lappin-Scott et al., 1988a; Lappin-Scott et al., 1988b; MacLeod et al. 1988). Starvation of bacteria results in (radical) size reduction (Bryers and Sanin, 1994; Sanin, 1996) and a rapid decrease in metabolic activity until the bacteria approach complete dormancy (Kjelleberg, 1993).

The significance of starvation on bacterial transport is illustrated in a recent study by Gerlach et al. (1998) in which starved and active cells of *Shewanella alga* BrY were injected into porous media columns (length: 10 ft = 3.048 m; diameter: 1 in. = 2.54 cm) filled with 40 mesh quartz sand. *S. alga* BrY is a facultative anaerobic bacterium, which has been extensively studied and characterized (Caccavo et al., 1996a; Caccavo et al., 1992; Caccavo et al., 1996b; Caccavo et al., 1997). Caccavo et al. (1996a) showed that starvation of BrY results in a gradual decrease in the mean cell volume from 0.48 μm^3 to 0.2 μm^3 and a dramatic decrease in endogenous metabolic activity. The column transport experiments were carried out in an up-flow mode at flow rates of approximately 10 mL/min, resulting in interstitial fluid velocities of approximately 11.5 cm/min (nutrients were removed from the influent, and influent cell concentration was continuously monitored to ensure no-growth conditions for the duration of the experiment). The studies compared the porous media transport of starved BrY cells to their active counterpart as shown in Figures 5.7 and 5.8. The cell breakthrough data shown in Figure 5.7 show that the passage through 3.048 m of quartz sand leads to almost a 5-log removal of active cells. Starved cells, however, break through the column with about a 1.5-log removal. The influent concentrations of starved and active cells were approximately equal at 5×10^7 cfu/ml. The quantification of cells (as described in Gerlach et al. (1998)) sorbed to the quartz sand along the flow-path shown in Figure 5.8 supports the breakthrough data. Active cells tended to adhere more to the porous medium than the starved cells (Fig. 5.8).

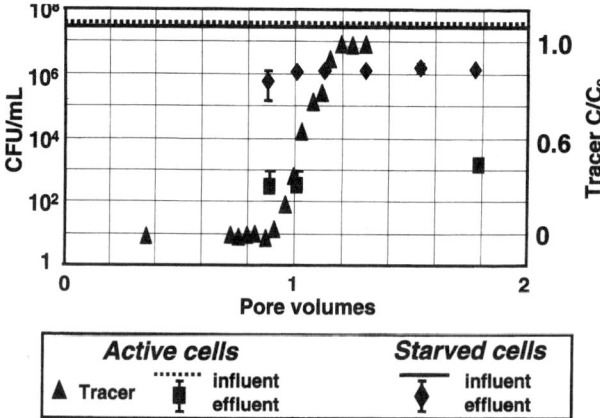

Figure 5.7 Breakthrough of starved and active *Shewanella alga* BrY cells through 10 ft columns filled with 40 mesh quartz sand. Starved cells break through in concentrations and numbers that are orders of magnitude higher than active cells.

5.3 BIOFILM ACCUMULATION

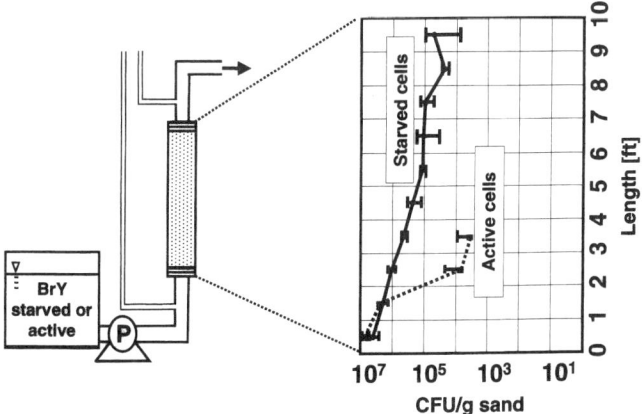

Figure 5.8 Transport of starved and active cells of *Shewanella alga* BrY (a dissimilatory metal reducing bacterium) through 10 ft columns filled with 40 mesh quartz sand. Approximately the same number and concentration of starved and active bacteria were injected. Starved cells are less retarded and more evenly distributed throughout the column.

High concentrations of adsorbed (adherent) active cells within the first few feet of the point of injection were observed. Starved cells, in contrast, were more evenly distributed throughout the system. The detection limit for culturable cells was 10^3 CFU/g sand for sorbed cells and 10^2 CFU/mL for planktonic cells. Results clearly indicate that starvation of *Shewanella alga* BrY provides a means for enhancing bacterial transport in quartz sand porous media.

5.3 BIOFILM ACCUMULATION

The microscale phenomena that govern biofilm accumulation and activity (i.e., chemical and hydrodynamic gradients) are difficult to measure in a porous media environment in comparison to other common reactor systems such as flasks, tanks, reservoirs, and pipelines. Biofilm growth, for example, is complicated by the nature of fluid and nutrient transport which, in porous media, occurs along tortuous flow paths of variable geometry. Similarly, the wide distribution of pore velocities introduces considerable variation in the processes of transport, adhesion, and detachment on a microscale. Consequently, most investigations to date have focused on quantifying the effects of biofilm development rather than on the behavior of individual biofilm processes.

In porous media flow systems, biofilm accumulation generally decreases with distance along the flow path, especially if substrate concentration is depleted through the medium. Laboratory (Taylor and Jaffe, 1990c) and field (van Beek, 1984) observations illustrate the plug flow nature of biofilm accumulation in porous media. In a plug flow system, as discussed in Chapter 2, the concentration of substrate and oxygen (or other electron acceptors) decreases in the downstream flow direction. Thus, cell growth also diminishes at some downstream point where substrate and/or oxygen become limiting to the growth process. The potential for both deposition and detachment are likewise highest in the upstream location experiencing maximum cell growth rates.

5.3.1 Hydrodynamics of Porous Media

The hydrodynamic principles presented here provide the necessary background for subsequent discussion of biofilm processes in porous media. For a more comprehensive treatment of flow through porous media, the reader should consult a standard groundwater text such as Bear (1979), H. Bouwer (1978), Freeze and Cherry (1979), or Todd (1980). These references also provide an overview of mass and momentum transport in porous media. In porous media, mass transport determines the capacity of the flow regime to transport dissolved or suspended substances (i.e., tracers, chemicals, nutrients, and microbial cells) while momentum transport determines the resistance to flow through the porous matrix. Mass transport processes influence the net accumulation of biofilm and the rates of degradation of organic contaminants, and they control the migration of contaminated groundwater. Momentum transport processes are likewise of importance to the analysis of biofouling in packed bed filters and porous media.

5.3.1.1 Darcy's Law.
Consider flow through a cylinder of cross-sectional area A (L^2) filled with a granular porous medium. Analysis of this experimental system provides the basis for illustrating fundamental porous medium flow principles. If water is introduced into the cylinder until all the pores are filled, the inflow rate Q is equal to the outflow rate. If an arbitrary datum is chosen, the elevations of the column inlet and outlet water levels relative to the datum are h_1 and h_2. Then,

$$Q = \frac{-KA \, \Delta h}{\Delta L} \tag{5.1}$$

where ΔL = column length or distance traveled by the water (L), $\Delta h = h_1 - h_2$ (L), K = proportionality coefficient (Lt^{-1}).

Expressed in differential terms,

$$Q = -KA\frac{dh}{dL} \tag{5.2}$$

or

$$v = \frac{Q}{A} = -K\frac{dh}{dL} \tag{5.3}$$

where Q = volumetric flow rate ($L^3 t^{-1}$), v = specific discharge (Lt^{-1}), K = hydraulic conductivity (Lt^{-1}), dh/dL = hydraulic gradient (−).

The preferred units for K and v are m day^{-1}, which result in values ranging from essentially zero for highly impermeable material (sandstone, clay, rock) to as much as 1000 m day^{-1} for very permeable media (gravel, fractured limestone). The negative sign indicates that the flow of water is in the direction of decreasing head. Equation 5.3 is known as Darcy's law and states that the specific discharge v equals the product of hydraulic conductivity and hydraulic gradient. Darcy's law is experimentally based, and its range of validity is determined by the magnitude of the porous medium Reynolds number

$$Re = \frac{vd\rho}{\mu} \tag{5.4}$$

where d = grain diameter (L), μ = fluid viscosity ($ML^{-1}t^{-1}$), ρ = fluid mass density (ML^{-3}).

5.3 BIOFILM ACCUMULATION

d is the mean grain diameter, although alternative quantities such as d_{10} (the diameter such that 10% by weight of the grains are smaller) are used in the literature. According to Bear (1979), despite the various definitions for d, Darcy's law can be considered valid as long as Re does not exceed 1.

Darcy's law is of fundamental importance in the analysis of groundwater flow; it is also of importance in many other applications of porous medium flow. As indicated by Freeze and Cherry (1979), Darcy's law describes the flow of soil moisture and is used by soil physicists, agricultural engineers, and soil mechanics specialists. It describes the flow of oil and gas in deep geological formations and is used by petroleum reservoir analysts. It is relevant to the design of filters by chemical engineers. It has also been used by bioscientists to describe the flow of bodily fluids across porous membranes in the body.

5.3.1.2 Permeability. The hydraulic conductivity K is a function of porous medium properties and hydraulic properties; it provides a measure of the ability of a porous medium to conduct water. However, it is often useful to express the permeability of a porous medium as a property of the medium independent of the density and viscosity of the fluid. Thus, the intrinsic permeability is defined as

$$k = \frac{K\eta}{\rho g} \qquad (5.5)$$

where k = permeability (L^2), g = acceleration due to gravity (Lt^{-2}).

Expressing g in cm s^{-2}, ρ in g cm^{-3}, and K in cm s^{-1}, then k is expressed in cm^2. The customary unit of k is the darcy, which is 0.987×10^{-8} cm^2. Intrinsic permeability is mainly applied in the petroleum and natural gas industries, which deal with underground flow of fluids of varying viscosity and density.

Because water movement occurs through the pores and cracks of the aquifer material only, the actual pore velocity of the water is greater than the specific discharge. If the effects of tortuosity are neglected, the relation between the actual pore velocity v_p and the specific discharge v is

$$v_p = \frac{A}{A_{\text{eff}}} v \qquad (5.6)$$

where v_p = pore velocity (Lt^{-1}), A = total area normal to the flow direction (L^2), A_{ef} = effective area available for flow (L^2)

or

$$v_p = \frac{v}{\varepsilon} \qquad (5.7)$$

where $\varepsilon = A_{\text{eff}}/A$ = porosity (–).

Equation 5.7 is theoretically correct only for the condition where the effective pore area is the same as the total pore area A_p. Also, if tortuosity is considered, the true microscopic velocities will be larger than v_p, since the fluid must travel along irregular flow paths that are longer than the linearly averaged path represented in v_p. However, according to H. Bouwer (1978), Eq. 5.7 reasonably estimates the actual velocity in most types of soil or aquifer materials.

5.3.2 Influence of Initial Bacteria Deposition

Laboratory-scale column (50 mm i.d. × 300 mm packed with 3 mm borosilicate glass beads) experiments were carried out to demonstrate the effect of the level of initial bacterial deposition on the rate of biofilm accumulation and organic substrate removal (Martin, 1990). Since the extent of substrate removal depends on the amount of active biomass, the measurement of substrate concentration along the column and over time provided an indirect indication of the spatial and temporal development of the biofilm. Each experiment comprised three phases designated "deposition," "flush," and "growth." In the deposition phase, the column was seeded with *P. aeruginosa* at a volumetric flow rate of 10 mL/min. The basic approach was to apply bacterial suspensions to the column and monitor the concentrations of cells in the effluent following the same procedure used for column deposition experiments described by Martin et al. (1992). Two trials were completed satisfactorily for which the concentration of cells applied to the column during the deposition phase differed by about 3 orders of magnitude. The average inlet cell concentration during the 60-min deposition phase for the low inoculum experiment was 70 cfu/mL and for the high inoculum experiment was 65,000 cfu/mL. During the deposition phase, the average effluent cell concentrations were 85% and 90% of the feed concentrations for the low and high inoculum experiments, respectively, indicating minimal changes in bulk liquid cell concentrations along the length of the columns. After 60 min, the column was flushed with sterile media containing 0.2 mg/L acetate at the same flow rate for a period of 120 min. The growth phase began at a total elapsed time of 180 min when sterile media containing 3 mg/L acetate was fed to the column. The concentration of acetate and cells in the bulk fluid were measured as a function of time and distance along the column.

The two different levels of cell concentrations applied to the column resulted in a pronounced difference in the time required to reach a steady state column effluent acetate concentration (Fig. 5.9). Data for the high inoculum experiment are represented by the squares and show between 5 and 6 days were required to reach 90% removal of the inlet acetate. For

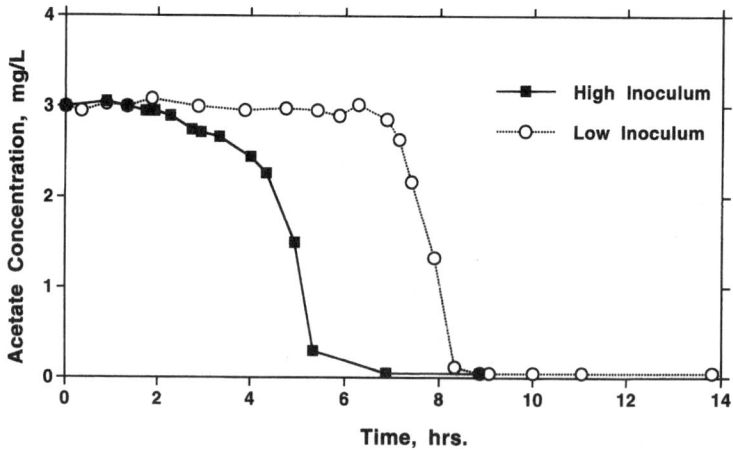

Figure 5.9 Variation in the effluent acetate concentration over time for both low and high inoculum biofilm growth experiments. The influent acetate concentration was 3.0 mg/L.

the low inoculum trial, indicated by the open circles, this time was just over 8 days. The steady state effluent acetate concentration was about 0.1 mg/L, representing about 97% removal. This level was reached within 7 and 9 days for the high and low inoculum trials, respectively. The implications of these laboratory findings to field-scale accumulation of biofilms for bioremediation of contaminants depend on the substrate removal kinetics. For rapidly biodegradable substrates, such as acetate, differences of a few days in the time scale to achieve steady state substrate biodegradation do not necessitate using a high inoculum over a low inoculum in the field. However, for slow contaminant biodegradation reactions, the time scale indicated in Figure 5.9 is markedly extended so a high inoculum greatly shortens the time needed to accumulate sufficient biofilm for successful bioremediation. Additional ramifications of a slow-growing species for application to biofilms utilizing toxic compounds are discussed in Chapter 7.

5.3.3 Influence on Porous Media Hydrodynamics and Permeability

Cunningham et al. (1991) carried out visual observations of biomass accumulation patterns occurring in specially designed packed-bed reactors under high substrate loading conditions (feed contained 25 mg/L glucose). Average biofilm thickness, as determined from 10 separate microscopic measurements of biofilm thickness on the exposed edges of media particles, was the variable used to estimate net accumulation. The experimental system consisted of parallel, horizontally mounted rectangular porous media reactors. Nutrients were supplied to the reactors under gravity flow from a constant head mixing chamber. A constant piezometric head drop of 0.5 cm/cm was maintained across the system throughout each experiment by locating the downstream end of each reactor effluent tube at an elevation ranging between 2.7 cm and 16 cm below the level in the constant head mixing chamber. Flow rate through each reactor varied in proportion to reactor permeability, which decreased substantially as biofilm accumulated during the experiments.

The progression of biofilm thickness for porous media of various sizes and composition followed a sigmoidal curve, reaching a maximum thickness after about 5 days (Fig. 5.10). The ultimate biofilm thickness varied directly with media pore space size and was not influenced significantly by media composition. The following interactions gave rise to the sigmoidal shape. As biofilm thickness increased, the diffusional path length within the biofilm increased, thereby decreasing nutrient concentrations in the base film. Since the piezometric head gradient remained constant, increased thickness also resulted in decreased interstitial pore velocity (decreased pore velocities were observed using nigrosine dye before and after biofilm development). Decreased pore velocities reduced both advective and dispersive transport, thereby lowering nutrient concentrations at the film–water interface, which subsequently reduced growth rate. Decreased pore velocities also reduced shear stress within the biofilm matrix. After biofilm thickness reached quasi-steady state (plateau between 5 and 10 days in Fig. 5.10), the volume of effective pore space (permeability) appeared to stabilize and remain constant.

Wanner et al. (1995) investigated microscale biofilm accumulation in a packed bed biofilm reactor inoculated with a pure culture of *P. aeruginosa* and operated under high substrate loading (feed concentration of 7 to 16 g C/m^3) and constant flow rate conditions. The 3.1 cm diameter cylindrical reactor was 5 cm in length and packed with 1 mm glass beads. Daily observations of biofilm thickness, influent and effluent glucose substrate concentration, and effluent dissolved and total organic carbon were made during the 13-day experiment. A published biofilm process simulation program (AQUASIM) was used to analyze the experimental data (Reichert, 1994). Biofilm accumulation appeared to reach a quasi-steady state condition (based on thickness measurements) after 10 days. Analysis of the experimental data using AQUASIM identified three distinct phases of biofilm accumulation: (1) an initial phase in which substrate removal was determined by biofilm thickness and mass transfer resistance between biofilm and bulk fluid, (2) a growth phase in which thickness, mass transfer resistance, and area of the biofilm–bulk fluid interface were important, and (3) a mature biofilm phase in which the area of the interface was the impor-

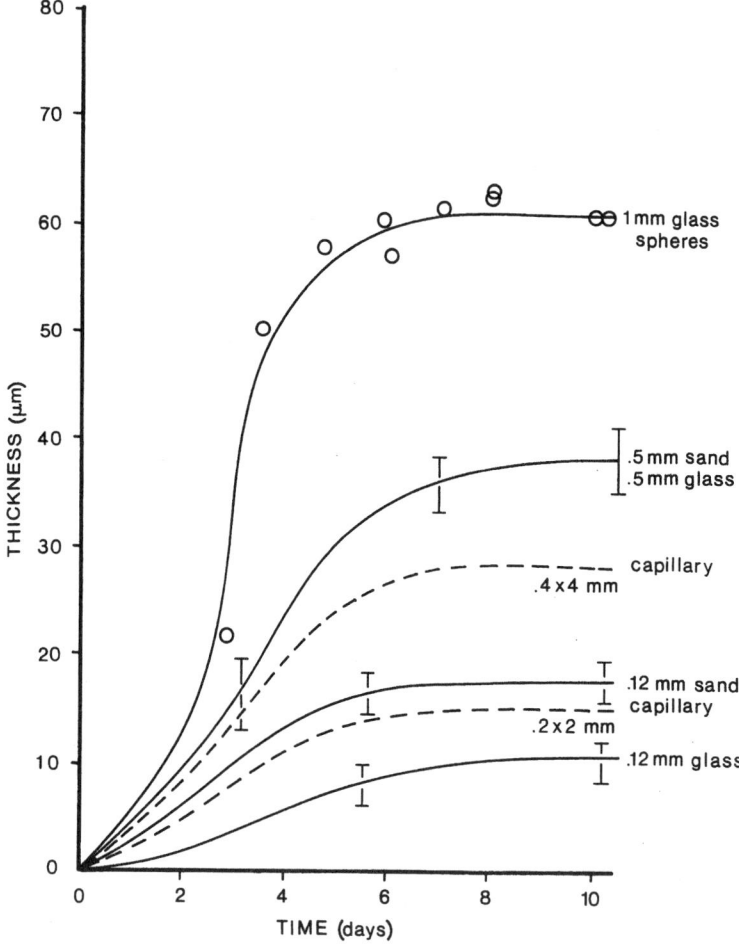

Figure 5.10 Progression of biofilm thickness (*P. aeruginosa*) for media of different diameter and composition. All reactors were run in parallel under a piezometric gradient of 0.5. Data indicate a quasi-steady state thickness is reached after about 5 days of reactor operation. Maximum biofilm thickness values of 63 μm, 40 μm, and 9–14 μm were observed for media particle diameters of 1 mm, 0.54 mm, and 0.12 mm. Clean surface permeability values of 2.1×10^{-5}, 2.17×10^{-6}, and 9.7×10^{-7} cm^2 correspond to these particle diameters, suggesting a direct relationship between media permeability and maximum biofilm thickness.

5.3 BIOFILM ACCUMULATION

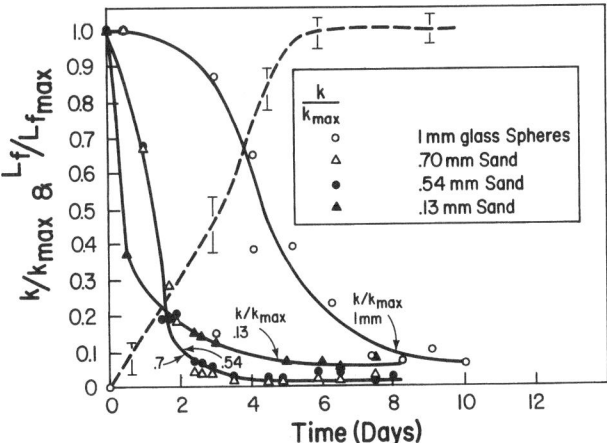

Figure 5.11 Porous media permeability decrease corresponding to increased biofilm thickness. L_f/L_{fmax} is a single composite dimensionless curve representing all thickness curves from Figure 5.10. K_{max} values are 2.1×10^{-5} cm^2 (1 mm glass spheres), 3.19×10^{-6} cm^2 (0.70 mm sand), 2.17×10^{-6} cm^2 (0.54 mm sand), and 9.7×10^{-7} cm^2 (0.12 mm sand). After 5 days of reactor operation, the media/biofilm permeability stabilized and remained essentially constant in the range of 3 to 7×10^{-8} cm^2.

tant parameter. In phase 3, the surface area of the biofilm in contact with the fluid had decreased significantly compared to phases 1 and 2. Substrate flux into the biofilm was utilized entirely within the outer layers of biofilm and thus the total thickness did not correlate with the substrate utilization rate. AQUASIM simulations also showed that biofilm detachment correlated directly with shear stress at the biofilm surface.

Investigations documenting permeability reduction due to biofilm accumulation in porous media have been widely reported (see review by Jaffe and Taylor, 1998). There appear to be two distinct mechanisms by which porous media biofilms reduce permeability. In case one, continuous films are formed, resulting in pore space reduction, causing both porosity and permeability to decrease, as demonstrated by Cunningham et al. (1991). An example of the permeability decrease, corresponding to increasing biofilm thickness for a laboratory porous media reactor system, is shown in Figure 5.11. Here it is seen that biofilm accumulation results in a 95–99% reduction in the original clean surface permeability. Medium porosity decreased between 50% and 96%. A minimum permeability (3 to 7×10^{-8} cm^2) persisted after biofilm thickness had reached a maximum value. These observations suggest that the biofilm accumulation process stabilizes and maintains a minimum permeability within the porous media system. No measurements have been made of "biofilm matrix permeability" yet.

In the second case, permeability is reduced by patchy biofilm aggregates accumulating mainly in pore throats. Under this condition, relatively small amounts of biomass (relative to free pore volume) can cause large reductions in permeability. According to Rittmann (1993), substrate flux to the biofilm is mainly responsible for determining whether case 1 or case 2 occurs. Rittmann developed an index variable, the normalized surface loading, which can be used to predict the occurrence of patchy versus continuous biofilm accumulation in porous media.

5.4 REMEDIATING SUBSURFACE CONTAMINATION

Contamination of groundwater and soils with hazardous organic compounds is widespread (National Research Council, 1994). Exploiting the metabolic capabilities of indigenous or introduced microorganisms in regions of subsurface contamination offers the prospect of converting dissolved and sorbed contaminants to harmless products. This represents a major new challenge for the application of biofilms in porous media. This section addresses some important issues concerning the design and implementation of in situ bioremediation.

5.4.1 Microbial Metabolism of Organic Contaminants

The metabolic capabilities of subsurface microorganisms are quite diverse. For growth of microorganisms, the presence of electron donors and acceptors, a carbon source, and essential nutrients is required. Besides natural compounds, many contaminants can provide these growth requirements. Most organic contaminants can typically be categorized as either aliphatic or aromatic compounds that contain different functional groups, such as -OH, -Cl, $-NH_2$, $-NO_2$, and $-SO_3$. As electron donors, these chemicals are oxidized during microbial metabolism to yield energy for growth and, in the best case, the organic carbon is converted entirely to CO_2. Some of the breakdown intermediates may also be assimilated as a carbon source for microbial growth. Functional groups (e.g., $-NH_2$, $-NO_2$, and $-SO_3$) may either be used as nutrients or cleaved from the carbon skeleton when the compound is reduced or oxidized. Oxidation can take place aerobically (in the presence of oxygen) or anaerobically (in the absence of oxygen). Oxygen serves two distinct functions. It can act as terminal acceptor of electrons that are released during the oxidation of electron donors, or it can react directly with the organic molecule. As an electron acceptor, oxygen can be replaced by other oxidized inorganic compounds such as nitrate, metal ions (e.g., Fe(III) or Mn(IV)), sulfate, or carbon dioxide, although the energy gains to the microorganism are then smaller. These alternate electron acceptors are reactants in anaerobic microbial processes, although they cannot substitute for the function of oxygen as a direct reactant (Schink, 1988).

Classes of organic contaminants known to be biotransformed by subsurface microorganisms are listed in Table 5.3 along with an indication of the required electron acceptor. In some instances, the compounds are the primary energy and carbon supply for the microorganisms (upper portion of Table 5.3). For other compounds, the biotransformation occurs as cosubstrate utilization where enzymes involved in the metabolism of one substrate are also able to degrade the contaminant (lower port of Table 5.3). Petroleum hydrocarbons and halogenated compounds are among the most prevalent organic contaminants at waste sites (Plumb, 1991). Petroleum hydrocarbons are easily degraded by certain aquifer microorganisms under aerobic conditions, but the anaerobic biodegradation of these compounds is much more variable and depends highly upon the compound, the terminal electron acceptor, and the specific site (Nales et al., 1998; Wilson and Bouwer, 1997). The aerobic co-oxidation of chlorinated solvents by methanotrophic bacteria while using methane as a cosubstrate (Table 5.3) is under intensive investigation for in situ bioremediation applications. In the absence of molecular oxygen (anaerobic conditions), halogenated organic compounds may be reductively dehalogenated during their degradation or be used as terminal electron acceptor for growth of microorganisms (Holliger et al., 1992; Mohn and Tiedje, 1991). In reductive dehalogenation, the halogenated compound becomes an electron acceptor; and in this process, a halogen is removed and is replaced with a hydrogen atom. Detailed information on biotransformation of halogenated compounds is presented elsewhere (Adriaens and Vogel, 1995; Norris et al., 1994; Vogel et al., 1987).

TABLE 5.3. Some Important Classes of Organic Contaminants Susceptible to In Situ Bioremediation

Chemical Class	Frequency of Occurrence	Favorable Electron Acceptor(s)	Notes
Primary Metabolism			
Monoaromatic hydrocarbons	Very Common	Oxygen[a], anaerobic[b]	Difficult to degrade if >4–5 rings
Polyaromatic hydrocarbons	Common	Oxygen[a], Nitrate[b]	
Aliphatic hydrocarbons	Common	Oxygen[a]	
Phenols	Infrequent	Oxygen[a], anaerobic[b]	
Nitroaromatics	Common	Oxygen[b], anaerobic[b]	
Alcohols, ketones, esters	Common	Oxygen[a], anaerobic[a]	
Some chlorinated solvents (CH_2Cl_2, CH_3CH_2Cl, $CH_2=CHCl$)	Common	Oxygen[a]	Biodegradable under a narrow range of conditions
Less halogenated aromatics	Common	Oxygen[a]	
Less chlorinated polychlorinated biphenyls	Infrequent	Oxygen[a]	Biodegradable under a narrow range of conditions
Cosubstrate Metabolism			
Highly halogenated aliphatics	Very Common	Oxygen[a], anaerobic[a]	Aerobic cometabolism by methanotrophs in special cases; anaerobic cometabolism by many bacteria
Less halogenated aliphatics	Very Common	Oxygen[a], anaerobic[a]	Aerobic cometabolism by methanotrophs; anaerobic cometabolism by many bacteria
Highly halogenated aromatics	Common	Aerobic[a], anaerobic[a]	

[a]Many observations, confirmed in laboratory and supported by field evidence.
[b]Demonstrated, variable effectiveness in the field.

5.4.2 Treatment Approaches

The most important principle of bioremediation is that microorganisms (mainly bacteria) can be used to destroy hazardous contaminants or transform them to less harmful forms. The microorganisms act against the contaminants only when they have access to a variety of materials—compounds to help them generate energy and nutrients to build more cells. In some cases the natural conditions at the contaminated site provide sufficient quantities of essential materials that bioremediation can occur without human intervention—a process called *natural* or *intrinsic bioremediation*. In most cases, bioremediation requires the application of engineered systems to supply microbe-stimulating materials—a process called *engineered bioremediation*. Engineered bioremediation relies on accelerating the desired biodegradation reactions by encouraging the growth of more organisms, as well as by optimizing the environment in which the organisms must carry out the detoxification reactions. Consequently, bioremediation can be considered as a continuum, extending from completely natural at one end to fully engineered at the other. The common and distinct issues for these two broad types of in situ bioremediation are discussed in the following two sections. An expanded discussion of the design considerations for in situ bioremediation of organic contaminants is given by Bouwer et al. (1998).

5.4.2.1 Intrinsic Bioremediation.

Intrinsic bioremediation (also termed natural attenuation) relies on the innate capabilities of naturally occurring microbial populations to convert environmental pollutants to harmless forms. Intrinsic bioremediation occurs at many sites, sometimes at a rate significant enough to prevent further movement of the pollutant with the flowing groundwater. There is increasing interest in relying on intrinsic bioremediation for control of all or some of the contamination at waste sites (National Research Council, 1993; Rifai, 1998). Intrinsic bioremediation is attractive economically because it is relatively passive, requiring only a demonstration (via extensive site characterization) that natural biological processes are destroying contaminants in situ. Examples of sites where intrinsic bioremediation has been shown to play a significant role in attenuating organic contaminants in groundwater have been presented by Madsen et al. (1991), Godsy et al. (1992), Klecka et al. (1990), Davies et al. (1994), Cox et al. (1995), Bosma et al. (1997), Ravi et al. (1998), Kennedy et al. (1998), Mondello et al. (1998), Rijnaarts et al. (1998a, 1998b), and Rijnaarts (1998b).

Before intrinsic bioremediation can be considered as a legitimate cleanup method for a given site, the capacity of the indigenous microorganisms to metabolize the contaminants must be documented through laboratory and field testing (Chapelle and Bradley, 1998). Furthermore, the effectiveness of intrinsic bioremediation must be proven with a site-monitoring program to confirm the progress of contaminant biodegradation. Chemical analyses of contaminants, terminal electron acceptors, and/or other reactants and products indicative of biodegradation processes should be performed. Consequently, employing intrinsic bioremediation is in contrast to no-action alternatives.

For intrinsic bioremediation to be effective as a stand alone approach to aquifer or vadose zone restoration, the naturally occurring hydrogeochemical conditions at the site must allow the rate of biodegradation to exceed the rate of contaminant migration. Site conditions that favor intrinsic bioremediation include presence of microorganisms adapted to contaminant degradation, consistent and known groundwater flow throughout the year to ensure that contamination is not spreading with the flowing groundwater, presence of carbonate minerals to buffer acidity produced during biodegradation, adequate supply of electron acceptors (for oxidative degradation), electron donors (for primary substrate and for reductive dehalogenation), and nutrients to meet the microbial growth requirements, and an absence of compounds that might be toxic to microorganisms (e.g., Hg, cyanide). Intrinsic bioremediation of petroleum hydrocarbons and monochlorinated compounds is especially applicable to sites where contamination is in the vadose zone and oxygen is more readily available. In areas where aquifers generally have low redox conditions (like in the Netherlands (Rijnaarts, 1998b)), these compounds are often not readily degraded by intrinsic processes. Microbial populations in such systems have not yet become adapted to the anaerobic degradation of petroleum hydrocarbons (e.g., benzene) which typically pose the greatest health risks and are the chemicals of most concern at the sites (van Heiningen, 1999). In contrast, highly chlorinated compounds (e.g., PCE, TCE, and hexachlorocyclohexanes) are often found to be naturally degraded by reductive processes in porous media with low redox conditions (Gerritse et al., 1999; Langenhoff et al., 1999; Rijnaarts et al., 1998b). This intrinsic bioremediation potential can be realized only when the contaminant load does not exceed the natural electron donor capacity.

5.4.2.2 Engineering In Situ Bioremediation. Although the potential for intrinsic bioremediation is substantial at many sites, it can often be unacceptably slow as a sole remediation strategy due to poorly adapted microorganisms, the limited availability of electron acceptors/donors and nutrients, cold temperatures, high concentrations of contaminants (NAPL), and mass transfer limitations in the subsurface. When site conditions are not favorable for natural biotransformation, bioremediation requires construction of engineering systems to supply microbe-stimulating materials. Engineered bioremediation relies on accelerating the desired biodegradation reactions by encouraging the growth of more organisms, as well as by optimizing the environment in which the organisms must carry out the detoxification reactions. In some cases, bioaugmentation with adapted microbial consortia may be considered. Examples include the initiation of the anaerobic degradation of PCE and its degradation products cis-1,2-DCE and vinyl chloride (Morgan, 1998) or the stimulation of the anaerobic conversion of benzene to innocuous products (van Heiningen et al., 1999, in press).

Frequently, the necessary stimuli for microbial growth in the subsurface are oxygen and other electron acceptors (such as nitrate or sulfate) and nutrients (such as nitrogen and phosphorus) or organic substrates as electron donors. Typical engineered bioremediation systems therefore perfuse electron acceptors/donors and nutrients through the contaminated region as shown in Figure 5.2. Engineering in situ bioremediation near the land surface can be achieved by using infiltration galleries that allow water amended with nutrients and electron acceptors/donors to percolate through the soil. When contamination is deeper, in situ bioremediation systems inject the amended water through wells. Some in situ bioremediation systems use extraction and injection wells in combination to control the flow of electron acceptors/donors and nutrients and to hydraulically isolate the contaminated area. Another approach to engineered in situ bioremediation is extraction of contaminated groundwater combined with above-ground bioreactor treatment and subsequent reinjection of nutrient-spiked effluent (Bouwer et al., 1998; National Research Council, 1993).

The application of engineered bioremediation for aerobic cometabolic biodegradation of trichloroethylene (TCE) was recently demonstrated in field scale at the Edwards Air Force Base in California using toluene as primary substrate (McCarty et al., 1998). Groundwater contaminated with 500 to 1200 µg/L trichloroethylene was circulated between two contaminated aquifers through two treatment wells located 10 m apart. Groundwater circulation was achieved by pumping groundwater from the 8 m thick upper aquifer to the 5 m thick lower aquifer in one well, while the second well pumped groundwater from the lower to the upper aquifer using flowrates between 25 and 38 L/min. Toluene (7 to 13.4 mg/L), oxygen, and hydrogen peroxide were injected into the recirculating groundwater to stimulate biofilm formation (bioactive zones) in the exit flow regions surrounding each well. With each pass through the bioactive zone in the upper aquifer, 87 ± 8% of the TCE in the contaminated groundwater was biodegraded. In the lower aquifer bioactive zone, TCE removals were as high as 83 ± 16% with each passage of contaminated water. The groundwater recirculation on a regional scale (multiple passes through the two bioactive zones) achieved 97–98% total removal of TCE. The average downgradient residual toluene concentration was 1.1 ± 1.6 µg/L, indicating that 99.98% of the primary substrate was biodegraded in the engineered system.

5.4.3 Bioavailability

Biodegradation rates in the field can be significantly slower than in the laboratory because of lower field temperatures and reduced bioavailability (Sturman et al., 1995). Bioavailability is generally defined as the availability of a chemical to biological transformation, and it is determined by the extent to which a chemical is exposed to organisms (Hamelink et al., 1994). Hydrophobic organic contaminants tend to partition among solid, liquid, and gas phases in the subsurface. For example, hydrophobic organic compounds can be found as a nonaqueous phase liquid, dissolved in the aqueous phase, sorbed to soil and sediment materials, and/or associated with colloids and other large organic molecules (e.g., natural organic matter) in water. Bioavailability of hydrophobic organic contaminants in the subsurface is affected by sorption/desorption in two important ways (Zhang et al., 1998). First, sorption diminishes the organic concentration in bulk water such that only a small fraction of the compound may actually be in the water phase. Most evidence indicates that microorganisms are most efficient in utilizing substrates dissolved in the aqueous phase (van Loosdrecht et al., 1990). Sorption causes high organic concentrations in microporous regions and impermeable zones to which bacterial access is obstructed. Separation of organic compounds by sorption from the aqueous phase is likely to reduce the rate and extent of biotransformation in the subsurface (Mihelcic et al., 1993). Second, because desorption and immobile zone diffusion must occur before biodegradation can proceed, the overall rate of bioremediation can be limited or even controlled by these mass transfer processes, not by the activity of the degrading microorganisms (Chung et al., 1993). The practical impact of reduced bioavailability is to decrease the rate of removal of the contaminants, thereby increasing the time required to achieve cleanup and the amount of chemicals that must be added to sustain microbial activity.

The influence of soil on the rate and extent of benzene, toluene, and naphthalene biodegradation has been demonstrated through a series of batch experiments (Zhang and Bouwer, 1997). Nearly complete aerobic naphthalene biodegradation (1.28 mg/L) by indigenous soil bacteria occurred within 60 h in aqueous solution (soil-free) while it took 2 weeks to degrade the same amount in the presence of 0.47 kg soil/L. The batch data also showed that the slower the desorption, the lower the biodegradation rate. Large soil particles exhibited a slower approach to equilibrium than smaller particles, implying intraparticle diffusion limited biodegradation. The rate of biodegradation was also observed to decrease with increasing organic compound hydrophobicity, soil/water ratio, and soil organic carbon content.

5.4 REMEDIATING SUBSURFACE CONTAMINATION

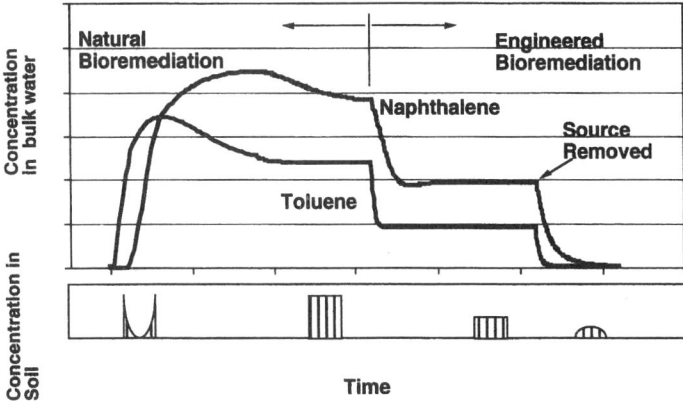

Figure 5.12 Simulations of sorption and biodegradation of toluene and naphthalene in the subsurface that illustrate the qualitative behavior of hydrophobic pollutants during bioremediation. The lower portion depicts the concentration profile along the centerline of the soil particle for different times.

A qualitative response of organic concentrations in aqueous and solid phases as a function of time during bioremediation in the subsurface is shown in Figure 5.12. Biotransformation is initially insignificant as the amount of biomass present is very small. Sorption removes a large portion of the organic compounds from the bulk water. As concentration gradients within soil aggregates diminish, organic concentrations in the bulk water increase quickly and finally level off and reach a steady state with biomass accumulated from utilization of the organic contaminant. As the contact time between the contaminant and the solid phase increases, the amount of the contaminant available for desorption or biodegradation tends to decrease. The mechanisms responsible for these contaminant aging effects are poorly understood (Luthy et al., 1997).

With engineered bioremediation, microbial activity is enhanced, which brings about a rapid decrease in the bulk water concentration (Fig. 5.12). This lowering of the bulk water concentration promotes desorption of the contaminants by increasing the local concentration gradient for diffusion (Rijnaarts et al., 1990). The desorbing contaminants can be biodegraded as they pass through the attached biomass, which keeps the bulk water concentrations low. Thus, the net rate of desorption is accelerated in the presence of biotransformation. However, if the rate of desorption is much slower than the biodegradation rate, stimulating microbial growth only has a minimal impact on accelerating the rate of soil decontamination. Furthermore, upgradient sources of contamination must be removed (e.g., removal and containment of separate phase organic liquids) in order to achieve effective in situ bioremediation of the contaminant plume.

5.4.4 Bioactivated Zones and Biobarriers

Often one of the first steps for remediation of groundwater pollution is containment of the dissolved contaminant plume. Bacteria can be injected into porous media to create either a bioactive zone to biodegrade contaminants in the flowing groundwater or a biobarrier by plugging of the formation (Cunningham et al., 1997). By reducing the overall permeability and mass transport properties of the aquifer matrix, the creation of subsurface biofilm barriers (i.e., biobarriers) serves as an alternative technology for controlling the migration of contaminants from hazardous waste sites. Biobarriers may also be useful as a means of funneling contaminated groundwater through subsurface treatment systems (Fig. 5.13).

The first successful attempt at selectively plugging permeable strata with microbial biomass was conducted with ultramicrobacteria (UMB) (cell diameters from 0.3 to 0.5 μm) (MacLeod et al., 1998). The UMB are not motile, and therefore they followed the path of injected water in the subsurface environment. They penetrated wherever pore sizes were permissive, with a significant number retained in each pore structure by trapping and/or deposition. When a solution containing a suitable nutrient mixture for UMB resuscitation was subsequently injected into the porous medium, these nutrients initiated the relatively rapid process of resuscitation. The UMB returned to their active 1.0 to 1.4 μm form and began to reproduce. Immediately following their resuscitation, the active bacteria derived from these UMB began to grow, reproduce, adhere to surfaces, and secrete exopolysaccharides (EPS). The simultaneous nutrient-driven processes of reproduction and EPS secretion rapidly filled the available pore space (Lappin-Scott et al., 1988a). The processes of resuscitation and plugging could be controlled by selecting nutrients that were either fast acting or slow acting and exhibited different extents of assimilation by the resuscitating organisms and by the rate at which these nutrients were pumped into the porous medium. Rapid re-

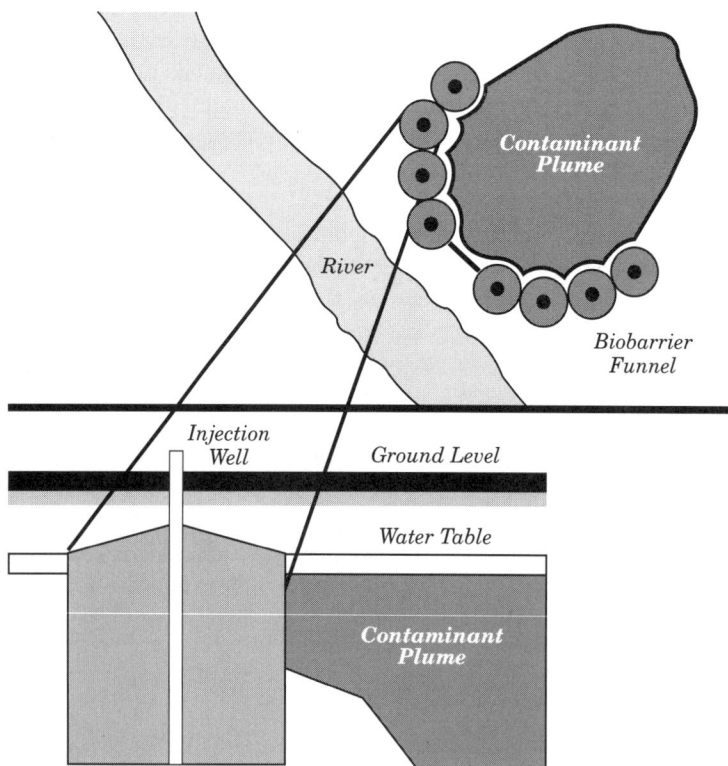

Figure 5.13 Schematic illustrating use of biofilm barriers to channel the flow of a dissolved contaminant plume through a zone of active treatment.

suscitation of UMB and slow rates of nutrient injection caused plugging very close to the injection point and limited the depth of the placed biobarrier.

Recent experiments with a series of packed-bed column and small tank reactors were conducted to evaluate biobarrier formation and persistence (Cunningham et al., 1997). An initial reduction in the hydraulic conductivity to less than 10% of the original value was observed in 91.4 cm long packed sand columns operated in a vertical mode. A reduction in hydraulic conductivity to less than 0.0001% of the initial hydraulic conductivity was obtained with simulated biobarrier formation in the small tanks (sand bed 91.4 cm wide, 30.4 cm deep, and 122 cm long). Biobarrier persistence in the columns and tanks was significantly different. Starvation conditions in the 91.4 cm long packed sand columns had a deleterious effect on biobarrier effectiveness. The sand column required continuous nutrient addition for barrier maintenance, while the tank barriers were able to persist for a considerable period without nutrient addition. The persistence of the biobarriers in the tanks compared to the column experiments may be due to the higher head pressure continually present in the column reactor design. The difference is approximately 122 cm of head above the top of the 91.4 cm column compared to approximately 3.6 cm of head across the 122 cm length of the tank. The higher head provides a higher flow rate and pore velocity through the columns, which may increase the detachment rate. The performance and formation of the biobarrier in the sand columns was not altered by the presence of 1 ppm strontium and 1 ppm cesium for periods up to 120 days. The column biobarrier challenged with 100 mg/L carbon tetrachloride for extended periods and with 300 mg/L carbon tetrachloride for short periods did not have any measurable effect on biobarrier stability as measured by changes in hydraulic conductivity. No detectable amounts of strontium or carbon tetrachloride were detected in the column effluent during these biobarrier challenge experiments.

Biobarriers are attractive for environmental purposes because of their potential to preferentially plug zones of high permeability. Also, they are conceptually not limited to near-surface application (i.e., 30 m or less) as is the case for sheet piling and grout curtains. Biobarriers are currently being considered for secondary oil recovery by water flooding (Cusack et al., 1992), protection of rivers and lakes from pollutant plumes in emerging groundwater (Cunningham, 1997), and for sealing of earthen dams, berms, and hazardous waste landfill liners (Lappin-Scott and Costerton, 1992).

Bioactivated zones constructed using strains that produce very little EPS would provide less resistance to groundwater flow. Bacterial cells within these "loose" biobarriers could be capable of pollutant degradation and/or metal reduction as groundwater passes through them. In some cases, a microbial population needs to be introduced to establish a certain biodegradation capability. In other cases, the indigenous microorganisms can be stimulated to degrade or immobilize contaminants. For petroleum hydrocarbons, chlorinated solvents, and chlorinated pesticides, feasibility and pilot test have been performed and implementation at full scale at several large contaminated sites is expected (Rijnaarts, 1998a). The hydraulic conductivity within bioactivated zones can be manipulated by seeding with bacterial strains with the desired clogging or nonclogging ability. This procedure was demonstrated by Koene and Rijnaarts (1996) in aerobically toluene degrading packed bed biofilm reactors under continuous flow conditions (Fig. 5.14). The reactors were packed with activated carbon (0.25–0.50 mm grain diameter), and parallel reactors were inoculated either with a pure culture of *Rhodococcus sp C. 125* (high blocking factor and unfavorable cell–cell interactions yielding a low α and low tendency to clog), an undefined mixed culture, or a pure culture of *Pseudomonas putida mt2* (low blocking factor and favorable cell–cell interactions yielding a high α and high tendency to clog). After 50 days of operation, hydraulic conductivity decreased from about 100 cm/day

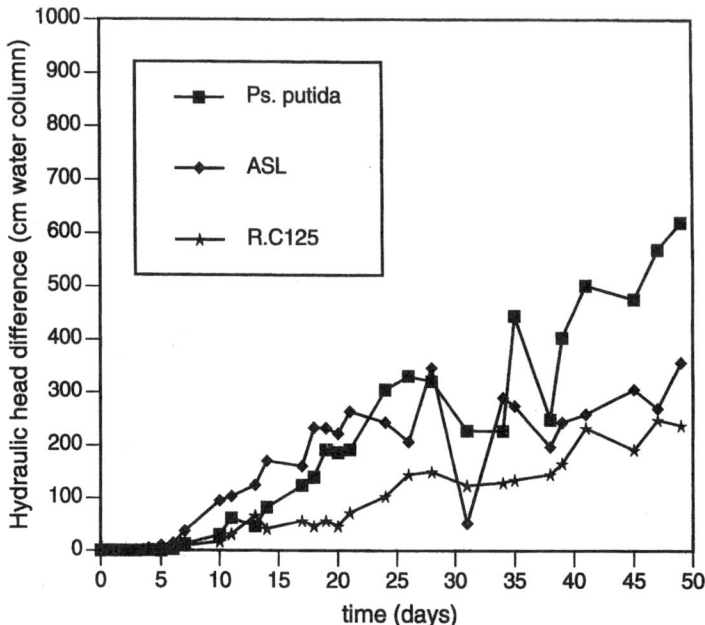

Figure 5.14 Pressure buildup in aerobic activated carbon-packed bed columns inoculated with a strongly blocking pure culture (*Ps. putida*), a weakly blocking pure culture (*R. C125*), and a mixed culture (termed ASL) and fed with toluene as the substrate.

to values of 24, 13, and 8 cm/day, for *R. C125*, the mixed culture, and *Ps. putida*, respectively. This finding shows that the nature of the cell–cell interactions of the dominant microbial population influences the hydraulic performance of a bioactivated zone or biobarrier.

5.5 CONCLUDING REMARKS

Biofilm accumulation in porous media is controlled by the processes of transport, adhesion, microbial growth, and detachment. Interactions between bacteria and solid surfaces strongly influence the behavior of bacteria in natural and engineering environments. Subsurface microorganisms are capable of transforming many organic contaminants. This has led to great interest in exploiting biological processes for in situ treatment including intrinsic bioremediation, engineered bioremediation, and biobarriers. More attention must be directed to the role of biofilms in contaminant transformations because biofilms predominate in many natural aquatic systems and have several advantages in engineered treatment systems. Future research should be especially directed toward developing methods to improve the bioavailability of hydrophobic organic contaminants.

ACKNOWLEDGEMENTS

The authors acknowledge the support of the National Science Foundation (NSF) through Cooperative Agreement ECD-8907039 between NSF and Montana State University.

REFERENCES

Absolom, D. R., F. V. Lamberti, A. Policova, W. Zingg, C. J. van Oss, and A. W. Neumann. 1983. Surface thermodynamics of bacterial adhesion. *Appl. Environ. Microbiol.* 46: 90–97.

Adriaens, P., and T. M. Vogel. 1995. Biological treatment of chlorinated organics. In *Microbial Transformation and Degradation of Toxic Organic Chemicals*, L. Y. Young and C. E. Cerniglia, eds. New York: John Wiley, pp. 435–486.

Albrechtsen, H. J., and A. Winding. 1992. Microbial biomass and activity in subsurface sediments. *Microbial Ecol.* 23: 303–317.

Barbaro, S. E., H. J. Albrechtsen, B. K. Jensen, C. I. Mayfield, and J. F. Baker. 1994. Relationships between aquifer properties and microbial populations in the Borden Aquifer. *Geomicrobiol. J.* 12: 203–219.

Bear, J. 1964. *Chemistry of Soil*. New York: Reinhold Publishing.

Bear, J. 1979. *Hydraulics of Groundwater*. New York: McGraw-Hill.

Beloin, R. M., J. L. Sinclair, and W. C. Ghiorse. 1988. Distribution and activity of microorganisms in subsurface sediments of a pristine study site in Oklahoma. *Microbial Ecol.* 16: 85–97.

Bitton, G., N. Lahav, and Y. Henis. 1974. Movement and retention of *Klebsiella aerogenes* in soil columns. *Plant and Soil* 40: 373–380.

Bosma, T. N. P., M. A. van Alast, H. H. M. Rijnaarts, J. Taat, and J. Bovendeur. 1997. Intrinsic dechlorination of 1,2-dichloroethane at an industrial site. In *In Situ and On-Site Bioremediation: Volume 3*, B. C. Alleman and A. Leeson, eds. Columbus, OH: Battelle Press, pp. 155–159.

Bouwer, H. 1978. *Groundwater Hydrology*. New York: McGraw-Hill.

Bouwer, E. J., N. D. Durant, L. P. Wilson, and W. Zhang. 1998. Design considerations for in situ bioremediation of organic contaminants. In *Biological Treatment of Hazardous Waste*, G. A. Lewandowski and L. J. DeFilippi, eds. New York: John Wiley, pp. 237–269.

Bryers, J. D., and S. L. Sanin. 1994. Resuscitation of starved ultramicrobacteria to improve in situ bioremediation. *Ann. NY Acad. Sci.* 745: 61–76.

Busscher, H. J., M. N. Bellon-Fontaine, N. Mozes, H. C. van der Mei, J. Sjollema, O. Cerf, and P. G. Rouxhet. 1990a. Deposition of *Leuconostoc mesenteroides* and *Streptococcus thermophilus* to solid substrata in a parallel plate flow cell. *Biofouling* 2: 55–63.

Busscher, H. J., M. N. Bellon-Fontaine, J. Sjollema, and H. C. van der Mei. 1990b. Relative importance of surface free energy as a measure of hydrophobicity in bacterial adhesion to solid surfaces. In *Microbial Cell Surface Hydrophobicity*, R. J. Doyle and M. Rosenberg, eds. Washington, DC: American Society for Microbiology Press, pp. 335–359.

Busscher, H. J., M. M. Cowan, and H. C. van der Mei. 1992. On the relative importance of specific and non-specific approaches to oral microbial adhesion. *FEMS Microbiol. Rev.* 88: 199–210.

Caccavo, F., Jr., R. P. Blakemore, and D. R. Lovley. 1992. A hydrogen-oxidizing Fe(III)-reducing microorganisms from the Great Bay Estuary, New Hampshire. *Appl. Environ. Microbiol.* 58: 3211–3216.

Caccavo, F., Jr., N. Birger Ramsing, and J. W. Costerton. 1996a. Morphological and metabolic responses to starvation by the dissimilatory metal-reducing bacterium *Shewanella alga* BrY. *Appl. Environ. Microbiol.* 62: 4678–4682.

Caccavo, F., Jr., B. Frolund, F. van Ommen Kloeke, and P. H. Nielsen. 1996b. Deflocculation of activated sludge by dissimilatory Fe(III)-reducing bacterium *Shewanella alga* BrY. *Appl. Environ. Microbiol.* 62: 1487–1490.

Caccavo, F., Jr., P. C. Schamberger, K. Keiding, and P. H. Nielsen. 1997. Role of hydrophobicity in adhesion of the dissimilatory Fe(III)-reducing bacterium *Shewanella alga* to amorphous Fe(III) oxide. *Appl. Environ. Microbiol.* 63: 3837–3843.

Camper, A. K., J. T. Hayes, P. J. Sturman, W. L. Jones, and A. B. Cunningham. 1993. Effects of motility and adsorption rate coefficients on transport of bacteria through saturated porous-media. *Appl. Environ. Microbiol.* 59(10): 3455–3462.

Chapelle, F. H., and P. M. Bradley. 1998. Selecting remediation goals by assessing the natural attentuation capacity of groundwater systems. *Bioremediation J.* 2(3–4): 227–238.

Chapelle, F. H., J. L. Zelibor, D. J. Grimes, and L. L. Knobel. 1987. Bacteria in deep coastal plain sediments of Maryland: A possible source of $CO2$ to ground water. *Water Resources Research* 23(8): 1625–1632.

Chung, G. Y., B. J. McCoy, and K. M. Scow. 1993. Criteria to assess when biodegradation is kinetically limited by intraparticle diffusion and sorption. *Biotech. Bioeng.* 41: 625–632.

Colwell, F. S. 1989. Microbiological comparison of subsurface soil and unsaturated subsurface soil from a semiarid high desert. *Appl. Environ. Microbiol.* 55: 2420–2423.

Corapcioglu, M. Y., and A. Haridas. 1984. Transport and fate of microorganisms in porous media: A theoretical investigation. *J. Hydrology* 72: 149–169.

Cox, E., E. Edwards, L. Lehmicke, and D. Major. 1995. Intrinsic biodegradation of trichloroethene and trichloroethane in a sequential anaerobic-aerobic aquifer. In *Intrinsic Bioremediation, Third International In Situ and On-Site Bioreclamation Symposium*, R. E. Hinchee, J. T. Wilson, and D. C. Downey, eds. Columbus, OH: Battelle Press, pp. 223–231.

Cunningham, A. B. 1997. Building biobarriers to control the spread of hazardous wastes. *Centerpoint* (a publication of the USEPA Hazardous Substance Research Centers).

Cunningham, A. B., W. G. Characklis, F. Abedeen, and D. Crawford. 1991. Influence of biofilm accumulation on porous media hydrodynamics. *Environ. Sci. Technol.* 25(7): 1305–1311.

Cunningham, A., B. Warwood, P. Sturman, K. Horrigan, G. James, J. W. Costerton, and R. Hiebert. 1997. *Biofilm Processes in Porous Media—Practical Applications*. In *The Microbiology of the Terrestrial Deep Subsurface*, P. S. Amy and D. L. Haldeman, eds. Boca Raton, FL: Lewis Publishers, pp. 325–344.

Cusack, F., S. Singh, C. McCarthy, J. Grecco, M. De Rocco, D. Nguyen, H. M. Lappin-Scott, and J. W. Costerton. 1992. Enhanced oil recovery—Three dimensional sandpack simulation of ultramicrobacteria resuscitation in reservoir formation. *J. Gen. Microbiol.* 138: 647–655.

Davis, J. W., N. J. Klier, and C. L. Carpenter. 1994. Natural biological attentuation of benzene in ground water beneath a manufacturing facility. *Ground Water* 32: 215–226.

Dawson, M. P., B. Humphrey, and K. C. Marshall. 1981. Adhesion: A tactic in the survival strategy of a marine Vibrio during starvation. *Current Microbiol.* 6: 195–198.

Durant, N. D., L. P. Wilson, and E. J. Bouwer. 1995. Microcosm studies of subsurface PAH-degrading bacteria from a former manufactured gas plant. *J. Contaminant Hydrol.* 17(3): 213–237.

Ehlers, L. J., and E. J. Bouwer. 1999. RP4 plasmid transfer among strains of pseudomonas in a biofilm reactor. *Water Science and Technol.* 39(7): 163–171.

Fletcher, M. 1976. The effects of proteins on bacterial attachment to polystyrene. *J. Gen. Microbiol.* 94: 400–404.

Fletcher, M., and G. D. Floodgate. 1973. An electron-microscopic demonstration of an acidic polysaccharide involved in the adhesion of a marine bacterium to solid surfaces. *J. Gen. Microbiol.* 74: 325–334.

Fletcher, M., and G. L. Loeb. 1979. Influence of substratum charactertistics on the attachment of a marine *Pseudomonad* to solid surfaces. *Appl. Environ. Microbiol.* 37: 67–72.

Fletcher, M., and K. C. Marshall. 1982. Bubble contact angle method for evaluating substratum interfacial characteristics and its relevance to bacterial attachment. *Appl. Environ. Microbiol.* 44: 184–192.

Fontes, D. E., A. L. Mills, G. M. Hornberger, and J. S. Herman. 1991. Physical and chemical factors influencing transport of microorganisms through porous media. *Appl. Environ. Microbiol.* 57(9): 2473–2481.

REFERENCES

Freeze, R. A., and J. A. Cherry. 1979. *Groundwater*. Englewood Cliffs, NJ: Prentice-Hall.

Gannon, J., Y. Tan, P. Baveye, and M. Alexander. 1991a. Effect of sodium chloride on transport of bacteria in a saturated aquifer material. *Appl. Environ. Microbiol.* 57(9): 2497–2501.

Gannon, J. T., V. B. Manilal, and M. Alexander. 1991b. Relationship between cell surface properties and transport of bacteria through soil. *Appl. Environ. Microbiol.* 57(1): 190–193.

Geesey, G. G., and J. W. Costerton. 1979. Microbiology of a northern river: Bacterial distribution and relationship to suspended sediment and organic carbon. *Can. J. Microbiol.* 25: 1058–1062.

Gerlach, R., A. B. Cunningham, and F. Caccavo, Jr. 1998. Formation of redox-reactive subsurface barriers using dissimilatory metal-reducing bacteria. In *Hazardous Waste Research—Bridging Gaps in Technology and Culture*, L. E. Ericksen and M. M. Rankin, eds. Great Plains/Rocky Mountain Hazardous Substance Research Center, Kansas State University, Manhattan, KS, Snowbird, UT, pp. 209–223.

Gerritse, J., A. Borger, E. van Heiningen, H. H. M. Rijnaarts, T. N. P. Bosma, J. Taat, B. van Winden, J. Dijk, and J. A. M. de Bunt. 1999. Assessment and monitoring of 1,2-dichloroethane dechlorination, in situ and on-site. In: Engineered approaches for in situ bioremediation of chlorinated solvent contamination, A. Leeson and B. C. Alleman (eds.). *Bioremediation, the fifth international symposium*. Battelle Press, Columbus, OH, pp. 77–79.

Gilbert, P., D. J. Evans, I. G. Duguid, and M. R. W. Brown. 1991. Surface characteristics and adhesion of *Escherichia coli* and *Staphylococcus epidermidis*. *J. Appl. Bacteriol.* 71: 72–77.

Godsy, E. M., D. F. Goerlitz, and D. Grbic-Galic. 1992. Methanogenic biodegradation of creosote contamination in natural and simulated ground water ecosystems. *Ground Water* 30: 232–242.

Goldberg, S., Y. Konis, and M. Rosenberg. 1990. Effect of cetylpyridinium chloride on microbial adhesion to hexadecane and polystyrene. *Appl. Environ. Microbiol.* 56: 1678–1682.

Grady, C. P. L., Jr., G. T. Daigger, and H. C. Lim. 1999. *Biological Wastewater Treatment*. New York: Marcel Dekker, p. 1076.

Grasso, D., B. F. Smets, K. A. Strevett, B. D. Machinist, C. J. van Oss, R. F. Giese, and W. Wu. 1996. Impact of physiological state on surface thermodynamics and adhesion of *Pseudomonas aeruginosa*. *Environ. Sci. Technol.* 30(12): 3604–3608.

Gross, M. J., and B. E. Logan. 1995. Influence of different chemical treatments on transport of *Alcaligenes paradoxus* in porous media. *Appl. Environ. Microbiol.* 61(5): 1750–1756.

Hamelink, J. L., P. F. Landrum, H. L. Bergman, and W. H. Benson. 1994. *Bioavailability: Physical, Chemical, and Biological Interactions*. Chelsea, MI: Lewis Publishers.

Hanna, M. L., and R. T. Taylor. 1996. Attachment/detachment and trichloroethylene degradation-longevity of a resting cell methylosinus trichosporium OR3b filter. *Biotechnol. Bioeng.* 51(6): 659–672.

Harvey, R. W. 1997. In situ and laboratory methods to study subsurface microbial transport. In *Manual of Environmental Microbiology*, C. J. Hurst, G. R. Knudsen, M. J. McInerney, L. D. Stetzenback, and M. V. Walter, eds. Washington, DC: ASM Press, pp. 586–599.

Harvey, R. W., and S. P. Garabedian. 1991. Use of colloid filtration theory in modeling movement of bacteria through a contaminated aquifer. *Environ. Sci. Technol.* 25: 178–185.

Harvey, R. W., D. W. Metge, N. Kinner, and N. Mayberry. 1997. Physiological considerations in applying laboratory-determined buoyant densities to predictions of bacterial and protozoan transport in groundwater: Results of in-situ and laboratory tests. *Environ. Sci. Technol.* 31: 289–295.

Hicks, R. J., and J. K. Fredrickson. 1989. Aerobic metabolic potential of microbial populations indigenous to deep subsurface environments. *Geomicrobiol. J.* 7: 67–77.

Holliger, C., G. Schraa, A. J. M. Stams, and A. J. B. Zehnder. 1992. Enrichment and properties of an anaerobic mixed culture reductively dechlorinating 1,2,3-trichlorobenzene to 1,2-dichlorobenzene. *Appl. Environ. Microbiol.* 58: 1636–1644.

Jackson, A., D. Roy, and G. Breitenbeck. 1994. Transport of a bacterial suspension through a soil matrix using water an an anionic surfactant. *Water Research* 28: 943–949.

Jaffe, P. R., and S. W. Taylor. 1998. Assessment of the potential for clogging and its mitigation during in situ bioremediation. In *Biological Treatment of Hazardous Waste*, G. A. Lewandowski and L. J. DeFilippi, eds. New York: John Wiley, pp. 215–235.

Jenkins, M. B., and L. W. Lion. 1993. Mobile bacteria and transport of polynuclear aromatic hydrocarbons in porous media. *Appl. Environ. Microbiol.* 59(10): 3306–3313.

Jenneman, G. E., M. J. McInerney, and R. M. Knapp. 1985. Microbial penetration through nutrient-saturated Berea Sandstone. *Appl. Environ. Microbiol.* 50: 383–391.

Jewett, D. G., T. A. Hilbert, B. E. Logan, R. G. Arnold, and R. C. Bales. 1995. Bacterial transport in laboratory columns and filters: Influence of ionic strength and pH on collision efficiency. *Water Research* 29(7): 1673–1680.

Johnson, W. P., and B. E. Logan. 1996. Enhanced transport of bacteria in porous media by sediment-phase and aqueous-phase natural organic matter. *Water Research* 30: 923–931.

Kennedy, L. G., J. W. Everett, K. J. Ware, R. Parsons, and V. Green. 1998. Iron and sulfur mineral analysis methods for natural attenuation assessments. *Bioremediation J.* 2(3–4): 259–276.

Kinoshita, T., R. C. Bales, M. T. Yahya, and C. P. Gerba. 1993. Bacteria transport in a porous medium: Retention of *Bacillus* and *Pseudomonas* on silica surfaces. *Water Research* 28: 1295–1301.

Kjelleberg, S. 1984. Effects of interfaces on survival mechanisms of copiotrophic bacteria in low-nutrient habitats. In *Current Perspectives in Microbial Ecology*, M. J. Klug and C. A. Reddy, eds. Washington, DC: American Society for Microbiology, pp. 151–159.

Kjelleberg, S., ed. 1993. *Starvation in Bacteria*. New York: Plenum Press.

Klecka, G. M., J. W. Davis, D. R. Gray, and S. S. Madsen. 1990. Natural bioremediation of organic contaminants in groundwater: Cliffs-Dow Superfund site. *Ground Water* 28: 534–543.

Koene, J. J. A., and H. H. M. Rijnaarts. 1996. In-situ activated bioscreens: A feasibility study (in Dutch, with English summary). R 96/072, TNO-MEP, Apeldoorn, The Netherlands.

Langenhoff, A. A. M., H. C. van Liere, M. H. Harkes, C. G. J. M. Pijls, G. Schraa, and H. H. M. Rijnaarts. 1999. Combined intrinsic and stimulated in situ biodegradation of hexachlorocyclohexane (HCH). In: Phytoremediation and innovative strategies for specialized remedial applications. A. Leeson and B. C. Alleman (eds.). *In situ and on-site Bioremediation, the fifth international symposium*. Batelle Press, Columbus, OH, pp. 81–87.

Lappin-Scott, H. M., and J. W. Costerton. 1992. Ultramicrobacteria and their biotechnological applications. *Curr. Opin. Biotechnol.* 3: 283–285.

Lappin-Scott, H. M., F. Cusack, and J. W. Costerton. 1988a. Nutrient resuscitation and growth of starved cells in sandstone cores: A novel approach to enhanced oil recovery. *Appl. Environ. Microbiol.* 54(6): 1373–1382.

Lappin-Scott, H. M., F. Cusack, F. A. MacLeod, and J. W. Costeron. 1988b. Starvation and nutrient resuscitation of *Klebsiella pneumoniae* isolated from oil well waters. *J. Appl. Bacteriol.* 64: 541–549.

Lindqvist, R., and C. G. Enfield. 1992. Cell density and non-equilibrium sorption effects on bacterial dispersal in groundwater microcosms. *Microbial Ecol.* 24: 25–41.

Luthy, R. G., G. R. Aiken, M. L. Brusseau, S. D. Cunningham, P. M. Gschwend, J. J. Pignatello, M. Reinhard, S. J. Traina, W. J. Weber, Jr., and J. C. Westall. 1997. Sequestration of hydrophobic organic contaminants by geosorbents. *Environ. Sci. Technol.* 31(12): 3341–3347.

MacLeod, F. A., H. M. Lappin-Scott, and J. W. Costerton. 1988. Plugging of a model rock system by using starved bacteria. *Appl. Environ. Microbiol.* 54: 1365–1372.

Madsen, E. J., J. L. Sinclair, and W. C. Ghiorse. 1991. In situ biodegradation: Microbiological patterns in a contaminated aquifer. *Science* 252: 830–833.

Martin, R. E. 1990. Quantitative description of bacterial deposition and initial biofilm development in porous media. Ph.D. Thesis, Johns Hopkins University, Baltimore, 321p.

REFERENCES

Martin, R. E., E. J. Bouwer, and L. M. Hanna. 1992. Application of the clean-bed filtration theory to bacterial deposition in porous media. *Environ. Sci. Technol.* 26: 1053–1058.

McCarty, P. L., M. N. Goltz, G. D. Hopkins, M. E. Dolan, J. P. Allan, B. T. Kawakami, and T. J. Carrothers. 1998. Full-scale evaluation of *in situ* cometabolic degradation of trichloroethylene in groundwater through toluene injection. *Environ. Sci. Technol.* 32(1): 88–100.

McCaulou, D. R., R. C. Bales, J. F. McCarthy, and R. G. Arnold. 1994. Use of short-pulse experiments to study bacteria transport through porous media. *J. Contaminant Hydrol.* 15: 1–14.

McCaulou, D. R., R. C. Bales, and R. G. Arnold. 1995. Effect of temperature-controlled motility on transport of bacteria and microspheres through saturated sediment. *Water Res. Research* 31(2): 271–280.

McDowell-Boyer, L. M., J. R. Hunt, and N. Sitar. 1986. Particle transport through porous media. *Water Res. Research* 22(13): 1901–1921.

Mihelcic, J. R., D. R. Lueking, R. J. Mitzell, and M. Stapleton. 1993. Bioavailability of sorbed- and separate-phase chemicals. *Biodegradation* 4: 141–153.

Mills, A. L., J. S. Herman, G. M. Hornberger, and T. H. de Jesus. 1994. Effects of solution ionic strength and iron coatings on mineral grains on the sorption of bacterial cells to quartz sand. *Appl. Environ. Microbiol.* 60(9): 3300–3306.

Mohn, W. W., and J. M. Tiedje. 1991. Evidence for chemiosmotic coupling of reductive dechlorination and ATP synthesis in desulfomonile tiedjei. *Arch. Microbiol.* 157: 1–6.

Mondello, F. J., D. A. Abramowicz, and J. R. Rhea. 1998. Natural restoration of PCB-contaminated Hudson River sediments. In *Biological Treatment of Hazardous Waste*, G. A. Lewandowski and L. J. DeFilippi, eds. New York: John Wiley, pp. 303–326.

Morgan, P., M. W. Holmes, R. J. Buchanan, D. E. Ellis, C. L. Bartlett, G. J. Hanson, J. M. Odom, M. D. Lee, E. J. Lutz, M. A. Heitkamp, M. A. Harkness, K. A. DeWeerd, J. L. Spivak, J. M. Davis, G. M. Klecka, D. L. Pardieck, and M. J. Bell. 1998. Demonstration of accelerated anaerobic bioremediation for trichloroethylene (TCE) contaminated groundwater at Dover Air Force Base. *Contaminated Soil '98*. London, UK: Thomas Telford, pp. 571–574.

Mueller, R. F. 1996. Bacterial transport and colonization in low nutrient environments. *Water Research* 30: 2681–2690.

Mueller, R. F., W. G. Characklis, W. L. Jones, and J. T. Sears. 1992. Characterization of initial events in bacterial surface colonization by two *Pseudomonas* species using image analysis. *Biotechnol. Bioeng.* 39: 1161–1170.

Nales, M., B. J. Butler, and E. A. Edwards. 1998. Anaerobic benzene biodegradation: A microcosm survey. *Bioremediation J.* 2(2): 125–144.

National Research Council. 1993. *In Situ Bioremediation When Does It Work?* Washington, DC: National Academy Press. 208 p.

National Research Council. 1994. *Alternatives for Groundwater Cleanup*. Washington, DC: National Academy Press. 315 p.

Norris, R. D., R. Hinchee, R. Brown, P. L. McCarty, L. Semprini, J. Wilson, D. Kampbell, M. Reinhard, E. J. Bouwer, R. Borden, T. Vogel, M. Thomas, and H. Ward. 1994. *Handbook of Bioremediation*. Boca Raton, FL: Lewis Publishers, 257 p.

Paul, J. H., and W. H. Jeffrey. 1984. The effect of surfactants on the attachment of estuarine and marine bacteria to surfaces. *Can. J. Microbiol.* 31: 224–228.

Peyton, B. M., and W. G. Characklis. 1993. A statistical analysis of the effect of substrate utilization and shear stress on the kinetics of biofilm detachment. *Biotechnol. Bioeng.* 41: 728–735.

Plumb, R. H. 1991. The occurrence of Appendix IX organic constituents in disposal site groundwater. *Ground Water Monitoring Remediation* 11: 157–164.

Ravi, V., J. S. Chen, J. T. Wilson, J. A. Johnson, W. Gierke, and L. Murdie. 1998. Evaluation of natural attenuation of benzene and dichloroethanes at the KL landfill. *Bioremediation J.* 2(2–4): 239–258.

Reichert, P. 1994. AQUASIM—A Tool for Simulation and Data Analysis of Aquatic Systems. *Water Sci. Technol.* 30(2): 21–30.

Reynolds, P. J., P. K. Sharma, G. E. Jenneman, and M. J. McInerney. 1989. Mechanisms of microbial movement in subsurface materials. *Appl. Environ. Microbiol.* 55: 2280–2286.

Rifai, H. S. 1998. One hundred years of natural attenuation. *Bioremediation J.* 2(3–4): 217–219.

Rijnaarts, H. H. M. 1998a. Bioprocesses in treatment walls: Bioscreens. In *NATO/CCMS Pilot Study: Evaluation of demonstrated and emerging technologies for the treatment of contaminated land and groundwater, special session on treatment walls and reactive barriers*. S.S. US EPA—Environmental Management Support Inc., Maryland, USA, ed. Vienna: NATO Committee on the challenges of modern society (CCMS) and US-EPA, pp. 44–47.

Rijnaarts, H. H. M. 1998b. Natural in situ bioprocesses as a basis for cost-effective remediation: Examples of 16 contaminated sites in The Netherlands. *Groundwater Quality: Remediation and Protection*. Tübingen, Germany: IAHS, pp. 283–288.

Rijnaarts, H. H. M., A. Bachmann, J. C. Jurnelet, and A. J. B. Zehnder. 1990. Effect of desorption and intraparticle mass transfer on the aerobic biomineralization of α-hexachlorocyclohexane in a contaminated calcareous soil. *Environ. Sci. Technol.* 24(9): 1349–1354.

Rijnaarts, H. H. M., W. Norde, E. J. Bouwer, J. Lyklema, and A. J. B. Zehnder. 1993. Bacterial adhesion under static and dynamic conditions. *Appl. Environ. Microbiol.* 59(10): 3255–3265.

Rijnaarts, H. H. M., W. Norde, E. J. Bouwer, J. Lyklema, and A. J. B. Zehnder. 1995. Reversibility and mechanism of bacterial adhesion. *Colloids and Surfaces B-Biointerfaces* 4: 5–22.

Rijnaarts, H. H. M., W. Norde, E. J. Bouwer, J. Lyklema, and A. J. B. Zehnder. 1996a. Bacterial deposition in porous media related to the clean bed collision efficiency and to substratum-blocking by attached cells. *Environ. Sci. Technol.* 30(10): 2869–2876.

Rijnaarts, H. H. M., W. Norde, E. J. Bouwer, J. Lyklema, and A. J. B. Zehnder. 1996b. Bacterial deposition in porous media: Effects of cell-coating, substratum hydrophobicity, and electrolyte concentration. *Environ. Sci. Technol.* 30(10): 2877–2883.

Rijnaarts, H. H. M., W. Norde, J. Lyklema, and A. J. B. Zehnder. 1999. DLVO and steric contributions to bacterial deposition in media of different ionic strengths. *Colloids and Surfaces B-Biointerfaces* 14(1–4): 179–195.

Rijnaarts, H. H. M., M. A. van Aalst-van Leeuwen, E. van Heiningen, H. van Buijzen, A. Sinke, H. C. van Liere, M. Harkes, R. Baartmans, T. N. P. Bosma, and H. J. Doddema. 1998a. Intrinsic and enhanced bioremediation in aquifers contaminated with chlorinated and aromatic hydrocarbons in The Netherlands. *Contaminated Soil '98*. London, Edinburgh, UK: Thomas Telford, pp. 119–112.

Rijnaarts, H. H. M., J. H. De Best, H. C. van Liere, and T. N. P. Bosma. 1998b. *Intrinsic Biodegradation of Chlorinated Solvents: From Thermodynamics to Field*. TNO-MEP-R 98/130, TNO-MEP, Apeldoorn, The Netherlands.

Rittmann, B. E. 1982. The effect of shear stress on biofilm loss rate. *Biotechnol. Bioeng.* 24: 501–506.

Rittmann, B. E. 1993. The significance of biofilms in porous media. *Water Res. Research* 29(7): 2195–2202.

Sanin, S. L. 1996. Hydrodynamic and physiological effects on bacterial transport and biofilm formation in porous media. Ph.D. Thesis, Duke University, Durham, NC. 149 p.

Sarkar, A. K., G. Georgiou, and M. M. Sharma. 1994. Transport of bacteria in porous media: 1. An experimental investigation. *Biotechnol. Bioeng.* 44: 489–497.

Schäfer, A., H. Harms, and A. J. B. Zehnder. 1998a. Bacterial accumulation at the air-water interface. *Environ. Sci. Technol.* 32(23): 3704–3712.

Schäfer, A., P. Ustohal, H. Harms, F. Stauffer, T. Dracos, and A. J. B. Zehnder. 1998b. Transport of bacteria in unsaturated porous media. *J. Contaminant Hydrol.* 33(1–2): 149–169.

REFERENCES

Scheuermann, T. R., A. K. Camper, and M. A. Hamilton. 1998. Effects of substratum topography on bacterial adhesion. *J. Colloid and Interface Sci.* 208: 23–33.

Schink, B. 1988. Principles and limits of anaerobic degradation: Environmental and technological aspects. In *Biology of Anaerobic Microorganisms*, A. J. B. Zehnder, ed. New York: Wiley, pp. 771–846.

Scholl, M. A., and R. W. Harvey. 1992. Laboratory investigations on the role of sediment surface and groundwater chemistry in transport of bacteria through a contaminated sandy aquifer. *Environ. Sci. Technol.* 26: 1410–1417.

Scholl, M. A., A. L. Mills, J. S. Herman, and G. M. Hornberger. 1990. The influence of mineralogy and solution chemistry on the attachment of bacteria to representative aquifer material. *J. Contaminant Hydrol.* 6: 321–336.

Sharma, P., and M. J. McInerney. 1994. Effect of grain-size on bacterial penetration, reproduction, and metabolic activity in porous-glass bead chambers. *Appl. Environ. Microbiol.* 60(5): 1481–1486.

Sharma, M. M., Y. I. Chang, and T. F. Yen. 1985. Reversible and irreversible surface charge modifications of bacteria for facilitating transport through porous media. *Colloids and Surfaces* 16: 193–206.

Sharp, R. R., R. Gerlach, and A. B. Cunningham. 1999. Bacterial transport issues related to subsurface biobarriers. Engineered approaches for in sito bioremediation of chlorinated solvent contaminants. In *Fifth International Symposium on In Situ and On-Site Bioremediation*, A. L. Leeson, and B. C. Alleman, eds. Columbus, OH: Battelle Press, pp. 211–216.

Sinclair, J. L., S. J. Randtke, J. E. Denne, L. R. Hathaway, and W. C. Ghiorse. 1990. Survey of microbial populations in buried-valley aquifer sediments from Northeastern Kansas. *Ground Water* 28: 369–377.

Smith, M. S., G. W. Thomas, R. E. White, and D. Ritonga. 1985. Transport of *Escherichia coli* through intact and disturbed soil columns. *J. Environ. Qual.* 14: 87–91.

Speitel, G. E., and F. A. DiGiano. 1987. Biofilm shearing under dynamic conditions. *J. Environ. Eng. Division, ASCE* 113(3): 464–475.

Sturman, P. J., P. S. Stewart, A. B. Cunningham, E. J. Bouwer, and J. H. Wolfram. 1995. Engineering scale-up of in situ bioremediation processes: A review. *J. Contaminant Hydrol.* 19: 171–203.

Taylor, S. W., and P. R. Jaffe. 1990a. Biofilm growth and the related changes in the physical properties of a porous medium: 1. Experimental investigation. *Water Res. Research* 26: 2153–2159.

Taylor, S. W., and P. R. Jaffe. 1990b. Biofilm growth and the related changes in the physical properties of a porous medium: 3. Dispersivity and model verification. *Water Res. Research* 26: 2171–2180.

Taylor, S. W., and P. R. Jaffe. 1990c. Substrate and biomass transport in a porous medium. *Water Res. Research* 26: 2181–2194.

Te Stroet, C. B. M., E. R. Melger, A. A. M. Langenhoft, and H. H. M. Rijnaarts. 1999. Modelling of Bacterial Transport in Porous Media. TNO-Research Report R99/939. TNO Environment Energy and Process Innovation. The Netherlands.

Thomas, J. M., and C. H. Ward. 1989. In situ biorestoration of organic contaminants in the subsurface. *Environ. Sci. Technol.* 23(7): 760–766.

Todd, D. K. 1980. *Groundwater Hydrology*. New York: Wiley.

Trevors, J. T., J. D. van Elsas, L. S. van Overbeek, and M. E. Starodub. 1990. Transport of a genetically engineered *Pseudomonas fluorescens* strain through a soil microcosm. *Appl. Environ. Microbiol.* 56(2): 401–408.

van Beek, C. G. E. M. 1984. Restoring well yield in The Netherlands. *J. Am. Water Works Assoc.* 76(10): 66–72.

Vandevivere, P., and P. Baveye. 1992a. Effect of bacterial extracellular polymers on the saturated hydraulic conductivity of sand columns. *Appl. Environ. Microbiol.* 58: 1690–1698.

Vandevivere, P., and P. Baveye. 1992b. Saturated hydraulic conductivity reduction caused by aerobic bacteria in sand columns. *J. Soil Sci. Soc. Am.* 56: 1–13.

van Heiningen, E., A. A. M. Nipshagen, J. Griffioen, A. G. Veltkamp, A. A. M. Langenhoff, and H. H. M. Rijnaarts. 1999. Intrinsic and enhanced biodegradation of benzene in strongly reduced aquifers. In: In situ bioremediation of Petroleum hydrocarbons and other organic compounds. B. C. Alleman and A. Leeson (eds.) *In situ and on-site Bioremediation, The fifth international symposium*. Battelle Press, Columbus, OH, pp. 481–485.

van Loosdrecht, M. C. M., J. Lyklema, W. Norde, G. Schraa, and A. J. B. Zehnder. 1987a. Electrophoretic mobility and hydrophobicity as a measure to predict the initial steps of bacterial adhesion. *Appl. Environ. Microbiol.* 53(8): 1898–1901.

van Loosdrecht, M. C. M., J. Lyklema, W. Norde, G. Schraa, and A. J. B. Zehnder. 1987b. The role of bacterial cell wall hydrophobicity in adhesion. *Appl. Environ. Microbiol.* 53(8): 1893–1897.

van Loosdrecht, M. C. M., J. Lyklema, W. Norde, and A. J. B. Zehnder. 1990. Influence of interfaces on microbial activity. *Microbiol. Rev.* 54(1): 75–87.

Vogel, T. M., C. S. Criddle, and P. L. McCarty. 1987. Transformations of halogenated aliphatic compounds. *Environ. Sci. Technol.* 21: 722–736.

Wan, J., T. K. Tokunaga, and T. Chin-fu. 1995. Bacterial sedimentation through a porous medium. *Water Res. Research* 31(7): 1627–1636.

Wanner, O., A. B. Cunningham, and R. Lundman. 1995. Modeling biofilm accumulation and mass transport in a porous medium under high substrate loading. *Biotechnol. Bioeng.* 47: 703–712.

Weiss, T. H., A. L. Mills, G. M. Hornberger, and J. S. Herman. 1995. Effect of bacterial cell shape on transport of bacteria in porous media. *Environ. Sci. Technol.* 29: 1737–1740.

Williams, V., and M. Fletcher. 1996. *Pseudomonas fluorescens* adhesion and transport through porous media are affected by lipopolysaccharide composition. *Appl. Environ. Microbiol.* 62: 100–104.

Wilson, L. P., and E. J. Bouwer. 1997. Biodegradation of aromatic compounds under mixed oxygen/denitrifying conditions: A review. *J. Ind. Microbiol. Biotechnol.* 18: 116–130.

Wood, W. W., and G. G. Ehrlich. 1978. Use of baker's yeast to trace microbial movement in ground water. *Ground Water* 16: 398–403.

Zhang, W., and E. J. Bouwer. 1997. Biodegradation of benzene, toluene, and naphthalene in soil-water slurry microcosms. *Biodegradation* 8: 167–175.

Zhang, W., E. J. Bouwer, and W. P. Ball. 1998. Bioavailability of hydrophobic organic contaminants: Effects and implications of sorption-related mass transfer on bioremediation. *Ground Water Monitoring and Remediation* 18(1): 126–138.

INNOVATIVE BIOFILM TREATMENT TECHNOLOGIES FOR WATER AND WASTEWATER TREATMENT

VALENTINA LAZAROVA
JACQUES MANEM
Societé Lyonnaise des Eaux, Le Pecq, France

6.1 INTRODUCTION

Biofilms have been successfully used in water and wastewater treatment for over a century (Atkinson, 1981). The first biological process applied to the water field was a slow sand filter for drinking water treatment in London, developed by English engineers in 1829 (Imbeaux, 1935). Also in England, the first work began on wastewater filtration in the early 1860s. The biological mechanisms of sewage purification in filtration columns were demonstrated in 1865 in Germany by A. Mueller and in 1868 in England by Sir E. Frankland (Peters and Foley, 1983). The first full-scale operation of intermittent biofilters (trickling filters) for wastewater treatment was carried out in the early 1880s in Wales and several years later in the United States. At the beginning of the twentieth century, the rotating disc contactors and activated sludge processes were also developed. However, it was not until the early 1980s that the numerous advantages of biofilm reactors became a focus of interest for a considerable number of researchers and engineers, not only in the field of water technology, but also in many other areas of biotechnology.

The main factors promoting the development of new intensive biofilm technologies for water and wastewater treatment are the increasing volume of wastewater, limited space availability, and progressively tightening standards and quality control (Table 6.1). In order to satisfy these requirements, new processes have to be developed, characterized by higher

---

Biofilms II: Process Analysis and Applications, Edited by James D. Bryers.
ISBN 0-471-29656-2 Copyright © 2000 Wiley-Liss, Inc.

compactness, flexibility, and stability of operation at acceptable costs. The technological solutions proposed include improvements in both biological performances (reaction rates, biofilm control, biomass age, and population dynamics) and in reactor hydrodynamics (mass transfer, mixing).

The main advantages of such systems using biofilm processes are their ability to increase biological reaction rates through accumulation of high concentration of active biomass and the high resistance of this attached biomass to overloading and toxic compounds (Lazarova and Manem, 1994). Moreover, biofilms enable us to maintain high biomass age, which favors the selective development of specific slowly growing bacteria such as nitrifiers and reduces their washout from the system.

Significant advances in basic science, including both modeling of biofilm process hydrodynamics and kinetics (Furumai and Rittmann, 1994; Hermanowicz, 1998; Lewandowski et al., 1994) and the characterization of biofilm structure, composition, activity, and population dynamics (Bishop, 1997; Godon et al., 1997; Lazarova et al., 1994a; Lazarova and Manem, 1995; Lazarova et al., 1998a; Raskin et al., 1995; Zahid and Ganczarczyk, 1994; Zhang et al., 1994) favor the industrial development of new biofilm reactors and enhance their control and operation.

TABLE 6.1. Factors Promoting the Development of Innovative Biofilm Water and Wastewater Treatment Processes and Technological Solutions Developed to Achieve the Needed Process Intensification

Driving Forces and Needs of Water Industry	Process Characteristic	Technological Solutions	Process Intensification
Increasing wastewater volume	High capacity	High biomass concentrations	Increased volumetric rates
Space constraints Plant upgrading Low space availability Soil characteristics	Compactness	High biomass activity (biofilm control) Improved mass transfer Integrated phase separation (cell/liquid)	Improved specific conversion rates Higher oxygen consumption rates Improved effluent quality and biomass recovery
Noise and odors Stringent standards Lower concentration limits Higher requirements for sample conformity Treatment of runoff water	Covered units Stable and flexible operation	Attached biomass growth Improved mixing Multiple reactions in a single reactor	Higher resistance to toxic compounds Biofilm age and quantity independent of throughput Maintenance of low growing or specialized bacterial culture in particular nitrifying bacteria Low sensibility to higher mass and hydraulic loads

Numerous laboratory, bench scale, and industrial studies have consistently illustrated the technical advantages of biofilms. In 1989, 1993, and 1996, specialized international conferences sponsored by the International Association of Water Quality (IAWQ) were held in Nice (France), Paris (France), and Copenhagen (Denmark) respectively, focusing specifically on the advances in biofilm reactor technologies for water and wastewater treatment.

This chapter presents an overview of the state-of-the-art and advanced biofilm technologies applied to the treatment of drinking water, municipal sewage, and industrial wastewater. Emphasis is placed on the factors and techniques ensuring effective control of biofilm activity and higher operation reliability. A brief historical review of the development of the two main groups of biofilm reactors (fixed and moving beds) is presented together with information for process design, operation, and industrial applications.

6.2 PRINCIPLES AND CHARACTERISTICS OF INNOVATIVE BIOFILM PROCESSES

Biofilm processes used in water and wastewater treatment can be divided into two main groups according to the state of the support media (Fig. 6.1): (1) fixed beds and (2) mobile beds whose medium is kept continuously moving through the driving force of the liquid, air, or mechanical stirring. Depending on oxidoreduction conditions, these processes can operate as aerobic (aeration), anoxic (no oxygen but presence of nitrogen oxidized forms, NO_3—N or NO_2—N), or anaerobic processes (no air, no nitrogen oxidized forms).

The biofilm reactors are, in principle, less complex to operate compared with activated sludge systems, obviating the need for the unit processes of sludge settling and recycling which eliminates the problems of sludge bulking and rising. New designs for biofilm reactors are becoming an attractive technological solution that offer numerous features and advantages.

6.2.1 Fixed Bed Processes

The fixed beds include all processes in which the biological reaction takes place in biofilms developed on various types of static media (rocks, schist, plastic profiles, sponges, granular carrier, membranes). The transport of oxygen and nutrients is ensured by the liquid flow through the fixed medium. The main applications of fixed bed biofilm reactors are for secondary treatment (carbon removal or simultaneous carbon and nitrogen removal) and tertiary treatment (nitrification or postdenitrification) after appropriate pretreatment by screening, oil and grease removal, and primary settling or as polishing step after conventionally activated sludge systems (Fig. 6.2).

The first biofilm systems applied to wastewater treatment were trickling filters using rock medium. Due to the relatively low reaction rates, the low specific area for biofilm development, and clogging problems, plastic-medium trickling filters were developed (Duddles et al., 1974) and have been used for combined carbon removal and nitrification (Parker and Richards, 1986; Daigger et al., 1994). The main advantages of trickling filters are their low capital and operating costs. The main disadvantage, however, is that trickling filters do not allow optimal oxygen supply to the biofilm. This is the main reason for low nitrification rates and strong dependence on carbon loads. Several other disadvantages, such as the high sensitivity to shock loading and mass transfer limitations, keep their application essentially to carbon removal.

Some of these disadvantages have been eliminated with the development of rotating biological contactors (RBCs). RBCs with plastic discs were developed in the 1950s in Germany and further optimized in the United States with the Biosurf process and in Switzerland with the Biospiral process (Jerrard, 1989). RBCs are predominantly used for carbon removal, and more recently for nitrification. The nitrification stage uses "high density" media with up to 50% higher specific surface area.

In the 1970s, fixed bed reactors with granular media were developed in France, characterized by a higher surface area for biofilm attachment, better oxygen transfer, and higher nitrification rates (Pujol, 1991; Rogalla et al., 1991; Dauthuille et al., 1992). The main characteristic and advantage of this technology is its capacity to do both biological treatment and suspended solid removal simultaneously. Boller et al. (1997) demonstrated in a comparative study of different industrial processes that granular biofilters can ensure volumetric nitrification rates up to 3.5 and 8 times higher than RBC and trickling filter (plastic media), respectively. For a given degree of treatment, granular biofilters require three times less aeration volume than activated sludge units, and 20 times less than trickling filters (Smith and Hardy, 1992). Pujol et al. (1998) showed the advantage and possibility to increase the up-flow velocity to 35 m h^{-1} in order to improve mass transfer and treatment capacity.

6.2.1.1 Submerged Biofilter with Heavy Granular Media.
The principle of operation of submerged biological filters is similar to that of high-rate sand filters except for the injection of air (aerated biofilters) or external carbon source (denitrifying biofilters). Moreover, specially designed carriers were used whose characteristics are well adapted to the specific water or wastewater treatment application. Compared with the trickling filters and rotating biological contactors, submerged biofilters are considered to be advanced biofilm reactors with the following advantages and innovative elements: (1) higher specific areas for biofilm growth ensuring high biomass concentration and increased volumetric rates; (2) improved biofilm thickness control through periodic backwashing; (3) better ef-

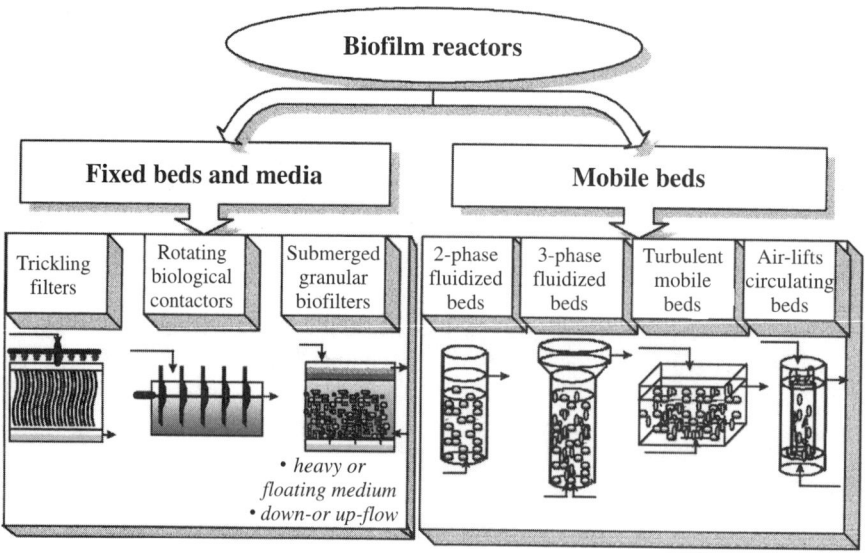

Figure 6.1 Classification of biofilm processes as a function of the state of the support medium.

6.2 PRINCIPLES AND CHARACTERISTICS OF INNOVATIVE BIOFILM PROCESSES

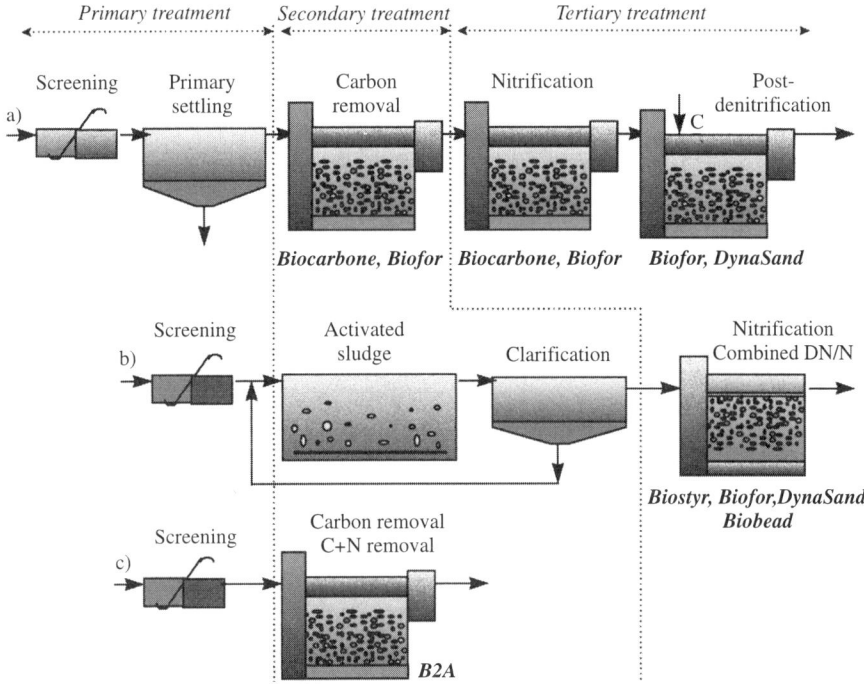

Figure 6.2 Main treatment processes and innovative commercially developed fixed bed bioreactors used for municipal wastewater treatment.

fluent quality due to the simultaneous removal of dissolved and suspended matter; (4) intensification of nitrification rates.

Biofilter processes with heavy carriers have been studied and developed using inorganic, porous filling materials, such as expanded clay with an average density of 1.5 g cm^{-3} and a diameter of 2–6 mm, and puzzolane particles with a diameter of 3–5 mm (Pujol, 1991; Rogalla et al., 1991; Villaverde et al., 1997).

Several examples of industrial processes are shown in Table 6.2. The reactors may work as down-flow or up-flow processes. The first full-scale biological aerated biofilters (BAF) with optimized granular media were developed at the end of the 1970s in France, and are known as Biofor (Degrémont company) and Biocarbone (OTV company) processes. Their compactness and high efficiency enabled the incorporation of these bioreactors into covered industrial buildings (Fig. 6.3). More details on process design and industrial applications of fixed biofilm reactors are provided in Section 6.3.

Figure 6.4 illustrates the principle of the up-flow filtration (Biofor, Pujol et al., 1994; Peladan et al. 1997). This process was developed and operated at full scale for almost all the treatment processes shown in Figure 6.2a: secondary and tertiary treatment for carbon removal, nitrification, and pre- or postdenitrification. The feed water flows from the bottom to the top in a cocurrent flow with the air or the carbon source in the case of tertiary denitrification. The carrier used is 2–5 mm expanded clay known as Biolite medium. The bed depth is usually 3 m (from 2 to 4 m depending on the application) and the contact time is in the range of 3–60 min. Filtration velocity varies depending on the application with a maximum value up to 35 m h^{-1}. Loading rates above 18 kg COD m^{-3} d^{-1} and 1.5 kg N-NH$_4$

TABLE 6.2. Design Parameters of several innovative fixed bed processes developed in industrial scale

Type	Process (Flow direction)	Country	Manufacturer	Medium[a]	Applications[b]	Volumetric Loads (kg m^{-3} d^{-1})	Hydraulic Loads (m^3 m^{-2} h^{-1})	References
Fixed Bed Biofilters with Heavy Carriers	Biofor (UF)	France	Degrémont	Expanded clay Biolite, 2–5 mm	MWW-CR, MWW-NF, MWW-postDN	5–18 (COD) 0.5–1.5 (N-NH$_4$) 5 (N-NO$_3$)	Up to 35	Pujol et al., 1994, 1998 Hamon et al., 1993
	Biocarbone (DF)	France	OTV	Expanded clay Biodamine, 2-5 mm	MWW-CR,NF IWW	5–10 (COD)	Up to 10	Rogalla et al., 1991
	B2A (UF)	France	OTV	Multi-media, 3 layers 1–40 mm	MWW-CR, NF, NF/DN	25 (COD) 0.7 (N-NH$_4$)		Bigot et al, 1998
	Nitrazur (UF)	France	Degrémont	Biolite L, 1.7–2.7 mm	DW-DN	3.1–9.5 (NO$_3$)	8.2	Richard and Thébault, 1992
	Biodenit (DF)	France		Biodagene, 3–6 mm	DW-DN	3–5 (NO$_3$)		Ravarini et al., 1988
	Sulfur/Limestone (UF)	Netherlands	OTV	2–6 mm sulfur and limestone	DW-DN		9.75	Kruithof et al., 1988
Biofilters with Continuous Washing	DynaSand (UF)	US	Parkinson CO	Sand 1.2–2 mm	MWW-postDN MWW-NF	2.7 (N-NO$_3$); 1.6 (N-NH$_4$)	5.4–24.5	Koopman et al., 1990; Sanz et al., 1996
	(UF)	France	SOGEA	Sand	MWW-postDN	3–5 (N-NO$_3$)		Larané, 1993
Fixed Beds Biofilters with Organic and Floating Media	Biostyr (UF)	France	OTV	Expanded PS, 1–2 mm	MWW-CR MWW-NF/DN	5 (COD), 0.5–1.5 (N-NH$_4$)	10	Rogalla et al., 1991, Boller et al., 1997
	Denitropur (UF)	Germany	Sulzer	Plastic elements Mellapak	DW-DN	0.5 (N-NO$_3$)		Rutten and Schnoor, 1992
	Denipor (UF/DF)	Germany	Preussag AG	Expanded PS, 2–3 mm	DW-DN	0.7–1.0 (N-NO$_3$)		Roennefahrt, 1986
	Biobead (UF)		Brightwater Eng.	PE or PP, 2–5 mm	MWW-NF	0.35 (N-NH$_4$)		Meaney and Strickland, 1994

[a] PE-polyethylene; PS-polystyrene; PP-polypropylene
[b] MWW-municipal wastewater; IWW-industrial wastewater; DW-drinking water treatment; CR-carbon removal, NF-nitrification, DN-denitrification; NF/DN-combined nitrification-denitrification

6.2 PRINCIPLES AND CHARACTERISTICS OF INNOVATIVE BIOFILM PROCESSES

$m^{-3}\,d^{-1}$ can be applied. Depending on the industrial application, the backwashing cycles vary from 12 to 48 h. More than 100 municipal and industrial wastewater treatment Biofor units are under operation with an installed total capacity of up to 1,200,000 population equivalents (p.e.) (Ninassi et al., 1998).

The granular aerated biofilter can also be operated in a contracurrent flow, such as the Biocarbone process, which has been installed in numerous countries for carbon and nitrogen removal from municipal wastewater for treatment capacities ranging from 20,000 to 200,000 p.e. (Rogalla et al., 1990, 1991; Kleiber et al., 1994). The main processes carried out in the Biocarbone BAF are carbon removal, simultaneous carbon and nitrogen removal, and nitrification (Fig. 6.2a).

Generally, the BAF are installed after the conventional or lamellar primary settler to minimize bed clogging. More recently, a new generation of multimedia up-flow B2A filters (Fig. 6.5) have been developed in France (Bigot et al., 1998). The main advantage of the process is the possibility of treating raw nonsettled sewage (see Fig. 6.2c). To achieve high filtration efficiency, the B2A process consists of a 3 to 4 m deep bed with three media layers: a fine media layer with particle size from 1 to 4 mm and average and coarse media layers (4–40 mm). The backwashing is performed in two steps by a gravity flush to remove the particles from the coarse media, followed by the use of air and water to remove the excess biomass. This biofilm process is designed to achieve secondary and tertiary treatment for carbon removal or simultaneous carbon and nitrogen removal, as well as simultaneous nitrification-denitrification in the same reactor. The maximum reported loading rates of the B2A process are up to 0.7 kgTKN $m^{-3}\,d^{-1}$ and 25 kgCOD $m^{-3}\,d^{-1}$.

The main advantage of aerated biofilters is that they can operate simultaneously as biological reactors and filters, which means that secondary settling is not required. Specific energy consumption is similar to that of activated sludge tanks (approximately 1.0–1.4 kWh kg^{-1} COD removed), but the loading rates are much higher (Pujol, 1991). These reactors are competitive especially when land is limited and/or tertiary treatment is required (Pujol, 1991; Rogalla et al., 1991; Dauthuille et al., 1992; Pujol et al., 1992). The main constraint for operation is the frequent biofilter backwashing (one each 24 or 48 h) due to rapid clog-

Figure 6.3 View of the covered wastewater treatment plant in Clos de Hilde, France, with the compact Biofor processes for nitrogen removal (capacity 150,000 p.e.).

ging of the bed. However, it is important to stress that all backwashing procedures are fully automatic.

New continuous backwashing reactors are being developed in France and the United States (Hanus and Bernard, 1988; Koopman et al., 1990; Larané, 1993). The principle of operation of these systems is illustrated in Figure 6.6 by the DynaSand process: The up-flow filtration is carried out through a bed of sand (1.2–2 mm particle size). The sand bed is drawn slowly downward into the suction of an air-lift pump, cleaned by the turbulence, and returned to the top of the bed. The tested bed heights were 3.5–4.8 m and the hydraulic loads varied from 5.4 to 24.5 m h^{-1} (Hultman et al., 1994). Up to 1.6 kg N-NH$_4$ m^{-3} d^{-1} nitrification rates were measured at 12–14°C and filter velocity of 14.3 m h^{-1} (Sanz et al., 1996). For tertiary denitrification with methanol, volumetric removal rates up to 2.7 kg N-NO$_3$ m^{-3} d^{-1} were reached (Koopman et al., 1990).

6.2.1.2 Submerged Biofilters with Light and Floating Media.
Floating beds are the most recent application of submerged biofilters. The main innovative element is the use of light granular particles enabling the reduction of the specific energy input needed for backwashing.

The floating carriers used have a specific density from 0.03 g cm^{-3} (Rogalla et al., 1991; Rols and Capdeville, 1992; Vital et al., 1990) to 0.9 g cm^{-3} (Meyer, 1993). These processes already have several industrial applications (see Table 6.2): Biostyr, OTV (Rogalla et al., 1991); Denipor, Preussag AG (Roennefahrt, 1986; Meyer, 1993); Biobead (Meaney and Strickland, 1994).

The up-flow floating bed Biostyr (Fig. 6.7) was designed both for tertiary nitrification-denitrification in the same unit or for tertiary nitrification after secondary treatment for carbon removal (see Fig. 6.2b). The filter is filled with very light 1–2 mm diameter expanded polystyrene beads with a specific surface area of about 1200 m^2/m^{-3}. In the first configuration, air is injected into the filter media separating an unaerated zone for prefiltration and anoxic denitrification from the aerated upper zone for nitrification. The floating media facilitate the backwashing through countercurrent flushing. Nitrogen removal between 70% and 80% can be achieved with a detention time of 2–3 h and loading rates below 1 kg

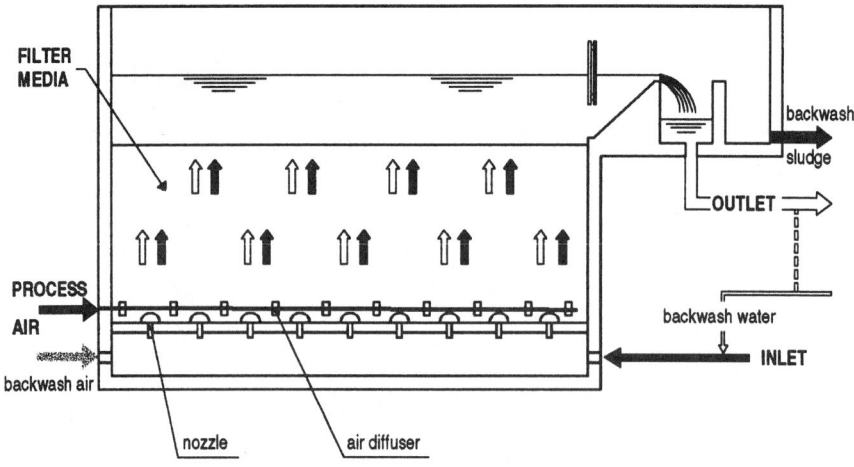

Figure 6.4 Biofor process diagram, Degrémont (France).

6.2 PRINCIPLES AND CHARACTERISTICS OF INNOVATIVE BIOFILM PROCESSES

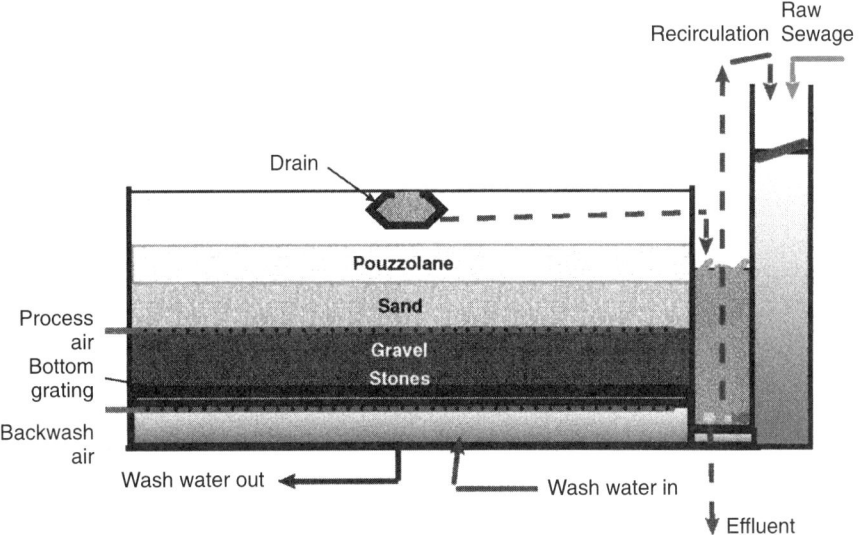

Figure 6.5 Principle of the multimedia B2A process, OTV (France).

N-NH$_4$ m^{-3} d^{-1} and 5 kg COD m^{-3} d^{-1}. The maximum reported nitrification rates in conditions of tertiary treatment are up to 1 kg N-NH$_4$ m^{-3} d^{-1} (Rogalla et al., 1992; Le Tallec et al., 1997) and 1.48 kg N-NH$_4$ m^{-3} d^{-1} (Boller et al., 1997). Toettrup et al. (1994) observed similar kinetic parameters in laboratory-scale and industrial prototypes.

Another type of up-flow floating bed known as Biobead has been developed in the United Kingdom (Meaney and Strickland, 1994). The carrier used is a plastic (polyethylene or polypropylene) granular medium with a density close to that of water (0.9–0.95 g cm^{-3}) and a size between 2 and 5 mm. During the operation, a zone at the bottom of the reactor is free of media. This volume above the bed-retaining mesh provides the backwash water for a daily bed cleaning. This process has been applied for tertiary nitrification and suspended solids removal at full scale since 1991. The nitrification rate reported is 0.35 kg N-NH$_4$ m^{-3} d^{-1}. Denitrification using settled sewage as a carbon source has also been investigated.

The biological performance of fixed bed processes on granular media with densities higher or lower than water are similar. However, higher hydraulic loads have been achieved in up-flow biofilters with a heavy carrier. Submerged granular biofilters offer several advantages over conventional processes such as compactness, ease of operation, and ability to treat higher applied loads. The main design-related needs are a homogenous flow and air distribution and efficient backwashing.

6.2.2 Moving Beds

This group of biofilm reactors covers all biofilm processes with continuously moving media (stirred, expanded, fluidized, turbulent, and circulating) maintained by high air and/or water velocity or by mechanical stirring (see Fig. 6.1). The main innovative element is the continuous motion of the biofilm-covered particles. This mode of operation has the following advantages and beneficial effects: improves mass transfer, reduces biofilm diffusion limitations, accelerates biochemical reactions by increasing the biofilm/liquid transfer area, and uses very fine granular (0.2–2.0 mm) or very porous (plastic foams or sponges) materials to increase the specific surface area available for bacterial growth. The autocontrol of the biofilm thickness is one of the major characteristics of this group of biofilm reactors, allowing them to maintain thin and active biofilms (Fig. 6.8). Moreover, no clogging problems were observed. These advantages explain the better performances of mobile bed systems (Table 6.3) compared with some fixed biofilm reactors such as high carbon and nitrogen removal rates through large amounts of fixed biomass and with low hydraulic retention times (Cooper, 1981; Audic et al., 1985; Elmaleh et al., 1987; Heijnen et al., 1989; Sutton and Mishra, 1990).

Compared with fixed bed reactors, moving bed processes did not require intensive pretreatment or primary settling allowing raw sewage feeding (Fig. 6.9a). Some of the innovative moving bed reactors have been successfully used for retrofitting and upgrading of existing activated sludge plants (Fig. 6.9b).

6.2.2.1 Two-Phase Liquid–Solid Fluidized Beds. Two-phase liquid-solid fluidized beds have been reported to be the most efficient biological treatment systems, as they have the highest volumetric rates for anaerobic carbon removal (70 kg COD m^{-3} d^{-1}, Holst et al., 1997), nitrification (5 kg N-NH$_4$ m^{-3} d^{-1}, Deboosere et al., 1984), and nitrate removal (10 kg N-NO$_3$ m^{-3} d^{-1}, Cooper and Wheeldon, 1982). This high efficiency is due to the large biomass concentration of up to 30–90 kg VSS m^{-3}, with average values of 20–40 kg

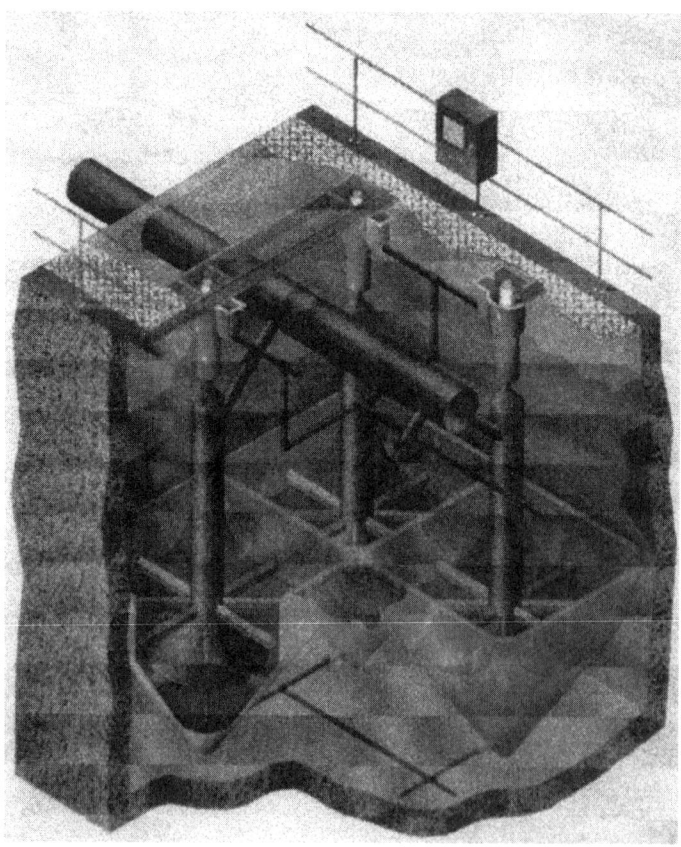

Figure 6.6 Cutaway view of the DynaSand filter, Parkson Corporation (United States).

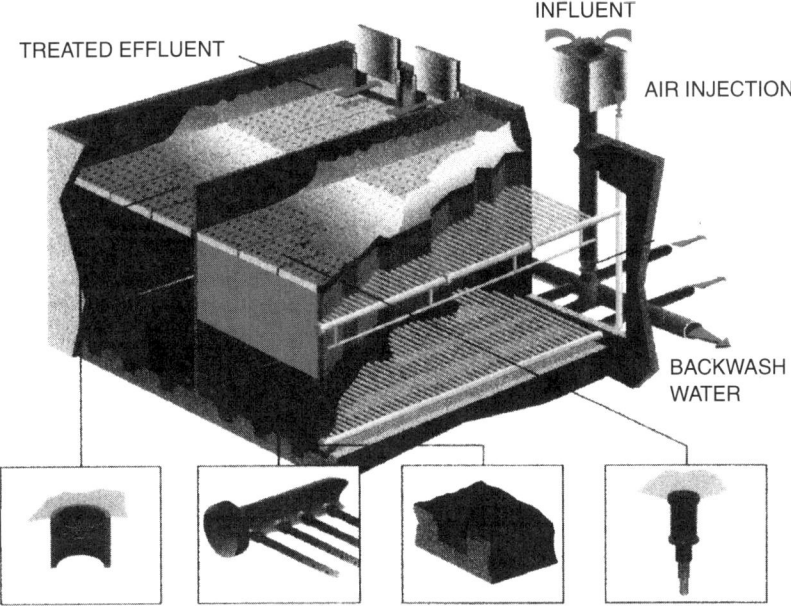

Figure 6.7 Principle of the Biostyr process, OTV (France).

VSS m^{-3}, and good mass transfer characteristics. The application, design, and development of fluidized bed technologies have been reviewed by Cooper (1981), Mishra and Sutton (1991), and Heijnen et al. (1989). Europe has been the leader in applying advanced anaerobic technologies with far more installations than in the USA (McCarty and Smith, 1986).

Fluidized bed reactors (FBR) differ from other mobile beds due to the homogeneous expansion of the medium, maintained by the upward liquid velocity created by the feed and recirculation flow rates. The expanded bed particles provide a large surface area for biofilm growth, ensuring very high biomass concentrations compared with activated sludge systems. The high settling particle velocity (e.g., sand) enables the application of high liquid velocities from 10 to 30 m h^{-1}. A three-phase separator, located at the top of the reactor, ensures the separation of the gas, effluent, and biofilm-covered solids. When the biofilm becomes very thick, the particle density decreases, enabling the extraction of these solids and the biofilm removal in a separate mechanical device.

Numerous hydraulic models, based on the principles of solid-liquid fluidization, have been developed to predict system dynamic and the theoretical value of the FBR biomass (Hermanowicz and Ganczarczyk, 1984; Rittmann et al., 1992; Shieh et al., 1981). The process design parameters are defined by the combination of a hydrodynamic model and a model for biofilm kinetics. Although knowledge of the kinetic and hydrodynamic parameters is critical, proper design of mechanical devices is of major importance for the operation and performances of fluidized bed reactors helping to control liquid distribution, oxygen transfer, bed height, and biofilm thickness (Holst et al., 1997; Mishra and Sutton, 1991). Finally, in order to guarantee process stability and avoid clogging of the distribution system, especially in the case of industrial wastewater, an efficient pretreatment is necessary to remove grease and suspended solids up to < 500 mg SS L^{-1}.

Although the history of this technology for water and wastewater treatment can be traced back to the 1940s in the United Kingdom, real development of fluidized biofilm

TABLE 6.3 Mobile Bed Bioreactors: Summary of Some Design Parameters Determined on the Basis of Recent Pilot Plant and Full-Scale Studies

Process and Type of Reactor

Type	Process	Country	Scale	Manufacturer	Medium[a]	Applications[b]	Volumetric loads[c] (kg m^{-3} d^{-1})	Biomass[d] (g SS L^{-1})	References
2-Phase Aerobic Fluidized Beds	Oxitron system	US, UK	Full scale	Dorr-Oliver/Ecolotrol	Sand/GAC	MWW-NF	0.6–1.3 (N-NH$_4$)	5.5–22.0	Cooper, 1981; Owsley et al., 1989
	Rex process	US	Full scale	Envirex/Ecolotrol	Sand, 0.4–0.6 mm	MWW-NF IWW-NF	1 (N-NH$_4$)	2.0–12.0	Sutton and Mishra, 1990, 1994
2-Phase Anerobic or Anoxic Fluidized Beds	Anitron system	US	Full scale	Dorr-Oliver/Ecolotrol	Sand/GAC	MWW-DN, IWW-CR	5–10 (N-NO$_3$)		Cooper, 1981; Sutton and Mishra, 1990
	Rex process	US	Full scale	Envirex/Ecolotrol	Sand, 0.4–0.6 mm	MWW-DN, IWW-CR, DN	9 (N-NO$_3$)	11.0–22.0	Cooper, 1981; Owsley et al., 1989
	Anaflux reactor	France	Full scale	Degrémont	Biolite (expanded clay)	IWW-CR	70 (COD)		Holst et al., 1997
	Biobed	Netherlands	Full scale	Gist-Brocades	Sand 0.1–0.3 mm	IWW-CR	30 (COD)	20.0	Heijnen et al., 1989
3-Phase Fluidized Beds	Biolift	France	Prototype	OTV	Sand 0.2–0.6 mm	MWW-NF	1 (N-NH$_4$)		Badot et al., 1994
		Japan	Pilot		Anthracite 0.6 mm	MWW-NF			Hosaka et al., 1991
		Spain	Pilot		Pumice 0.35–0.43 mm	MWW-NF	<0.2 (N-NH$_4$)		Fdez-Polanco et al., 1994
		France	Lab scale		Clay, plastics	MWW-CR	24 (COD)	2.0–3.0	Elmaleh et al., 1987, 1992
		France	Lab scale		Plastics OSBG 2.7 mm	MWW, IWW	6.4–15.5 (COD)	0.1–0.5	Legile et al., 1988
Inverse Fluidized Beds		US	Lab scale		PE 2–5 mm	MWW-DN			Turner et al., 1990
		Bulgaria	Prototype		Expanded PS	MWW-CR	12–35 (COD)	5.0–20.0	Nikolov et al., 1990
Moving and Turbulent Beds	Linpor	Germany	Full scale	Linde AG	PU cubes	MWW-CR/NF IWW-CR	2–4(BOD)/0.36(N-NH$_4$) 5–10 (BOD)	7.0–16.0 15.0	Reimann, 1984; Morper, 1994
	Captor	UK, US	Full scale	Simon-Hart, Ltd	PU cubes	MWW-CR/NF	1.4–4.0(BOD); 0.6(N-NH$_4$)	6.0–12.0	Cooper et al., 1984; Banerji and Liu, 1993
	MBBR	Norway	Full scale	Kaldness Miljoteknology	Patented, PE, 7–10 mm	IWW-CR, MWW-NF,DN	15–22 (COD); 0.45 (N-NH$_4$)	3.0–4.0	Rusten, 1997
	Pegasus/Pegasur	Japan/France	Full scale	Hitachi/Degrémont	PEG pellets	MWW-NF	0.2–1 (N-NH$_4$)		Emori et al., 1998 Jackson et al., 1998
	Circox/BAS	Netherlands	Full scale	Pacques/Gist-Brocades	Basalt, 0.2 mm	MWW-CR/NF IWW-CR	0.2–0.8 (N-NH$_4$) 3–5 (COD); 1.2 (N-NH$_4$)	15.0–40.0	Heijnen et al., 1992; Tijuis et al., 1992; Benthum 1997a
Air-lifts and Circulating Beds	Turbo N	France	Full scale	Degrémont	PE granules, 1–3 mm	MWW-NF	20 (COD) 0.6–1.2 (N-NH$_4$)	5.0–7.0	Lazarova and Manem, 1996
	Mixazur	France	Full scale	Degrémont	Clay, 0.1–0.3 mm	MWW-DN	0.6 (N-NO$_3$)		Chudoba et al., 1998
	Air-lift	Germany	Lab scale		Sand, 0.2 mm	MWW-NF	4.2 (N-NH$_4$)		Bennemann et al., 1991

[a]GAC-granulated active carbon; OSBG-optimized support of biological growth; PE-polyethylene; PS-polystyrene; PU-polyurethane; PEG-polyethylene glycol
[b]MWW-municipal wastewater; IWW-industrial wastewater; DW-drinking water treatment; CR-carbon removal; NF-nitrification; DN-denitrification
[c]Loads per m^3 of total reactor volume
[d]SS-suspended solids

6.2 PRINCIPLES AND CHARACTERISTICS OF INNOVATIVE BIOFILM PROCESSES

reactors did not occur until the early 1970s (Mishra and Sutton, 1991). One of the first fluidized bed patents credited for water treatment was granted in 1974 in the United States to Ecolotrol Inc. (Cooper, 1981; Sutton and Mishra, 1990). The first full-scale application for denitrification of municipal wastewater treatment occurred in Pensacola, Florida, in the mid-1970s (Mishra and Sutton, 1991). A couple of years later, the first fluidized bed reactor was set up for the anaerobic treatment of industrial wastewater from a soft drink bottling plant in Alabama. In Europe, the first industrial FBRs were built in 1984 in Delft, the Netherlands, and in 1985 in Prouvy, France, by Gist-Brocades Company (Heijnen et al., 1989). In 1986 the French company Degrémont constructed the first Anaflux reactor in France for the anaerobic treatment of brewery wastewater (Holst et al., 1997).

There are currently more than 80 two-phase industrial fluidized bed plants that have been built in the United States and Europe (Sutton and Mishra, 1994). Table 6.3 summarizes the characteristics of some of the full-scale FBR constructed by different companies. Most of the existing industrial fluidized bed reactors are two-phase liquid-solid systems, applied for anaerobic, anoxic (denitrification), and aerobic treatment. The common approach to satisfy the oxygen requirements in a two-phase aerobic system is the use of pure oxygen dissolved in the influent flow (Fig. 6.10, Mishra and Sutton, 1991).

In aerobic treatment, one-stage configuration has usually been applied, while in anaerobic treatment two-stage treatment is recommended. The main advantages of two-step anaerobic treatment are the higher specific methonogenic activity (up to 3 times), better process stability, easier pH control, less alkalinity production, and easier biofilm thickness control (Heijnen et al., 1989).

The conventional support is inert media: 0.1–0.6 mm sand (Heijnen et al., 1989; Mishra and Sutton, 1991) or expanded clay (< 0.5 mm, Holst et al., 1997). Smaller particles favor biofilm growth because of the lower shear stress and high specific surface area. Under well-controlled operating conditions, this media allows rapid bacterial adhesion and provides a large specific area for the development of an active biofilm after a relatively short period of seeding and startup. Recently, inert media has started to be displaced by active supports: GAC or granulated sludge. The use of activated carbon (GAC) provides the synergy effect of both biotreatment and adsorption and is an attractive process for the elimination of complex

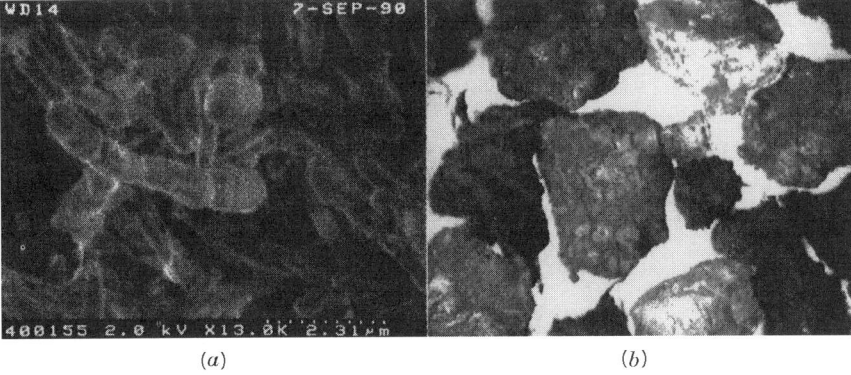

(a) (b)

Figure 6.8 View of thin and active biofilms developed in mobile bed reactors: *a*) scanning electron microscopy view of a denitrifying biofilm from a fluidized bed reactor; *b*) binocular microscopy view of the active biomass after 2-(p-iodophenyl)-3-(*p*-nitrophenyl)-5-phenyltetrazolium chloride (INT) staining of a nitrifying biofilm from a circulating floating bed reactor. Figure also appears in Color Figure section.

organics from municipal and industrial wastewater (Suidan et al., 1983; Sutton and Mishra, 1994; Wang et al., 1986; Chang and Rittmann, 1987; Fox et al., 1990; Nakhla et al., 1990).

The main application of the fluidized bed technology in Europe and North America is industrial wastewater treatment, with more than 65% and 90% of the existing industrial units respectively (Mishra and Sutton, 1991). Several full-scale plants are operated for drinking water denitrification (Hall et al., 1985) and for postdenitrification of municipal wastewater (Mishra and Sutton, 1991; Cooper and Williams, 1990; MacDonald, 1990).

Two-phase fluidized beds are recommended as an attractive, competitive process in the following treatments: (1) aerobic or anoxic treatment of industrial water with low levels of organic pollutants (Sutton and Mishra, 1990), (2) tertiary treatment with nitrification (Cooper and Williams, 1990), and (3) methanization of food industry effluents (Ehlinger, 1992).

The main advantages of fluidized beds are their high treatment capacity, lack of clogging problems, less sensibility to shock overloading, and high compactness. Despite these advantages, the FBR design, its potential to scale-up and its industrial development have been limited by numerous problems and technical constraints, the biggest obstacles being the control of bed expansion and biofilm thickness, influent distribution devices, and oxygen saturation systems (Cooper, 1981; Sutton and Mishra, 1990). Most of these technical constraints have been overcome in the more recent designs of anaerobic reactors (Heijnen et al., 1989; Holst et al., 1997).

6.2.2.2 Three-Phase Gas-Liquid-Solid Fluidized Beds.
Three-phase fluidized bed reactors can be divided into process groups according to the medium used and the driving force applied for fluidization (Fig. 6.11). Depending on the carrier density, two types of

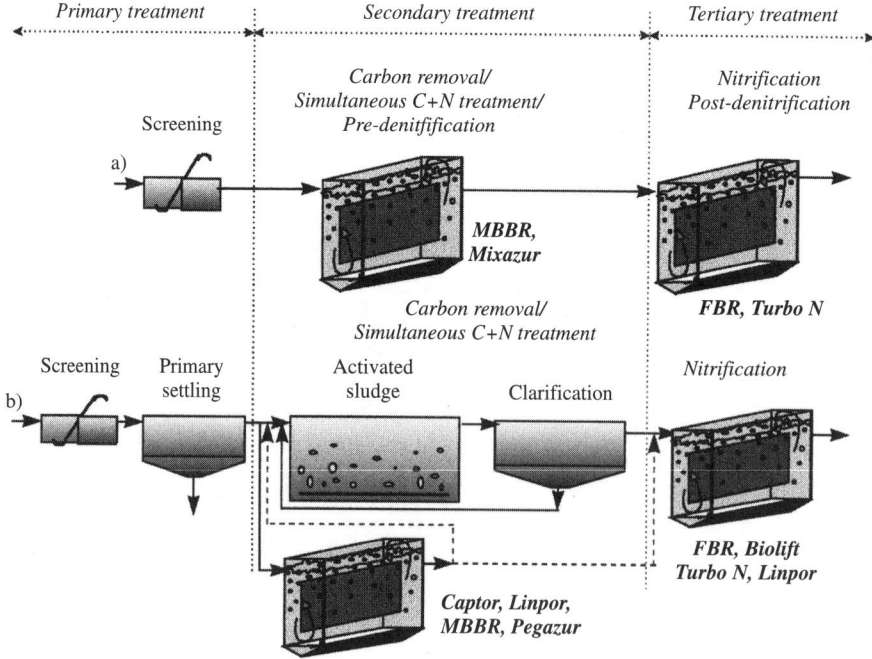

Figure 6.9 Place and role of innovative mobile bed processes in the municipal wastewater treatment schemes.

6.2 PRINCIPLES AND CHARACTERISTICS OF INNOVATIVE BIOFILM PROCESSES

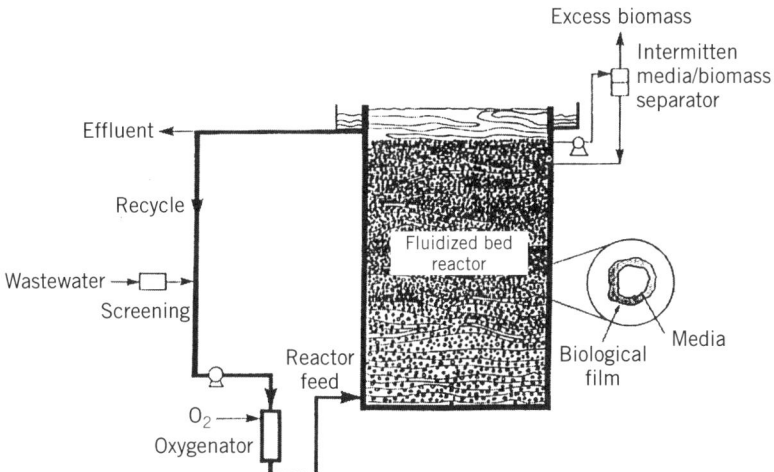

Figure 6.10 Principle of the Oxytron fluidized bed reactor.

fluidized bed technologies can be defined: (1) up-flow fluidized beds (conventional FBR) with heavy support media and (2) inverse FBR using a floating medium. The driving force used for bed expansion enables the distinction of another two groups of reactors: (1) column type (conventional FBR), where the bed expansion is ensured by the liquid velocity and (2) draft-tube fluidized bed bioreactors, where the air injection induces the liquid circulation and the bed expansion.

Three-phase fluidized beds partly solve the problems of oxygen saturation in the conventional FBR by using simultaneous gas and liquid injection. This improves liquid/solid mass transfer and induces high shear stress that controls biofilm thickness (Trinet et al., 1991; Lazarova and Manem, 1994). However, little information is available concerning their operation and industrial applications (see Table 6.3). Most of the results are from laboratory studies of up-flow high density granular medium beds (Elmaleh et al., 1987, 1992; Koch et al., 1991; Trinet et al., 1991; Puhakka and Järvinen, 1992; Rittmann et al., 1992; Zheng et al., 1995) and pilot scale studies with mineral (Hosaka et al., 1991; Badot et al., 1994; Fdez-Polanco et al., 1994) and GAC media (Ouyang and Liaw, 1994). The major advantage is that these processes do not require equipment to control biofilm thickness. However, they still present problems in terms of controlling bed expansion and homogeneous flow distribution.

Several models have been developed to predict the bed dynamics as a function of particle density, biofilm thickness, and gas and liquid velocities (Chang and Rittmann, 1994; Chen et al., 1995; Zheng et al., 1995; Yu and Rittmann, 1997).

According to Elmaleh et al. (1992), energy consumption in three-phase FBR reactors is around 1 kWh kg^{-1} COD removed, similar to that in activated sludge processes and lower than in two-phase fluidized bed reactors with preoxygenation. The main drawback of these processes is the very great technical difficulty in simultaneously producing fluidization and air injection (coalescence of air bubbles, bed turbulence, etc.). Consequently, there has been no successful scale-up of the three-phase FBR.

Fan et al. (1984) and Tang et al. (1987) developed a draft-tube fluidized bed reactor by replacing the driving force created by the liquid distribution rate with the driving force of an air-lift type gas injection. The biological and hydraulic performance of this reactor has thus been improved.

Down-flow or inverse fluidized beds form another group of reactors and have been developed to produce a more reliable three-phase process by using a low-density granular medium (floating medium). This medium can be fluidized using two principles: the driving force of the liquid (column type reactor) or the driving force of the gas (draft-tube reactor). The main technical constraint is that homogeneous distribution of liquid flow and effluent recirculation are required. The rate of expansion of the floating bed is difficult to control and there is a high risk of bed clogging (Turner et al., 1990).

Addition of a draft-tube enables all the advantages of fluidized beds to be achieved without complex technical equipment (Nikolov and Karamanev, 1990; Nikolov et al., 1990). Internal liquid circulation, carried out using the air-lift principle, expands the floating bed to the bottom of the reactor. During development of the biofilm, bioparticle density increases and bioparticles are led into the internal column (three-phase part) where the overgrowth biofilm is stripped by shear stress. Inverse fluidized beds are characterized by high carbon and nitrogen removal rate, 10–17 times higher than for conventional processes (Karamanev and Nikolov, 1988; Nikolov et al., 1990). However, no industrial development has been reported due to the difficulties in controlling and maintaining homogenous bed expansion.

6.2.2.3 Turbulent and Moving Bed Reactors.
Turbulent or moving beds are defined as column type reactors where the biofilm-covered particles are continuously moving under the driving force of a moderate or high gas velocity and mixers. To reduce the energy consumption required to create and maintain high bed turbulence, the medium materials are chosen to have a low density close to that of water. High shear stress keeps the biofilm very thin and active (Rusten et al., 1992; Lazarova and Manem, 1994). Consequently, these reactors combine biofilm-suspended culture systems where 25–35% of the total biomass is suspended. The main advantages of these processes are good oxygen transfer, autoregulation of biofilm thickness, and simpler distribution of liquid flow that enables raw, unsettled wastewater to be treated directly. The possibility of easily upgrading existing activated sludge tanks to treat higher organic loads or to enhance nitrogen removal is another significant advantage of these systems.

Three types of turbulent beds were developed to full scale: the Captor sponge biofilm process developed in the United Kingdom (Cooper et al., 1984) and studied in the United States by the U.S. Environmental Protection Agency (EPA) (Heidman et al., 1988), the Linpor sponge biofilm process developed by Linde AG in Germany (Reimann, 1984; Morper, 1994), and the more recent moving bed reactor that uses a plastic carrier developed

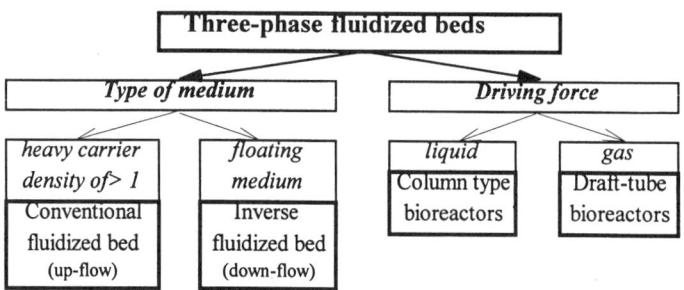

Figure 6.11 Classification of three-phase fluidized bed processes.

in Norway by Kaldness Miljøteknology (KMT) company (Ødegaard and Rusten, 1990; Rusten et al., 1997).

The use of porous media (polyurethane or polyether foam) in the Captor and Linpor processes with both attached and suspended growth provides a large specific surface area for bacterial retention (1000–5000 $m^2\ m^{-3}$). Typical biomass concentrations of the mixed liquor vary from 7 to 12 kg m^{-3}. The filling rate of porous pads is in the range of 10–40%. These porous sponge particles require continuous biomass stripping in specially designed rollers. The main problem with these systems is the need to keep the porous cubes circulating and to prevent their floatation (Cooper et al., 1984; Reimann, 1984). A packed bed Captor process was studied by Banerji and Liu (1993) to compare the efficiency of different Captor sponge media and other carriers. At loading rates up to 0.6 $kgN\text{-}NH_4\ m^{-3}\ d^{-1}$, the open cell Captor media performed better than gravel media, plastic trickling filter media, and two other types of Captor media. The main industrial use reported for the Captor process is for upgrading of existing wastewater treatment plants (WWTP) for enhanced carbon and nitrogen removal (see Fig. 6.9b).

The Kaldness moving bed biofilm reactor (MBBR) was developed and tested at laboratory-, pilot-, and full-scale over the last 5 years (Ødegaard and Rusten, 1990; Rusten et al., 1992; Rusten et al., 1997; Pastorelli et al., 1997). Currently, there are more than 75 installations worldwide sized from small community or multihome units to large municipal plants (Neu and Rusten, 1998). The MBBR is filled with small carrier elements about 10 mm in diameter, made of polyethylene with a density of 0.92–0.96 g cm^{-3} (Rusten et al., 1997). The maximum filling of about 70% corresponds to a specific area of 350 $m^2\ m^{-3}$. The typical biomass concentration reported is 3–4 kgSS m^{-3}. According to Ødegaard et al. (1998), the normal and maximum design loading rates for carbon removal are 5 and 8 $kgBOD_7\ m^{-3}\ d^{-1}$ at 67% standard fill. These values correspond to surface loads of 15 and 25 g $BOD_7\ m^{-2}\ d^{-1}$, which is in the same range as in RBC processes. In conditions of simultaneous C and N removal, significant nitrification started at organic volumetric loads below 2–2.2 kg BOD $m^{-3}\ d^{-1}$. The maximum design values for nitrification are 0.15 and 0.35 kg $N\text{-}NH_4\ m^{-3}\ d^{-1}$ (0.45 and 1 g $N\text{-}NH_4\ m^{-2}\ d^{-1}$) for residual ammonia values below and over 3 mg L^{-1} respectively (Ødegaard et al., 1998). The MBBR process has also been applied for denitrification with design values for predenitrification and postdenitrification of 0.3 and 0.7 $kgN\text{-}NO_3\ m^{-3}\ d^{-1}$ respectively (0.9 and 2 $gN\text{-}NO_3\ m^{-2}\ d^{-1}$). In this case, the agitation of the media is ensured by mechanical stirring. Greater flexibility of the system makes it possible to switch between different operating modes. Three different process combinations have been tested and documented, including pre- and postdenitrification.

6.2.2.4 Air-Lifts and Circulating Bed Reactors.
Air-lift reactors are used in different fields of biotechnology for production of beer, vinegar, citric acid, yeasts, and other bioproducts (Siegel and Robinson, 1992). Particular advantages of air-lifts compared with other types of bioreactors are the very good mixing capacities and enhanced mass transfer. Biofilm air-lifts have recently been developed to treat industrial and urban wastewater and for nitrification (see Table 6.3).

The excellent capacity and efficiency of air-lift bioreactors are due to the good hydrodynamic characteristics of the reactors and the high biomass concentrations of up to 40 kg m^{-3} (Bennemann et al., 1991; Tijhuis et al., 1992). The media used in the first air-lift processes were very fine granular minerals: sand and basalt with a size of 0.2–0.4 mm (Bennemann et al., 1991; Heijnen et al., 1991). Due to the large biofilm surface area of up to 2000–3000 $m^2\ m^{-3}$ (reactor volume), high volumetric conversion rates can be achieved.

Recently, draft tube air-lift reactors with floating organic media were successfully operated at laboratory and pilot scale (Lazarova and Manem, 1996; Tsubone et al., 1992).

Three groups of industrial scale air-lift reactors have been developed: (1) the tall reactors with suspended activated sludge developed by ICI in 1975 in the United Kingdom known as Deep-Shaft process (Fields, 1972) and Tower Biology (Bayer AG) developed in Germany; (2) the Biofilm Air-lift Suspension Reactor, BAS, developed by Gist-Brocades in the Netherlands in the early 1990s and commercialized by Paques B.V. as the Circox process using fine mineral media (Heijnen et al., 1991; Frijters et al., 1997); and (3) the TURBO N circulated bed reactor developed by Degrémont, France, using floating granular media (Lazarova and Manem, 1996).

The first group of air-lift reactors were operated predominantly as activated sludge processes, with only a few examples of the use of biofilms formed on powdered activated carbon or organic carriers (Leistner et al., 1997; Diesterweg et al., 1980; Pascik and Mann, 1984; Austrup, 1985; Fields, 1992; Fouhy, 1992).

The Biofilm Air-lift Suspension Reactor, BAS (Fig. 6.12) has been developed on the principle of internal air-lift reactors with a draft-tube (Heijnen et al., 1991). The fine basalt carrier (0.1–0.3 mm, average density of 3.3 g cm^{-3}) ensures a high specific biofilm area (up to 4000 m^2 m^{-3}). Its separation from the effluent is performed by means of a special device, the phase separator, situated in the top of the reactor. Numerous laboratory-scale studies have been carried out to quantify the biofilm performances of formation and detachment (Benthum et al., 1997a; Heijnen et al., 1992; Tijhuis et al., 1996) demonstrating also the possibility of achieving high nitrification rates of up to 5–6 kg N-NH$_4$ m^{-3} d^{-1} (Garrido et al., 1997; Tijhuis et al., 1992).

The principle of the circulating bed TURBO N is based on internal air-lift circulation (Lazarova and Manem, 1996; Lazarova et al., 1997, 1998b,c). The industrial scale reactor has a rectangular configuration with two compartments (Fig. 6.13). The carrier used is a floating plastic granular medium with a density of 0.9 g cm^{-3} and a size of 1–3 mm. The injection of air into one compartment induces a homogenous circulation of the liq-

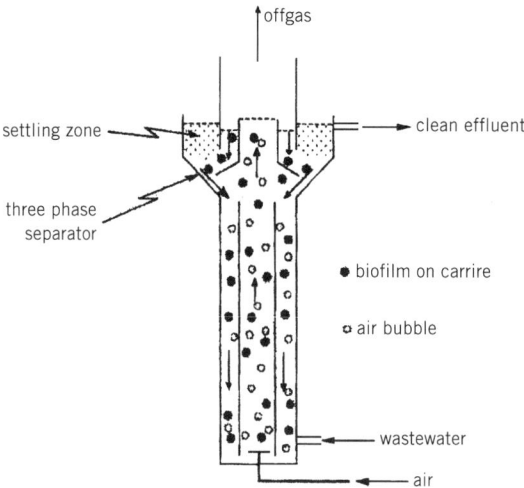

Figure 6.12 Principle of the air-lift suspension reactor (BAS).

6.2 PRINCIPLES AND CHARACTERISTICS OF INNOVATIVE BIOFILM PROCESSES

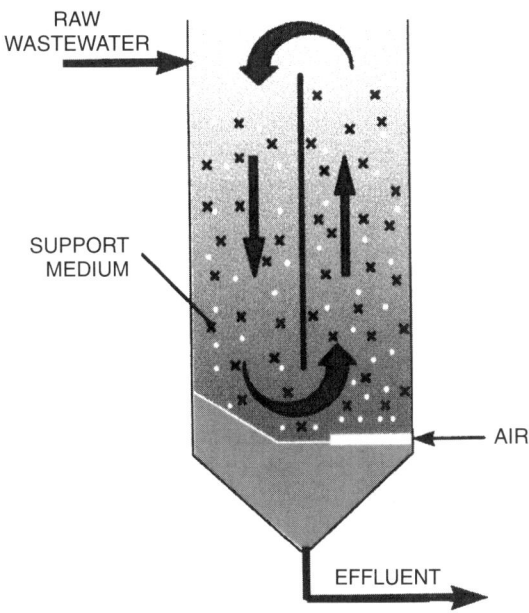

Figure 6.13 Principle of the circulating floating bed TURBO N reactor, Degrémont (France).

uid, solid, and gas bubbles with a flow velocity of up to 35 m s^{-1}. High nitrification rates and operation stability have been reported both in tertiary (up to 2 kg N-NH$_4$ m^{-3} d^{-1}) and secondary nitrification (up to 0.6 kgN-NH$_4$ m^{-3} d^{-1}), as well as in coupling with a predenitrification for total nitrogen removal (up to 1.0 kg N-NH$_4$ m^{-3} d^{-1}). Two full-scale plants have been constructed for total nitrogen removal in the Sahurs WWTP, France (1200 p.e.), and tertiary nitrification for upgrading the Würsellen WWTP in Germany (24,000 p.e.).

An anoxic circulating bed reactor known as the Mixazur process has been developed by Degrémont (France) for denitrification of municipal wastewater (Chudoba et al., 1998). The carrier used is fine mineral clay particles with a size of about 0.1–0.2 mm with very low concentrations below 5%. The reactor volume is separated in two zones: a circulating bed with mechanical stirring and an external plug-flow zone for separation of biofilm-covered particles from the effluent (Fig. 6.14). Denitrification rates of up to 0.6 kg N-NO^3 m^{-3} h^{-1} have been achieved in predenitrification without an external carbon source.

6.2.2.5 Immobilized Cells in Natural and Synthetic Gels.
Synthetic or natural gels enable the selective immobilization of enriched microbial cultures (e.g., nitrifying bacteria) thus increasing the specific reaction rates (Wijffels and Tramper, 1995). The treatment capacity of such systems is due to the activity of both the immobilized strains and the external spontaneously growing biofilm. The main advantage of natural gels (alginate, κ-carrageenan) is the survival of a great part of the entrapped cells during the immobilization procedure. However, the low mechanical stability of natural gels and especially their biodegradability limit their industrial application to a few types of fermentation and food industries.

Synthetic gels formed by polymerization or cross-linking are more resistant to shear stress. However, bacterial activity losses are over 90% and nutrient mass transport rates are lower. Different synthetic resins have been studied, such as polyvinyl alcohol, polyurethane, and polyethylene glycol by Tanaka et al. (1991) and Wijffels and Tramper (1995). Tsubone et al. (1992) compared the specific BOD removal activity of photopolymerizable resin and polyvinyl alcohol carriers with and without entrapped activated sludge cultures. No difference was observed in the treatment efficiency, although the respiration activity of gels with immobilized cells was higher. These results support the hypothesis that most of the reaction took place not in the gels, but in the biofilm formed on the surface of the carriers.

Recently, a new generation of moving bed reactors known as the Pegasus process, which uses immobilized bacteria cells in polyethylene glycol pellets with an average size of 3 mm has been developed in Japan (Mikawa et al., 1996). Tanaka et al. (1996) reported maximum nitrification rates of up to 0.8 kgN-NH_4 m^{-3} d^{-1} and significantly higher numbers of nitrifying bacteria in the pellets compared with activated sludge. The recommended filling rate is 7.5% and the reported ammonia loading is 4.2 kgN-NH_4 m^{-3} pellets d^{-1} with 85% removal efficiency (Emori et al., 1996). The process has two major applications for nitrification and single stage predenitrification-nitrification treatment. Three full-scale WWTP have been operated in Japan since 1990. This process has been commercialized in Europe by Degrémont under the name of Pagazur process with a first industrial application for the upgrading of a 120,000 p.e. WWTP in Jersey island, United Kingdom (Jackson et al., 1998).

6.2.3 Design of Innovative Biofilm Reactors

The optimum design of advanced biofilm processes depends on numerous factors including the type of wastewater (composition, variability, and treatability), type of reactor and biofilm support media, existing treatment standards, pretreatment processes, environmental conditions (temperature, oxygen concentration), as well as soil constraints, energy consumption, and other specific local conditions. The major factors influencing biofilm reactor design are interrelated and have an impact on capital and operating costs. Tables 6.2 and 6.3

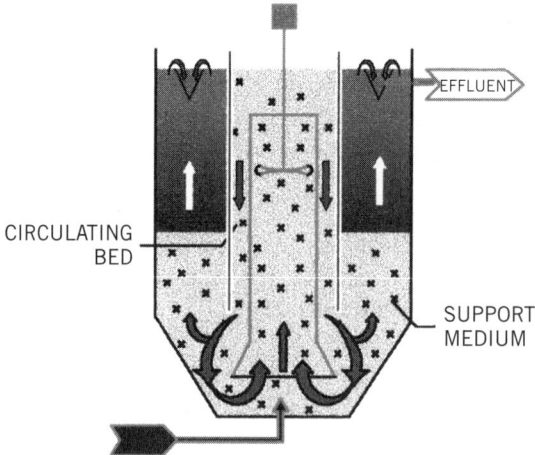

Figure 6.14 Schematics of the anoxic Mixazur process, Degrémont (France).

Figure 6.15 View of biofilm carriers with large industrial application. *a*) Expanded clay Biodagene (Biocarbone process); *b*) expanded polystyrene beads (Biostyr process).

summarize some experimental data reported for the performances of full-scale biofilm processes. More details on innovative biofilm reactor design for treatment of drinking, municipal, and industrial wastewaters are presented in Section 6.3.

Taking into account the highly dynamic development of advanced biofilm processes, as well as the specifities of each application for water or wastewater treatment in given local conditions, direct application of the reported values for design purposes is not recommended.

6.2.3.1 Biofilm Carriers. The choice of biofilm carrier is very important not only because it represents a significant part of the initial capital costs, but also because it greatly affects process efficiency and operating costs (backwashing). The most important carrier characteristics include size, porosity, density, attrition resistance, and capability for biofilm attachment (Gjaltema et al., 1997; Wijffels and Tramper, 1995; Sumino et al., 1992).

Two main types of support media were used for the development of innovative high-rate biofilm reactors (Fig. 6.15): (1) mineral particles such as fine sand or basalt (< 0.6 mm) and expanded clays (2–5 mm) and (2) plastic low-density materials such as polystyrene, polyethylene, or polyurethane (2–10 mm). Most of these carriers are specially designed (size, shape) to improve the biofilm attachment and activity (Tsubone et al., 1992; Ødegaard et al., 1998). Table 6.4 presents comparative information on some carriers used in recent advanced biofilm reactors.

6.2.3.2 Loading Rates. Some of the basic design parameters of biofilm reactors are the hydraulic, organic, and nitrogen loading rates. Hydraulic loading is a very important parameter for the design of submerged biofilters and represents the volume of treated liquid per unit of reactor cross-sectional area per hour, $m^3 \, m^{-2} \, h^{-1}$.

The volumetric organic, nitrification, or denitrification rates are currently expressed as kg BOD_5, COD, $N-NH_4$, or $N-NO_3$ applied per total volume of the biofilm reactor per day, kg $m^{-3} \, d^{-1}$. It is important to underline that some researchers present the performances of biofilm reactors on the basis of organic or nitrogen loads per volume of granular carrier. Very little information exists on the loads applied per unit of surface area in the recent innovative biofilm reactors. In some cases, specific in situ reaction rates were computed by dividing the volumetric rates by the biomass concentration.

Table 6.5 lists design loads reported for carbon and nitrogen removal in some commercial innovative biofilm reactors demonstrated on the basis of a full-scale experience. It is worth stressing the significantly higher hydraulic and organic loads of these processes

TABLE 6.4. Specifications of Some Carriers Used in Recent Advanced Biofilm Reactors

Type of Carrier	Size[a] (mm)	Density[b] (g cm^{-3})	Shape/ Porosity	Specific Surface Area[c] (m^2 m^{-3})	Process (Solid hold-up[d])
Mineral media					
Sand	0.1–0.6	2.7	Uniform/No	3000	3-Phase fluidized bed (5–20%)
Expanded clays	1.7–6	1.5–1.8	Round or irregular/ Rough	700–1500	Biofilters
Organic media					
Expanded polystyrene	1–3	0.3	Found		Biofilter
Plyethylene granules	1–5	0.9–0.95	Irregular/Rough	1500	Air-lift (40%); biofilters
Customized plastic elements (MBBR)	7–15	0.95	Cylinder/Low	350	Moving Bed (70%)

[a]Depending on the specific application
[b]Real desnity
[c]m^2 of surface area available for biofilm development per m^3 of reactor volume
[d]Solid hold-up or maximum filling rate: volume of carrier per reactor volume

TABLE 6.5. Summary of Loading Criteria for Innovative Biofilm Reactors in Full-Scale Operations for Municipal Wastewater Treatment

Type of Biofilm Reactor (*Manufacturer*)	Type of Treatment	Maximum Installed Capacity[a] [m^3 h^{-1} (*p.e.*)[b]]	Mass Loading, Rates (kg m^{-3} d^{-1})	Hydraulic Loading Rates (m^3 m^{-2} h^{-1})
Submerged biofilter Biofor (*Degrémont*)	Carbon removal		18 (COD)	16
	Nitrification	43,200	1.5 (N-NH$_4$)	20
	Postdenitrification	(*1,200,000*)	6 (N-NO$_3$)	35
Submerged biofilter Biostyr (*OTV*)	Carbon removal			10
	Combined	43,200		
	Nitrification/ Denitrification	(*1,200,000*)	1 (N-NH$_4$); 5 (COD)	6
	Nitrification		1.5 (N-NH$_4$)	10
Moving bed MBBR (*Kaldness Miljøtechnology*)	Carbon removal		2.5–8 (BOD$_7$)	NA
	Nitrification	3585 (*70,000*)	0.15–0.35 (N-NH$_4$)	
	Predenitrification		0.3 (N-NH$_4$)	
	Postdenitrification		0.7 (N-NO$_3$)	

[a]Peak wet weather flow
[b]Population equivalents
NA = Not applicable

compared to conventional trickling filters and rotating biological contactors. High hydraulic loads up to 35 m^3 m^{-2} h^{-1} have been reported for the Biofor process (Pujol et al., 1998) compared with the average design value of 0.4 m^3 m^{-2} h^{-1} for trickling filters (American Society of Civil Engineering Manual). Higher volumetric loading and removal rates have also been reported in innovative fixed- and moving-bed reactors for treatment of municipal wastewater as well as for the aerobic and anaerobic treatment of industrial wastewater. Because of the great diversity and specificity of industrial wastewaters, loading rates vary within a large range and are discussed in Section 6.3.

6.2.3.3 Parameters Influencing Nitrification Performances of Biofilm Reactors.
Sharma and Ahlert (1977) provided a comprehensive review of the main parameters and characteristics of nitrification. Harremoës et al. (1998) discussed the main factors influencing the design of nitrogen removal plants. The analysis of the biological mechanisms and kinetics of this process and numerous experimental studies demonstrated that biofilm systems offer better prospects in terms of achieving complete nitrification, as they favor the attachment and selective development of slow-growing nitrifying bacteria.

In fixed bed systems, diffusion rates within the biofilm are usually reaction rate limiting. The mass transport of substrate and oxygen to the attached cells and the elimination of metabolic products must be considered for the design and operation of the reactor (Harremoës, 1978).

The bulk oxygen concentration is the main operating parameter for real time control of nitrification rates (Lazarova et al., 1998b,c). The oxygen equilibrium in the biofilm nitrifying reactors depends on both bacterial respiration and mass transfer rates. In the case of biological processes with high oxygen consumption rates such as nitrification, oxygen transfer is the limiting parameter. In general, the average oxygen penetration depth in biofilms is around 100 to 200 μm (De Beer et al., 1993). For this reason, the biofilm thickness should be preferably in the same range and a high biofilm surface area is required in order to achieve higher volumetric reaction rates.

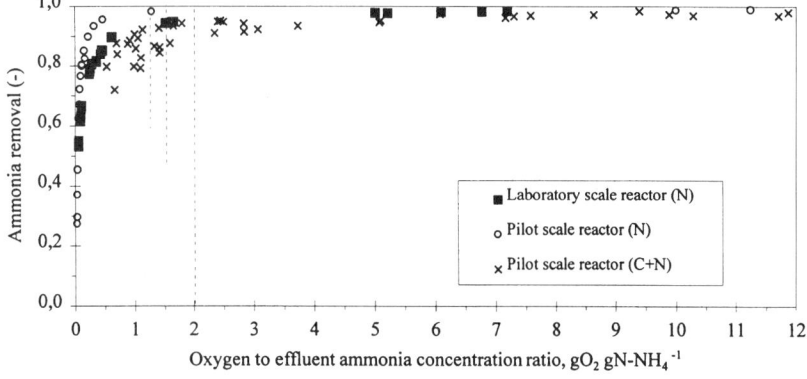

Figure 6.16 Influence of bulk oxygen to ammonia concentration ratio on nitrification efficiency in the circulating bed reactor TURBO N for simultaneous carbon and nitrogen removal (C+N) and tertiary nitrification (N).

Nitrification kinetics depend either on ammonia or oxygen bulk concentration. It is important to underline that the transition values for the transition from zero to half or first-order reaction are significantly higher for oxygen compared with ammonia concentrations (0.5–2 mgN-NH_4 L^{-1} compared to 3–6 mgO_2 L^{-1}). A commonly used indicator for the transition from the ammonia rate limited to the oxygen-limited nitrification rate is the oxygen to ammonia ratio. The reported values are within the range of 2.5 to 4 gO_2 gN-NH_4^{-1} for moving beds, rotating discs, and biofilters (Hem et al., 1994; Çeçen and Gönenç, 1994; Gönenç and Harremoës, 1985). Lower values of about 1.2–2 gO_2 gN-NH_4^{-1} were reported by Nogueira et al. (1998) for the circulating bed reactor TURBO N both at laboratory and pilot-scale (Fig. 6.16). This difference could be due to the hydrodynamic characteristics of the circulating bed, which results in a thinner biofilm, thus minimizing the external and internal resistances to oxygen mass transfer. In full-scale biofilm systems these criteria were hardly ever achieved and the nitrification process was always oxygen limited.

Another important parameter influencing biofilm nitrification rates is the redox potential (Lazarova et al., 1998c). The redox potential in wastewater treatment systems is strongly influenced by the daily variations in loading rates (Fig. 6.17). During the daily peak loading, the redox values fall sharply. Higher values were observed by night or during the weekends, respectively 400–440 mV/ENH compared to 300 mV/ENH during the peak loads. Figure 6.17 also shows that in the same range of nitrification loading rates, the values of the redox potential vary with the bulk oxygen concentration. The decrease in the redox potential is 20, 40, and 60 mV/ENH, respectively for dissolved oxygen concentrations of 6, 5, and 3 mgO_2 L^{-1}. It is important to underline that the critical value of the redox potential reported for biofilm reactors, for example 340 mV/ENH for the circulating bed TURBO N, is very low compared to the values reported for activated sludge above

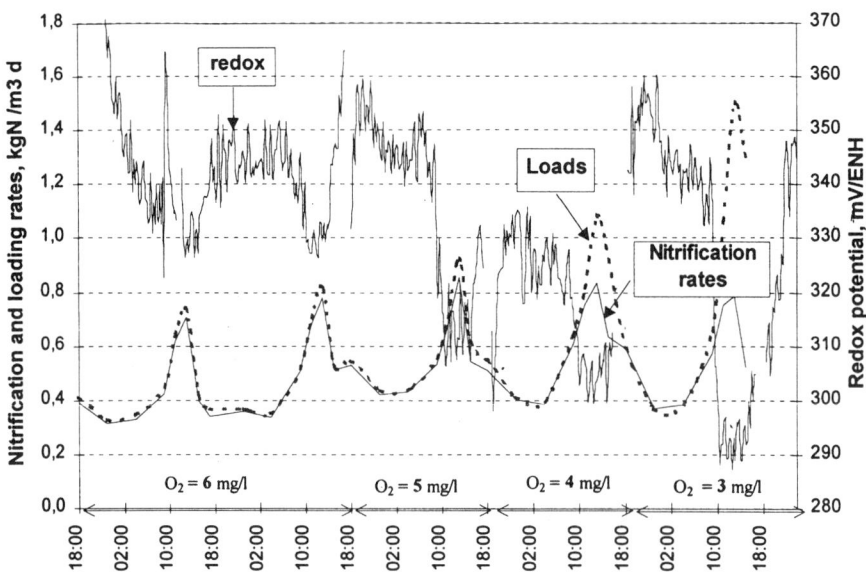

Figure 6.17 Influence of daily shock loads on redox potential for different bulk oxygen concentrations: results obtained in a nitrifying circulating loed reactor TURBO N.

6.3 INDUSTRIAL APPLICATIONS

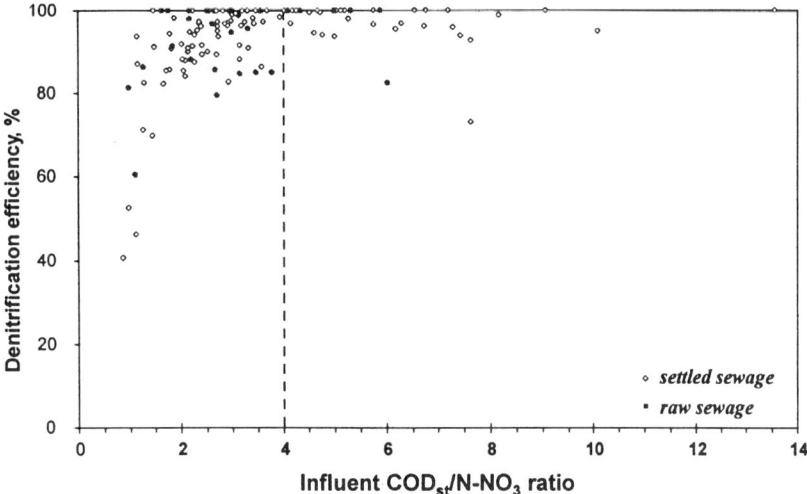

Figure 6.18 Influence of the variations in the $COD_s/N\text{-}NO_3$ ratio on predenitrification efficiency (floating bed denitrifying reactor).

400 mV/ENH. This fact can be explained by the higher oxidation state of the fixed biomass (Lazarova et al., 1998a).

As a rule, biofilm reactors for nitrification are designed on the basis of the "lowest temperature" under winter conditions. The rate of growth of nitrifiers generally triples with each 10°C and is around 0.86 d^{-1} at 20°C (Wijffels et al., 1995; Harremoës et al., 1998).

6.2.3.4 Parameters Influencing Denitrification Performances of Biofilm Reactors.
The main parameter influencing the predenitrification removal rate is the carbon source availability or the C/N ratio. The biodegradable carbon in raw wastewater is hardly ever sufficient to ensure maximum denitrification rates. Under the given loading rates, an increase in the denitrification efficiency is observed for the $COD_s/N\text{-}NO_3$ ratios below 4–6 (Fig. 6.18) or the $COD_{tot}/N\text{-}NO_3$ ratios below 10–12. Moreover, the oxygen rich water recycled from the nitrification step to the denitrifying reactor leads to the fast consumption of organic matters for oxygen respiration thus reducing the available amount for denitrification. Consequently, the predenitrification volumetric rates are rather lower than those reported for postdenitrification with an external carbon source addition.

The control of denitrification and consequently the level of nitrogen removal depends on the recycling ratio (the ratio between the recycled and treated flow) from the nitrifying to the denitrifying reactor. The commonly used values vary from 100 to 600%.

6.3 INDUSTRIAL APPLICATIONS

6.3.1 Drinking Water Treatment

Biofilm reactors have been studied both at laboratory and full scale for removing nitrates from drinking water in Europe and to a more limited extent in North America. Despite the large development of biological denitrification for wastewater treatment over many years, the transfer of this technology to water treatment was relatively slow due to concerns over possible bacterial contamination, increasing organic content and chlorine demand. Existing experience in this field has been reviewed by Gayle et al. (1989), Rittmann and Huck (1989), Richard and Thébault (1992), and Kapoor and Viraraghavan (1997).

In water treatment, nitrates are reduced to nitrogen gas by heterotrophic or autotrophic bacteria under anoxic conditions. Heterotrophic denitrification utilizes external organic carbon such as ethanol, acetic acid, and methanol with C/N ratios 1.05, 1.32, and 0.93, respectively. In autotrophic denitrification, hydrogen serves as a substrate and carbon dioxide or bicarbonate ensure organic carbon for cell synthesis.

The most important requirements in water treatment are high treatment efficiency and the high hygienic quality of the effluent. The first objective can be successfully attained by the application of compact biofilm reactors operating at low temperature (lower limit 2–6°C) with a high nitrate removal capacity from 3.1–4.4 kg NO_3^- m^{-3} d^{-1} (Richard and Partos, 1986; Roennefahrt, 1986) to 9–12 kg NO_3^- m^{-3} d^{-1} (Eppler and Eppler, 1986; Liessens et al., 1993). The clogging problems of biofilters related to gas bubbles or biomass accumulation were avoided by the application of high conversion rate fluidized bed reactors (Gregory and Sheiham, 1981; Hall et al., 1985; Hollo and Czako, 1987; Lazarova et al., 1994b). The second requirement concerning effluent quality posed some difficulties and has been resolved by an intensive posttreatment, for example granular activated carbon (GAC) filtration (Richard and Thébault, 1992; Liessens et al., 1993).

A critical analysis of the literature indicates the wider application at full scale of heterotrophic denitrification in comparison with autotrophic denitrification, especially on a full-scale level. It is important to underline also that fixed beds are applied predominantly for nitrate removal from groundwater while fluidized beds are used for treatment of surface waters as this process is less sensitive to variations in water quality (turbidity, suspended solids, organic matters).

6.3.1.1 Heterotrophic Denitrification

6.3.1.1.1 Fixed Bed Processes. Two heterotrophic fixed bed technologies with a heavy carrier known as Nitrazur and Biodenit processes, respectively up-flow and down-flow processes, have been operated in Europe since 1981, when the first Nitrazur reactor was installed in Château-Landon, France. The filtering media of the Nitrazur reactor is 1.7–2.7 mm expanded clay Biolite specially designed for improved bacteria development (Richard and Thébault, 1992). The main carbon source was acetic acid and ethanol with denitrification rates in the range of 2.8 to 9.5 kg NO_3^- m^{-3} d^{-1}. Capital and operating costs depend on plant size and vary from $0.8 to $1.3 million and from $0.06 to $0.13 m^{-3} for water treatment plant capacities from 50 to 835 m^3 h^{-1}, respectively (Richard and Thébault, 1992).

The down-flow Biodenit process operates under pressure using 3–6 mm expanded clay Biodagene (Ravarini et al., 1988). Denitrified effluent is treated in a Biocarbone filter filled with 0.8–1.2 mm sand and 1.7–3.4 mm CAG media.

A floating bed bioreactor called Denipor has been developed by Preussag AG in Germany and uses expanded polystyrene for biofilm support and ethanol as its carbon source (Roennefahrt, 1986). Excess biomass is easily removed by downward flushing. The denitrified effluent is further treated by two aerobic filters. The reported Denipor capacity was in the range of 0.7–1.0 kg $N-NO_3^-$ m^{-3} d^{-1} and the effluent TOC - 1 mg L^{-1}. Two full-scale plants were built in Germany.

6.3.1.1.2 Fluidized Bed Processes. Fluidized bed reactors were constructed in the early 1980s for surface water denitrification in Great Britain (Bucklesham and Ipswich). Gregory and Sheiham (1981) demonstrated that this technology is more cost-effective than fixed beds for large water treatment plants (45,000–450,000 $m^3 d^{-1}$) and costs remain comparable for small plants of 4500 $m^3 d^{-1}$. The main carbon source applied is methanol. The media used is sand with an average diameter of 0.35 mm, which enables the biomass to achieve a concentration of up to 15 kg m^{-3} (Hall et al., 1985). A constant flow velocity of 22 m h^{-1} is maintained by recirculation of the effluent. High specific denitrification rates of up to 5.5 g N-NO_3 kg^{-1} VSS h^{-1} have been reached at the low temperature of 2°C. The sludge production reported is 0.2 kg SS kg^{-1} of eliminated carbon. The backtrough of methanol in the effluent is minimized by the operation under conditions of carbon source limitation, ensuring an average nitrate concentration in the effluent of 3 mg N-NO_3 L^{-1}. A posttreatment is necessary for the removal of suspended solids.

6.3.1.2 Autotrophic Denitrification.

An autotrophic up-flow fixed bed industrial process, Denitropur, has been developed by Sulzer company in Switzerland and Germany using hydrogenotrophic denitrification (Gros et al., 1986; Rutten and Schnoor, 1992). The main advantages of this process compared with heterotrophic denitrification are (1) the lack of a need for posttreatment for organic substrate removal and (2) the low sludge production, due to the slowly growth autotrophic biofilm. The main disadvantage is the safety requirements related to hydrogen. The Denitropur plant at Mönchengladbach, Germany, was constructed in 1986 to treat 100 $m^3 h^{-1}$ groundwater containing about 40 mg L^{-1} of nitrate. The process operates under pressure (4–6 bar) with Sultzer-mixing plastic elements as a support medium with a three dimensional structure known as Mellpark. The hydrogen is added before the reactor system with a consumption rate of 11 g H_2 per 100 g nitrate. The denitrification rate was 0.25 kg N-NO_3^- $m^{-3} d^{-1}$ with a sludge production of 0.1–0.3 kg $kg^{-1} d^{-1}$. The total cost for design capacities of 50 and 200 m h^{-1} are \$1.14 and \$0.76 m^{-3}, respectively, 50% of which are operating costs (Rutten and Schnoor, 1992).

A sulfur/limestone denitrification fixed bed is operated at full scale in Monterferland, Holland, for nitrate removal from groundwater (Kruithof et al., 1988). The plant consists of a vacuum degasser, a slow up-flow sulfur/limestone filter, a cascade tray, and an infiltration pond with a capacity of 35 $m^3 h^{-1}$. One of the main disadvantages of this process is the increase in sulfate concentration and water hardness.

Kurt et al. (1987) reported the results from hydrogenitrophic denitrification in a fluidized bed reactor. A residence time of 4.5 h was required for the removal of 25 mgN-NO_3 L^{-1}.

The capital and operating costs for nitrate removal using autotrophic denitrification was found to be higher than the costs of heterotrophic denitrification (Kapoor and Viraraghavan, 1997).

6.3.2 Municipal Wastewater Treatment

The principal contaminants of urban wastewater are suspended solids (ss), rganic matter expressed by a COD of about 200–1000 mg L^{-1}, and ammonia nitrogen with average concentrations of 30–100 mgN-NH_4 L^{-1}. Consequently, the major challenges of the treatment are carbon removal, nitrification, and denitrification for relatively high flow rates of up to 300,000 $m^3 d^{-1}$ in a relatively economic way. Stringent standards for wastewater discharge lead to requirements of higher process efficiency and reliability of operation. The importance of biofilm reactors for urban wastewater treatment is widely recognized.

Figure 6.19 shows a general view of the new wastewater treatment plant in Colombes near Paris, France, where the influent from 1,200,000 p.e. is treated until total nitrogen removal by Biostyr and Biofor biofilters in covered buildings. Space constraints in the densely populated area near Paris favored the implementation of this more compact WWTP.

During the last decades, various types of innovative biofilm reactors have been developed for carbon removal, nitrification, pre- and postdenitrification in order to comply with the new standards for wastewater discharge (residual concentration as low as 1 mg L^{-1} ammonia nitrogen and phosphorus and 10 mg L^{-1} BOD, SS, and total nitrogen). The biofilm design depends on the specific application. Figure 6.20 illustrates the main schemes and processes used for municipal wastewater treatment.

6.3.2.1 Carbon Removal

6.3.2.1.1 Fixed Bed Processes. Numerous papers have reported the successful application of biofilm reactors for carbon removal from urban wastewater. Daigger et al. (1994) discuss the practical experience from three full-scale trickling filter installations in the United States operating with organic loading in the range of 0.2–1.0 kg BOD_5 m^{-3} d^{-1}. Simultaneous nitrification was achieved under low carbon loads. Canler and Perret (1994) presented an assessment of the performances of 12 French up-flow and down-flow biological aerated filters with a capacity from 7500 to 150,000 p.e. treating predominantly domestic sewage. The study demonstrated a good reliability of operation and a high efficiency to remove the organic carbon and suspended solids. At applied loads under 7 kg COD m^{-3}

Figure 6.19 View of the compact and covered wastewater treatment plant in Colombes near Paris, France, using the fixed bed biofilm processes Biostyr and Biofor (design capacity 1,200,000 p.e., dry weather flow 2.8 m^3 s^{-1} and wet flow rate 8.5 m^3 s^{-1}).

6.3 INDUSTRIAL APPLICATIONS

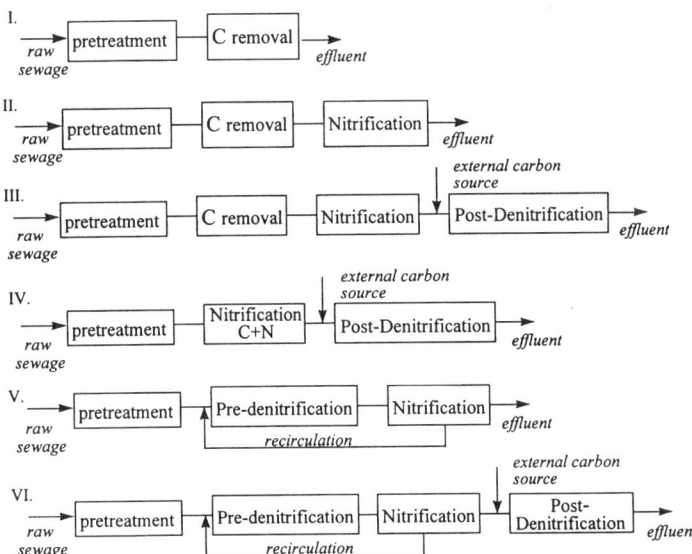

Figure 6.20 Main treatment schemes and processes applied for municipal wastewater treatment.

d^{-1}, the required effluent quality of less than 90 mg COD L^{-1} was attained. Some plants, not designed to treat nitrogen, showed more than 80% nitrification efficiency with an average elimination rate of 0.36 kg N-NH_4 m^{-3} filter medium d^{-1}. The specific sludge production was 0.41 kg SS kg^{-1} COD_{elim} d^{-1}. The average specific energy consumption measured was 1.16 kWh kg^{-1} COD_{elim} d^{-1}.

More than 80 Biofor WWTP are in operation in France, Scandinavia, and North America (Pujol et al., 1994; Peladan et al., 1997; Ninassi et al., 1998). The carbon removal is achieved usually after a primary settling or physicochemical treatment step and sometimes after high-rate activated sludge. More than 60–70% COD removal is ensured with loading rates of up to 18 kg COD m^{-3} d^{-1}. Suspended solids removal remained above 80% with residual concentration less than 10 mg SS L^{-1}.

A great number of aerated biofilters Biocarbone, have been installed in Europe, Japan, and North America for carbon removal from municipal wastewater, ensuring that effluent quality conforms to the standard of 15 and 30 mg L^{-1} BOD and suspended matters, respectively (Rogalla et al., 1991; Kleiber et al. 1994). Plant capacities vary from 35,000 to 550,000 p.e. and up to 7.87 kg COD m^{-3} d^{-1} organic loads have been applied. The air and energy consumptions were 15–20 Nm^3 kg^{-1} COD applied and 0.8–1.2 kWh kg^{-1} COD removed, respectively, and the sludge production—0.38 kg SS kg^{-1} COD applied.

6.3.2.1.2 Mobile Bed Processes.
The Captor sponge process has been selected for the upgrading of the Moundsville/Glen Dale WWTP in West Virginia, United States, for secondary treatment with average and peak flows of 8707 and 30,284 m^3 d^{-1}, respectively (Golla et al., 1994). The treatment consists of primary settling, the Captor process, and activated sludge, forming secondary treatment, secondary clarification, and ultraviolet disinfection. The Captor process was designed for high rate BOD reduction and the activated sludge as a treatment to obtain the BOD permit. However, during a 3-year full-scale operation it was

found that the Captor process also achieved a high rate of nitrification. With a 5.5 kg SS m^{-3} biomass concentration and 50–100 min detention time, the average effluent soluble BOD was 5.9 ± 1.4 mg L^{-1} (influent concentration 85–135 mg L^{-1}) and 70–75% of the ammonia loading was removed.

One of the first full-scale applications of the Linpor sponge process in 1984–1986 was also for upgrading several overloaded existing WWTP (Morper, 1994). The good nitrification capacity led to the development and implementation of Linpor for simultaneous carbon and nitrogen removal, as well as for ammonia removal as a polishing step (<1 mgN-NH$_4$ L^{-1} discharge).

6.3.2.2 Total Nitrogen Removal. Total nitrogen removal can be achieved by different treatment schemes by coupling nitrification with pre- or postdenitrification biofilm reactors (see Fig. 6.20, schemes III and VI). The treatment schemes with nitrification and postdenitrification have as a major disadvantage the use of external carbon source (methanol). To avoid this constraint, other schemes of predenitrification and nitrification have been developed. In this case, carbon limitation is the main constraint: Carbon demands are in the range of 3.5–4.5 gCOD$_s$ per g N-NO$_3$ reduced.

6.3.2.2.1 Fixed Bed Processes. Total nitrogen removal in compact units can be achieved by the coupling of nitrifying and denitrifying biofilters. Kleiber et al. (1994) reported the performances of a full-scale Biocarbone WWTP in Hundested, Scandinavia, with an average capacity of 20,000 p.e. and a restrictive permit discharge of 8 mg N$_{tot}$ L^{-1} and 1.5 mg P$_{tot}$ L^{-1}.

One of the higher biofilter plant capacities is installed in the Oslo WWTP, Norway: 450,000 p.e., flow rate 7.5 m^3 s^{-1}, under operation since 1994 (Pujol et al., 1998). The plant consists of eight treatment lines located in rock caverns and each line is composed of a sedimentation tank, four nitrifying and four denitrifying Biofor filters each of 87 and 65 m^2, respectively (see Fig. 6.20, scheme IV). The pilot plant and full-scale studies showed the possibility of treating maximum nitrification loads up to 0.6 and 0.8 kg N-NH$_4$ m^{-3} d^{-1} at <10 and 10–15°C, respectively, with residual ammonia concentrations below 2 mg L^{-1}. The maximum denitrification rates with methanol addition are up to 5 kg N-NO$_3$ m^{-3} d^{-1}.

The largest WWTP designed for total nitrogen removal and treatment of runoff water is the plant of Colombes near Paris, France, in operation since July 1998. The plant was designed to treat the wastewater from 1,200,000 p.e. with a maximum flow capacity of 12 m^3 s^{-1} (heavy rains). The plant design consists of physicochemical primary treatment, followed by three stages of biofilters and provides very high flexibility of operation: almost total removal of carbon, nitrogen, and phosphorus pollution (25 mgBOD L^{-1}, 20 mgSS L^{-1}, 10 mgN L^{-1} and 1 mgP L^{-1}) for dry weather flow rates (2.8 m^3 s^{-1}) with the operation of biofilters in series and carbon and nitrogen removal from wet weather of 8.5 m^3 s^{-1} (30 mgBOD L^{-1}, 30 mgSS L^{-1}, 15 mgTKN L^{-1}) after change in biofilter operation to "in parallel" mode. The carbon removal and postdenitrification are performed in Biofor reactors, while the nitrification is ensured by the Biostyr process. All the installation is covered and the air is chemically treated in four treatment chains, each with an installed capacity of 200,000 m^3 h^{-1}.

A full-scale WWTP for 85,000 p.e. for total nitrogen removal without addition of an external carbon source (Fig. 6.20, scheme V) has been operating since the beginning of 1998 in Ahlen, Germany (Ninassi et al., 1998). The system consists of eight predenitrifying Biofor of 47 m^2 and eight nitrifying Biofor of 52 m^2. The outlet nitrogen concentrations meet the standards of 5 mgN-NH$_4$ L^{-1} and 13 mgN-NO$_3$ L^{-1}. The pilot plant study results showed maximum denitrification rates of up to 1.2 kg N-NO$_3$ m^{-3} d^{-1} at flow velocities in the range of 8–21.5 m h^{-1} and recirculation rates of up to 350%.

The French floating fixed bed Biostyr ensures almost total nitrogen removal in the same reactor. This process is usually designed for tertiary treatment after preliminary carbon removal in activated sludge systems. Gonçalves et al. (1994) reported the results from pilot plant experiments with simultaneous total nitrogen removal and biological phosphorus elimination resulting in residual concentrations of 10 mg N-NO$_X$ L^{-1} and 2 mg P$_{tot}$ L^{-1}. Since 1994 in Denmark, according to Borregaard (1997), four plants have been in operation for nitrification-denitrification with the addition of methanol and one for nitrification of municipal wastewater with a design capacity from 2000 to 13,000 m^3 d^{-1}.

6.3.2.2.2 Three-Phase Fluidized Bed Processes. The technical-economic feasibility of three-phase fluidized beds was studied in Japan by Hosaka et al. (1991) for the nitrification of municipal wastewater in a 50 m^3 d^{-1} pilot plant in three reactors with a total volume of 16 m^3, that were applied after conventional a activated sludge treatment containing anoxic zone. The FBR diameter and height were 1.2 and 4 m, respectively. Granular anthracite with a diameter of 0.6 mm was used as a support medium. The liquid volumetric flow rate was 13 m^3 h^{-1} and the air flow rate was 0.6 m^3 min^{-1}. After a 3-month startup period, the hydraulic retention time was maintained at 2.1 h. The total nitrogen removal efficiency reported was 56–73% with less than 15 mg L^{-1} total nitrogen concentration in the effluent.

Badot et al. (1994) reported the results of an industrial prototype of the three-phase Biolift process applied for tertiary wastewater treatment at the Maxeville WWTP, France, and coupled with an anoxic predenitrification step with activated sludge (Fig. 6.20, scheme V). The design parameter were as follows: diameter 3.8 m, height 10 m, volume treated 100 m^3 h^{-1}, internal recirculation flow 500–1000 m^3 h^{-1}, air-flow rates 100–400 Nm^3 h^{-1}, 5–20% filling with sand with a size of 0.2–0.6 mm. One characteristic of this system is the use of air both for process needs (injection into the fluidized bed) and for recirculation (airlift system). Up to 5 kg N-NH$_4$ m^3 of support d^{-1} nitrification rates have been reached.

A 6.5 m^3 anaerobic/aerobic fluidized bed with 0.35–0.43 mm pumice stone media and air injection into the medium of the bed has been studied by Fdez-Polanco et al. (1994) for municipal wastewater treatment. The results showed a short startup time, good stability despite variations in organic and nitrogen loads and a good recovery after process upsets. With a detention time of 24 h based on the volume of the bed (bed expansion 20%), over 80% and 95% of COD$_s$ and ammonia were removed at organic and nitrogen loading rates under 1.2 and 0.2 kg m^{-3} d^{-1}, respectively.

6.3.2.2.3 Moving Bed Processes. Moving bed reactors have been tested for total nitrogen removal from the early 1990s at full scale at several small wastewater treatment plants (< 2000 p.e.) in Norway (Ødegaard et al., 1994). Two process schemes were studied: (1) predenitrification for N removal and postprecipitation for P removal or (2) preprecipitation and postdenitrification. The second scheme is more suitable for wastewater with a small fraction of easily biodegradable organic matter and can meet higher N-removal efficiency >85% (compared to <70% for the first scheme). Reduction of external carbon source consumption could be achieved in a scheme based both on pre- and postdenitrification (Fig. 6.20, scheme VI). One of the largest WWTP under operation was installed in Lillehammer (70,000 p.e.) with dry weather capacity of 26,000 m^3 d^{-1} and a peak flow of 43,000 m^3 d^{-1}. The total installed MBBR volume is of 3840 m^3 with specific biofilm area of 325 m² m⁻³ and water depth of 5.5 m. At cold temperatures of 6.5°C, the nitrification rate was 0.165 kg N-NH$_4$ m⁻³ d⁻¹ at dissolved oxygen levels of 6.0–6.5 mg L⁻¹. The final objective is a 70% nitrogen removal.

6.3.2.2.4 Air-Lift and Circulating Bed Processes. Air-lift reactors have also been studied in pilot plants and applied at full scale for urban wastewater treatment with nitrogen removal. Frijters et al. (1997) reported the results of a two-stage 4.56 m³ pilot located in a WWTP in the Netherlands. The second stage air-lift reactor was with integrated denitrification. A nitrification rate up to 1.5 times higher was reached in the aerobic zone in the reactor with integrated denitrification compared to the completely aerobic first step, 0.9 and 1.32 kg N-NH$_4$ m⁻³ d⁻¹, respectively. In both reactors, nitrification was complete at temperature above 15°C and COD conversion rates were more than 2.5 kg COD m⁻³ d⁻¹.

Lazarova et al. (1998c) reported experimental results obtained for an industrial-scale prototype of the new circulating bed reactor TURBO N coupled with a fixed floating bed denitrification reactor (see Fig. 6.20, scheme V). The new biofilm system ensured a total nitrogen removal rate of 65–80% with residual nitrogen concentration below 12 mgN-NO$_3$ L⁻¹. Under conditions of daily shock loading, nitrate removal remained above 90% within the whole range of loading from 0.2 to 0.55 kgN-NO$_3$ m⁻³ d⁻¹ and recirculation rates of 200 and 400%. The circulating bed TURBO N also showed high tolerance to the daily shock loading ensuring nitrification efficiency above 93–95% (Fig. 6.21). The maximum nitrification rates reached 0.8 kgN-NH$_4$ m⁻³ d⁻¹ with very low residual ammonium concentrations. The reaction rate constants found from modeling and respirometric kinetic studies indicated that the apparent specific nitrification rates achieved by the circulating bed biofilm were comparable to the maximum intrinsic rate. Real time control of dissolved oxygen enabled savings of up to 44% of energy costs and led to a twofold improvement of oxygen transfer rate.

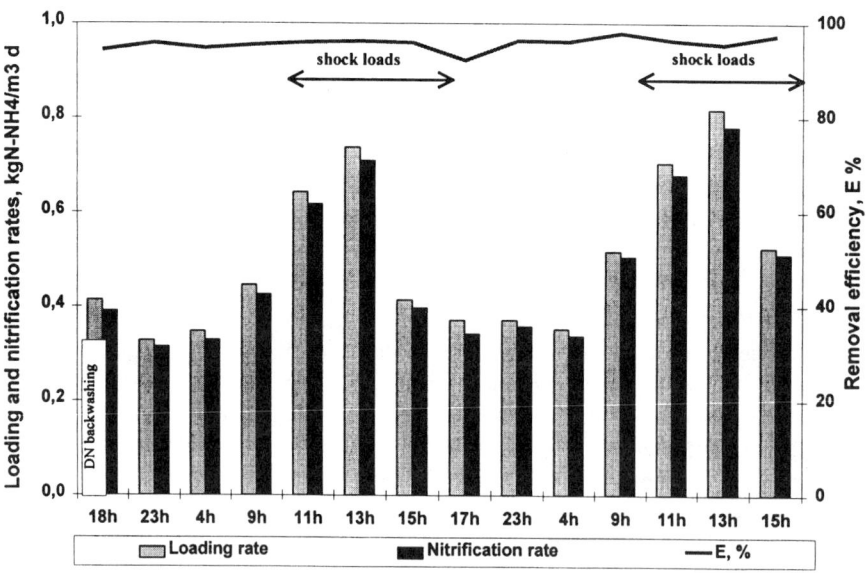

Figure 6.21 Influence of daily shock loads on nitrification rates in the circulating bed reactor TURBO N (T = 22°C, 6 mgO$_2$ L⁻¹, hydraulic residence time, HRT 37 and 30 min).

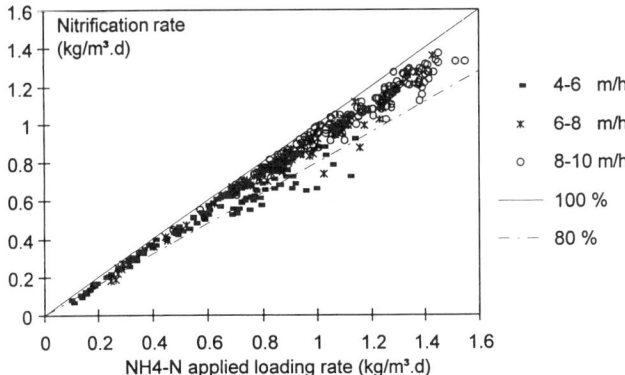

Figure 6.22 Performances of an industrial-scale Biofor biofilter in tertiary nitrification: relationship between applied ammonia loads and nitrification rates for different hydraulic conditions. Nitrification efficiency between 80 and 100%.

6.3.2.2.5 Immobilized Cell Processes. The Pegasus process using immobilized cells in PEG pellets has been operated in Japan since 1990 in three full-scale facilities for nitrification coupled with a predenitrification step with activated sludge (see Fig. 6.20, scheme V) in Kawagoe City (3000 m^3 d^{-1}) and Munakata City (11,300 m^3 d^{-1}), as well as in Osaka for single nitrification of 960 m^3 d^{-1} (Tanaka et al., 1996). A demonstration study at Takinoshita WWTP (3000 m^3 d^{-1}) showed that the retrofitting of the conventional activated sludge with the Pegasus process for nitrification ensured a stable removal of the total nitrogen around 70% with an average hydraulic retention time of 8 h (Emori et al., 1996).

The first full-scale plant in Europe with immobilized cells, the Pegazur process of Degrémont, is under construction at Bellozanne WWTP on Jersey Island, United Kingdom, for upgrading the existing plant with a capacity to treat the influent from 120,000 p.e. (Jackson et al., 1998). The pilot plant studies showed nitrification rates from 0.2 to 0.8 kg N-NH$_4$ m^{-3} d^{-1} at applied BOD loads of 0.1–0.6 kg BOD m^{-3} d^{-1} at 20°C.

6.3.2.3 Tertiary Nitrification.
Tertiary nitrification, as well as total nitrogen removal and upgrading of existing WWTP, is one of the most important and beneficial applications of the advanced biofilm reactors (see Fig. 6.20, scheme II). Biofilm processes guarantee efficient nitrification at very low hydraulic residence times and thus reach very high volumetric loading rates.

6.3.2.3.1 Fixed Bed Processes. The monitoring of more than 10 full-scale Biofor aerated filter plants demonstrated nitrification efficiency in the range of 80–90% for loading of up to 1.4 kg N-NH$_4$ m^{-3} d^{-1} (Pujol et al., 1994). The maximum nitrification rates reached at pilot scale was 2.7–4.3 kg N-NH$_4$ m^{-3} d^{-1} (Peladan et al., 1996, 1997, 1998). The common design hydraulic load is 10–20 m h^{-1} with maximum values of up to 35 m h^{-1}. An increase in nitrification rates has been observed with increasing water velocity. The analysis of these results shows that the water velocity is a positive factor rather than a limiting factor for the nitrification as long as the nitrification conditions are not limiting (temperature, aeration, backwashing, etc.). Similar results (Fig. 6.22) have been observed at industrial scale in a 144 m^2 Biofor reactor installed at the WWTP of Achères near Paris in

France (Pujol et al., 1998). Moreover, the increase in water velocity does not have a negative impact on suspended solids removal.

The tertiary nitrification in biofilters has been successfully combined with chemical phosphorus removal. For example, the Köln facility (Germany), operated since 1992 with a total capacity of 1,450,000 p.e., was designed for upgrading the existing activated sludge plant for simultaneous nitrification and tertiary phosphorus removal with 48 Biofor reactors, each with a unit surface area of 73 m^2 (Pujol et al., 1995). The operation results demonstrated removal rates above 95% for nitrification and 70% for P removal with residual concentrations of less than 0.2 mg $N-NH_4$ L^{-1} and 0.5 mg P_{tot} L^{-1}.

6.3.2.3.2 Fluidized Bed Processes. Fluidized beds were investigated for tertiary nitrification at pilot scale at Beckton and Horley WWTPs, United Kingdom (Green and Hardy, 1985; Williams et al., 1986). Full nitrification was achieved in a 4 m tall FBR in Beckton in less than 30 min residence time with an average biomass concentration of 15–20 kg m^{-3}. The Horley's FBR has a total volume of 12 m^3, design up-flow velocity of 28 m h^{-1}, hydraulic retention time of 30 min, and biomass concentration of 12 kg m^{-3}. The time required to achieve a fully nitrifying population was 10 weeks. It was shown that at 250 and 400 m^3 d^{-1}, the process operated satisfactorily removing above 90% of the ammonia despite the low biomass concentration of 2.8 kg m^{-3}. Financial analysis showed that the capital investment required to construct a fluidized bed is under half that needed for the extensions to an activated sludge plant. Running costs, however, were 70% greater due to the higher energy consumption. Finally, the overall costs are approximately 10% lower over a period of 30 years.

6.3.2.4 Tertiary Postdenitrification with Methanol Addition. In the United States, denitrification in tertiary filtration with methanol addition has been successfully operated for more than 15 years (see Fig. 6.20, schemes III and IV). The main advantages of biofilm reactors for upgrading existing municipal WWTPs are space conservation, a low hydraulic retention time, fast and easy construction without retrofitting of the existing plant, and the flexible control of nitrate reduction.

6.3.2.4.1 Fixed Bed Processes. Several pilot and demonstration studies have been carried out with the Biofor DN process and methanol addition showing up to 5 kg m^{-3} d^{-1} denitrification rates at 11°C and flow velocity of 10 m h^{-1} (Hamon et al., 1993). The influence of different external carbon sources of a packed bed biofilm reactor Biofor was studied by Æsøy et al. (1998). A maximum denitrification rate was achieved with ethanol and hydrolyzed sludge with C/N ratios of 4.5 and 8–10 g COD g^{-1} $N-NO_3$, respectively. Acetic acid gave a lower and more variable rate below 2 kg $N-NO_3$ m^{-3} d^{-1}.

The technical and economic feasibility of tertiary denitrification with methanol was studied in a full-scale filter with a 1.2 m expanded slate at the WWTP of Zürich-Werdhölzli, Switzerland (Koch and Siegrist, 1997). At peak flow rate, the denitrification rates were 1 and 2 kg $N-NO_3$ m^3 d^{-1}, in winter and summer, respectively, with an effluent nitrogen concentration of less than 3 mg N L^{-1}. The estimated cost of the tertiary nitrification was $2.50/kg N, which is approximately 50% less expensive than extended denitrification in activated sludge.

Combined suspend solids and nitrogen removal were investigated in a pilot- and full-scale study of a continuous up-flow sand filter DynaSand (Koopman et al., 1990; Hultman et al., 1994; Sanz et al., 1996). Combined tertiary denitrification and phosphorus removal were tested in full scale in Sweden in a DynaSand filter with a cross-sectional area of

6.3 INDUSTRIAL APPLICATIONS

4.7 m^2, bed height of 6 m, and 1.2–2 mm sand (Hultman et al., 1994). Methanol addition was controlled with the measurement of the residual nitrate concentration. No operational disturbances resulted from the formation of nitrogen gas bubbles, which was probably favored by the continuous bed cleaning. Influent nitrate concentrations up to 20 mg N-NO$_3$ L^{-1} were decreased to 0.5–2 mg N-NO$_3$ L^{-1}. Phosphorus may limit the denitrification rate for concentrations below 0.1 mg L^{-1} soluble phosphorus in the filter effluent.

6.3.2.4.2 Fluidized Bed Processes. Tertiary denitrification is also the main application of the fluidized bed technology for wastewater treatment. The first industrial units were constructed in the United States in the early 1980s in Pensacola, Florida, and Reno, Nevada (Mishra and Sutton, 1991). In 1990, the EPA recommended this application as an efficient and cost-effective solution, and new plants have been implemented in the 7500 m^3 d^{-1} Rancho California Water Reclamation Plant (MacDonald, 1990) with an effluent nitrogen standard <2.5 mg L^{-1}.

A full-scale demonstration of the Envirex fluidized bed process (flow rate 1325 m^3 d^{-1}, bed expansion height 2.9–3.4 m, velocity 36 m h^{-1}) was carried out in 1993–1996 in Long Island Sound near New York. The experiments illustrated the possibility of achieving total effluent nitrogen in the 3–5 mg/L range at temperatures ranging from 11°C to 25°C (Sadick et al., 1996; Semon et al., 1997). Effluent total suspended solids (TSS) and BOD remained well below the 30-mg L^{-1} permit requirement. The process was characterized as relatively simple to operate and maintain.

In Europe, the first fluidized bed reactor for the tertiary denitrification of urban wastewater has been operating in Sweden since 1997 (Dahlberg, 1997). The WWTP serving 240,000 p.e. has been upgraded to achieve near zero nitrogen discharge. The main objective of the separation of the activated sludge system for BOD removal from the biofilm denitrification with a FBR was to enhance the stability of operation, to avoid bulking and foaming problems in the activated sludge system, and to achieve treatment levels difficult to obtain with conventional treatment schemes. Four FBR with sand media have been constructed and operated with a liquid velocity of 39 m h^{-1} and methanol addition (2.5 g g^{-1} N-NO$_3$). The expected denitrification rate and the effluent nitrogen concentration are respectively 5.6 kg m^{-3} d^{-1} and less than 5 mg N$_{tot}$ L^{-1}.

The FBR has also been investigated in Germany for the upgrading of the Frankfurt-Niederrad WWTP (1,3000,000 p.e.) to ensure denitrification. An Oxitron pilot plant of Dorr-Oliver GmbH (Pöpel and Kristeller, 1997) has been operated for postnitrification with methanol at a filtration velocity of up to 30 m h^{-1}. The maximum denitrification rate was 2.6–5.2 kg m^{-3} d^{-1}. The results obtained and the process reliability demonstrated the flexibility of the FBR and appeared to be a more economic solution than retrofitting of the existing plant. The estimated additional cost for postdenitrification was between \$0.04 and \$0.065 m^{-3} depending on energy and methanol costs.

6.3.3 Industrial Wastewater Treatment

In general, industrial wastewaters pose greater problems for treatment as compared to urban wastewater. Industrial wastewater effluents are highly polluted, with significant daily and seasonal fluctuations, they are not balanced with respect to nutrients, and may contain toxic substances in high concentrations. This means that the biofilm reactor design is more difficult and case-specific.

6.3.3.1 Anaerobic Biofilm Processes

6.3.3.1.1 Fixed Bed Processes. Oliva et al. (1995) presented some important design parameters for anaerobic filters applied for the treatment of food processing industries in Brazil. At 15–30°C, COD removal efficiency was above 80% for applied organic loading rates lower than 5 kg COD m^{-3} d^{-1}.

Expanded bed anaerobic GAC reactors have been used successfully in the treatment of inhibitory wastewaters from the coal gasification industry (Fox et al., 1990; Suidan et al., 1983; Wang et al., 1986) and landfill leachate (Imai et al., 1993). The efficiency of the active GAS medium in the purification of toxic wastewater was attributed to its excellent microbial attachment properties and high adsorption capacity for toxic organic compounds.

6.3.3.1.2 Fluidized Bed Processes. Anaerobic fluidized beds are commonly used for the treatment of food processing industrial wastewater, the petrochemical industry, and denitrification of industrial sewage. In the anaerobic digestion of industrial wastewater, a two-step system offers many advantages, such as the possibility of optimizing the steps of hydrolysis and methanogenesis, improving reaction kinetics, and increasing the stability of the reactor operation. Simple, low cost, mixed acidogenesis reactors have been applied successfully in France in coupling with Anaflux FBR for the treatment of food processing and paper wastewaters (Holst et al., 1997).

More than 17 full-scale Anaflux reactors were installed between 1986 and 1996 for the treatment of food processing wastewaters (breweries, grape juice, starch, soft drinks (Fig. 6.23), milk/whey, chocolate, citric acid, jam/preserves and corn mill industries), as well as for two pulp and paper factories and one perfume industry (Holst et al., 1997). The FBR diameters varied from 2.4 to 6.1 m and the treatment capacity from 670 to 50,000 kg COD d^{-1}. An optimized inlet distribution system in the form of a grid with distribution holes facing downward ensures uniform influent distribution. The specially designed three-phase separator enables the efficient separation of the biogas even with high gas velocities and the recirculation of particles that can be settled. This system was characterized by the highest volumetric treatment capacity, up to 70 kg COD m^{-3} d^{-1}.

6.3.3.1.3 Air-Lift Processes. Gist Brocades constructed 3 two-stage plants in Delft, the Netherlands, and Prouvy, France, for the treatment of fermentation process wastewaters (Heijnen et al., 1989). The fluidized bed heights were 13 and 11 m and the reactor volumes 215 and 80 m^3, respectively. Biofilm control is ensured by the gas/liquid turbulence. The treatment efficiency reported was 60–70% on COD, while the effluent contained 100 mg L^{-1} fatty acids (influent COD about 3.2 g L^{-1}). The maximum capacity reached was 30 kg COD m^3 d^{-1} with a biomass concentration of 20 kg m^{-3}. Large fluctuations in pH (3–10) and peaks of nitrates did not cause any deterioration of the process operation, except for a temporary stop in gas production.

Paques Water Systems has recently installed eight full scale airlift reactors in Europe, Shanghai (Shanghai Fosters), and Brazil (Brahma Navigantes) for the treatment of industrial effluents containing organic pollutants. Reactor volumes lie in the range of 120–420 m^3. In most of these systems COD conversion capacities of 3–5 kg COD m^3 d^{-1} together with nitrification rates of up to 1.2 kg N m^3 d^{-1} have been reported. One of these systems concerns a new development in the form of a denitrifying airlift reactor with an integrated anoxic compartment.

Figure 6.23 View of the Anaflux fluidized bed reactor for treatment of wastewater from the soft drink industry.

6.3.3.2 Aerobic Biofilm Processes. On the basis of laboratory-scale experiments, Arcangeli and Arvin (1995) demonstrated that several commonly found aromatic pollutants (aromatic hydrocarbons, phenol, methylphenol, chlorophenols, nitrophenols, chlorobenzenes, and aromatic nitrogen-, sulfur- and oxygen-containing heterocyclic compounds) can be biodegraded with relatively high rates in biofilm systems in a large range of bulk concentrations from just a few to 500 mg L^{-1}.

6.3.3.2.1 Fixed Bed Processes. Aerated biofilters Biofor have been studied in more than 15 pilot plants and successfully operated in more than 12 full-scale installations in Canada, France, Germany, and Italy for high-rate secondary treatment of pulp and paper mill wastewater of paper recycling units and polishing treatment of less concentrated effluents generated by the manufacture of fine papers (Rovel et al., 1994). Consistent SS and BOD removal rates above 75% were observed. Moreover, complete toxicity removal was observed from paper mill effluents.

Polypropylene nonwoven fabric carriers with a specific surface area of 600 m^2 m^{-3} and 55 × 55 mm size were used for the upgrading of existing aeration tanks to floating fixed bed reactors for the treatment of wastewater from oily food processing and chemical factories in Japan (Koyama et al., 1998). Overall BOD removal efficiency was 98% at 0.83 and 1.12 kg BOD m^{-3} d^{-1} loading, respectively for the oil food and chemical factory effluents.

6.3.3.2.2 Fluidized Bed Processes. The aerobic fluidized bed reactors are successfully applied for treatment of wastewater from coke plants (Nutt et al., 1984; Marvan et al., 1992), metal works (Sutton et al., 1986), and wastewaters containing high concentrations of nitrogen (Cooper and Williams, 1990; Owsley et al., 1989) and chlorinated compounds (Holst et al., 1991). General Motors Corporation represents the largest single industrial end-user of the aerobic fluidized bed technology with 12 aerobic fluidized bed reactors installed in the 1980s for carbon removal and nitrification from automotive manufacturing operations (Mishra and Sutton, 1991).

Several aerobic and anoxic FBR with GAC media have been operated in the USA for removal of petroleum hydrocarbons from groundwater, as well as for the aerobic treatment of chemical plants (Sutton and Mishra, 1994). Improved performances of FBRs compared to conventional suspended growth processes result from the stability of operation in conditions of shock loading. Long solid retention times of the biofilm enable the acclimation of attached biomass to less biodegradable organic compounds.

Marvan et al. (1992) reported the performances of aerobic fluidized beds for the removal of phenolic compounds from coking plants in Algoma, Ontario, Canada (2 reactors each 65.7 m^2 and 8.53 m height) and Hoogovens, Holland (2 reactors each 84.3 m^2 and 8.7 m height). Within a hydraulic residence time of 4.14 h, the concentration of phenolics was reduced from 200 to less than 6.6 mg L^{-1}.

Inverse fluidized bed (Olem and Unz, 1980) and rotating disc biofilm reactors are tested for treatment of acid mine drainage wastewater. Under extreme conditions of low temperature and low pH from 2.18 to 5.5, continuous biooxidation of Fe^{2+} to Fe^{3+} was accomplished with a treatment efficiency above 90%.

6.3.3.2.3 Moving Bed Processes. Recent pilot plant studies showed the feasibility of the application of the new Norwegian moving biofilm bed reactor MBBR for the treatment of industrial wastewater from the dairy industry (Rusten et al., 1992), corrugated paper mill wastewater (Broch-Due et al., 1994), and potato chips wastewater (Ødegaard et al., 1994). The highest volumetric carbon removal rates were achieved during the treatment of paper mill wastewater—about 22 kgCODs m^3 d^{-1}.

Several pilot plant and full-scale experiences for the application of the Linpor process for the treatment of wastewater from milk, distillery, textile, potato, pulp, paper, and petroleum industries have been reported by Reimann (1984) and Morper (1994). These bioreactors with sponge carriers ensured a high volumetric loading of 5–10 kg BOD m^3 d^{-1} at biomass concentrations greater than 15 kg m^{-3}.

6.3.3.2.4 Air-Lift Processes. Two 300 m^3 BAS reactors have been operated in Delft, the Netherlands, for posttreatment of anaerobically treated industrial wastewater (Heijnen et al., 1991). High volumetric conversion rates were achieved, 20 kg COD m^3 d^{-1}. The BAS alone or coupled with a predenitrifying suspended culture was recommended for the treatment of concentrated ammonia containing effluents above 500 mg L^{-1} (Benthum et al., 1997b).

6.4 CONCLUSION AND PROSPECTS

Successful resolution of environmental protection problems requires the development of new intensive biological technologies. Biofilm processes offer successful technological solutions for the new challenges facing the water industry for compactness, high specific conversion rates, and stable operation. Various innovative biofilm processes have been developed during the past decades for both water and wastewater treatment. Their main advantages could be summarized as follows:

Advantages Concerning Treatment Performances

1. Simultaneous biological treatment and suspended solids removal (submerged granular biofilters).
2. High biomass concentrations through the ability to use small granular media (submerged granular biofilters, mobile and fluidized beds), high solid hold-up (fluidized beds), and moderate shear stress (air-lift circulating beds, two-phase liquid-solid fluidized beds).
3. Higher biomass activity through simple, effective control of biofilm thickness (three-phase reactors), high biomass retention times, biochemical reactions accelerated by better phase mixing, and a larger surface area for mass transfer (expansion or turbulence of the support media).
4. Better mass transfer due to fewer limitations of external liquid/biofilm surface diffusion (turbulence, high liquid velocity), removal of internal diffusion limitations in the biofilm (thin biofilms), high transfer coefficients similar to those in tower bioreactors (three-phase fluidized beds, turbulent beds, air-lift circulating beds).
5. Stable operation with variations in substrate and toxin concentrations, temperature, and hydraulic loads (in particular the turbulent and air-lift circulating beds).

Advantages Concerning Design and Operation

1. No sludge recycling and bulking problems.
2. Design without complex technical devices: (a) easier effluent distribution (turbulent beds, air-lift circulating beds); (b) simple and economical aeration system (draft-tube air-lift reactors); (c) autocontrol of biofilm thickness (three-phase biofilm reactors).
3. Higher aeration efficiency (tall reactors).
4. No backwashing requirement (three-phase biofilm reactors).
5. Compactness and no noise and odor emissions (closed biofilm reactors).

The greater activity of the fixed biomass and its stable operation with variations in operating and environmental conditions are a unique advantage specific to advanced biofilm processes. Biofilm thickness control has been found to be an important parameter ensuring high efficiency of advanced biofilm processes.

Several hundreds of submerged granular biofilters have already been applied successfully throughout the world to remove nitrogenous and carbon-containing compounds from drinking (Nitrazur, Biodenit, Denitropur, etc.), municipal (Biofor, Biostyr, B2A, etc.) and industrial wastewater (Biofor). In the case of the upgrading of existing wastewater treatment plants, turbulent and moving bed reactors have the largest application (Kaldness moving bed, Linpor and Captor sponge moving beds, Pegasus and Pegazur immobilized beds, TURBO N and Mixazur circulating beds).

Industrial wastewaters from food processing are successfully treated by anaerobic fluidized bed processes. Aerobic fluidized beds are mostly applied in coke, chemical, and metal works industries, as well as for high stretch nitrogen wastewaters.

It is important to stress the high dynamics in the development of these innovative biofilm processes. For this reason, reactor configurations and design parameters are characterized by a very fast evolution. The reported data and design values must be used by design engineers with great care and after consultation of the most recent case studies for a given water or wastewater application in similar environmental conditions.

To take advantage of all features of the biofilm reactors, further studies are needed for a better understanding of process hydrodynamics and biofilm properties and to optimize reactor design, scale-up and biofilm activity control. Such an approach, which integrates the latest advances in the field of biological science and chemical engineering, is necessary for the development and application of new innovative biofilm reactors.

ACKNOWLEDGMENTS

The authors want to acknowledge all their colleagues from Degrémont and Anjou Recherche, France, as well as Prof. Mark Van Loosdrecht from the Technical University of Delft, The Netherlands, for their help in completing the information concerning design parameters and for providing illustrations for industrial biofilm processes. The technical review, comments, and suggestions of Prof. M. Suidan, Prof. B. Rittmann, Prof. J. Bryers, and Dr. P. Chudoba were very helpful in completing this chapter and are gratefully acknowledged.

REFERENCES

Arcangeli, J. P., and E. Arvin. 1995. Biodegradation rates of aromatic contaminants in biofilm reactors. *Wat. Sci. Tech.* 31(1): 117–128.

Æsøy, A., H. Ødegaard, K. Bach, R. Pujol, and M. Hamon. 1998. Denitrification in a packed bed biofilm reactor (Biofor)—Experiments with different carbon sources. *Wat. Res.* 32(5): 1463–1470.

Audic, J. M., D. Amar, G. M. Faup, and J. Partos. 1985. The use of biomass attached on fine granular medium for water treatment application to aerobic processes. *Deutsch-franzosisches kolloquium*, 18–19 April 1985, Hamburg, Germany.

Austrup, R. 1985. Reinigung von Chemie-Abwässern mit BIOHOCH-Reaktoren. *Korresp. Abwasser* 11: 948–953.

Atkinson, B. 1981. Immobilized biomass—A basis for process development in wastewater treatment. In *Biological Fluidized Bed Treatment of Water and Wastewater*, P. F. Cooper and B. Atkinson, eds. Chichester: Ellis Harwood, pp. 22–34.

Badot, R., T. Coulom, N. de Longeaux, M. Badard, and J. Sibony. 1994. A fluidized-bed reactor: The Biolift process. *Wat. Sci. Tech.* 29(10/11): 329–338.

Banerji, S. K., and M. H. Liu. 1993. Comparison of Captor system with other fixed-film media nitrification systems. *Water Poll. Res. J. Canada* 28(2): 289–310.

Bennemann, H., M. Feldmann, and D. C. Hempel. 1991. Nitrifikation mit immobilisierten Bakterien. *Wasser-Abwasser* 132(12): 686–689.

REFERENCES

Benthum, W. A. J. van, M. D. M. van Loosdrecht, and J. J. Heijnen. 1997a. Control of heterotrophic layer formation on nitrifying biofilms in a biofilm airlift suspension reactor. *Biotechn. Bioeng.* 53(4): 397–405.

Benthum, W. A. J. van, M. D. M. van Loosdrecht, and J. J. Heijnen. 1997b. Process design for nitrogen removal using nitrifying biofilm and denitrifying suspended growth in a biofilm airlift suspension reactor. *Wat. Sci. Tech.* 36(1): 119–128.

Bigot, B., X. Le Tallec, and M. Badard. 1998. A new generation of compact multi-media biological filter. *Proc. EWPCA/WEF/CIWEM Conf. Treatment Innovation For the Next Century*, Cambridge University, UK, July 7–10, 9p.

Bishop, P. L. 1997. Biofilm structure and kinetics. *Wat. Sci. Tech.* 36(1): 287–294.

Boller, M., M. Tschui, and W. Gujer. 1997. Effects of transient nutrient concentrations in tertiary biofilm reactors. *Wat. Sci. Tech.* 36(1): 101–109.

Borregaard, V. R. 1997. Experience with nutrient removal in a fixed-film system at full-scale wastewater treatment plants. *Wat. Sci. Tech.* 36(1): 129–137.

Broch-Due, A., R. Andersen, and O. Kristoffersen. 1994. Pilot plant experience with an aerobic moving bed biofilm reactor for treatment of NSSC wastewater. *Wat. Sci. Tech.* 29(5/6): 283–294.

Canler, J. P., and J. M. Perret. 1994. Biological aerated filters: Assessment of the process based on 12 sewage treatment plants. *Wat. Sci. Tech.* 29(10/11): 13–22.

Çeçen, F., and I. E. Gönenç. 1994. Nitrogen removal characteristics of nitrification and denitrification filters. *Wat. Sci. Tech.* 29(10/11): 409–416.

Chang, H. T., and B. E. Rittmann. 1987. Mathematical modeling of biofilm on activated carbon. *Environ. Sci. Technol.* 21: 231–241.

Chang, H. T., and B. E. Rittmann. 1994. Predicting bed dynamics in three-phase fluidized bed biofilm reactor. *Wat. Sci. Tech.* 29(10/11): 273–279.

Chen, Z., C. Zheng, and Y. Feng. 1995. Distributions of flow regimes and phase holdups in three-phase fluidized beds. *Chem. Eng. Sci.* 50(13): 2153–2159.

Chudoba, P., M. Pannier, A. Truc, and R. Pujol. 1998. A new fixed-film mobile bed bioreactor for denitrification of wastewaters. *Wat. Sci. Tech.* 38(8/9): 233–240.

Cooper, P. F. 1981. The use of biological fluidized beds for the treatment of domestic and industrial wastewater. *The Chem. Eng.* (8/9): 373–376.

Cooper, P. F., and D. H. V. Wheeldon. 1982. Complete treatment of sewage in a two-stage fluidized-bed system. *Wat. Pollut. Control* 81: 447–464.

Cooper, P. F., and S. C. Williams. 1990. High-rate nitrification in a biological fluidized bed. *Wat. Sci. Tech.* 22(1/2): 431–442.

Cooper, P., H. E. Crabtree, E. P. Austin, and M. K. Green. 1984. Some recent developments in sewage treatment in UK with CAPTOR and biological fluidised bed. In La biomasse fixée dans le traitement des eaux, *37es Journées Int.*, CEBEDEAU, Liège, 23–25 mai, pp. 307–339.

Dahlberg, A. G. 1997. Fluid idea. *WQI*, May/June, 22–23.

Daigger, G. T., T. A. Heinemann, G. Land, and R. S. Watson. 1994. Practical experience with combined carbon oxidation and nitrification in plastic media trickling filters. *Wat. Sci. Tech.* 29(10/11): 189–196.

Dauthuille, P., X. Kandel, and M. Druesne. 1992. Association de réacteurs à cultures fixées pour l'élimination de la pollution carbonée et azotée. *T.S.M. L'eau* 87(4): 177–185.

De Beer, B., J. C. Van der Heuvel, and S. P. P. Ottengraf. 1993. Microelectrode measurements of the activity distribution in nitrifying bacteria aggregates. *Appl. Environ. Microbiol.* 59: 573–579.

Deboosere, S., J. Gemoets, and W. Verstraete. 1984. Activated nitrification in a shallow fluidized bed reactor. *37es Journées Int.*, CEBEDEAU, Liège, 23–25 mai.

Diesterweg, G., I. Pascik, and J. F. Lawson. 1980. Tower-Biology and its application for the nitrification/denitrification of ammonia-rich wastewater. *Proc. 40th Ind. Waste Conf.*, Purdue Univ., Butterworth Publishers, 535–542.

Duddles, G. A., S. E. Richardson, and E. F. Barth. 1974. Plastic-medium trickling filters for biological nitrogen control. *J. Water Pollut. Control Fed.* 46(5): 937–946.

Ehlinger, F. 1992. La fermentation méthanique en lits fluidisés en France. *L'eau, L'ind., Les Nuisances* (155): 48–50.

Elmaleh, S., S. Papaconstantinou, G. M. Rios, and A. Grasmick. 1987. Organic carbon conversion in a large-particle spouted bed. *The Chem. Eng. J.* 34: B29–B34.

Elmaleh, S., S. Papaconstantinou, and G. M. Rios. 1992. Biological wastewater treatment in a high compacting multiphasic reactor. In *Récents progrès en génie des procédés*, Vol. 6, (20), 1–9, Toulouse.

Emori, H., K. Mikawa, M. Hamaya, T. Yamaguchi, K. Tanaka, and T. Takeshima. 1996. Pegasus—Innovative biological nitrogen removal process using entrapped nitrifiers. In Immobilised cells: Basics and applications. *Progress in Biotechnology* 11: 546–555.

Eppler, D., and A. Eppler. 1986. Neus Verfahren zur biologischen Denitrifikation von Grundwasser. *Wasserwirtschaft* 76: 492–494.

Fan, L. S., K. Fujie, and T. R. Long. 1984. Some remarks on gas-liquid mass transfer and biological phenol degradation in a draft tube gas-liquid-solid fluidized bed bioreactor. *A.I.Ch.E. Symp. Series* 80(241): 102–109.

Fdez-Polanco, F., F. J. Real, and P. A. Garcia. 1994. Behavior of an anaerobic/aerobic pilot scale fluidized bed for the simultaneous removal of carbon and nitrogen. *Wat. Sci. Tech.* 29(10/11): 339–346.

Fields, P. R. (1992). Deep or semi-deep shaft technology. *Env. Prot. Bull.* 20: 14–22.

Fouhy, K. 1992. Wastewater treatment reaches new heights. *Chem. Eng.* 12: 101–102.

Fox, P., M. T. Suidan, and J. T. Pfeffer. 1990. Hybrid expanded GAC reactor for treating inhibitory wastewater. *J. Environ. Eng.* 116: 438–453.

Frijters, C. T. M. J., D. H. Eikelboom, A. Mulder, and R. Mulder. 1997. Treatment of municipal wastewater in a Circox® airlift reactor with integrated denitrification. *Wat. Sci. Tech.* 36(1): 173–181.

Furumai, H., and B. Rittmann. 1994. Evaluation of multiple-species biofilm and floc processes using a simplified aggregate model. *Wat. Sci. Tech.* 29(10/11): 439–446.

Garrido, J. M., J. L. Campos, R. Méndez, and J. M. Lema. 1997. Nitrous oxide production by nitrifying biofilms in a biofilm airlift suspension reactor. *Wat. Sci. Tech.* 36(1): 157–163.

Gayle, B. P., G. D. Boardman, J. H. Sherreard, and R. E. Benoit. 1989. Biological denitrification in water. *J. Envir. Eng.*, ASCE, 115(5): 930–943.

Gjaltema, A., N. Van der Marel, M. C. M. van Loosdrecht, and J. J. Heijnen. 1997. Adhesion and biofilm development on suspended carriers in airlift reactors: Hydrodynamic conditions versus surface conditions. *Biotech. Bioeng.* 55(6): 880–889.

Godon, J. J., E. Zumstein, P. Dabert, F. Habouzit, and R. Moletta. 1997. Microbial 16S rDNA diversity in an anaerobic digester. *Wat. Sci. Tech.* 36(6/7): 49–55.

Golla, P. S., M. P. Reddy, M. K. Simms, and T. J. Laken. 1994. Three-years of full-scale Captor® process operation at Moundsville WWTP. *Wat. Sci. Tech.* 29(10/11): 175–181.

Gonçalves, R. F., L. Le Grand, and F. Rogalla. 1994. Biological phosphorus uptake in submerged biofilters with nitrogen removal. *Wat. Sci. Tech.* 29(10/11): 135–143.

Gönenç, E., and P. Harremoës. 1985. Nitrification in rotating disc Systems-I, criteria for transition from oxygen to ammonia rate limitation. *Water Res.* 19(9): 1119–1127.

Green, M. K., and P. J. Hardy. 1985. The development of a high-rate nitrification fluidized-bed. *Wat. Pollut. Control* 84: 44–55.

REFERENCES

Gregory, R., and I. Sheiham. 1981. Biological fluidized bed denitrification of surface water. The economics of a remedy for nitrate in drinking water. In *Biological Fluidized Bed Treatment of Water and Wastewater*, P. F. Cooper and B. Atkinson, eds. Chichester: Ellis Harwood, pp. 329–361.

Gros, H., G. Schnoor, and P. Rutten. 1986. Nitrate removal from ground water by autotrophic microorganisms. *Water Supply* 4: 11–21.

Hall, T., R. A. Walker, and T. F. Zabel. 1985. Nitrate removal from drinking water-process selection and design. *Proc. Conf. Nitrates dans les Eaux*, October 22–24, Ecole National Superieure de Chimie de Rennes, Paris, vol. 1, pp. 51–57.

Hamon, M. 1993. From pilot to full scale plant nitrogen removal with biofiltration process. *Proc. 9th Eruopean Sewage Conf.*, Munich, 11–14 May 1993, pp. 283–296.

Hanus, F., and C. Bernard. 1988. Dénitrification des eaux potables dans un réacteur biologique à élimination continue de la biomasse en excès: Un procédé inédit. *T.S.M.—L'eau* 83(4): 243–246.

Harremoës, P. 1978. Biofilm kinetics. In *Water Pollution Control Microbiology*, vol. 2, R. Mirchel, ed. New York: John Wiley, pp. 82–109.

Harremoës, P., A. Haarbo, M. Winther-Nielsen, and C. Thirsing. 1998. Six years of pilot plant studies for design of treatment plants for nutrient removal. *Wat. Sci. Tech.* 38(1): 219–226.

Heidman, J. A., R. C. Brenner, and H. J. Shah. 1988. Pilot-plant evaluation of porous biomass supports. *J. Environ. Eng.* 114: 1077–1096.

Heijnen, J. J., A. Mulder, W. Enger, and F. Hoeks. 1989. Review on the application of anaerobic fluidized bed reactors in wastewater treatment. *Chem. Eng. J.* 41: B37–B50.

Heijnen, J. J., A. Mulder, R. Weltevrede, J. Hols, and H. Van Leeuwen. 1991. Large scale anaerobic-aerobic treatment of complex industrial waste water using biofilm reactors. *Wat. Sci. Tech.* 23: 1427–1436.

Heijnen, J. J., M. C. M. van Loosdrecht, A. Mulder, and L. Tijhuis. 1992. Formation of biofilms in a biofilm air-lift suspension reactor. *Wat. Sci. Tech.* 26(3–4): 647–654.

Hem, L. J., B. Rusten, and H. Ødegaard. 1994. Nitrification in a moving bed biofilm reactor. *Wat. Sci. Tech.* 28(6): 259–272.

Hermanowicz, S. 1998. A model of two-dimensional biofilm morphology. *Wat. Sci. Tech.* 37(4/5): 219–222.

Hermanowicz, S., and J. J. Ganczarczyk. 1984. Dynamics of nitrification in a biological fluidized bed reactor. *Wat. Sci Tech.* 17: 351–366.

Hollo, J., and L. Czako. 1987. Nitrate removal from drinking water in a fluidized bed biological denitrification bioreactor. *Acta Biotechnol.* 7: 417–423.

Holst, J., B. Martens, H. Gulyas, N. Greiser, and I. Sekoulov. 1991. Aerobic biological regeneration of dichloromethane-loaded activated carbon. *J. Envir. Eng.*, ASCE, 117(2): 194–208.

Holst, T. C., A. Truc, and R. Pujol. 1997. Anaerobic fluidized beds: Ten years of industrial experience. *Wat. Sci. Tech.* 36(6–7): 415–422.

Hosaka, Y., T. Minami, and S. Nasuno. 1991. Fluidized-bed biological nitrogen removal. *Wat. Env. Tech.* (8): 48–51.

Hultman, B., K. Jönsson, and E. Plaza. 1994. Combined nitrogen and phosphorus removal in a full-scale continuous up-flow sand filter. *Wat. Sci. Tech.* 29(10/11): 127–134.

Imai, A., N. Iwami, K. Matsushige, Y. Inamori, and R. Sudo. 1993. Removal of refractory organics and nitrogen from landfill leachate by the microorganism-attached activated carbon fluidized bed. *Wat. Res.* 27(1): 143–145.

Imbeaux, I. D. 1935. *Qualité de l'eau et moyens de corrections*. Paris, France: Ed. Dunod.

Jackson, G., R. Mene, E. de Jong, and P. Chodoba. 1998. Upgrading and expansion using a Pegazur combined activated sludge system: Experience from a pilot scale trial on the island of Jersey. *Proc. WEFTEC '98*, October 3–7, Orlando, Florida, vol. 1, pp. 703–721.

Jerrard, R. 1989. RBCs in concrete: A new generation of sewage treatment. *Water and Waste Treatment*, May, pp. 46–49.

Kapoor, A., and T. Viraraghavan. 1997. Nitrate removal from drinking water-review. *J. Envir. Eng.*, ASCE, 123(4): 371–380.

Karamanev, D. G., and L. N. Nikolov. 1988. Influence of some physicochemical parameters on bacterial activity of biofilm: Ferrous iron oxidation by *Thiobacillus ferrooxidans*. *Biotech. Bioeng.* 31: 295–299.

Kleiber, B., G. Roudon, B. Bigot, and J. Sibony. 1994. Assessment of aerated biofiltration at industrial scale. *Wat. Sci. Tech.* 29(10/11): 197–208.

Koch, B., M. Osterman, H. Höke, and D. C. Hempel. 1991. Sand and activated carbon as biofilm carriers for microbial degradation of phenols and nitrogen-containing aromatic compounds. *Wat. Sci. Tech.* 25(1): 1–8.

Koch, G., and H. Siegrist. 1997. Denitrification with methanol in tertiary filtration at wastewater treatment plant Zürich-Werdhölzli. *Wat. Sci. Tech.* 36(1): 165–172.

Koopman, B., C. M. Stevens, and C. A. Wonderlick. 1990. Denitrification in a moving bed upflow sand filter. *J. Water Pollut. Control Fed.* 62(3): 239–245.

Koyama, T., Y. Sunaga, and H. Satoh. 1998. Application of floating bed reactor using non-woven fabrics carriers for biological wastewater treatment. *Proc. WEFTEC Asia TechN Conf. and Exhibition*, vol. 1, pp. 227–233.

Kruithof, J. C., C. A. van Bennelhom, H. A. L. Dierx, W. A. M. Hijnen, J. A. M. van Paassen, and J. C. Schippers. 1988. Nitrate removal from ground water by sulfur/limestone filtration. *Water Supply* 6: 207–217.

Kurt, M., I. J. Dunn, and J. R. Bourne. 1987. Biological denitrification of drinking water using autotrophic organisms with H_2 in a fluidized bed biofilm reactor. *Biotech. Bioeng.* 29: 493–501.

Larané, A. 1993. Un biofiltre où les bactéries travaillent 24 heures sur 24. *Ind. Techniques*, vol. 737, avril, p. 89.

Lazarova, V., and J. Manem. 1994. Advances in biofilm aerobic reactors ensuring effective biofilm control. *Wat. Sci. Tech.* 29(10/11): 319–327.

Lazarova, V., and J. Manem. 1996. An innovative process for wastewater treatment: The circulating floating bed reactor. *Wat. Sci. Tech.* 34(9): 89–99.

Lazarova, V., V. Pierzo, D. Fontvieille, and J. Manem. 1994a. Integrated approach for biofilm characterization and biomass activity cotnrol. *Wat. Sci. Tech.* 29(7): 345–354.

Lazarova, V., B. Capdeville, and L. Nikolov. 1994b. Influence of seeding conditions on nitrite accumulation in a denitrifying fluidized bed reactor. *Wat. Res.* 28(5): 1189–1197.

Lazarova, V., R. Nogueira, J. Manem, and L. Melo. 1997. Control of nitrification efficiency in a new biofilm reactor. *Wat. Sci. Tech.* 36(1): 31–41.

Lazarova, V., D. Bellahcen, D. Rybacki, B. Rittmann, and J. Manem. 1998a. Population dynamics and biofilm composition in a new three-phase circulating bed reactor. *Wat. Sci. Tech.* 37(4/5): 149–158.

Lazarova, V., R. Nogucira, J. Manem, and L. Melo. 1998b. Influence of dissolved oxygen on nitrification kinetics in a circulating bed reactor. *Wat. Sci. Tech.* 37(4–5): 189–193.

Lazarova, V., F. Ferre, L. Duval, and J. Manem. 1998c. Innovative approach for high rate nitrogen removal: Real time control of the coupling of TURBO DN and TURBO N biofilm reactors. *Proc. WEFTEC '98 Conf.*, Orlando, Florida, vol. 1, Water Environment Federation, pp. 99–108.

Leistner, G., G. Müller, G. Sell, and A. Bauer. 1979. Der Bio-Hochreaktor—Eine biologische Abwasserreinigungsanlage in Hochbauweise. *Chem. Ing. Tech.* 51(4): 288–294.

Le Tallec, X., S. Zeghal, A. Vidal, and A. Lesouëf. 1997. Effect of effluent quality variability on biofilter operation. *Wat. Sci. Tech.* 36(1): 111–117.

REFERENCES

Lewandowski, Z., P. Stoodley, S. Altobelli, and E. Fukushima. 1994. Hydrodynamics and kinetics in biofilm systems—Recent advances and new problems. *Wat. Sci. Tech.* 29(10/11): 223–229.

Liessens, J., R. Germonpré, S. Beernaert, and W. Verstraete. 1993. Removing nitrate with a methylotrophic fluidized bed: Technology and operating performance. *J. AWWA* 85(4): 144–154.

MacDonald, D. V. (1990). Denitrification of fluidized biofilm reactor. *Wat. Sci. Tech.* 22(1/2): 451–461.

Marvan, I. J., F. Craig, and P. M. Sutton. 1992. Treatability evaluation of coking plant effluent. *Int. Biodeterioration and Biodegradation* 30: 313–329.

McCarty, P. L., and D. P. Smith. 1986. Anaerobic wastewater treatment. *Environ. Sci. Technol.* 20(12): 1200–1206.

Meaney, B. J., and J. E. T. Strickland. 1994. Operating experience with submerged filters for nitrification and denitrification. *Wat. Sci. Tech.* 29(10/11): 119–125.

Meyer, J. M. 1993. Radioscopie de la mini-station d'épuration de Cergy-Neuville. *L'Usine Nouvelle* (2394): 54–55.

Mikawa, K., H. Emori, T. Takeshima, E. Ishiyama, and K. Tanaka. 1996. High rate and compact two-stage post-denitrification process with single sludge pre-denitrification. *Wat. Sci. Tech.* 34(1/2): 467–475.

Mishra, P. N., and P. M. Sutton. 1991. Biological fluidized beds for water and wastewater treatment: A state of the art review. In *Biodeterioration and Biodegradation 8*, H. W. Rossmoore, ed. New York: Elsevier Applied Science, pp. 340–357.

Morper, M. R. 1994. Upgrading of activated sludge systems for nitrogen removal by application of the LINPOR-CN process. *Wat. Sci. Tech.* 29(12): 167–176.

Nakhla, G. F., M. T. Suidan, and J. T. Pfeffer. 1990. Control of anaerobic GAC reactor for treating inhibitory wastewater. *Res. J. Water Pollution Control Fedn.* 62: 65–72.

Neu, K. E., and B. Rusten. 1998. The Kaldnes moving biofilm reactor (MBBR) process for on-site and small flow wastewater treatment or plant upgrades. *Proc. Conf. WEFTEC '98*, October 3–7, Orlando, FL, pp. 265–276.

Nikolov, L., and D. Karamanev. 1990. Change of microbial activity after immobilization of microorganisms. In *Physiology of Immobilized Cells*, J. A. M. DeBont et al., eds. Amsterdam, The Netherlands: Elsevier, pp. 461–466.

Nikolov, L., D. Karamanev, T. Penev, and D. Dimitrov. 1990. Full-scale inverse fluidized bed biofilm reactor for waste water treatment. *2nd Int. Biotechnology Conf.*, May 6–9, Seoul, p. 12.

Ninassi, M. V., J. G. Peladan, and R. Pujol. 1998. Pre-denitrification of municipal wastewater: The interest of up-flow biofiltration. *Proc. Conf. WEFTEC '98*, October 3–7, Orlando, Florida, vol. 1, pp. 445–466.

Nogueira, R., V. Lazarova, J. Manem, and L. Melo. 1998. Influence of dissolved oxygen on the nitrification kinetics in a circulating bed biofilm reactor. *Bioprocess Eng.* 19: 441–449.

Nutt, S. G., H. Melcer, and J. H. Pries. 1984. Two-stage biological fluidized bed treatment of coke plant wastewater for nitrogen control. *J. Water Pollut. Control Fed.* 56(7): 851–857.

Ødegaard, H., and B. Rusten. 1990. Upgrading of small municipal wastewater treatment plants with heavy dairy loading by introduction of aerated submerged biological filters. *Wat. Sci. Tech.* 22(7/8): 191–198.

Ødegaard, H., B. Rusten, and T. Westrum. 1994. A new moving bed reactor—Applications and results. *Wat. Sci. Tech.* 29(10/11): 157–165.

Ødegaard, H., B. Rusten, and J. Siljudalen. 1998. The development of the moving bed biofilm process—From idea to commercial product. *Proc. Conf. Treatment Innovation for the Next Century, Innovation 2000*, July 7, Cambridge, UK.

Olem, H., and R. F. Unz. 1980. Rotating disc biological treatment of acid mine drainage. *J. Water Pollut. Control Fed.* 52(2): 257–269.

Oliva, L. C. H. V., M. Zaiat, and E. Foresti. 1995. Anaerobic reactors for food processing wastewater treatment: Established technology and new developments. *Wat. Sci. Tech.* 32(12): 157–163.

Ouyang, C. F., and C. M. Liaw. 1994. The optimum medium of the suspended bio-medium aeration contactor process. *Wat. Sci. Tech.* 29(10/11): 183–188.

Owsley, D. S., J. S. Jeris, and R. Owens. 1989. Ammonia removal allows effluent reuse at fish hatchery using fluidized bed reactors. *43rd Purdue Ind. Waste Conf. Proc.* Chelsea, MI: Lewis Publ., pp. 449–457.

Parker, D. S., and T. Richards. 1986. Nitrification in trickling filters. *J. Water Pollut. Control Fed.* 58: 896–903.

Pascik, I., and T. Mann. 1984. The two-step nitrification of ammonia-rich waste water. *Wat. Sci. Tech.* 16: 215–223.

Pastorelli, G., G. Andreottola, C. Canziani, C. Darriulat, E. de Fraja Frangipane, and A. Rozzi. 1997. Organic carbon and nitrogen removal in moving-bed biofilm reactors. *Wat. Sci. Tech.* 35(6): 91–99.

Peladan, J. G., H. Lemmel, and R. Pujol. 1996. High nitrification rate with upflow biofiltration. *Wat. Sci. Tech.* 34(1/2): 347–353.

Peladan, J. G., H. Lemmel, and R. Pujol. 1997. Improved nitrification rate using high water velocity on upflow biofilters. *Proc. of Environ. Biotechnology International Symposium*, Oostende, Germany, April 21–23, part II, pp. 147–150.

Peladan, J. G., H. Lemmel, S. Tarallo, S. Tattersall, and R. Pujol. 1998. A new generation of upflow biofilters with high water velocities. *Proc. Int. Conf. On Advanced Wastewater Treatment Processes*, Leeds, UK, September 8–11, 10 pp.

Peters, R. W., and V. L. Foley. 1983. Fixed-film wastewater treatment systems: Their history and development as influenced by medical, economic, and engineering factors. In *Fixed-Film Biological Processes for Wastewater Treatment*, Y. C. Wu and E. D. Smith, eds. Park Ridge, NJ: Noyes Data Company, pp. 1–52.

Pöpel, H. J., and W. Kristeller. 1997. Post-denitrification at the Frankfurt-Niederrad wastewater treatment plant by fluidized-bed technology. *Wat. Sci. Tech.* 35(10): 95–102.

Puhakka, J. A., and K. Järvinen. 1992. Aerobic fluidized-bed treatment of poly-chlorinated phenolic wood preservative constituents. *Wat. Res.* 26(6): 765–770.

Pujol, R. 1991. L'épuration par biofiltration. Premiers constants. *Etude inter agences*, n°2, CEMAGREF.

Pujol, R., J. P. Canler, and A. Iwema. 1992. Biological aerated filters: An attractive and alternative biological process. *Wat. Sci. Tech.* 26(3–4): 693–702.

Pujol, R., M. Hamon, X. Kandel, and H. Lemmel. 1994. Biofilters: Flexible, reliable biological reactors. *Wat. Sci. Tech.* 29(10/11): 33–38.

Pujol, R., J. G. Peladan, and K. H. Buchl. 1995. Nutrient removal by upflow filtration. *Proc. 7th Int. Conf. On Design and Operation of Large Wastewater Treatment Plants*, Vienna, August 27–September 1, vol. 1, pp. 245–253.

Pujol, R., H. Lemmel, and M. Gousailles. 1998. A keypoint of nitrification in an upflow biofiltration reactor. *Wat. Sci. Tech.* 38(3): 43–49.

Raskin, L., R. Amann, L. K. Puolsen, B. Rittmann, and D. Stahl. 1995. Use of ribosomal RNA-based molecular probes for characterization of complex microbial communities in anaerobic biofilms. *Wat. Sci. Tech.* 31(1): 261–272.

Ravarini, P., J. Couttelle, and F. Damez. 1988. L'Usine de Dennemont. Une unité de dénitrification-nitrification à grande échelle. *TSM-L'eau* 83(4): 235–240.

Reimann, H. 1984. The LINPOR-process with fixed biomass on plastic foam—Practical aspects and results. In La biomasse fixée dans le traitement des eaux, *37es Journées Int.*, CEBEDEAU, Liège, 23–25 mai, pp. 353–361.

Richard, Y., and J. Partos. 1986. Elimination biologique des nitrates en vue de la production d'eau potable. *TSM-L'eau* 81: 141–147.

Richard, Y., and P. Thébault. 1992. Biological removal of nitrates-report on 7 years of operation and progress. *Water Supply* 10(3): 151–160.

Rittmann, B. E., and P. M. Huck. 1989. Biological treatment of public water. *CRC Critical Rev. Envir. Control* 19(2): 119–184.

Rittmann, B. E., F. Trinet, D. Amar, and H. T. Chang. 1992. Measurement of the activity of a biofilm: Effects of surface loading and detachment on a three-phase liquid-fluidized-bed reactor. *Wat. Sci. Tech.* 26(3–4): 585–594.

Roennefahrt, K. W. 1986. Nitrate elimination with heterotrophic aquatic microorganisms in fixed bed reactors with buoyant carriers. *Aqua* 5: 283–285.

Rogalla, F., M. Payraudeau, G. Becquet, M. M. Bourbigot, J. Sibony, and P. Gilles. 1990. Nitrification and phosphorus precipitation with biological aerated filters. *J. Water Pollut. Control Fed.* 62(2): 169–176.

Rogalla, F., B. Lacamp, G. Bacquet, and F. Hansen. 1991. "Ten years after": Les biofiltres aérés à l'heure européenne. *L'eau, L'Ind., Les nuisances* 147: 49–52.

Rogalla, F., M. Badard, F. Hansen, and P. Dansholm. 1992. Upscaling a compact nitrogen removal process. *Wat. Sci. Tech.* 26(5/6): 1067–1076.

Rols, J. L., and B. Capdeville. 1992. Un nouveau type de filtre biologique: Le lit flottant. In *Récents progrès en génie des procédés*, Vol. 6, n°20, 37–45, Toulouse.

Rovel, J. M., J. P. Trudel, P. Lavallée, and I. Schroeter. 1994. Paper mill effluent treatment using biofiltration. *Wat. Sci. Tech.* 29(10/11): 217–222.

Rutten, P., and G. Schnoor. 1992. Five years' experience of nitrate removal from drinking water. *Water Supply* 10(3): 183–190.

Rusten, B., H. Odegaard, and A. Lundar. 1992. Treatment of dairy wastewater in a novel moving bed biofilm reactor. *Wat. Sci. Tech.* 26(3–4): 703–711.

Rusten, B., O. Kolkinn, and H. Odegaard. 1997. Moving bed biofilm reactors and chemical precipitation for high efficiency treatment of wastewater from small communities. *Wat. Sci. Tech.* 35(6): 71–79.

Sadick, T., J. Semon, D. Palumbo, P. Kennan, and G. T. Daigger. 1996. Fluidized-bed denitrification. *Wat. Env. Tech.*, August, pp. 81–85.

Sanz, J. P., M. Freund, and S. Hother. 1996. Nitrification and denitrification in continuous upflow filters—Process modelling and optimization. *Wat. Sci. Tech.* 34(1/2): 441–448.

Semon, J., T. Sadick, D. Palumbo, M. Santoro, and P. Kennan. 1997. Biological upflow fluidized bed denitrification reactor demonstration project—Stamford, CT, USA. *Wat. Sci. Tech.* 36(1):139–146.

Sharma, B., and R. Ahlert. 1977. Nitrification and nitrogen removal. *Wat. Res.* 11(1): 897–925.

Shieh, W. K., P. M. Sutton, and P. Kos. 1981. Predicting reactor biomass concentration in a fluidized bed system. *J. Water Pollut. Control Fed.* 53: 1574–1584.

Siegel, M. H., and C. W. Robinson. 1992. Application of air-lift gas-liquid-solid reactors in biotechnology. *Chem. Engrg. Sci.* 47(13/14): 3215–3229.

Smith, A. T., and J. P. Hardy. 1992. High rate sewage treatment using biological aerated filters. *JIWEM* 3(2): 154–167.

Suidan, M. T., C. E. Strubler, S. W. Kao, and J. T. Pfeffer. 1983. Treatment of goal gasification wastewater with anaerobic filter technology. *J. Water Pollut. Control Fed.* 55(10): 1263–1270.

Sumino, T., H. Nakamura, N. Mori, and Y. Kawaguchi. 1992. Immobilization of nitrifying bacteria by polyethylene glycol polymer. *J. Fermen. Bioeng.* 73(1): 37–42.

Sutton, P. M., and P. N. Mishra. 1990. Fluidized bed biological wastewater treatment: Effects of scale-up on system performance. *Wat. Sci. Tech.* 22(1/2): 419–430.

Sutton, P. M., and P. N. Mishra. 1994. Activated carbon based biological fluidized beds for contaminated water and wastewater treatment: A state of the art review. *Wat. Sci. Tech.* 29(10/11): 309–317.

Sutton, P. M., P. N. Mishra, and L. Hachigian. 1986. Using biotechnology to treat metalworking wastewater. *Pollution Eng.* 18(4): 46–49.

Tanaka, K., T. Minori, T. Kimata, S. Harada, Y. Fujii, T. Mizuguchi, N. Mori, and H. Emori. 1991. Development of new nitrogen removal system using nitrifying bacteria immobilized in synthetic resin pellets. *Wat. Sci. Tech.* 23: 681–690.

Tanaka, K., T. Sumino, H. Nakamura, T. Ogasawara, and H. Emori. 1996. Application of nitrification by cells immobilized in polyethylene glycol. In *Immobilized Cells: Basics and Applications, Progress in Biotechnology* 11: 622–632.

Tang, W. T., K. Wisecarver, and Fan Liang-Shih. 1987. Dynamics of a draft tube gas-liquid-solid fluidized bed bioreactor for phenol degradation. *Chem. Eng. Sci.* 42(9): 2123–2134.

Tijhuis, L., M. C. M. van Loosdrecht, and J. J. Heijnen. 1992. Nitrification with biofilms on small suspended particles in airlift reactors. *Wat. Sci. Tech.* 26(9–11): 2207–2211.

Tijhuis, L., B. Hijman, M. D. M. van Loosdrecht, and J. J. Heijnen. 1996. Influence of detachment, substrate loading and reactor scale on the formation of biofilms in airlift reactors. *Appl. Microbiol. Biotechnol.* 45: 7–17.

Toettrup, H., F. Rogalla, A. Vidal, and P. Harremoës. 1994. The treatment trilogy of floating filters: From pilot to prototype to plant. *Wat. Sci. Tech.* 29(10/11): 23–32.

Trinet, F., R. Heim, D. Amar, H. T. Chang, and B. E. Rittmann. 1991. Study of biofilm and fluidization of bioparticles in a three-phase liquid-fluidized bed reactor. *Wat. Sci. Tech.* 23: 1347–1354.

Tsubone, T., Y. Ogaki, Y. Yoshy, and M. Takahashi. 1992. Effects of biomass entrapment and carrier properties on the performance of an air-fluidized-bed biofilm reactor. *Wat. Environ. Res.* 64(7): 884–889.

Turner, C. D., M. J. Koehmstedt, and J. R. Gallagher. 1990. A comparison of coupled upflow and coupled downflow fluid bed reactors treating synfuel wastewater. *44th Purdue Ind. Waste Conf. Proc.* Chelsea, MI: Lewis Publ., pp. 475–484.

Villaverde, S., P. A. García-Encina, and F. FDZ-Polanco. 1997. Influence of pH over nitrifying biofilm activity in submerged biofilters. *Wat. Res.* 31(5): 1180–1186.

Vital, J. L., H. Lemmel, and M. P. Gaudin. 1990. Un nouvel appareil de filtration sur lit flottant. *L'eau, L'Ind., Les nuisances* 135(3): 57–58.

Wang, Y. T., M. T. Suidan, and B. E. Rittmann. 1986. Kinetics of methanogens in an expanded-bed reactor. *J. Environ. Engr.* 112: 155–170.

Wijffels, R. H., and J. Tramper. 1995. Nitrification by immobilized cells: A review. *Enzyme and Microbial Technol.* 17: 482–492.

Wijffels, R. H., G. Englund, J. H. Hunik, E. J. T. M. Leenen, A. Bakketun, A. Günther, J. M. Obón de Castro, and J. Tramper. 1995. Effects of diffusion limitation on immobilized nitrifying microorganisms at low temperatures. *Biotech. Bioeng.* 45: 1–9.

Williams, S. C., D. W. Harrington, P. F. Cooper, and J. J. Quinn. 1986. High-rate nitrification in a biological fluidized bed at Horley STW—An interim report. *Wat. Pollut. Control* 85: 81–89.

Yu, H., and B. E. Rittmann. 1997. Predicting bed expansion and phase holdups for three-phase fluidized bed reactors with and without biofilm. *Wat. Res.* 31: 2604–2616.

Zahid, W. M., and J. J. Ganczarczyk. 1994. Fractal properties of the RBC biofilm structure. *Wat. Sci. Tech.* 29(10/11): 271–279.

Zhang, T. C., Y. C. Fu, and P. L. Bishop. 1994. Competition in biofilms. *Wat. Sci. Tech.* 29(10/11): 263–270.

Zheng, C., Z. Chen, Y. Feng, and H. Hofmann. 1995. Mass transfer in different flow regimes of three-phase fluidized beds. *Chem. Eng. Sci.* 50(10): 1571–1578.

BIOFILMS APPLIED TO HAZARDOUS WASTE TREATMENT

BRUCE E. RITTMANN
Northwestern University, Evanston, Illinois

ALEX OTTO SCHWARZ
Catholic University of Chile, Santiago, Chile, and Northwestern University, Evanston, Illinois

PABLO BALDOMERO SAEZ
Catholic University of Chile, Santiago, Chile

7.1 INTRODUCTION

Biofilms play important roles in the detoxification of hazardous organic contaminants such as volatile aromatic hydrocarbons, chlorinated solvents, phenolics, and chlorinated aromatics. They dominate in situ bioremediation of contaminated groundwater and soil (e.g., National Research Council, 1993; Rittmann et al., 1994; Yager et al., 1997; Reinhard et al., 1997) and biofiltration of gas streams containing volatile organic compounds (VOCs) (e.g., Kirchner et al., 1989, 1991; Ottengraf, 1988; Speitel and Mclay, 1993; Tonga and Singh, 1994; Swanson and Loehr, 1997; Leson and Smith, 1997; Alonso et al., 1997). Biofilm processes also are used to treat aqueous streams containing hazardous organic compounds (e.g., Fox et al., 1990; Nakhla et al., 1990; Sáez et al., 1991; Arcangeli and Arvin, 1997; Shimomura et al., 1997; Khodadoust et al., 1997; Ballapragada et al., 1997; Melcer et al., 1995).

This chapter explores whether biofilms offer any special advantages when detoxification is the treatment goal. More specifically, do biofilms have features that make them especially well suited for biodegrading the hazardous organic chemicals commonly found in wastewaters, groundwaters, and soils?

Biofilms II: Process Analysis and Applications, Edited by James D. Bryers.
ISBN 0-471-29656-2   Copyright © 2000 Wiley-Liss, Inc.

Biodegradation of the common hazardous organic chemicals involves physiological phenomena that challenge microbiological systems in ways not present with more conventional applications, such as sewage treatment. Although these phenomena are not specific to biofilms per se, they define the ways in which biofilms might provide unique advantages. Five critical phenomena are:

1. The contaminants frequently are *inhibitory* or *toxic* to the microorganisms that degrade them. Many organic compounds are self-inhibitory at high enough concentration (Gantzer et al., 1991), and microorganisms wash out of the system if continually exposed to an inhibitory concentration. Because most hazardous waste conditions involve mixtures of many compounds (Rittmann et al., 1994), other compounds in the mixture can inhibit microorganisms capable of biodegrading one compound (Sáez and Rittmann, 1993).

2. Even when toxicity is avoided, the capable microorganisms accumulate only if they can extract enough energy. This situation is particularly acute when the target compound is biodegraded strictly through *cometabolism*, in which the cometabolite is transformed incidentally by an enzyme produced for another purpose (Rittmann et al., 1994). The incidental transformation yields no electrons, energy, or carbon to sustain microbial growth and maintenance. Thus, the biomass is present only when an appropriate primary substrate is available to sustain the microorganisms capable of transforming the target cometabolite.

3. Supporting sufficient capable biomass is not just a problem for cometabolites. Concentrations of target compounds frequently must be driven to very low concentrations in order to minimize risks to humans and the environment. When a primary (i.e., energy generating) substrate has a very low concentration, its biodegradation produces electron and energy flows that also are very low. Concentrations less than a threshold value defined as S_{min} (Rittmann, 1987) give biomass synthesis rates less than the loss rate due to maintenance, predation, and physical removal. A compound present at a concentration less than its S_{min} is called a *secondary substrate* and can be biodegraded only when the biomass is sustained by utilization of another primary substrate present at a concentration greater than its S_{min} (Namkung and Rittmann, 1987).

4. The microorganisms capable of biodegrading a target compound live in *microbial communities* in which they often comprise only a tiny fraction of the total biomass. To survive in the community, the bacteria must cooperate with other community members.

5. For many target compounds, biodegradation is only one of several competing sinks. Transfers to solid or gas phases can reduce the compound's availability to the microorganisms. Because biofilms exist attached to solid surfaces, solid phase transfers can be accentuated.

Section 7.2 describes key characteristics of biofilms and relates them to the five phenomena for detoxification. Then Sections 7.3 and 7.4 utilize mathematical modeling to illustrate different conditions in which biofilms can create an advantage for detoxification. All sections view biofilms within the context of treatment technology.

7.2 CHARACTERISTICS OF BIOFILMS

7.2.1 Biofilms as Microbial Aggregates

A biofilm is a form of microbial aggregation, a feature widespread in nature and environmental biotechnology. Other forms of microbial aggregates include suspended

7.2 CHARACTERISTICS OF BIOFILMS

flocs, marine snow, microbial mats, and colonies on solid growth media (Brusseau et al., 1997). At this time, biofilms and flocs are exploited in treatment technology and are the focus here.

Biofilms and flocs share common features of aggregation, but some differences are important.

- Within the aggregate, the microorganism density is relatively high.
- For readily biodegradable substrates, the potential to consume the substrate is very high per unit volume of the aggregate.
- Mass transfer within the aggregate is a limiting factor, because molecular diffusion is the main transport mechanism.
- When the two previous features occur, strong concentration gradients form. In other words, the concentration of substrate (or product) is drastically different between the inside and the outside of the aggregate. In particular, substrate concentrations inside the aggregate are much lower than in the bulk liquid—even zero.
- In general, microbial aggregates readily separate from the liquid phase. Flocs settle under quiescent conditions, and biofilms remain in the system as long as the solid substratum is retained. Thus, the retention time of the aggregates is much longer than the hydraulic detention time. Good cell retention is essential for stable process performance, selection of slow-growing species, and economic feasibility (Rittmann, 1987).
- The different microorganisms within an aggregate maintain more or less fixed spatial relationships. However, biofilms tend to have a more stable spatial architecture, because flocs often experience regular breakup and reflocculation cycles. While small subclusters may remain intact in the floc, microorganisms near the center of a floc at one time may find themselves near the outside of a reformed floc.
- Under certain conditions, biofilms experience virtually no physical removal. Substrata that are highly creviced, such as granular activated carbon (GAC), or have internal macropores shield the microorganisms from shear stress and turbulent forces that detach biofilm (Chang and Rittmann, 1988; Gantzer et al., 1989). Having no physical removal enhances selection of slow-growing species and reduces S_{min}.
- In some cases, the substratum to which the biofilm attaches is a strong sink for organic molecules. Activated carbon is the most important example, although resins, zeolites, and some soils exhibit significant sorption properties. Sorbable compounds that are not biodegraded in the biofilm are adsorbed by the substratum. The sorption sink becomes active during shock loading of readily degradable substrates for compounds that are poorly or not biodegradable and before significant biofilm has accumulated. Sorption sequesters compounds from the microorganisms. Sequestration is advantageous when the compound is an inhibitor (Fox et al., 1990; Nakhla et al., 1990; Sáez et al., 1991), but sequestration of primary substrate slows biomass growth, although it serves as a long-term reservoir of substrate once biofilm is established (Chang and Rittmann, 1988).
- In other cases, the substratum is a source for substrates. One dramatic example is the growth of a biofilm on a membrane through which gaseous substrates diffuse. Key gaseous substrates include oxygen, methane, and hydrogen. Another important example is the formation of a biofilm on a nonaqueous phase liquid, such as a pool or droplet of hydrocarbon (Rosenberg and Rosenberg, 1981; Rittmann et al., 1994).

7.2.2 Do Substrata Change the Microorganisms?

When biofilms form on living substrata, such as plant and animal tissue, very specific host–microorganism interactions take place (Jones, 1984; Dazzo, 1984). Cell-to-cell signaling and highly specialized adhesion sites cause the colonizing microorganisms to induce a set of surface-triggered genes whose expression markedly affects the microorganisms' physiology. These very strong host-induced responses with living substrata lead us to ask if biofilm formation on nonliving environmental surfaces brings about physiological changes in the biofilm microorganisms. Do such changes, if they occur, give the biofilm microorganisms some unique capabilities relevant to detoxification?

The results of scientific research are too sparse to come to a firm conclusion. In terms of gross, observable responses, the literature shows no obvious trend and often is contradictory. Van Loosdrecht et al. (1990) surveyed the literature and found that attachment to substrata increased and decreased growth and respiration rates, for example. On the other hand, biofilm formation seems to trigger signaling molecules, such as homoserine lactones, that cause physiological changes between planktonic and attached cells. Population density controls gene expression that regulates a range of behaviors, including sporulation, generation of aerial hyphae, release of hydrolytic enzymes, biofilm development, and luminescence (Kaiser and Losick, 1993; Fugua et al., 1994; Brint and Ohman, 1995; Clewell, 1993; Davies et al., 1998; Davies and Geesey, 1995; Davies et al., 1993). Earlier analyses (van Loosdrecht et al., 1990; Rittmann et al., 1994; Stal et al., 1989; Palenick et al., 1989) suggested that bacterial attachment to an inert substratum does not, by itself, alter the physiology of a bacterium in ways that directly affect detoxification; however, the final answer is far from obvious.

Regardless of whether attachment to inert surfaces alters the physiology of the microorganisms with regard to detoxification, dense aggregation can affect the environmental conditions to which the microorganisms are exposed. For instance, stable concentration gradients in biofilms offer a range of ecological *niches* over distances of only a few micrometers (Wanner and Gujer, 1986; Kissel et al., 1984; Rittmann and Manem, 1992; Furumai and Rittmann, 1994). Complex microbial communities can develop, because the consumption or production of materials by one type of microorganism makes possible the growth of another type just micrometers away. The fact that the biofilm's physical structure is relatively stable means that the different microbial types can select their most favorable location and maintain the synergistic associations. Through this ecological mechanism, biofilm formation indirectly influences the diversity of the microbial community and may enhance detoxification potential (Villaverde et al., 1997; Yu et al., 1998).

A second possible mechanism by which biofilm formation can enhance the biodegradation potential is through exchange of genetic information. Horizontal transfer of deoxyribonucleic acid (DNA) can occur by several pathways (Rittmann et al., 1990), but the mechanism most relevant in the detoxification context is conjugation. Conjugation involves plasmids, which are smaller, circular strands of DNA that exist independently from the cell's chromosomal DNA. Whereas the chromosome defines the cell genetically and is replicated when the cell reproduces, the plasmid is replicated when it is transferred from a plasmid-containing donor cell to a plasmid-free recipient cell. After conjugation occurs, the donor maintains the plasmid, and the recipient becomes a plasmid-bearing transconjugant. Thus, conjugation multiplies and proliferates the genetic information contained on the plasmid. Conjugation is an energy-demanding process that also requires direct donor-to-recipient contact (Smets et al., 1990, 1993, 1995; MacDonald et al., 1992).

Plasmid conjugation holds promise for using biofilms to carry out detoxification. First, genes for the initial steps of biodegradation often are found on plasmids (Leahy and

COLOR FIGURES

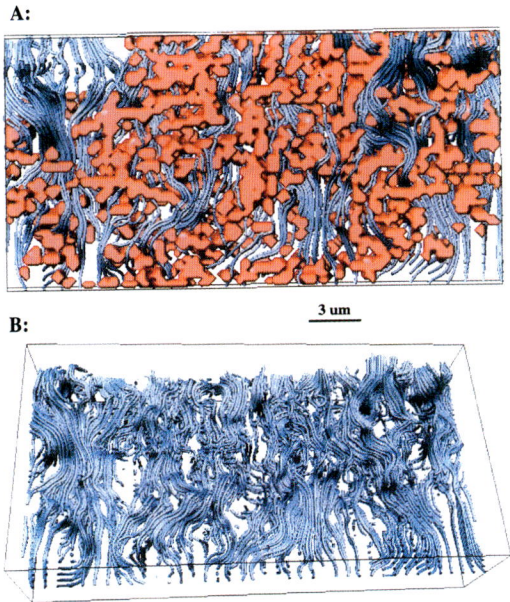

Figure 1.5 Propagational analysis of flow through a 3-D biofilm microstructure as recorded by confocal laser scanning microscopy. Flow is from front of picture to back (Singleton et al., 1997).

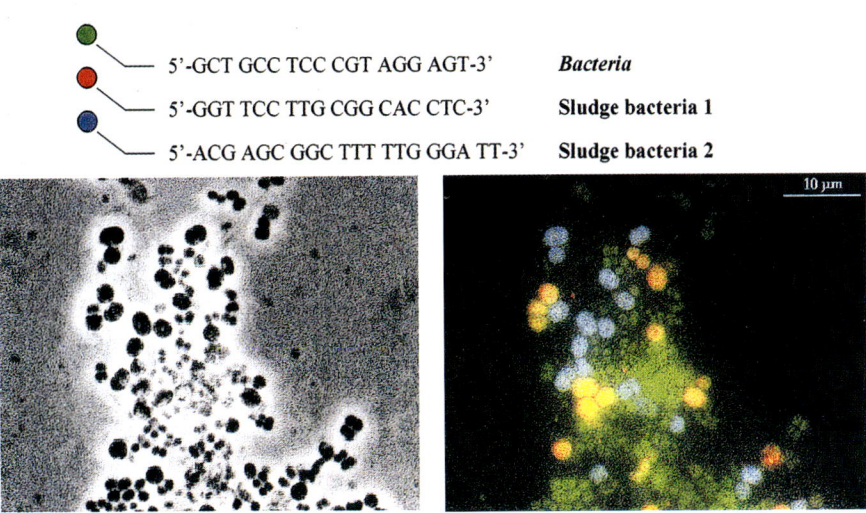

Figure 4.1 In situ rRNA hybridization for identification of organisms in an activated sludge system. In a deteriorated biological phosphorus removal reactor a novel group of bacteria have been identified, and specific rRNA probes were designed for their detection. The pictures show a representative sample from the sludge of the reactor, the left frame representing the phase contrast image from the microscope, and the right side presenting the fluorescence image. The coccus-shaped bacteria (red/yellow and blue cells) are new species identified from this reactor, whereas green cells belong to the *Bacteria*. (Reprinted from Nielsen et al., 1999)

COLOR FIGURES

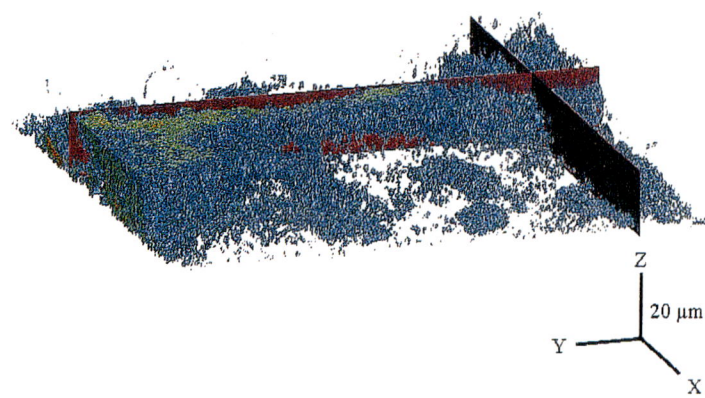

Figure 4.2 Structural organization of a composite biofilm. Confocal microscopy coupled with in situ rRNA hybridization and image analysis is a powerful method of revealing three-dimensional distributions of organisms and specific structural elements in the surface community. The image here shows that the biofilm is composed of densely populated regions as well as of void sectors most likely representing channels through which substrate and waste products may be transferred. It is seen that some cells form mounds penetrating from the substratum through the extracellular polysaccharide (EPS) matrix to the top of the biofilm.

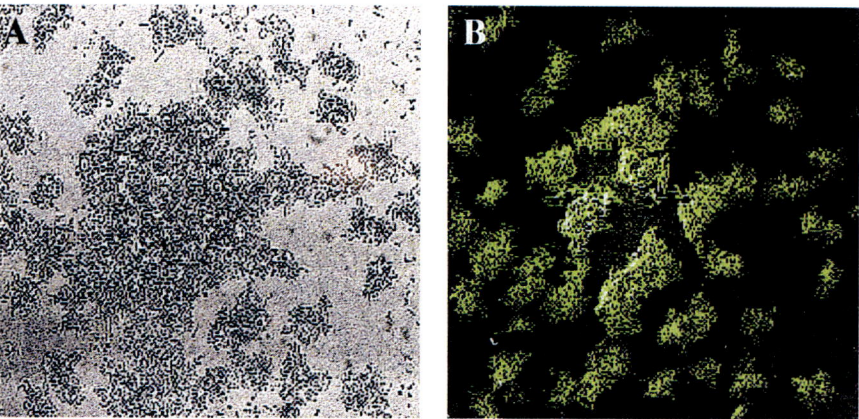

Figure 4.3 Metabolic interactions in complex biofilm communities. In a biofilm consisting of three bacterial strains, all capable of mineralizing toluene, a derivative of *P. putida* was introduced, which carries a gene fusion between the benzoate inducible TOL plasmid promoter, *Pm*, and the reporter gene, *gfp*. The community was fed benzyl alcohol as the sole carbon and energy source, and in a monoculture biofilm it was found that the specific strain of *P. putida* described showed no sign of induction of transcription from the *Pm* promoter (not shown). The presence of another community organism, *Acinetobacter spp.*, however, resulted in a significant induction of the *Pm* promoter in *P. putida* under conditions where the latter strains was colonizing the former (close proximity between the two organisms). The cells seen in *A* (phase contrast) form a dense microcolony composed of the two organisms, *P. putida* and *Acinetobacter*. The induction of green fluorescence, indicating activation of the *Pm* promoter, is coupled directly to the association between the two (*B*). The further away *P. putida* is from the *Acinetobacter* colony, the lower the Gfp signal. (Reprinted from Møller et al., 1998)

COLOR FIGURES

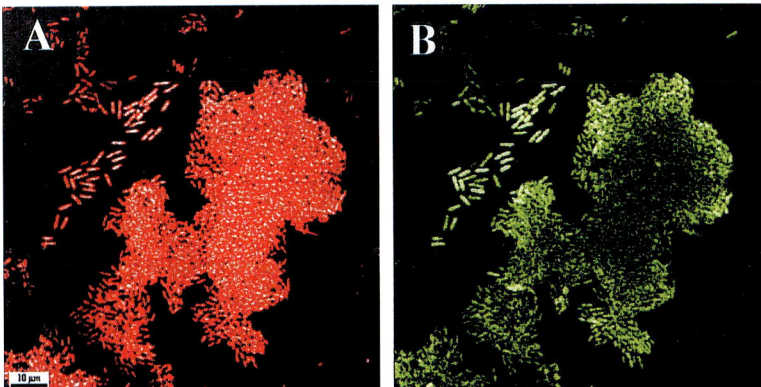

Figure 4.9 Growth activity monitored by reporter gene activity expressed from a fusion between a ribosomal promoter and the *gfp* gene. A fusion between the P1 rRNA promoter from *E. coli* and a gene encoding unstable Gfp was inserted in the chromosome of *P. putida* as part of a mini-transposon. The picture in the *A* frame shows a microcolony in the biofilm, and judging from the intensity of fluorescence it seems to be a homogenous collection of cells. The same colony was analyzed for its green fluorescence emitted from the unstable Gfp protein expressed from the gene fusion described, and it is clear from frame *B* that the cell population is fairly heterogenous with respect to green fluorescence. For full figure caption, see page 105.

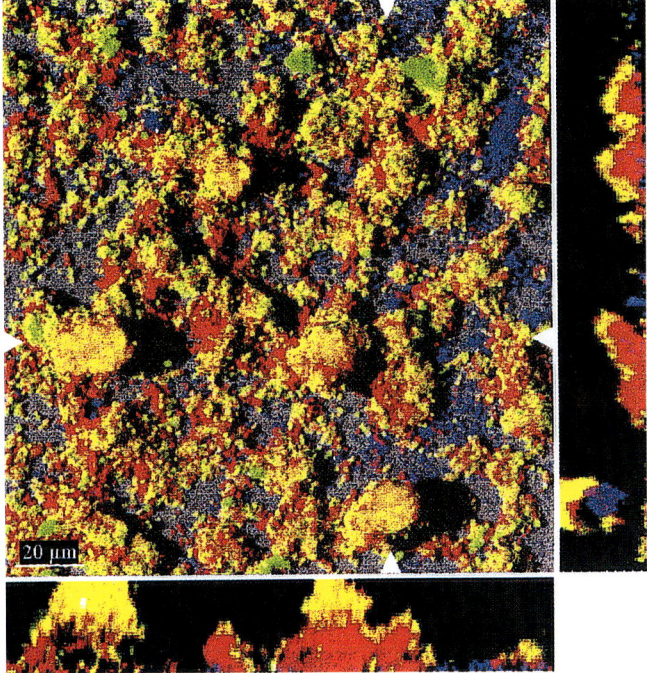

Figure 4.14 Direct monitoring of plasmid transfer in a complex biofilm community. Two organisms are identified with rRNA probes in the micrograph: *P. putida* (red cells) and *Acinetobacter* (blue cells). In addition, Gfp fluorescent cells represent transconjugants which are seen as either green or yellow cells. For full figure caption, see page 115.

COLOR FIGURES

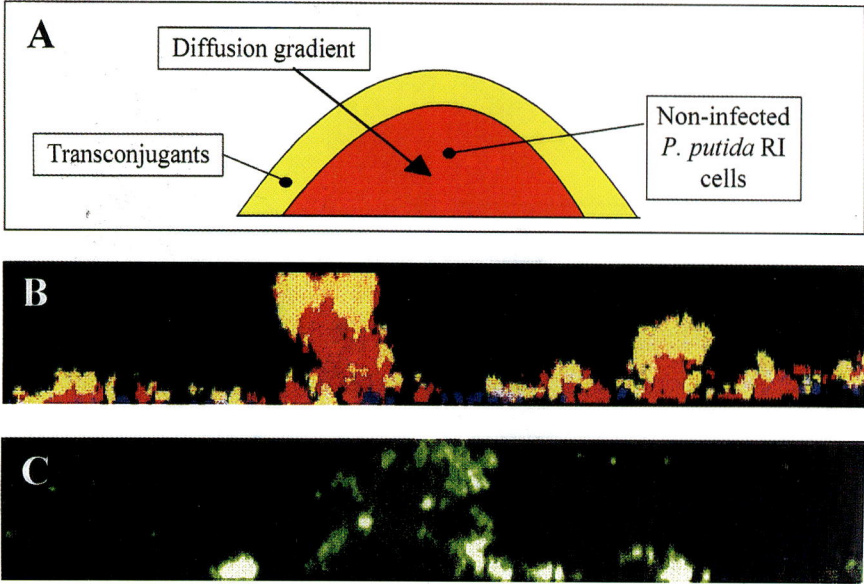

Figure 4.15 Correlation between active growth zones and plasmid conjugation zones in microcolonies in biofilms. It is likely that cells in microcolonies in biofilms face different nutritional conditions depending on their position in the colonies. *A*. A model based on nutrient diffusion gradients is suggested as an explanation for the lack of plasmid transfer to the potential recipient cells in the middle of the colony. *B*. A scan like the ones presented and explained in Figure 4.14, showing the position of transconjugant cells on the surface of a microcolony. *C*. The growth active cells are monitored through the fluorescence signal resulting from expression of an rRNA promoter/unstable Gfp fusion (cf. Fig. 4.9). There is a striking correlation between the growth active zone and the plasmid transfer active zone in these colonies. (Reprinted from Christensen et al., 1998)

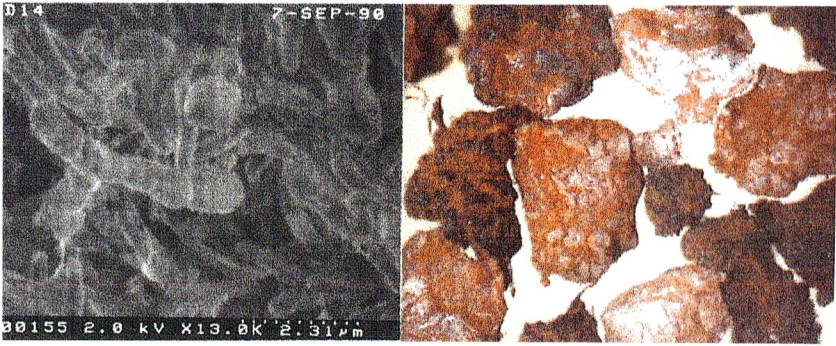

Figure 6.8 View of thin and active biofilms developed in mobile bed reactors: *a*) scanning electron microscopy view of a denitrifying biofilm from a fluidized bed reactor; *b*) binocular microscopy view of the active biomass after 2-(p-iodophenyl)-3-(*p*-nitrophenyl)-5-phenyltetrazolium chloride (INT) staining of a nitrifying biofilm from a circulating floating bed reactor.

7.2 CHARACTERISTICS OF BIOFILMS

Colwell, 1990; Rittmann et al., 1990, 1994). Apparently, sharing genes for the degradation of infrequently encountered molecules that are toxic to the microorganisms is an efficient way for the community to respond to their presence. Second, several researchers (Beaudoin et al., 1998a, 1998b; Bryers and Sharp, 1997; Ehlers and Bouwer, 1999) have shown recently that plasmid transfer occurs at significant rates with bacteria in biofilms. Third, the high cell density of the biofilm aggregate enhances one aspect of cell-to-cell contact, a prerequisite for conjugation. Whether biofilm formation significantly increases conjugation is a complicated issue. Whereas a high biomass density increases cell-to-cell contact in general, the cell's relatively fixed positions may prevent donors from coming into contact with plasmidless recipients. Smets et al. (1990) analyzed data on the kinetics of conjugation and found that the mixed, second-order rate coefficients for plasmid transfer were substantially lower for immobilized bacteria than for suspended and mixed bacteria. Hence, biofilm formation inevitably leads to one condition that speeds conjugation (high biomass density) and a second condition that slows conjugation (lack of cell mobility).

7.2.3 Protection in Biofilms

Although other mechanisms (like plasmid transfer) may also act, an overarching theme that connects the special challenges of detoxification to the characteristics of biofilms is *protection*. Being in a biofilm can protect sensitive and critical microorganisms from *inhibitory materials* and *excessive loss rates*.

7.2.3.1 Protection from Inhibition.
Protection from inhibitory materials, or toxicants, comes about in two ways. Perhaps the most obvious way is when the biofilm colonizes a strongly adsorptive surface, such as GAC. Rapid adsorption by GAC lowers the aqueous phase concentration of the toxicant, allowing the biofilm to survive a sudden shock and gradually degrade an inhibitory substrate. Figure 7.1 illustrates how active adsorption depresses the toxicant concentration to noninhibitory levels.

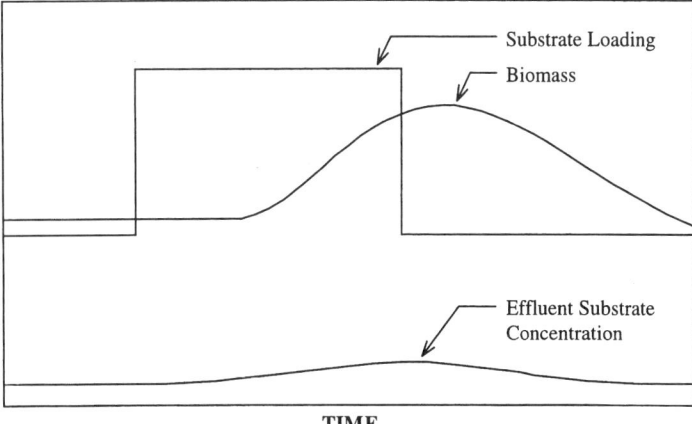

Figure 7.1 Schematic illustration of how active adsorption of an inhibitor into the micropores of an adsorbent substratum, such as GAC, reduces the inhibitor's concentration to below an inhibition threshold. Bacteria are able to function and grow near the substratum surface as long as the flux of inhibitor continues to draw down its concentrations.

Suidan and co-workers (e.g., Suidan et al., 1983; Nakhla et al., 1990; Fox et al., 1990; Khodadoust et al., 1997) performed extensive work demonstrating that using GAC as an adsorptive carrier for biofilms greatly reduces toxicity and allows successful removal of biodegradable components. A nonbiodegradable inhibitor is adsorbed in the GAC micropores, which are too small for entry by the microorganisms. Protection from a nonbiodegradable inhibitor works only when it is strongly adsorbable and continues to be sequestered from the microorganisms. Figure 7.1 illustrates that the flux of the inhibitor into the micropores draws down the inhibitor's concentration near the substratum surface, providing a location for microbial growth, regardless of whether the bulk liquid concentration is inhibitory. For extended operation with input of a nonbiodegradable toxicant, the GAC must be replaced regularly to maintain adsorption capacity and toxicant flux into the micropores.

A more subtle form of protection does not involve an adsorptive substratum, but requires that the inhibitor be biodegradable. In this case, the biofilm protects itself by taking advantage of the concentration gradients that normally develop. If the inhibitory substrate is utilized reasonably rapidly when inhibition is relieved, the interior of the biofilm can have a noninhibiting concentration that allows fast substrate utilization and positive biomass growth. The outer layer is exposed to higher, inhibitory concentrations, but protects interior portions of the biofilm by creating mass transport resistance that generates a strong concentration gradient. In effect, the microorganisms on the outside of the biofilm form a sacrificial shield for the microorganisms on the inside (Villaverde et al., 1997; Yu et al., 1998; Sáez et al., 1991).

Sáez et al. (1991) performed experimental and modeling studies on the anaerobic fermentation of phenol, a self-inhibitory primary substrate. An anaerobic, fluidized bed reactor having granular anthracite carriers was operated with a pH depressed to suppress methanogenesis and accentuate the effects of self-inhibition. Experimentally, Sáez et al. (1991) obtained stable phenol fermentation for phenol concentrations ranging from 24 to 640 mg/L in the bulk liquid. Under the same pH conditions, dispersed bacteria would have been inhibited and washed out at phenol concentrations about 76 mg/L.

Sáez et al. (1991) incorporated self-inhibition kinetics into the steady state-biofilm model (Rittmann and McCarty, 1980) and generated modeling results like those shown in Figure 7.2. At the outer surface of the biofilm, phenol is at a significantly inhibitory concentration, and the bacteria there have a net negative growth rate. In the middle of the biofilm, inhibition is largely relieved, and the bacteria have a net positive growth rate. Very deep in the biofilm, the phenol concentration is less than S_{min}, and the bacteria are in net decay.

For the steady state biofilm, the net positive growth in the middle zone just balances the net losses in the inner and outer zones. The substrate utilization in the middle zone drives the concentration gradient, and the mass transport resistance across the outer, inhibited zone allows the middle zone to experience little to no inhibition. Because the bacteria in the outer zone are inhibited and have a net negative growth rate, the middle zone continually "exports" bacteria to the outer zone. Thus, the steady state is a dynamic one in which the growing bacteria in the middle zone continually generate their protective layer.

The overall pattern is the same when a secondary substrate is the inhibitor. The modeling results of Figure 7.3 are based on 4-chlorophenol being an inhibitory secondary substrate for the primary substrate phenol (Sáez and Rittmann, 1991, 1993). The figure shows that a middle zone has a positive growth rate, and this zone begins once the concentration of the inhibitor (I_f) declines due to its flux through an outer inhibited layer. As in Figure 7.2, the middle zone grows bacteria and exports them to the outer layer, which is inhibited, but

7.2 CHARACTERISTICS OF BIOFILMS

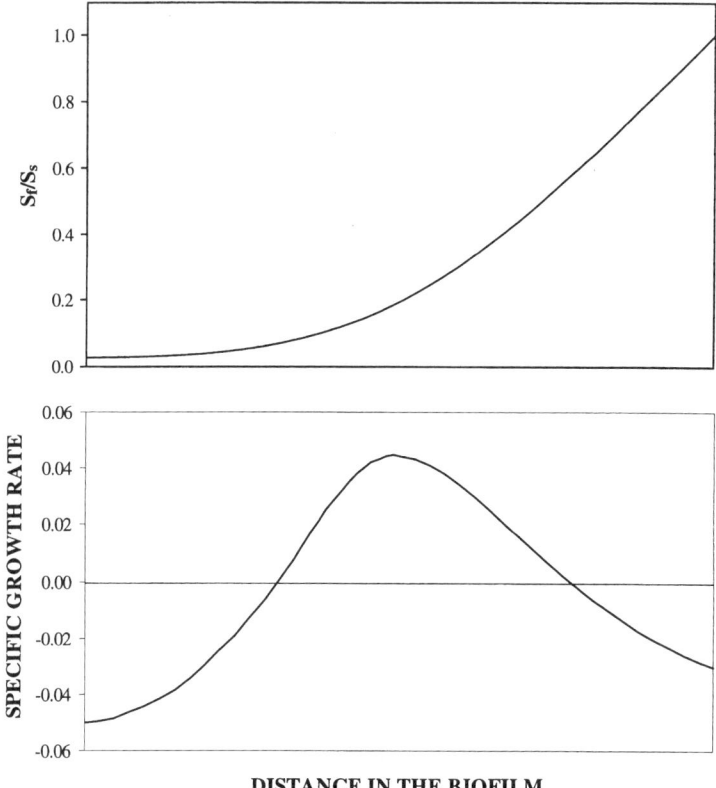

Figure 7.2 An example of how the concentration gradient for a self-inhibitory secondary substrate creates a positive growth zone at an intermediate location inside a steady state biofilm. The substratum is at the left, while the outer zone is at the right. S_f is the phenol concentration at any point in the biofilm, and S_s is the phenol concentration at the outer surface of the biofilm. This graphic is adapted from Sáez et al. (1991).

protecting. Protection from an inhibitory secondary substrate works best when the ratio S_s/I_s is large. The high ratio is particularly beneficial when the utilization kinetics for the secondary substrate are slow, since slow utilization creates a shallow concentration gradient. In such a case, ample utilization of primary substrate is needed in order to generate a thick enough protecting layer. Fortunately, a large S_s/I_s ratio is the classic situation for secondary utilization (Rittmann, 1987).

7.2.3.2 Protection from Excessive Loss Rate.
With a few exceptions, biofilm is lost from its outer surface (Gantzer et al., 1989). Most common is detachment by erosion, the continuous removal of small pieces of biofilm brought about by tangential shear stress or perpendicular stresses from turbulence or abrasion. Predation by protozoa or other higher forms also is concentrated at the biofilm's outer surface.

A biofilm loss rate due to detachment can be defined according to (Rittmann, 1982, 1989):

$$r_{\text{det}} = b_{\text{det}} X_f L_f \qquad (7.1)$$

where r_{det} = biomass loss rate per unit surface area ($M_x L^{-2} t^{-1}$) b_{det} = average, first-order rate coefficient for detachment (t^{-1}) X_f = average biofilm density ($M_x L^{-3}$) L_f = biofilm thickness (L) $X_f L_f$ = biofilm accumulation per unit surface area ($M_x L^{-2}$)

Although predation losses have not been well quantified, an average predation rate, r_{pred} ($M_x L^{-2} t^{-1}$) can be defined in parallel to the detachment rate:

$$r_{pred} = b_{pred} X_f L_f \tag{7.2}$$

where b_{pred} is the average, first-order rate coefficient for predation that depends on the number of types of predators present (t^{-1}).

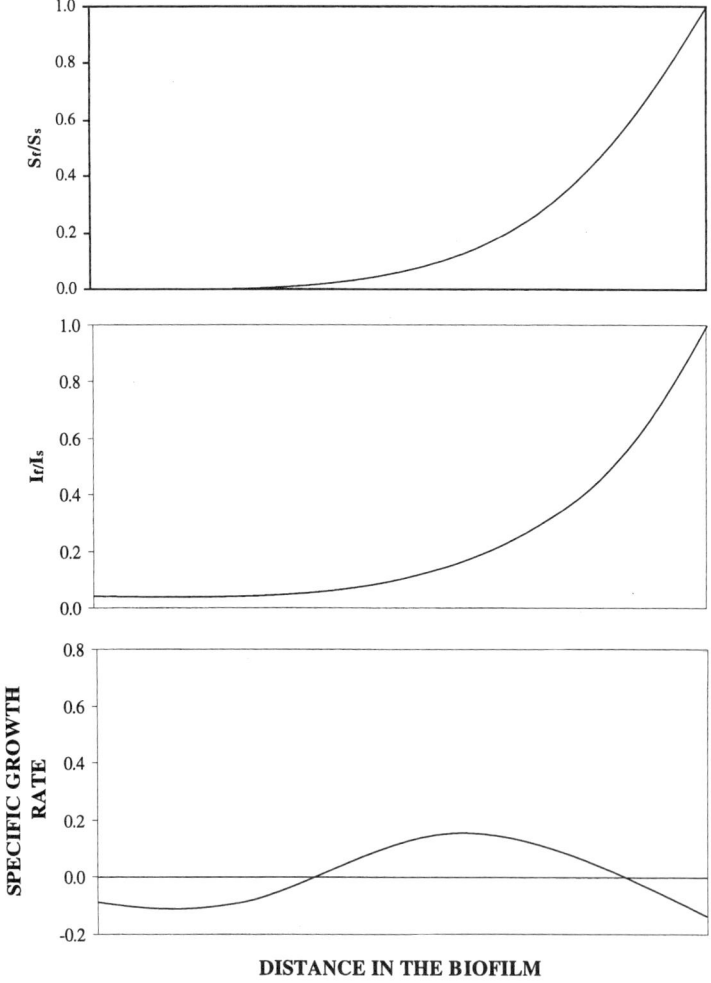

Figure 7.3 Modeling example of how biodegradation of an inhibitory secondary substrate (I) can be overcome by an outer protective layer created through primary substrate (S) utilization and growth inside the biofilm. I_f and I_s are the inhibitor concentrations inside the biofilm and at the outer surface, respectively.

7.2 CHARACTERISTICS OF BIOFILMS

The combined effects of detachment and predation, r_{loss} ($M_xL^{-2}t^{-1}$), is then described by

$$r_{loss} = r_{det} + r_{pred} = (b_{det} + b_{pred})X_f L_f \tag{7.3}$$

The average solids retention time, or sludge age, for a biofilm is defined as the reciprocal of the first-order loss rate coefficient (Rittmann, 1989). When detachment and predation occur, the average solids retention time is $(b_{det} + b_{pred})^{-1}$.

Section 7.2.1 described how certain creviced or macroporous carriers can reduce b_{det} to virtually zero. Likewise, creating conditions inimical to protozoa also can eliminate predation. Examples include strictly anaerobic conditions, very high liquid velocities, high salinity, and the presence of toxicants or antibiotics that act specifically against eucaryotes. The fluidized bed reactors exploited by Suidan and co-workers (Suidan et al., 1983; Wang et al., 1986; Nakhla et al., 1990; Fox et al., 1990; Khodadoust et al., 1997) had at least the first two conditions, which led to very low specific biofilm loss rates. The very low biofilm loss rates contributed importantly to the accumulation of slow-growing methanogens, despite the presence of toxicants. Thus, the ability to make $(b_{det} + b_{pred})$ extremely low in certain biofilm systems makes the average solids retention time very large and allows the colonization and maintenance of slow-growing species.

However, eliminating all predation and detachment is not feasible for many situations. In these cases, the fact that losses occur primarily from the outer surface offers the opportunity to protect sensitive microorganisms from loss rates that would otherwise wash them out of the system. Work coming from several research groups (Rittmann and Manem, 1992; Furumai and Rittmann, 1994; Wanner and Gujer, 1986; Rittmann et al., 1992, 1999; Kissel et al., 1984) shows that biofilms form "fuzzy layers" in which slower growing species accumulate more near the attachment surface, while faster growing species dominate near the outer surface. The layers are "fuzzy" because the transition of dominance from one microbial type to another is gradual.

The fuzzy layers are far from monocultures with strict boundaries. Instead, the relative amounts of the different species, averaged over some relevant surface area, shift gradually with depth. Figure 7.4 illustrates this gradual shifting in dominance for an aerobic biofilm comprised of aerobic heterotrophs, autotrophic nitrifiers, and inert (i.e., not metabolically active) biomass types. The heterotrophs are much faster growers than are the nitrifiers; hence, the heterotrophs dominate at the outer surface, where losses are greatest. The nitrifiers become most important in a middle zone. Inert biomass, which is formed as a decay by-product, is the "slowest growing" species and dominates near the attachment surface. Experimental work (Rittmann and Manem, 1992; van Loosdrecht et al., 1995; Watanabe et al., 1995; Okabe et al., 1996), while still preliminary, supports these general trends.

At the outer surface, where the losses occur, the biofilm is dominated by heterotrophs, which absorb the majority of the losses. Nitrifiers, on the other hand, are subject to only a fraction of the loss rate. As a rough approximation, we can assume that the loss rate of the nitrifiers is in proportion to their biomass fraction at the outer surface. For Figure 7.4, the nitrifiers have about 30% of the total biomass at the outer surface, and their first-order loss rate coefficient would be $0.3(b_{det} + b_{pred})$, giving an average solids retention time of $3.3/(b_{det} + b_{pred})$. In this way, slow-growing species are selectively retained, because they exist more in the protected layers away from the outer surface. In fact, this fuzzy zoning is a natural selection mechanism for multispecies biofilms (Rittmann and Manem, 1992).

In what situations are slow-growing populations particularly important for detoxification? The first situation occurs when a naturally slow growing species is critical for system performance. The second situation occurs when a key species is exposed to a toxicant, thereby slowing its growth rate. Of course, the combination of the two situations—a slow grower exposed to a toxicant—is the most severe challenge and one requiring maximum protection.

Important slow-growing species include:

- Nitrifiers, which are essential for nitrogen transformations and also can carry out cometabolic dechlorination of trichloroethylene (TCE) (Vanelli et al., 1991; Genestet, 1998).
- Methanogens, which stabilize organic oxygen demand to methane and indirectly are responsible for reductive dechlorination of halogenated aliphatics and aromatics (Wrenn and Rittmann, 1995, 1996; Nies and Vogel, 1990; Dolfing and Tiedje, 1986; Becker et al., 1998).
- Halorespiring anaerobes that garner energy from reductive dehalogenations (Fathepure et al., 1987; Gerritse et al., 1996; Hollinger et al., 1993; Hollinger and Schumacher, 1994; Scholz-Muramatsu et al., 1995; Maymo-Gatell et al., 1997; Krumholz et al., 1996; Wild et al., 1997; Christiansen and Ahring, 1996; Bouchard et al., 1996; Mohn and Tiedje, 1990; Dolfing and Tiedje, 1991).
- Any microorganisms that utilize as their primary electron donor only the target compound when that compound is at a very low concentration.

The situation in which inhibition slows the growth rate was discussed in Section 7.2.2 and is explored quantitatively in Sections 7.3 and 7.4. Like predation and detachment, the negative effects of inhibition occur most strongly at the outer surface when the inhibitor is input in an influent fluid. Regardless of whether the biomass at the outer surface grows there or is imported from the interior of the biofilm, protecting the sensitive species from an excessive loss rate is essential.

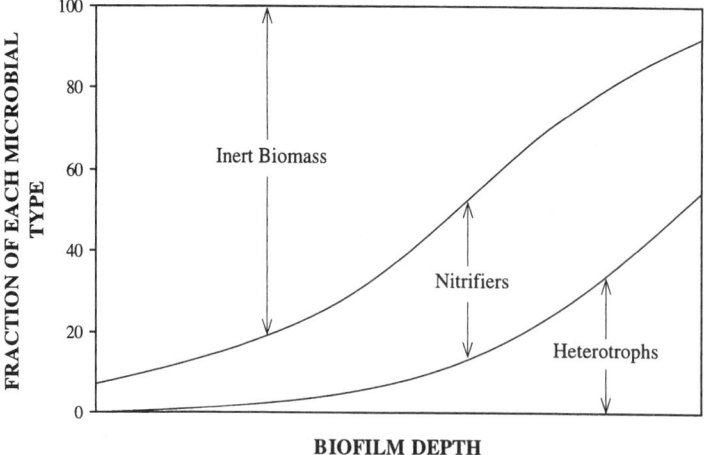

Figure 7.4 Illustration of fuzzy zoning in which the fast-growing heterotrophs dominate the outer surface, nitrifiers are most important toward the middle, and inert biomass dominated near the attachment surface. This schematic is adapted from Rittmann and Manem (1992).

7.3 PROTECTION FROM AN INHIBITORY SUBSTRATE FOR A STEADY STATE BIOFILM

The value of protection is illustrated tangibly through mathematical modeling, which quantifies the interactions among substrate utilization and diffusion, biomass growth and loss, and the effects of inhibition on substrate utilization and biomass growth. In this section, we first develop a mathematical model for a self-inhibitory substrate. We then solve the model for a steady state biofilm (Rittmann and McCarty, 1980), in which the growth of the biofilm due to substrate utilization is just balanced by its loss through decay and detachment. We show how inhibition places an upper limit on the substrate concentration that can be treated, but that biofilm growth can expand this range through protection, as described by Figure 7.2. Especially important in our elaboration of the effects of protection is the quantification of how the cells' maximum growth rate, inhibition, and detachment interrelate.

A primary goal of the modeling work is to identify relationships that ought to be observable and significant. These results can drive fundamental research and spur technological advancement by defining what is promising and how to quantify it.

7.3.1 Development of the Model

Sáez et al. (1991) and Gantzer (1989) presented models for steady state biofilms utilizing a self-inhibitory substrate according to the Haldane relationship. Our development follows the same patterns, but external mass transport resistance is neglected in order to simplify the presentation. Equations (7.4) to (7.8) are the mass balance and boundary condition equations that define the model: (a) the steady state substrate mass balance within a differential section of biofilm (Eq. (7.4)) that considers molecular diffusion (Fick's second law) and biological reaction (Haldane kinetics); (b) two boundary conditions (BCs) (Eq. (7.5)) at the substratum and Eq. (7.6) at the outer surface; (c) a steady state balance on total biofilm mass (Eq. (7.7)), including growth, decay, and detachment losses; and (d) the application of Fick's law to give the substrate flux into the biofilm (Eq. (7.8)).

$$\frac{\partial S_f}{\partial t} = D_f \frac{\partial^2 S_f}{\partial z^2} - \frac{q_m X_f S_f}{K_s + S_f + \frac{S_f^2}{K_i}} \qquad 0 \leq z \leq L_f \qquad (7.4)$$

$$\text{BC1:} \quad \frac{\partial S_f}{\partial z} = 0 \qquad \text{at } z = 0, \quad t \geq 0 \qquad (7.5)$$

$$\text{BC2:} \quad S_f = S_s \qquad \text{at } z = L_f, \quad t \geq 0 \qquad (7.6)$$

$$\frac{dL_f}{dt} = \int_0^{L_f} \left(\frac{Y q_m S_f}{K_s + S_f + \frac{S_f^2}{K_i}} - b' \right) dz \qquad (7.7)$$

$$J_b = D_f \frac{\partial S_f}{\partial z} \qquad \text{at } z = L_f \qquad (7.8)$$

where b' = total first order biofilm mass loss rate coefficient (t^{-1}), which includes decay (b) and detachment (b_{det}) D_f = molecular diffusion coefficient of the substrate within the biofilm (L^2t^{-1}) J_b = substrate flux into the biofilm ($M_sL^{-2}t^{-1}$) K_s = half-velocity concentration in absence of inhibition (M_sL^{-3}) K_i = inhibition coefficient (M_sL^{-3}) L_f = biofilm thickness (L) q_m = maximum specific rate of substrate utilization in absence of inhibition ($M_sM_x^{-1}t^{-1}$) S_f = rate-limiting substrate concentration in the biofilm (M_sL^{-3}) S_s = substrate concentration at the liquid/biofilm interface (M_sL^{-3}) t = time (t) X_f = cell density of the biofilm (M_xL^{-3}) Y = true yield coefficient ($M_xM_s^{-1}$) z = transverse coordinate in the biofilm (L)

Numerical solution of the model equations and concise interpretation of important concepts are greatly aided by a transformation of Eqs. (7.4) to (7.8) into a dimensionless domain. The following are the dimensionless variables, which are identified by an asterisk.

One main advantage of writing the model equations in the dimensionless domain is that, at steady state, $S_f^*(z^*)$, J_b^*, and L_f^* are functions only of S_s^*, K_i^*, and ϕ^*. Thus, the number of kinetic parameters of the model is reduced to only two dimensionless parameters: K_i^* and ϕ^*. Furthermore, the two dimensionless parameters hold special meaning and are crucial for interpreting the modeling results.

- K_i^*, the ratio of the Haldane inhibition concentration and the Monod half-maximum rate concentration, defines the *strength of the inhibition*. A large ratio indicates little inhibition, and a value of infinity means no inhibition. On the other hand, a small ratio (i.e., less than 1) indicates severe self-inhibition.
- ϕ^* is the ratio of the specific loss rate to the maximum specific growth rate. It is called the *growth potential* and has been denoted S_{min}^* in other places (e.g., Wirtel et al., 1992). We use ϕ^* in order to use a variation of S_{min}^* for another purpose. A small value of ϕ^* (i.e., much smaller than 1) means that the growth potential is high, since the maximum growth rate greatly exceeds the loss rate. On the other hand, a value of ϕ^* approaching or exceeding 1 indicates that the microorganisms can barely grow faster than their loss rate for the best conditions available to them.

Eqs. (7.4) to (7.8) now are written in the dimensionless domain as:

$$\frac{\partial S_f^*}{\partial t^*} = \frac{1}{L_f^{*2}} \frac{\partial^2 S_f^*}{\partial z^{*2}} - \frac{S_f^*}{1 + S_f^* + \frac{S_f^{*2}}{K_i^*}} = 0 \qquad 0 \leq z^* \leq 1 \qquad (7.9)$$

BC1: $\quad \dfrac{\partial S_f^*}{\partial z^*} = 0 \qquad\qquad\qquad\qquad\qquad\qquad$ at $z^* = 0$, $\quad t^* \geq 0 \quad (7.10)$

BC2: $\quad S_f^* = S_s^* \qquad\qquad\qquad\qquad\qquad\qquad\qquad$ at $z^* = 1$, $\quad t^* \geq 0 \quad (7.11)$

$$\frac{dL_f^*}{dt^*} = L_f^* \frac{YK_s}{X_f} \left\{ \int_0^1 \left(\frac{S_f^*}{1 + S_f^* + \frac{S_f^{*2}}{K_i^*}} \right) dz^* - \frac{\phi^*}{1 + \phi^*} \right\} = 0 \qquad (7.12)$$

$$J_b^* = \frac{1}{L_f^*} \frac{\partial S_f^*}{\partial z^*} \qquad\qquad\qquad\qquad\qquad \text{at } z^* = 1 \qquad (7.13)$$

Two special delimiting values can be obtained by solving Eq. (7.12) for the special case of $L_f^* = 0$. The two values are the two roots of the quadratic equation that results. The first

7.3 PROTECTION FROM AN INHIBITORY SUBSTRATE FOR A STEADY STATE BIOFILM

special value is $S_{min}*$, and it is the root with the negative sign on the radical term in the numerator:

$$S_{min}* = \frac{K_i^* - \sqrt{K_i^{*2} - 4K_i^*\phi^{*2}}}{2\phi^*} \qquad (7.14)$$

For no inhibition ($K_i^* = \infty$), Eq. (7.14) reduces to $S_{min}* = \phi*$, but for inhibited biofilms, $S_{min}* > \phi*$. The maximum value of $S_{min}*$ is $2\phi*$, which occurs when $K_i^*/\phi^{*2} = 4$. If $K_i^*/\phi^{*2} < 4$, the specific growth rate at any substrate concentration is less than the specific losses due to decay and shearing. Therefore, the biofilm loses mass continuously and cannot achieve a steady state condition. Consequently, $S_{min}*$ varies between $\phi*$ and $2\phi*$ for all feasible combinations of the parameters $\phi*$ and K_i^*.

The other solution of the quadratic equation is:

$$S_{max}*(L_f = 0) = \frac{K_i^* + \sqrt{K_i^{*2} - 4K_i^*\phi^{*2}}}{2\phi^*} \qquad (7.15)$$

$S_{max}*(L_f = 0)$ can be interpreted as the maximum bulk substrate concentration that allows a net positive growth rate when the biofilm is infinitely small. $S_{max}*(L_f = 0)$ varies between $2\phi*$ and ∞ for all possible combinations of $\phi*$ and K_i^*. For no inhibition, it has a value of ∞, which indicates that the microorganisms are never limited by inhibition at a high concentration. $S_{max}*(L_f = 0)$ defines the maximum concentration that dispersed microorganisms can tolerate; a higher concentration results in them having a net negative growth rate and being washed out.

The value of $S_{max}*(L_f = 0)$ is very important for the onset of biofilm growth, when the biofilm is physically thin and cannot provide protection through mass transport resistance. If a biofilm can grow to a thickness large enough that the concentration inside the biofilm is lower than it is at the outer surface, then the inner portion of the biofilm can have a net positive growth rate, producing biofilm overall net positive growth rates at concentrations higher than $S_{max}*(L_f = 0)$. For high enough substrate concentrations in the bulk liquid, however, the protection is overcome, and a steady state biofilm cannot exist. The substrate concentration at the biofilm outer surface at which inhibition is too great to allow a steady state biofilm is defined as $S_{smax}*$. It cannot be computed analytically, because the substrate concentration is changing inside the biofilm (recall Fig. 7.2). This substrate-concentration profile cannot be known a priori, which means that $S_{smax}*$ must be computed numerically.

7.3.2 Solving the Steady State Model

The dimensionless Eqs., (7.9) to (7.13), were solved using the numerical method of orthogonal collocation (Finlayson, 1980). Similar methods were used successfully in the analysis of biofilm (Sáez and Rittmann, 1988), activated carbon (Crittenden et al., 1980), and biofilm on activated carbon (Chang and Rittmann, 1987a). We used 32 collocation points in the biofilm.

Initial conditions were required for Eqs. (7.9) and (7.12). A bulk liquid concentration, S_s*, and values of $\phi*$ and K_i^* were imposed for each run. The model was then allowed to run until steady state values were obtained for the substrate concentration in the biofilm ($S_f*(z*)$), the substrate flux (J_b*), and the biofilm thickness (L_f*). The value of $S_{smax}*$ was determined as part of the exercise of varying S_s* for given values of $\phi*$ and K_i^*. In most cases, stable solutions for the global steady state resulted. In the cases with very low ϕ^* values, for large values of S_s^*, the solution showed oscillatory instability, but good results were obtained nevertheless.

7.3.3 Steady State Results

In order to represent the range of feasible situations, the variable ϕ^* was varied in the range of 0.01 to 1.0, while K_i^* was in the range of 0.1 to infinity. Figures 7.5 and 7.6 illustrate the major trends. In both figures, the substrate flux (J_b^*) is presented on the vertical axis as a function of the bulk-substrate concentration (S_s^*) on the horizontal axis and the strength of substrate inhibition (K_i^*) through the different lines. Figure 7.5 is for biomass with a very high growth potential ($\phi^* = 0.02$), while Figure 7.6 is for biomass with very poor growth potential ($\phi^* = 0.7$). The results in Figure 7.6 are relevant for intrinsically slow-growing microorganisms and/or those exposed to high detachment rates.

Three trends are common to both figures and illustrate the general response of biofilms utilizing self-inhibitory substrates. First, the substrate flux declines precipitously (and goes to zero) as S_s^* approaches S_{min}^*. This feature of steady state biofilms is well established (e.g., Rittmann and McCarty, 1980; Sáez and Rittmann, 1988, 1992) and means that the biofilm is severely limited by substrate availability. Second, the flux increases continually, but with a declining slope, for values of S_s^* greater than S_{min}^*. The continually declining slope reflects the increasing importance of mass transport resistance inside the biofilm. Third, the flux declines as the strength of inhibition increases, or K_i^* is smaller. Clearly, increasing self-inhibition reduces biomass accumulation and substrate flux.

Comparing Figures 7.5 and 7.6 points out a dramatic trend about how the growth potential interacts with inhibition: Figure 7.6 has many fewer lines. Whereas the microorganisms

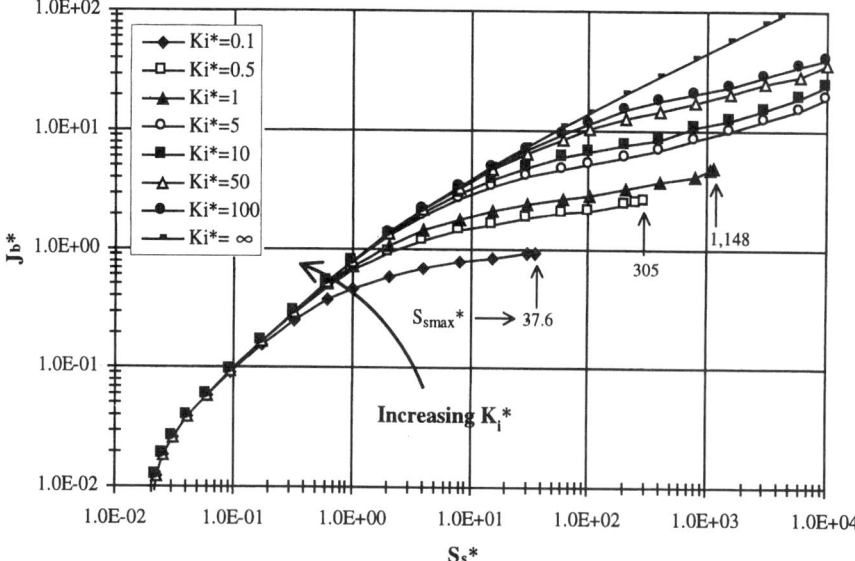

Figure 7.5 Effect of K_i^* on the relationship between substrate flux and substrate concentration for high growth potential, $\phi^* = 0.02$. The arrows show S_{smax}^* values that are less than 10^4.

7.3 PROTECTION FROM AN INHIBITORY SUBSTRATE FOR A STEADY STATE BIOFILM

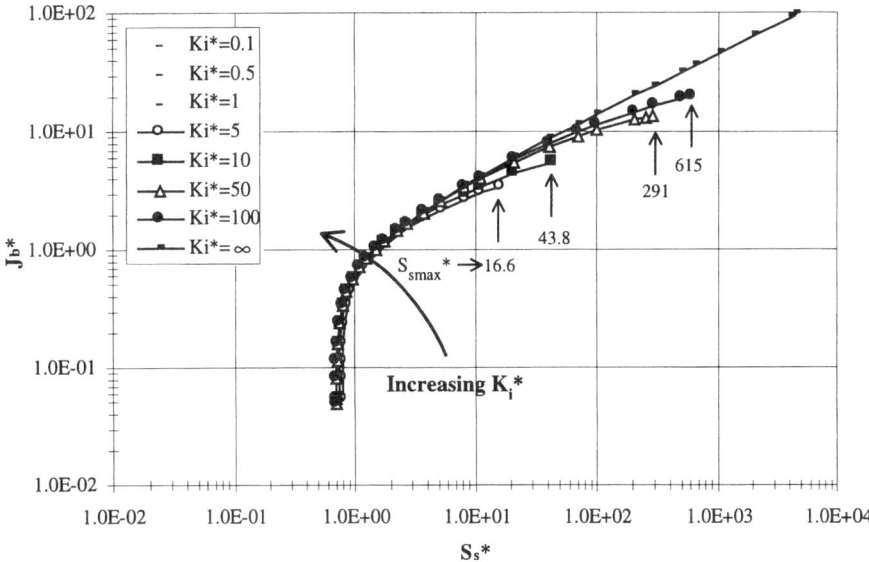

Figure 7.6 Effect of K_i^* on the relationship between substrate flux and substrate concentration for poor growth potential, $\phi^* = 0.7$. The arrows show S_{smax}^* values. For $K_i^* \leq 1$, no steady state biofilm exists, and no curve appears.

with the small ϕ^* can accumulate to form a steady state biofilm for even the most severe self-inhibition ($K_i^* = 0.1$), the microorganisms can develop a steady state biofilm only for modest to no inhibition ($K_i^* \geq 5$) when they have a poor growth potential. Quantitatively, having a larger value of ϕ^* makes S_{min}^* larger and S_{smax}^* smaller for any K_i^*. As inhibition increases, the distance between S_{min}^* and S_{smax}^* diminishes. When the growth potential is poor, and for $K_i^* < 5$, the biofilm cannot overcome inhibition and losses enough to create a protecting zone. Thus, no steady state biofilm exists for $K_i^* < 5$.

Table 7.1 summarizes the S_{smax}^* values for the whole range of ϕ^* and K_i^* values tested. A steady state biofilm is possible for any combination that yields an S_{smax}^* value, and a larger S_{smax}^* value indicates more biofilm accumulation and a higher substrate flux. The diagonal pattern of Table 7.1 demonstrates sharply the way in which enhancing the microorganisms' growth potential compensates for inhibition. Quantitatively, the same S_{smax}^* can be achieved by different combinations that trade off a low value of K_i^* versus a low value of ϕ^*. For example, severely inhibited microorganisms (i.e., $K_i^* = 0.1$) can form a steady state biofilm if the growth potential is good enough: $\phi^* \leq 0.1$. Therefore, Table 7.1 quantifies the limits on when substrate inhibition can be overcome by decreasing loss rates, particularly detachment.

Figure 7.7 emphasizes the way in which a low loss rate compensates for inhibition. For the example of strong inhibition ($K_i^* = 1$), Figure 7.7 shows that a steady state biofilm is possible for $\phi^* \leq 0.4$, but it is not possible for $\phi^* \geq 0.5$. Furthermore, the range of S_s^* values over which a steady state biofilm exists expands in both directions. The expansion to lower values of S_s^* is due to ϕ^* itself being smaller. Decreasing losses and ϕ^* also expands the S_s^* range in the higher direction, as greater growth potential allows protected layers inside the biofilm.

TABLE 7.1. S_{smax}* Values for Combinations of the Parameters K_i* and ϕ*

K_i*	\multicolumn{8}{c}{ϕ*}
	0.01	0.02	0.1	0.2	0.3	0.5	0.7	1
0.1	128[a]	38[a]	2.5	—[b]	—[b]	—[b]	—[b]	—[b]
0.5	—[c]	305[a]	24	7	—[b]	—[b]	—[b]	—[b]
1	—[c]	1148[a]	57	19	9	—[b]	—[b]	—[b]
5	—[c]	—[c]	348	133	73	32	17	6
10	—[c]	—[c]	740	288	162	75	44	22
50	—[c]	—[c]	4160	1590	926	463	291	175
100	—[c]	—[c]	8660	3270	1910	965	615	379
Infinity	Infinity	Infinity	Infinity	Infinity	Infinity	Infininity	Infinity	Infinity

[a] The numerical integrator delivers an oscillatory solution for high S_s* values, but a steady state biofilm exists for S_s* < S_{smax}*.
[b] A biofilm is not possible for this parameter combination, because the condition K_i*/ϕ*² ≥ 4 is not met.
[c] S_{smax}* > 10^5.

An example illustrates the trade-off. Sáez et al. (1991) found that for anaerobic biofilms grown on anthracite, the key parameters were: K_i* = 0.1 and ϕ* = 0.1. This is a case where a very strong self-inhibition (K_i* = 0.1) was compensated by a relatively strong growth potential (ϕ* ≤ 0.1). In this case, the high growth potential was achieved by having a negligible biofilm loss rate, since the anaerobic cells' growth rate was slow. For this example, Table 7.1 shows that S_{smax}* is 2.5, which was large enough to allow a steady state biofilm to form (Sáez et al., 1991).

7.4 PROTECTION INTERACTIONS WITH AN ADSORBING SUBSTRATUM

When the substratum is an adsorptive surface, such as GAC, any substrate or toxicant that penetrates the biofilm can be drawn to the substratum, where it binds. Adsorption becomes a sink that lowers the concentration of the penetrating material directly at the substratum interface and, depending on the balance between diffusion in the biofilm and the strength of the adsorbing driving force, can decrease the concentration inside the biofilm and even in the bulk liquid. When the adsorbed material is a substrate, adsorption sequesters substrate from the bacteria and slows their growth. However, adsorption of an inhibitor can relieve toxicity and increase microbial growth and activity.

The adsorption effects—positive or negative—are particularly strong for GAC, because it has a very large adsorption reservoir in its micropores, which are inaccessible to the bacteria. Thus, the substrate or toxicant is fully sequestered once it adsorbs into the micropores. The adsorbed material becomes available again only when it desorbs and travels back through the micropores and reaches the biofilm.

Any adsorbent substratum has a finite sorption capacity that eventually is exhausted by continued adsorption. Therefore, nonsteady state effects are essential. In this section, we develop a nonsteady state model in which biofilm is attached to a substratum that adsorbs a self-inhibitory substrate. It is directly appropriate for biofilm on GAC, although the model works equally well for substrata without micropores or adsorptive capacity. We first de-

7.4.1 The Nonsteady State Model for Biofilm on an Adsorptive Substratum

The model we used follows the prior work of Chang and Rittmann (1987a), who described biofilm on GAC. The GAC is described in spherical coordinates in which diffusion through micropores and equilibrium adsorption described by the Freundlich isotherm control the substrate concentration profile in the micropores. On the other hand, the biofilm processes (substrate diffusion and utilization, as well as biomass growth and loss) are described in a planar domain. The mass balance equations and the four boundary conditions are:

$$\frac{\partial \delta}{\partial t} = \frac{D_s}{r_s^2} \frac{\partial}{\partial r_s}\left(r_s^2 \frac{\partial \delta}{\partial r_s}\right) \qquad 0 \leq r_s \leq R \qquad (7.16)$$

$$\frac{\partial S_f}{\partial t} = D_f \frac{\partial^2 S_f}{\partial r_f^2} - \frac{q_m S_f}{K_s + S_f + \frac{S_f^2}{K_i}} X_f \qquad 0 \leq r_f \leq L_f \qquad (7.17)$$

$$\frac{d(L_f X_f)}{dt} = \int_0^{L_f}\left(\frac{Y q_m S_f}{K_s + S_f + \frac{S_f^2}{K_i}} - b'\right) X_f dr_f \qquad (7.18)$$

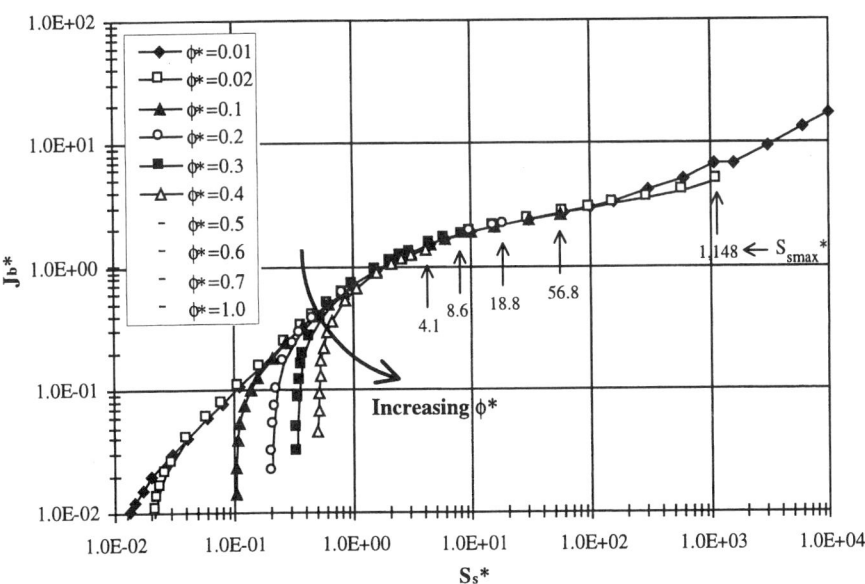

Figure 7.7 Decreasing ϕ^* expands the substrate-flux curve and the region in which a steady state biofilm exists. This example is for $K_i^* = 1$. For $\phi^* \geq 0.5$, no steady state biofilm exists, and no curve appears.

BC1: $\quad \dfrac{\partial \delta}{\partial rs} = 0 \quad r_s = 0, \quad t \geq 0 \quad (7.19)$

BC2: $\quad R^2 D_f \left. \dfrac{\partial S_f}{\partial r_f} \right|_{r_f = 0} = \rho_p \dfrac{\partial}{\partial t} \int_0^R \delta r_s^2 dr_s \quad t \geq 0 \quad (7.20)$

BC3: $\quad \delta_w = K_\delta (S_w)^{1/n} \quad t \geq 0 \quad (7.21)$

BC4: $\quad k_f(S_b - S_s) = D_f \left. \dfrac{\partial S_f}{\partial r_f} \right|_{rf = Lf} \quad t \geq 0 \quad (7.22)$

where b' = total first order biofilm mass loss-rate coefficient (t^{-1}), which includes decay (b) and detachment (b_{det}) δ = adsorption density of substrate at some position in the micropore ($M_s M_\delta^{-1}$) δ_w = adsorption density at the biofilm/GAC interface ($M_s M_\delta^{-1}$) D_f = substrate diffusion coefficient in the biofilm ($L^2 t^{-1}$) D_s = substrate surface diffusion coefficient in the activated carbon ($L^2 t^{-1}$) k_f = liquid-film mass transfer coefficient (Lt^{-1}) K_δ = Freundlich isotherm coefficient ($M_s^{1-1/n} L^{3/n} M_\delta^{-1}$) K_i = inhibition coefficient ($M_s L^{-3}$) K_s = half-velocity concentration in absence of inhibition ($M_s L^{-3}$) L_f = biofilm thickness (L) n = Freundlich isotherm exponent q_m = maximum specific rate of substrate utilization in absence of inhibition ($M_s M_x^{-1} t^{-1}$) ρ_p = apparent particle density of dry activated carbon ($M_\delta L^{-3}$) r_f = radial coordinate in the biofilm (L) r_s = radial coordinate in activated carbon (L) R = radius of the carbon particle (L) S_b = substrate concentration in the bulk liquid ($M_s L^{-3}$) S_f = rate-limiting substrate concentration in the biofilm ($M_s L^{-3}$) S_s = substrate concentration at the liquid/biofilm interface ($M_s L^{-3}$) S_w = substrate concentration at the biofilm/activated carbon interface ($M_s L^{-3}$) t = time (t) X_f = cell density of the biofilm ($M_x L^{-3}$) Y = true yield coefficient ($M_x M_s^{-1}$)

The four boundary conditions (Eqs. (7.19) to (7.22)) are significant, as they define the interactions between the biofilm and the GAC micropores and between the biofilm and the bulk liquid. Boundary condition 1 imposes symmetry of the concentration profile with respect to the center of the substratum particle. Boundary condition 2 states that the substrate mass entering the activated carbon from the biofilm must be equal to the increase in adsorbed substrate in the activated carbon. An adsorption isotherm is used to relate the concentration of substrate in the liquid phase to the adsorbed density at the biofilm/substratum interface (BC3). An isotherm of the same form is implicit in the mass balance inside the micropores (Eq. (7.16)). Boundary condition 4 states that the substrate flux entering the biofilm at its outer surface equals the flux entering the mass transport boundary layer from the bulk liquid.

The model also includes a mass balance on substrate in the bulk liquid of a single completely mixed reactor:

$$\dfrac{dS_b}{dt} = \dfrac{Q}{V\varepsilon}(S_0 - S_b) - \dfrac{3 X_w k_f (R + L_f)^2}{V \varepsilon \rho_p R^3}(S_b - S_s) \quad (7.23)$$

where ε = porosity of the reactor Q = flow rate of the feed solution ($L^3 t^{-1}$) S_0 = substrate concentration in the feed ($M_s L^{-3}$) V = empty bed volume of the reactor (L^3) X_w = weight of substratum in the reactor (M_δ).

We carried out the numerical solution of Eqs. (7.16) to (7.23) in a dimensionless domain. We do not repeat all of the dimensionless parameters and equations here. When presenting

7.4 PROTECTION INTERACTIONS WITH AN ADSORBING SUBSTRATUM

the results, we refer to the key dimensionless parameters of the steady state model—K_i^* and ϕ^*, which retain the same definitions. The dimensionless equations were solved simultaneously by orthogonal collocation (Finlayson, 1980; Chang and Rittmann, 1987a).

7.4.2 Model Parameters

Table 7.2 lists the parameter values for microbial kinetics, sorption, diffusion, and the physical system. The K_i^* value was 5 in all cases, giving a moderate inhibition. The only parameter varied was b_{det}. The ϕ^* values ranged from 0.02 for the lowest b_{det} to 0.7 for the highest b_{det}, thereby spanning from a very high growth potential to a poor growth potential. The acronyms BFAC and BFIS stand for biofilm on activated carbon and biofilm on an inert surface, respectively.

7.4.3 Nonsteady State Results and the Effects of an Adsorptive Substratum

The nonsteady state model was run until the solution reached steady states for the low and high values of ϕ^* and for the BFAC and BFIS modes. For the case of the high ϕ^*, the influent substrate concentration was increased in four equal steps in order to achieve the steady state. Without stepping, the inhibition was too great, and the biofilm gradually was lost. No stepping was needed for the low ϕ^*. Eventually, the BFAC and BFIS runs reached

TABLE 7.2. Parameters Used in the Nonsteady State Modeling

Parameter	Symbol	Unit	BFAC	BFIS
Substrate concentration in the feed	S_o	mg/cm³	0.0048	0.0048
Maximum specific rate of substrate utilization	a_m	mg/(mg of cell day)	13.7	13.7
Half-velocity concentration	K_s	mg/cm³	0.00024	00024
Haldane inhibition coefficient	K_i	mg/cm³	0.0012	0.0012
True yield of biomass	Y	mg of cell/mg	0.34	0.34
Biofilm decay coefficient	b	day⁻¹	0.068	0.068
Biofilm shear loss coefficient	b_{det}	day⁻¹	0.023–1.85	0.023–1.85
Biofilm density	X_f	mg of cell/cm³	6.44	6.44
Initial biofilm thickness	L_{fo}	cm	0.00025	0.00025
Substrate diffusion coefficient in the biofilm	D_f	cm²/day	0.584	0.584
Liquid-film mass transfer coefficient	K_f	cm/day	623.7	623.7
Surface diffusivity	D_s	cm²/day	0.0004	—[a]
Weight of adsorbent media in the reactor	X_w	mg	498.8	498.8
Freundlich isotherm coefficient	K_δ	—[b]	38.05	—[a]
Freundlich isotherm exponent	n	dimensionless	2.504	—[a]
Apparent particle density	ρ_p	mg/cm³	686	686
Radius of particle	R	cm	0.0249	0.0249
Empty bed volume of the reactor	V	cm³	12.38	12.38
Flow rate of the feed solution	Q	cm³/day	6000	6000
Porosity of the reactor	ε	dimensionless	0.4	0.4
Hydraulic detention time based on empty bed	θ	min	2.97	2.97

[a] The parameter is not needed.
[b] K_δ is in mg of sub$^{1-1/n}$L$^{1/n}$g of car^{-1}.

the same steady states. However, the time courses to reach the steady states were quite different when b_{det} and ϕ^* were large.

Figure 7.8 presents the bulk liquid substrate concentration (normalized to the final influent concentration) and the biofilm thickness for the modeling runs in which b_{det} and ϕ^* were at their maximum values, 1.85 day^{-1} and 0.7, respectively. Although both systems eventually reach the same steady states, the activated carbon substratum diminishes the peak substrate concentrations in the reactor immediately after a step increase in S_0. This effect is greatest near the beginning of the run, when the biofilm accumulation is small.

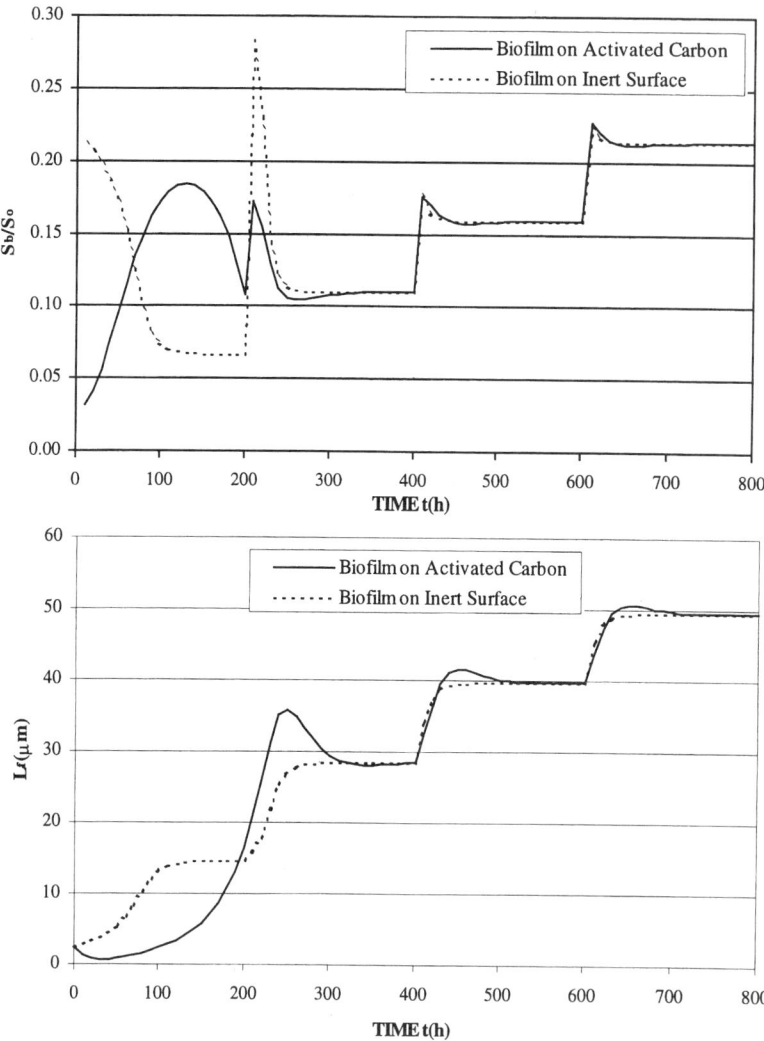

Figure 7.8 Evolution of the substrate concentration in the bulk liquid and the biofilm thickness for the BFAC and BFIS runs when $\phi^* = 0.7$. S_b is normalized to the final influent concentration (S_0 = 0.0048 mg/cm^3), and the spikes indicate the times at which the actual influent concentration was increased by 25% of S_0.

7.4 PROTECTION INTERACTIONS WITH AN ADSORBING SUBSTRATUM

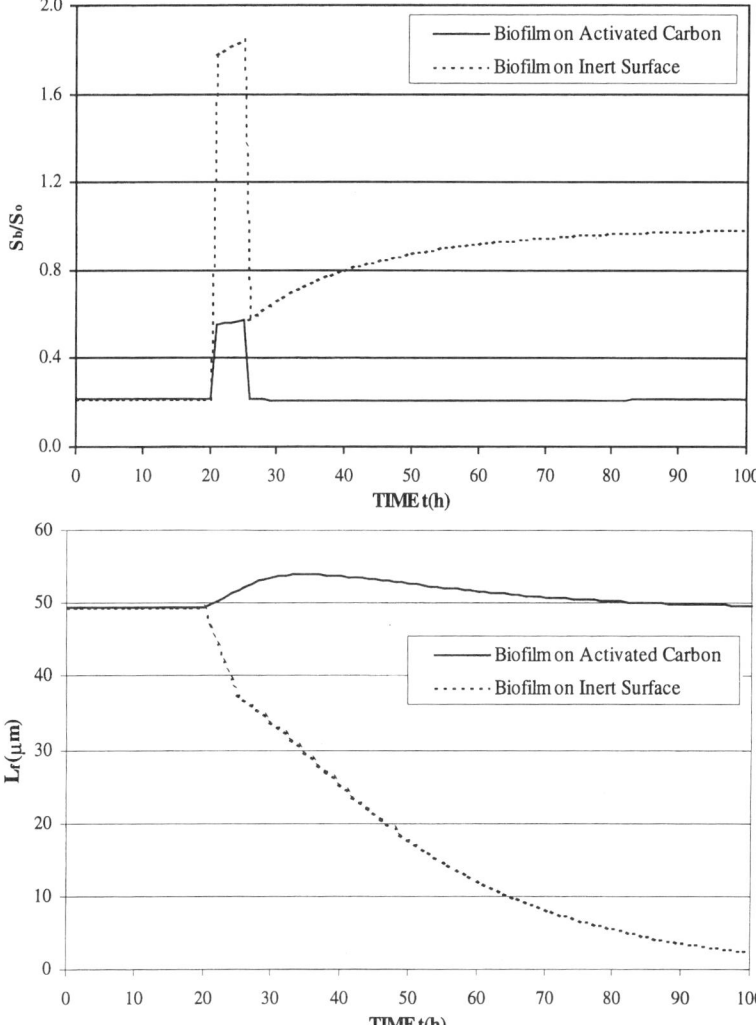

Figure 7.9 Evolution of the substrate concentration in the bulk liquid and the biofilm thickness when the BFAC and BFIS systems are exposed to a 100% shock-load increase in the influent substrate concentration for 20 to 25 h. $\phi^* = 0.7$. The substrate concentration is normalized to the original value of S_0.

Although sorption diminishes the effects of inhibition, the more dramatic effect is to defer biofilm growth. The top portion of Figure 7.8 illustrates that the substrate concentration is higher in the BFAC system from about 60 h until 200 h. The reason is that adsorption sequesters the substrate during the first 60 h and delays biofilm growth.

Once the biofilm in the BFAC begins to grow rapidly, after about 150 h, biofilm growth is very rapid. The biofilm accumulation in the BFAC eventually overtakes that in the BFIS reactor. This "spurt" in biofilm accumulation in the BFAC reactor comes about because the biofilm utilizes the substrate that had been adsorbed in the GAC micropores during the first 60 h. Hence, substrate sequestering spurs biofilm growth once the biofilm has grown enough to drive S_w to close to zero and force the substrate to back-diffuse out of the micropores (Chang and Rittmann, 1987b).

Once the steady states were attained, we exposed the four systems to a 100% increase in the influent concentration for 5 h. This severe shock loading of an inhibitory substrate identifies when the adsorbing substratum offers a protective advantage. For all the cases with the low ϕ^*, the differences between the BFAC and BFIS systems are minimal and not shown. The deep biofilms that occur with a small ϕ^* prevent substrate from reaching the substratum, and the adsorbing substratum is neither better nor worse than the inert substratum. When ϕ^* is large, 0.7 in this case, the adsorptive substratum helps protect the biomass in the biofilm from inhibition and washout. At the same time, adsorption helps keep the

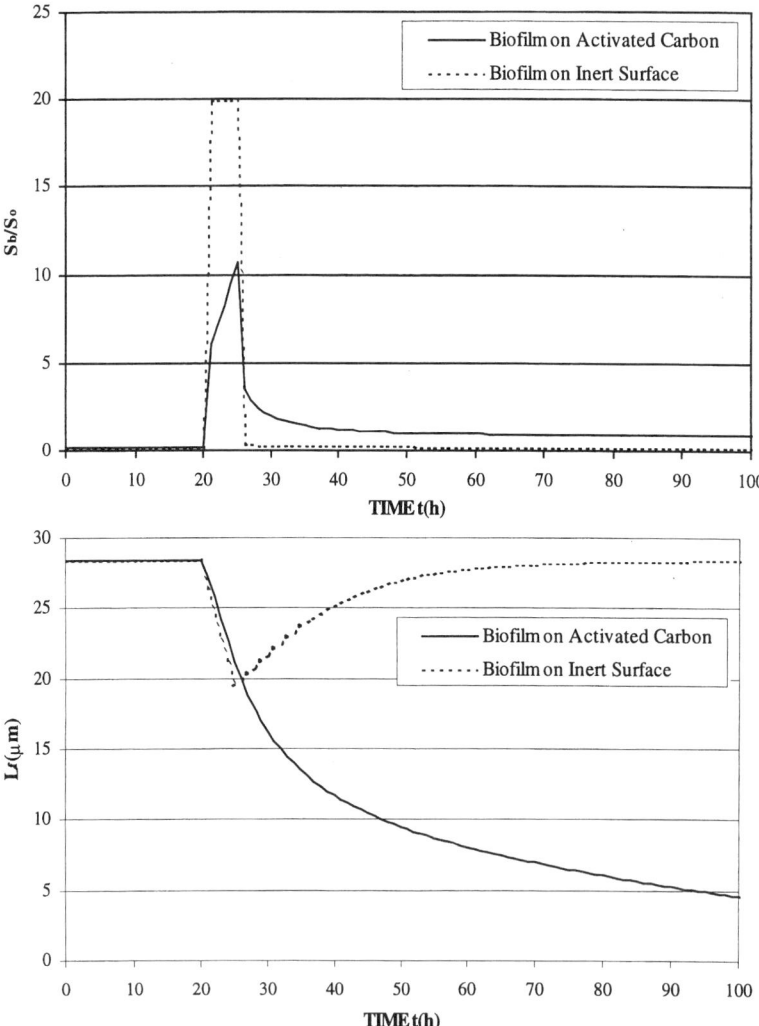

Figure 7.10 Response of the substrate concentration and biofilm thickness to a 20-fold shock increase in the influent substrate concentration for 5 h. $\phi^* = 0.7$. The steady state was established with $S_0 = 0.0024$ mg/cm^3.

effluent concentration low. The protective advantage of adsorption comes about because the biofilm is not deep when the growth potential is poor. Therefore, inhibitory substrate penetrates to the substratum, where it can be sequestered.

Figure 7.9 illustrates the protective effects of the adsorbing substratum when the growth potential is poor. Substrate adsorption keeps S_b near 0.5 S_0 during the shock, while it increases to about 1.8 S_0 for the inert surface (S_0 is the original influent concentration). By keeping the concentration of the inhibitory substrate relatively low in the bulk liquid, as well as in the biofilm, adsorption allows the biofilm to grow during and after the shock. Eventually, the biofilm in the BFAC system returns to its original steady state.

In contrast, the very high substrate concentration that occurs in the BFIS system causes a rapid loss of biomass during the 5-h shock. That loss of biomass reduces the normal degree of protection that had been established from the concentration gradient in the biofilm for S_0. Therefore, the biofilm continues to experience net loss after the shock, and it eventually washes out. This example demonstrates that biofilms exposed to inhibitors can be unstable when the inhibitor is applied with a large enough shock load and the growth potential is poor. A sorbing substratum can offer a major advantage in this case.

We saw that substrate adsorption has negative, as well as positive, effects during the initial growth of the biofilm (Fig. 7.8), and negative effects also can occur with shock loading of an inhibitory substrate. Adsorption to the substratum can be negative if the shock is very large, but the original concentration is not so inhibitory. Figure 7.10 illustrates that a 20-fold shock for 5 h washes out with an inert substratum, but the inhibition remains for a long time as the adsorbed substrate diffuses out of the substratum and into the biofilm after the shock has ended. The key to the successful recovery of the BFIS system is that S_0 is lower than in the previous examples, and this means that the biofilm is not too strongly inhibited once the shock has passed. For the BFAC system, however, the slow release of the inhibitory substrate maintains a negative growth condition, and the biofilm eventually washes out. The negative effect of substrate desorption is magnified when it diffuses out of the substratum, since inhibitory conditions are now imposed continually on the part of the biofilm that normally is protected from inhibition. Thus, the back diffusion eliminates the protected zone near the substratum, giving the biofilm a net negative growth rate overall.

7.5 SUMMARY

Biofilms can offer advantages in the detoxification of waters containing inhibitory materials when they provide zones protected from losses and/or inhibition. The need for substrates to diffuse into a biofilm means that the concentrations are lower inside the biofilm, a factor that can reduce inhibition by a substrate that is utilized by the biofilm microorganisms. Because detachment and predation occur primarily at the outer surface of the biofilm, microorganisms that have slow growth rates are able to accumulate in the inner protected zones. Microorganisms that transform toxic chemicals are among those slow growers, due to intrinsically slow kinetics, secondary utilization, cometabolism, or inhibition.

We developed and applied mathematical models for biofilms that utilize a self-inhibitory substrate. Modeling runs for steady state biofilms illustrate that microorganisms that have a low growth potential are at the greatest risk from inhibition. The growth potential is quantified by ϕ^*, and a value of ϕ^* approaching 1 indicates poor growth potential. Minimizing biofilm losses, mainly from detachment, can counteract substrate inhibition and extend the range of substrate concentrations tolerated. For example, decreasing ϕ^* from 0.7 to 0.1

increases the maximum tolerable substrate concentration more than 20-fold for moderate inhibition, or $K_i^* = 5$.

An adsorptive substratum, such as GAC, can provide transient protection from substrate inhibition, particularly when the normal operating conditions are near the upper limit on tolerable substrate concentration. However, an adsorptive substratum can have negative impacts, too. By sequestering substrate initially, it slows the onset of significant biofilm accumulation during startup. It also can prolong and intensify inhibition from short but very strong shock loads when the biofilm is not seriously inhibited for normal conditions.

REFERENCES

Alonso, C., M. T. Suidan, G. A. Sorial, F. L. Smith, P. Biswas, P. J. Smith, and R. C. Brenner. 1997. Gas treatment in trickle-bed biofilters: Biomass, how much is enough? *Biotechnol. Bioeng.* 54: 583.

Arcangeli, J., and E. Arvin. 1997. Modeling of the cometabolic biodegradation of trichloroethylene by toluene-oxidizing bacteria in a biofilm system. *Environ. Sci. Technol.* 31: 3044–3052.

Balapragada, B. S., H. D. Stensel, J. A. Puhakka, and J. F. Ferguson. 1997. Effect of hydrogen on reductive dechlorination of chlorinated ethenes. *Environ. Sci. Technol.* 31: 1728–1734.

Beaudoin, D. L., J. D. Bryers, A. B. Cunningham, and S. W. Peretti. 1998a. Mobilization of broad host range plasmid from *Pseudomonas putida* to established biofilm of *Bacillus azotoformans*. I. Experiments. *Biotechnol. Bioeng.* 57: 272–279.

Beaudoin, D. L., J. D. Bryers, A. B. Cunningham, and S. W. Peretti. 1998b. Mobilization of broad host range plasmid from *Pseudomonas putida* to established biofilm of *Bacillus azotoformans*. II. Modeling. *Biotechnol. Bioeng.* 57: 280–286.

Becker, J. G., G. Berardesco, B. E. Rittmann, and D. A. Stahl. 1998. Molecular and metabolic characterization of a 3-chlorobenzoate-degrading anaerobic microbial community. In *Natural Attenuation*, G. B. Wickramanayake and R. E. Hinchee, eds. Columbus, OH: Battelle Press, pp. 93–98.

Bouchard, B., R. Beaudet, R. Villemur, G. McSween, F. Lépine, and J.-G. Bisaillon. 1996. Isolation and characterization of *Desulfitobacterium frappieri sp. nov.*, an anaerobic bacterium which reductively dechlorinates pentachlorophenol to 3-chlorophenol. *Int. J. Syst. Bacteriol.* 46: 1010–1015.

Brint, J. M., and D. Ohman. 1995. Synthesis of multiple exoproducts in *Pseudomonas aeruginosa* is under the control of Rh1R-Rh1I, another set of regulators in strain PA01 with homology to the autoinducer-responsive LuxR-LuxI family. *J. Bacteriol.* 177: 7155–7163.

Brusseau, G. A., B. E. Rittmann, and D. A. Stahl. 1997. Addressing the microbial ecology of marine biofilms. In *Molecular Approaches to the Study of the Oceans*, K. E. Cooksey, ed. London, UK: Chapman and Hall Publishers, pp. 449–470.

Bryers, J. D., and R. R. Sharp. 1997. Retention and expression of recombinant plasmids between suspended and biofilm-bound bacteria degrading trichloroethene (TCE). *Water Sci. Technol.* 36(10): 1–8.

Chang, H. T., and B. E. Rittmann. 1987a. Mathematical modeling of biofilm on activated carbon. *Environ. Sci. Technol.* 21: 273–280.

Chang, H. T., and B. E. Rittmann. 1987b. Verification of the model of biofilm on activated carbon. *Environ. Sci. Technol.* 21: 280–288.

Chang, H. T., and B. E. Rittmann. 1988. Comparative study of biofilm kinetics on different adsorptive media. *J. Water Pollution Control Fedn.* 60: 362–368.

Christiansen, N., and B. K. Ahring. 1996. *Desulfitobacterium hafniense sp. nov.*, an anaerobic, reductively dechlorinating bacterium. *Int. J. Syst. Bacteriol.* 46: 442–448.

Clewell, D. B. 1993. Bacterial sex pheromone-induced plasmid transfer. *Cell* 73: 9–12.

Crittenden, J. C., B. W. C. Wong, W. E. Thacker, and V. L. Snoeyink. 1980. Mathematical model of sequential loading in fixed-bed adsorber. *J. Water Pollution Control Fedn.* 52: 2780–2795.

Davies, D. G., and G. G. Geesey. 1995. Regulation of the alginate biosynthesis gene *algC* in *Pseudomonas aeruginosa* during biofilm development in continuous culture. *Appl. Environ. Microb.* 61: 860–867.

Davies, D. G., A. M. Chakrabarty, and G. G. Geesey. 1993. Exopolysaccharide production in biofilms: Substratum activation of alginate gene expression by *Pseudomonas aeruginosa*. *Appl. Environ. Microb.* 59: 1181–1186.

Davies, D. G., M. R. Parsek, J. P. Pearson, B. H. Iglewski, J. W. Costerton, and E. P. Greenberg. 1998. The involvement of cell-to-cell signals in the development of a bacterial biofilm. *Science* 280: 295–298.

Dazzo F. B. 1984. Bacterial adhesion to plant root surfaces. In *Microbial Adhesion and Aggregation*, K. C. Marshall and W. G. Characklis, eds. Berlin: Springer-Verlag, pp. 85–94.

Dolfing, J., and J. M. Tiedje. 1986. Hydrogen cycling in a three-tiered food web growin on the methanogenic conversion of 3-chlorobenzoate. *FEMS Microbiol. Ecol.* 55: 293–298.

Dolfing, J., and J. M. Tiedje. 1991. Acetate as a source of reducing equivalents in the reductive dechlorination of 2,5-dichlorobenzoate. *Arch. Microbiol.* 156: 356–361.

Ehlers, L. J., and E. J. Bouwer. 1999. RP4 plasmid transfer among strains of *Pseudomonas* in a biofilm reactor. *Water Sci. Technol.*, in press.

Fathepure, B. Z., J. P. Nengu, and S. A. Boyd. 1987. Anaerobic bacteria that degrade PCE. *Appl. Environ. Microb.* 52: 2671–2674.

Finlayson, B. A. 1980. *Nonlinear Analysis in Chemical Engineering*. New York: McGraw-Hill.

Fox, P., M. T. Suidan, and J. T. Pfeffer. 1990. Hybrid expanded-bed GAC reactor for treating inhibitory wastewater. *J. Environ. Eng.* 116: 438–453.

Fugua, W. C., S. Winans, and E. P. Greenberg. 1994. Quorum sensing in bacteria: The LuxR-LuxI family of cell density-responsive transcriptional regulators. *J. Bacteriol.* 176: 269–275.

Furumai, H., and B. E. Rittmann. 1994. Evaluation of multi-species biofilm and floc processes using a simplified aggregate model. *Water Sci. Technol.* 29(10–11): 349–446.

Gantzer, C. J. 1989. Inhibitory substrate utilization by steady-state biofilms. *J. Environ. Eng.* 115: 302–319.

Gantzer, C. J., A. B. Cunningham, W. Gujer, B. Gutekunst, J. J. Heijnen, E. N. Lightfoot, G. Odham, B. E. Rittmann, E. Rosenberg, K. D. Stolzenbach, and A. J. B. Zehnder. 1989. Exchange processes at the fluid-biofilm interface. In *Structure and Function of Biofilms*. Chichester, UK: John Wiley, pp. 73–90.

Genestet, P. 1998. Biodégradation de solvants chlorés par une biomasse bactérienne nitrificante. Ph.D. thesis, Université Henri Poincaré, Nancy I, Nancy, France.

Gerritse, J., V. Renard, T. M. Pedro Gomes, P. A. Lawson, M. D. Collins, J. C. Gottschal. 1996. *Desulfitobacterium* sp. strain PCE1, an anaerobic bacterium that can grow by reductive dechlorination of tetrachloroethene or ortho-chlorinated phenols. *Arch. Microbiol.* 165: 132–140.

Holliger, C., and W. Schumacher. 1994. Reductive dehalogenation as a respiratory process. *Antonie van Leeuwenhoek* 66: 239–246.

Holliger, C., G. Schraa, A. J. M. Stams, and A. J. B. Zehnder. 1993. A highly purified enrichment culture couples the reductive dechlorination of tetrachloroethene to growth. *Appl. Environ. Microb.* 59: 2991–2997.

Jones, G. W. 1984. Adhesion to animal surfaces. In *Microbial Adhesion and Aggregation*, K. C. Marshall and W. G. Characklis, eds. Berlin: Springer-Verlag, pp. 71–84.

Kaiser, D., and R. Losick. 1993. How and why bacteria talk to each other. *Cell* 73: 873–875.

Khodadoust, A. P., J. A. Wagner, M. T. Suidan, and R. C. Brenner. 1997. Anaerobic treatment of PCP in a fluidized-bed GAC bioreactor. *Water Environ. Res.* 69: 180–187.

Kirchner, K., U. Schlachter, and H. J. Rehm. 1989. Biological purification of exhaust air using fixed bacterial monocultures. *Appl. Microb. Biotech.* 31: 629–632.

Kirchner, K., C. A. Gossen, and H. J. Rehm. 1991. Purification of exhaust air containing organic pollutants in a trickle-bed bioreactor. *Appl. Microb. Biotech.* 35: 396–400.

Kissel, J. C., P. L. McCarty, and R. L. Street. 1984. Numerical simulation of a mixed-culture biofilm. *J. Environ. Engr.* 110: 393–411.

Krumholz, L. R., R. Sharp, and S. S. Fishbain. 1996. A freshwater anaerobe coupling acetate oxidation to tetrachloroethylene dehalogenation. *Appl. Environ. Microbiol.* 62: 4108–4113.

Leahy, J. G., and R. R. Colwell. 1990. Microbial degradation of hydrocarbons in the environment. *Microbiol. Rev.* 54: 305–315.

Leson, G., and B. J. Smith. 1997. Petroleum environmental research forum field study on biofilters for control of volatile hydrocarbons. *J. Environ. Engr.* 123: 556–562.

MacDonald, J. A., B. Smets, and B. E. Rittmann. 1992. The effects of energy availability on the conjugative-transfer kinetics of plasmid RP4. *Water Res.* 26: 461–468.

Maymó-Gatell, X., Y.-T. Chien, J. M. Gossett, and S. H. Zinder. 1997. Isolation of a bacterium that reductively dechlorinates tetrachloroethene to ethene. *Science* 276: 1568–1571.

Melcer, H., W. J. Parker, and B. E. Rittmann. 1995. Modeling of volatile organic contaminants in trickling filter systems. *Water Sci. Technol.* 31(1): 95–104.

Mohn, W. W., and J. M. Tiedje. 1990. Strain DCB-1 conserves energy for growth from reductive dechlorination coupled to formate oxidation. *Arch. Microb.* 153: 267–271.

Mohn, W. W., and J. M. Tiedje. 1992. Microbial reductive dehalogenation. *Microbiol. Rev.* 56: 482–507.

Nakhla, G. F., M. T. Suidan, and J. T. Pfeffer. 1990. Control of anaerobic GAC reactors treating inhibitory wastewaters. *Res. J. Water Pollution Control Fedn.* 62: 65–72.

Namkung, E., and B. E. Rittmann. 1987. Evaluation of bisubstrate secondary utilization kinetics by biofilms. *Biotechnol. Bioeng.* 29: 335–342.

National Research Council. 1993. *In Situ Bioremediation: When Does it Work?* Washington, DC: National Academy Press.

Nies, L., and T. M. Vogel. 1990. Effects of organic substrates on dechlorination of Arochlor 1242 in anaerobic sediments. *Appl. Environ. Microbiol.* 56: 2612–2617.

Okabe, S., K. Hirata, Y. Ozawa, and Y. Watanabe. 1996. Spatial microbial distributions of nitrifiers and heterotrophs in mixed population biofilms. *Biotechnol. Bioeng.* 50: 24–35.

Ottengraf, S. P. P. 1988. Exhaust gas purification. In *Biotechnology*, H. J. Rehm and G. Reed, eds. Weinheim: VCH Verlagsgesellschaft, pp. 425–452.

Palenik, B. J.-C. Block, R. G. Burns, W. G. Characklis, B. E. Christensen, W. L. Chiorse, A. G. Cristina, F. M. M. Morel, W. W. Nichols, O. H. Tuovinen, G. J. Tuschewitzki, and H. A. Videla. 1989. Biofilms: Properties and processes. In *Structure and Function of Biofilms*, W. G. Characklis and P. A. Wilderer, eds. Chichester, UK: John Wiley, pp. 351–368.

Reinhard, M., S. Shang, P. K. Kitanidis, E. Orwin, G. D. Hopkins, and C. A. LeBron. 1997. In situ BTEX biotransformation under nitrate- and sulfate-reducing conditions. *Environ. Sci. Technol.* 31: 28–36.

Rittmann, B. E. 1982. The effect of shear stress on biofilm detachment rate. *Biotechnol. Bioeng.* 24: 501–506.

Rittmann, B. E. 1987. Aerobic biological treatment. *Environ. Sci. Technol.* 21: 128–136.

Rittmann, B. E. 1989. Detachment from biofilms. In *Structure and Function of Biofilms*, W. G. Characklis and P. A. Wilderer, eds. Chichester, UK: John Wiley, pp. 49–58.

Rittmann, B. E., and J. A. Manem. 1992. Development and experimental evaluation of a steady-state, multi-species biofilm model. *Biotechnol. Bioeng.* 39: 914–922.

Rittmann, B. E., and P. L. McCarty. 1980. Model of steady-state-biofilm kinetics. *Biotechnol. Bioeng.* 22: 2343–2357.

Rittmann, B. E., B. Smets, and D. A. Stahl. 1990. Genetic capabilities of biological processes. Part I. *Environ. Sci. Technol.* 24: 23–29.

Rittmann, B. E., F. Trinet, D. Amar, and H. T. Chang. 1992. Measurement of the activity of a biofilm: Effects of surface loading and detachment on a three-phase, liquid-fluidized-bed reactor. *Water Sci. Technol.* 26(3–4): 585–594.

Rittmann, B. E., E. Seagren, B. A. Wrenn, A. J. Valocchi, C. Ray, and L. Raskin. 1994. *In Situ Bioremediation*. Park Ridge, NJ: Noyes Publishers.

Rittmann, B. E., M. Pettis, H. W. Reeves, and D. A. Stahl. 1999. How biofilm clusters affect substrate flux and ecological selection. *Water Sci. Technol.* in press.

Rosenberg, M., and E. Rosenberg. 1981. Role of adherence in growth of Acinetobacter calcoaceticus RAG-1 on hexadecane. *J. Bacteriology* 148: 51–57.

Sáez, P. B., and B. E. Rittmann. 1988. An improved pseudo-analytical solution for steady-state-biofilm kinetics. *Biotechnol. Bioeng.* 32: 379–385.

Sáez, P. B., and B. E. Rittmann. 1988. The biodegradation kinetics of 4-chlorophenol, an inhibitory co-metabolite. *Res. J. Water Pollution Control Fedn.* 63: 838–847.

Sáez, P. B., B. E. Rittmann, and Q.-B. Zhang. 1991. Biodegradation kinetics of a self-inhibitory substrate by steady-state biofilms. *Proc. Purdue Industrial Waste Conf.* Ann Arbor, MI: Lewis Publ., pp. 273–279.

Sáez, P. B., and B. E. Rittmann. 1993. Biodegradation kinetics of a mixture containing a primary substrate (phenol) and an inhibitory co-metabolite (4-chlorophenol). *Biodegradation* 4: 3–23.

Scholz-Muramatsu, H., A. Neumann, M. Messmer, E. Moore, and G. Diekert. 1995. Isolation and characterization of *Dehalospirullum multivorans gen. nov.*, a tetrachloroethene-utilizing, strictly anaerobic bacterium. *Arch. Microbiol.* 163: 48–56.

Sharp, R. R., J. D. Bryers, W. G. Jones, and M. S. Shields. 1998. Activity and stability of a recombinant plasmid-borne TCE-degradative pathway in suspended cultures. *Biotechnol. Bioeng.* 57: 287–293.

Shimomura, T., F. Suda, H. Uchiyama, and O. Yagi. 1997. Biodegradation of trichloroethylene by *Methylocystis sp.* Strain M immobilized in gel beads in a fluidized-bed bioreactor. *Water Res.* 31: 2383–2386.

Smets, B., B. E. Rittmann, and D. A. Stahl. 1990. Genetic capabilities of biological processes. Part II. *Environ. Sci. Technol.* 24: 162–169.

Smets, B., B. E. Rittmann, and D. A. Stahl. 1993. The specific growth rate of *Pseudomonas putida* (TOL) influences the conjugal transfer rate of the TOL plasmid. *Appl. Environ. Microb.* 59: 3430–3437.

Smets, B., B. E. Rittmann, and D. A. Stahl. 1995. Quantification of the effect of substrate concentration on the conjugal transfer rate of the TOL plasmid in short-term batch mating experiments. *Letters in Appl. Microbiol.* 21: 167–172.

Speitel, G. E. J., and D. S. Mclay. 1993. Biofilm reactors for treatment of gas streams containing chlorinated solvents. *J. Environ. Eng.* 119: 658–678.

Stal, L. J., E. Bock, E. J. Bouwer, L. J. Douglas, D. L. Gutnick, K. D. Heckmann, P. Hirsch, J. M. Kölbel-Boelke, K. C. Marshall, J. I. Prosser, C. Shutt, and Y. Watanabe. 1989. Cellular physiology and interactions of biofilm organisms. In *Structure and Function of Biofilms*, W. G. Characklis and P. A. Wilderer, eds. Chichester, UK: John Wiley, pp. 269–288.

Suidan, M. T., C. E. Strubler, S. W. Kao, and J. T. Pfeffer. 1983. Treatment of coal-gasification wastewater with anaerobic filter technology. *J. Water Pollution Control Fedn.* 55: 1263–1279.

Swanson, W. J., and R. C. Loehr. 1997. Biofiltration: Fundamentals, design and operation principles, and applications of biological APC treatment. *J. Environ. Engr.* 123: 538–546.

Tonga, A. P., and M. Singh. 1994. Biological vapor-phase treatment using biofilter and biotrickling filter reactors: Practical operating regimes. *Environ. Prog.* 13: 94–97.

van Loosdrecht, M. C. M., J. Lyklema, W. Norde, and A. J. B. Zehnder. 1990. Influences of interfaces on microbial activity. *Microb. Rev.* 54: 75–87.

van Loosdrecht, M. C. M., L. Tijhuis, A. M. S. Wijdieks, and J. J. Heijnen. 1995. Population distribution in aerobic biofilms on small suspended particles. *Water Sci. Technol.* 56: 163–171.

Vannelli, T., M. Logan, D. M. Arciero, and P. B. Hooper. 1991. Degradation of halogenated aliphatic compounds by NH3-oxidizing bacterium *Nitrosomonas europeae*. *Appl. Environ. Microb.* 56: 1169–1171.

Villaverde, S., R. G. Mirpuri, Z. Lewandowski, and W. L. Jones. 1997. Physiological and chemical gradients in a *Pseudomonas putida* 54 G biofilm degrading toluene in a flat plate vapor phase reactor. *Biotechnol. Bioeng.* 56: 361–377.

Wang, Y. T., M. T. Suidan, and B. E. Rittmann. 1986. Anaerobic treatment of phenol by an expanded-bed reactor. *J. Water Pollution Control Fedn.* 58: 227–233.

Wanner, O., and W. Gujer. 1986. A multispecies biofilm model. *Biotechnol. Bioeng.* 28: 314–328.

Watanabe, Y., S. Okabe, K. Hirata, and S. Masuda. 1995. Simultaneous removal of organic materials and nitrogen by micro-aerobic biofilms. *Water Sci. Technol.* 31: 304–313.

Wild, A., R. Hermann, and T. Leisinger. 1997. Isolation of an anaerobic bacterium which reductively dechlorinates tetrachloroethene and tricholoroethene. *Biodegradation* 7: 507–511.

Wirtel, S. A., D. R. Noguera, D. T. Kampmeier, M. S. Heath, and B. E. Rittmann. 1992. Explaining widely varying biofilm-process performance with normalized loading curves. *Water Environ. Res.* 64: 706–711.

Wrenn, B. A., and B. E. Rittmann. 1995. A model for the effects of primary substrates on the kinetics of reductive dehalogenation. *Biodegradation* 6: 295–305.

Wrenn, B. A., and B. E. Rittmann. 1996. Evaluation of a mathematical model for the effects of primary substrates on reductive dehalogenation kinetics. *Biodegradation* 7: 49–64.

Yager, R. M., S. E. Bilotta, C. L. Mann, and E. L. Madsen. 1997. Metabolic adaptation and in situ attenuation of chlorinated ethenes by naturally occurring microorganisms in fractured dolomite aquifer near Niagara Falls, New York. *Environ. Sci. Technol.* 31: 2211–2217.

Yu, H., B. E. Rittmann, and B. Kim. 1998. *Kinetics of Removing Volatile Organic Compounds from Gas Streams and Nitroglycerin Using a Thru-Phase Circulating-Bed Biofilm Reactor*. Champaign, IL: Construction Engineering Research Laboratory.

PART 3

DETRIMENTAL ASPECTS OF BIOFILM SYSTEMS

8

BIOFOULING OF ENGINEERED MATERIALS AND SYSTEMS

GILL G. GEESEY

Department of Microbiology and Center for Biofilm Engineering, Montana State University, Bozeman, Montana

JAMES D. BRYERS

Center for Biomaterials, University of Connecticut Health Center, Farmington, Connecticut

8.1 INTRODUCTION

The operational definition of "fouling" is the undesirable formation of deposits on surfaces of equipment, which results in compromised performance and/or reduced life of the equipment. Fouling occurs on a variety of wetted surfaces, including pipes, heat exchangers, membranes, and medical implants. Fouling can occur through several mechanisms, involving biological, corrosion, particulate, and precipitation reactions (Videla and Characklis, 1992).

Biofouling is the result of extensive growth of organisms on surfaces. Biofouling occurs rapidly in flowing systems where nutrients are supplied on a regular basis. Extensive growth of microorganisms and their secretion of extracellular polymers lead to the formation of visible slimy deposits on solid surfaces. In spite of intense research over the past 30 years to develop a surface that is resistant to biofouling, no viable candidate material has yet emerged. Consequently, biofilm control still depends on judicious monitoring and regularly scheduled equipment surface cleaning that minimizes biofilm accumulation in order to achieve acceptable equipment performance. In this chapter, the process of biofouling and its impact is described for a wide variety of engineered systems. In addition, the engineering aspects of biofouling are revisited in light of recent discoveries in biofilm behavior. Finally, advances in monitoring biofilm behavior are described, and the applications and limitations discussed.

Biofilms II: Process Analysis and Applications, Edited by James D. Bryers.
ISBN 0-471-29656-2 Copyright © 2000 Wiley-Liss, Inc.

8.2 BIOFOULING OF STRUCTURES IN THE MARINE ENVIRONMENT

8.2.1 Oil and Gas Platforms

Man-made structures deployed in the offshore marine environment are subject to biofouling. Seawater contains the planktonic organisms that colonize submerged structures as well as the nutrients to support their growth and biofilm formation. Engineered structures of all types are prone to biofouling in seawater. In the southern North Sea the most severe fouling occurs in the tidal zone and on the first 20 m of structure below the water surface. Oil and gas platforms accumulate fouling organisms to an extent rarely observed in other industrial environments. Upon initial immersion, platform legs accumulate microscopic bacteria, including sulfur-reducing bacteria, algae, and fungi. Later, the microfouled surface becomes colonized by invertebrates such as mussels, barnacles, anemones, and seaweeds. In exceptional cases, fouling layers of 250 mm in thickness have been reported. Biofouling can develop to such an extent on platform legs that their structural properties change dramatically, exceeding the design allowance and accelerating fatigue. Costly, periodic cleaning, involving removal of the fouling layers by divers with hand tools, must be scheduled to avoid failure. Platform legs are so conducive to the growth of sessile organisms that they have been exploited for commercial shellfish rearing in some parts of the world.

The oil/water separation systems of an oil production operation provides a surface, an aqueous environment, and organic nutrients to support a rich anaerobic microbial biofilm that impedes the oil/water separation process, promotes corrosion, and plugs filters. Many bacteria in oil/water mixtures produce surfactants that act as emulsifiers. The emulsifiers form a separate phase that reduces the efficiency of the oil/water separation process. The biofilms formed in the oil/water separation systems periodically slough from the surfaces creating flocs in suspension that eventually become trapped in downstream filters. Plugging of the filters by the flocs also decreases the efficiency of the oil/water separation systems.

The seawater handling system of offshore oil production platforms treats, stores, and distributes water for various uses: to maintain pressure in the oil reservoir, fight fires, and provide water for heat exchangers and domestic services. The seawater handling system is a common site of biofouling. Biofouling of the water handling system contributes to injection well plugging, loss of pressure in fire water lines, and loss of heat exchange efficiency in heat exchangers.

Water produced with the oil during secondary production is often used as makeup water for reinjection into the oil reservoir. "Souring" of the oil reservoir occurs when sulfide-producing bacteria, slough from the anaerobic biofilms in the oil/water separation systems and are transported into the oil reservoir during water reinjection. Injection of raw seawater into the reservoir also promotes souring. Raw seawater contains relatively high concentrations of nutrients, electron donors, and sulfate, the electron acceptor used by sulfate-reducing bacteria (SRB) to form sulfide. Sour oil costs the industry in terms of corrosion and implementation of safety measures related to its toxicity at the production stage as well as in terms of air pollution mitigation during refining operations.

Biofilm bacteria sloughed into the makeup water also contribute to plugging at the site of water injection into the reservoir (Geesey et al., 1987). A thin but continuous biofilm develops at the interface of the well bore and porous reservoir sand, creating a back-pressure that requires more energy to pump the water into the reservoir. Eventually, the back-pressure caused by the membranelike biofilm becomes so great that the injection well must be taken off-line and treated with chemicals to remove the biofouling layer and restore reservoir permeability.

8.2.2 Biofouling of Ship Hulls

Fouling of ship hulls by marine organisms has been a problem for the maritime industry since shipbuilding began. Microbial biofilm formation is believed to precede macrofouling by multicellular algae and invertebrates. Among the macrofouling organisms, balanoid barnacles, mussels, polychaete worms, bryozoans, sponges, tunicates, and algae represent major fouling populations in the marine environment. The composition of the fouling community varies, however, depending on environmental conditions. The main macrofouling organisms off the Japanese coast in the East China Sea have been identified as *Balanus trigonus*, *Mytilus edulis*, and *Hormomya mutabilis* (Nakasono et al., 1993).

Indirect evidence suggests that surface growth of a wide diversity of bacteria and diatoms of the genera *Amphora*, *Amphiprora*, *Navicula*, and *Achnanthes* promote subsequent macrofouling (Cooksey and Wigglesworth-Cooksey, 1992). The accumulation of biomass on the hull increases turbulence at the hull surface, which in turn increases fluid frictional resistance and hydrodynamic drag to the ship as it moves through the water. It has been proposed that the nonrigid biofouling layer causes this increase in drag. Details of this phenomena can be found in Section 8.7.1.1. Microfouling alone has been shown in sea trials to increase drag by 5–15%. The drag-related fuel penalties have been estimated at $75–100 million per year (Alberte et al., 1992). The increased drag-related fuel costs incurred during the late 1970s forced the U.S. Navy to spend $360 million in dry dock costs for biofilm removal and prevention in 1981 (Alberte et al., 1992). Worldwide, the maritime industry spends an estimated $1 billion to control biofouling (Field, 1981).

The U.S. Navy applies antifouling coatings to hulls to reduce biofouling between scheduled hull overhaul dry-docking every 5–7 years. One coating developed by the Navy, RTV-11, now manufactured by General Electric Company, consists of a silicone-based polymer that presents a very unstable surface to colonizing marine organisms. The coating prevents ship hull biofouling as long as the ship is moving at moderate speed. The hull is subject to biofouling when it is anchored in port, however. Furthermore, the coating is easily abraided, resulting in exposure of the underlying hard surface to colonizing organisms, which can promote disbondment of the surrounding unabraided coating.

8.3 BIOFOULING OF HEAT EXCHANGERS

Heat exchangers are structures engineered to cool fluids. A schematic diagram of a typical tube-in-shell heat exchanger, such as that used to condense steam in the electric power industry is illustrated in Figure 8.1. As a heated fluid contacts a clean metal surface with high heat conductivity, the heat from the fluid is transferred to the metal. The metal then transfers the heat to another fluid, usually partitioned from the heated fluid by the metal itself. Copper-containing metal tubing often serves as the partition between the heated and cooled fluids in tubular heat exchangers. Heat transfer in fluid systems is controlled by (1) convective heat transfer through a gas, (2) conductive heat transfer from a fluid across the metal of the heat exchanger, and (3) advective heat transfer due to movement of a liquid across the metal surface (Marshall, 1992).

Deposits on the metal surface reduce heat transfer across the metal. Deposits may be either organic or inorganic or a combination of both. Calcium carbonate scales are an example of an inorganic deposit, whereas slimes or biofilms are an example of an organic deposit. It has been estimated that condenser tube failure through biological fouling is responsible for a 3.8% overall loss in availability of power plants generating 600MW or more in the United

States (Anson, 1977). The accumulation of a biological fouling layer on the metal surface decreases conductive heat transfer. The influence of a biological fouling layer on advective heat transfer, however, depends on the properties of the biofilm as described in Section 8.7.3.

8.3.1 Tubular Heat Exchangers

Fluid velocities in tube-in-shell heat exchangers such as that shown in Figure 8.1 generally range from 0.3–30 m/s and turbulent regime is the prevailing flow situation (Vieira et al., 1993). The buildup of a biofouling layer has several effects on hydrodynamics of the system. The buildup of a biofilm reduces the cross-sectional area of the water flow path, causing a pressure drop across the length of heat exchanger tubing. To force the same volume of water through a smaller cross-sectional area of tubing, one must increase flow velocity, which requires greater applied pressure. This results in increased turbulence at the tube wall, which can increase heat transfer efficiency. However, an additional expenditure of energy is required to force the water through the tubing, thus reducing overall operation efficiency. When the cross-sectional area of the water flow path is reduced by buildup of thicker fouling layers on the tube wall to a critical value that can no longer be compensated by increased pumping pressure, flow can change from turbulent to laminar regime. The reduction in the volume of water passing through the tubing per unit time and the change in flow regime will result in a reduction in heat transfer efficiency.

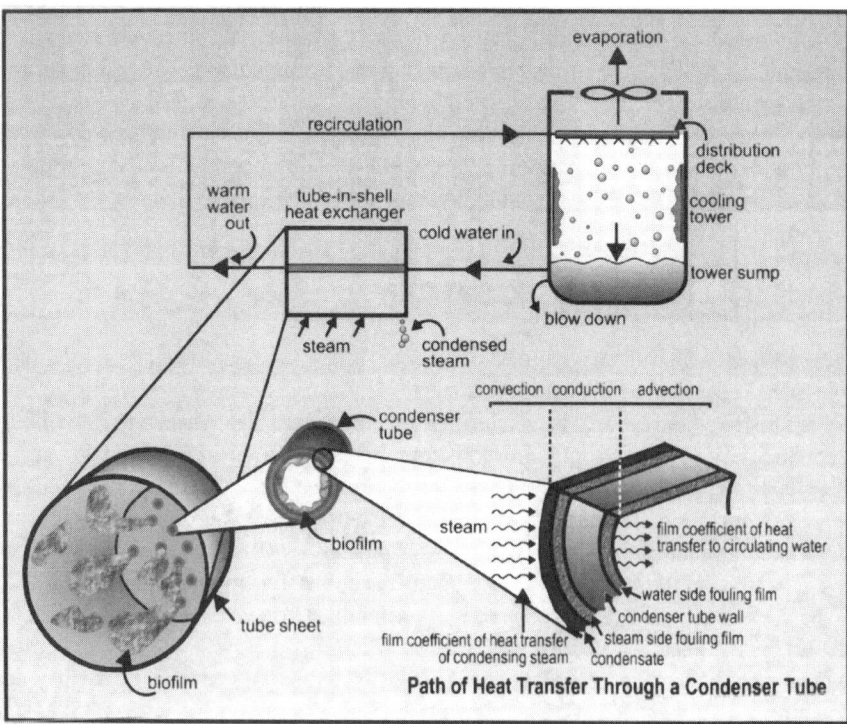

Figure 8.1 Schematic diagram of the tube-in-shell heat exchanger used in steam-generated power plant.

8.3 BIOFOULING OF HEAT EXCHANGERS

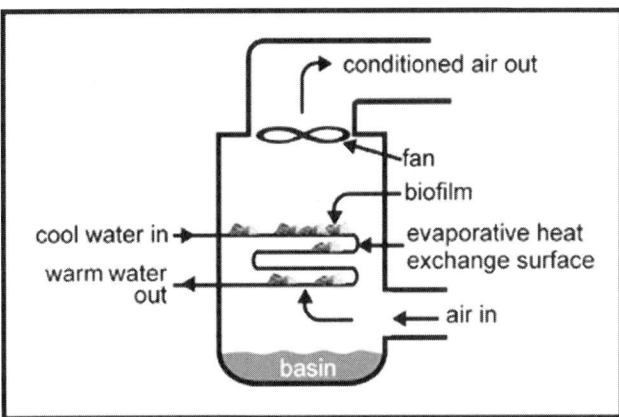

Figure 8.2 Schematic diagram of evaporative heat exchanger used in air-conditioning unit.

The presence of a biofouling layer on the surface of heat exchangers also alters the surface roughness. A biofilm layer converts the heat exchanger surface from a smooth rigid structure to one with a high degree of viscoelasticity, capable of deformation under the influence of the flowing fluid. This will increase surface roughness, which in turn increases fluid frictional resistance to flow near the tube wall. An increase in fluid frictional resistance at the heat exchanger surface can have different effects on heat transfer resistance. Fluid frictional resistance can decrease flow across the surface and thereby reduce heat transfer efficiency as described previously. Conversely, fluid frictional resistance increases turbulence at the heat transfer surface, thereby increasing heat transfer efficiency. The frictional effects of relatively thin biofilms (e.g., <100 μm thickness) are no larger than would be expected from the decrease in flow area. However, at some critical film thickness, frictional resistance increases sharply with thickness (Bott and Miller, 1983).

Characteristics of a heat exchanger system can influence the properties of the fouling layer. Within the range of fluid flow velocities encountered in tubular heat exchangers, hydrodynamics can significantly affect the physical and biological structure of the biofouling layer on the heat exchanger wall. The effect of fluid velocity on biofilm formation has been summarized as follows: higher fluid velocities (1) favor the transport of cells to the heat exchanger surface, (2) impede attachment of cells to the surface, (3) favor replication and production of extracellular polymeric substances (EPS) by the cells that do attach, (4) favor detachment of cells from the surface, and (5) favor denser, thinner, more cohesive biofilm structures with lower thermal resistance.

Internal diffusivity of nondegradable substances is lower in thinner, more compact biofilms. Diffusivity is therefore biofilm dependent and cannot be reduced to a common value (Vieira et al., 1993). Biofilm mass transfer coefficients can be interpreted as the quotient of the internal diffusivity and biofilm thickness. Mass transfer coefficients of biofilms formed over a range of fluid velocities typical of those in heat exchangers, do not differ significantly, however (Vieira et al., 1993).

Other system characteristics exert less of an influence on biofilm properties. Early studies demonstrated that substratum surface roughness has little influence on biofilm thickness (Harty and Bott, 1981). This is particularly relevant to weldments and the long-standing belief that grinding and polishing of all welds are necessary to achieve a surface roughness of less than 1 μm in order to eliminate sites for biofilm development on food preparation surfaces (Anonymous, 1993a). However, at other scales (i.e., roughness values of tens of microns or greater) surface roughness may influence properties such as biofilm thickness.

8.3.2 Evaporative Heat Exchangers

Tubing coils that circulate cooling water or other fluids may also have a surface in contact with a gas phase. Such evaporative heat exchangers are often used in air conditioning units as shown in Figure 8.2. The evaporative process results in the condensation of water from the air and the concentration of dissolved solids in the water. When this water condenses on the outer surface of the tubing, it promotes biofilm and scale formation on the tube surface. In addition to compromising heat transfer, biofouling of evaporative heat exchange surfaces poses a public health concern. Biofilms harbor infectious microorganisms such as *Legionella pneumophila*, which causes Legionnaires' disease, a type of pneumonia. The bacterium is infectious when inhaled as an aerosol. Evaporative cooling tower heat exchangers generate aerosols and thus create favorable conditions for the spread of infection, particularly among the aged and immuno-compromised human population. Dissemination of aerosol-borne opportunistic pathogens is likely to be widespread since many air-conditioning systems for residential and institutional buildings utilize this type of heat exchanger.

8.4 BIOFOULING AND BIOCONTAMINATION OF MEMBRANE SYSTEMS

Membrane systems are widely used in water purification. A number of membrane systems have been developed to remove particles, molecules, and ions from water streams. These include reverse osmosis (RO) membranes, ultrafiltration (UF) membranes, and microporous (MP) membranes, all of which are vulnerable to fouling (Fig. 8.3). In membrane

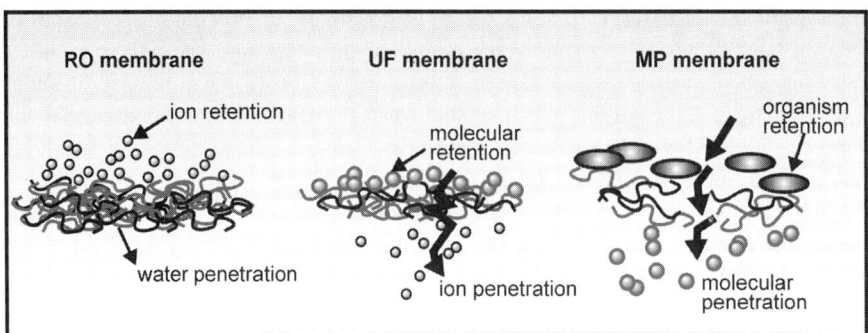

Figure 8.3 Schematic design of various types of membrane systems used to remove dissolved and particulate material from aqueous phase. RO, reverse osmosis; UF, ultrafiltration; MP, microporous membrane.

8.4 BIOFOULING AND BIOCONTAMINATION OF MEMBRANE SYSTEMS

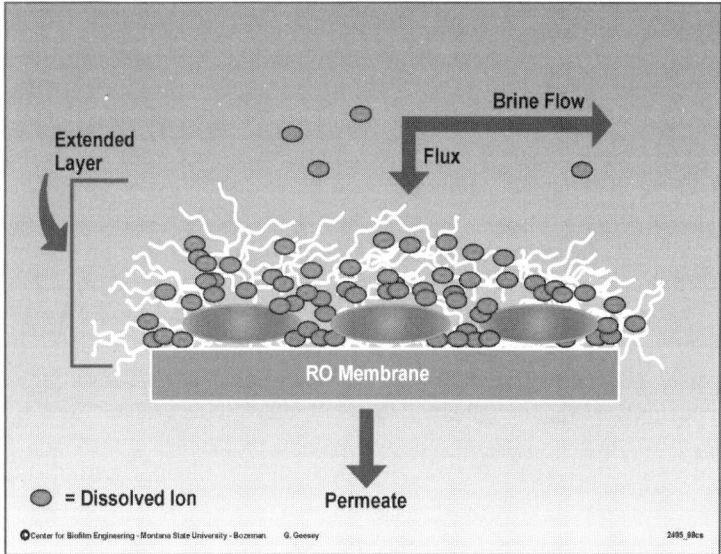

Figure 8.4 Schematic design of biofouling of reverse osmosis membrane. (Courtesy of Dr. Harry Ridgway.)

technology, the most important types of fouling are: crystalline fouling (deposition of inorganic minerals), organic fouling (deposition of humic substances, oils, and grease), particle and colloidal fouling, and microbiological fouling (accumulation of microorganisms as biofilms) (Flemming et al., 1994). Whereas the first three types of fouling can be controlled by reducing the foulant concentration in the water phase, biofouling of a membrane surface is not necessarily controlled by reducing the concentrations of microorganisms in the bulk water. Accumulation of a biofouling layer may occur by replication of the microorganisms adhering to the surface of the membrane. Bacterial attachment to and biofilm formation on membrane surfaces occur within the first hours of operation of a membrane unit if it is operated under nonsterile conditions. Although much effort has been directed to the development of membrane materials that resist bacterial adhesion, progress has been slow.

8.4.1 Reverse Osmosis Membranes

The presence of a biofilm on an RO membrane causes a transmembrane pressure drop ($\Delta p_{membrane}$ and a feed-brine pressure drop ($\Delta p_{feedbrine}$), which lead to a decline in flux of product water across the membrane (Fig. 8.4). The net result is a loss in membrane performance. The change in membrane performance is a gradual response to a gradually increasing accumulation of biofilm on the membrane surface.

The biofilm causes an increase in the concentration of ions or other dissolved substances on the raw water side of the RO membrane—a phenomenon referred to as "concentration polarization" (Fig. 8.4). This increases the pressure requirement for separation of the dissolved substances from the water. Concentration polarization also leads to precipitation of dissolved substances on the membrane as their solubility product is exceeded.

The RO membranes may accumulate surface biofilms ranging in thickness from 10–100 μm. The physical properties of the biofilm limit water transport across the membrane. The biofilm acts as a secondary membrane and impacts hydraulic resistance. When this hydraulic resistance leads to an unacceptable loss in permeability (membrane performance), it is referred to as "biofouling." Whereas water movement across a clean membrane is advection-controlled, a biofouled membrane is diffusion-controlled or advection-controlled, depending on the physical properties of the biofilm. Permeation across a biofouled membrane is controlled by the properties of the pore channels of the biofilm, not by the porosity of the membrane. The hydrodynamics of the water flowing across a biofouled RO membrane is no longer controlled by that of the bulk solution or the membrane spacers, but rather by the mass transport properties of the biofilm.

Flow porosity (p_f) is a term used to describe the movement of water through the biofilm matrix on a membrane (McDonogh et al., 1994). It refers to the available area for flow in an infinitesimal cross-section of the matrix, for example, percentage area of EPS. For any observed permeability, the resistance to water flow contributed by the biofilm matrix R_{bio} is a function of p_f.

$$R_{bio} = (p_f/\eta L_p) - R_{mem} \tag{8.1}$$

where L_p is the permeability of the filtration layer, R_{mem} is the resistance to flow by the membrane, and η is the viscosity of the flowing liquid.

The acids and exoenzymes excreted by biofilm microbial populations are concentrated in the EPS matrix at the membrane surface and can promote degradation of the membrane. The RO membranes fabricated of cellulose acetate reportedly are attacked and hydrolyzed by microbial products (Sinclair, 1982). Kutz et al. (1986) found evidence of bacteria of the genus *Selibera* in severely degraded areas of cellulose acetate membranes. In summary, biofouling contributes to a decline in membrane flux, a decrease in mineral rejection, an increase in transmembrane pressure, and possibly, membrane degradation.

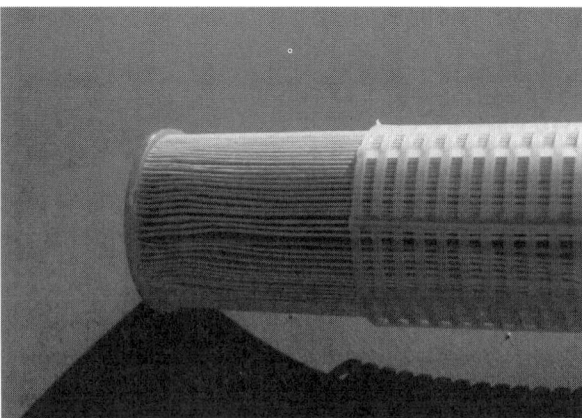

Figure 8.5 Ten-inch cellulose acetate cartridge filter (nominal pore size 0.2 μm) with plastic support removed from end of cartridge to reveal folded cellulose acetate media. (Courtesy of Marc Mittelman.)

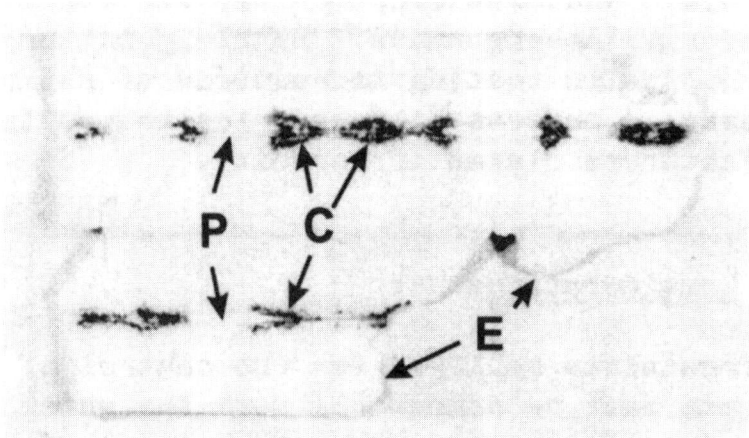

Figure 8.6 Cellulose acetate membrane media from cartridge filter similar to that shown in Figure 8.5 contaminated by bacteria. C is contaminate areas of cartridge localized at pleats (P). E shows edge of cut section of filter. (Courtesy of Marc Mittelman.)

8.4.2 Microporous Membranes

Although microporous membrane filters are designed to reduce or eliminate microorganisms in the bulk aqueous phase, their service life is intimately linked to the microbial biofilms, which develop on upstream surfaces. The sloughing and erosion of biofilm microorganisms into the bulk fluid phase is most often responsible for fouling of membrane surfaces downstream and the resultant transmembrane pressure increases that compromise system operating efficiency both in terms of increased energy demands and in membrane fatigue and failure. Perforation of membranes as a result of excess back-pressure due to biofouling leads to downstream contamination—a condition the membrane was intended to prevent.

8.4.3 Cartridge Filters

Cartridge filters are one of the most common types of filters used for solid/liquid separation (Fig. 8.5). Biofilm growth on cartridge filter surfaces also leads to filter deterioration (Fig. 8.6). Biofouling of these filters is a persistent problem in many industrial applications. The seawater handling systems for the offshore oil industry described earlier in this chapter, represent an important application of cartridge filters. Seawater filtration is usually carried out in two stages: coarse filtration, using 80–150 μm porosity filters and fine filtration using 2–10 μm porosity filters (Williams and Edyvean, 1995). The most important groups of organisms that contribute to blockage of filters used for seawater filtration are copepods, diatoms, dinoflagellates, and bacteria. Among the latter, iron- and sulfur-depositing bacteria are particularly problematic. Adhesive organic material, derived from the EPS of microbial biofilms, contribute to the blockage of filter elements. Cartridge filters used to filter seawater from the North Sea for oil field reservoir injection generally have a life span of anywhere from 3–40 days (Williams and Edyvean, 1995). The relative importance of particle loading and organic fouling in filter element blockage is an important question that has yet to be resolved for many of the applications of these types of filters.

8.5 ELECTRONICS-GRADE WATER SYSTEMS

Particulates constitute an important source of contamination in semiconductor device manufacturing deionized (DI) water lines. Particulates deposited on a chip surface during rinse procedures cause short-circuits and product rejection. With ever-increasing device densities, allowable levels of particulates must be reduced to enable manufacturing to maintain acceptable product yields. While all types of particulates cause this problem, biological particulates are the only ones that are self-replicating within the system. Whereas removal of nonliving particles from the water anywhere in the system reduce particle loads downstream, this is not necessarily the case with viable bacteria living on distribution line walls.

Conventional wisdom is that viable bacteria do not exist in DI water systems due to the lack of nutrients. Even in 18 megaohm water, where nutrient concentrations are extremely low or undetectable, bacteria propagate in biofilms on conduit, value, and various filter surfaces. Fluid flow velocities in these systems do not prevent bacterial biofilms from accumulating on these surfaces. In fact, the moving fluid promotes displacement of cells from the biofilm and their dissemination into the bulk liquid. The cells displace into the bulk liquid can serve as an inoculum for the colonization of other downstream surfaces, and ultimately the product (Fig. 8.7). Bacterial loading of the system is therefore a nonpoint source contamination phenomenon. The oligotrophic conditions associated with purified water, originally thought to prohibit bacterial growth and replication, actually promote bacterial attachment and colonization of surfaces, favoring the biofilm mode of growth that is so efficient in scavenging the trace amounts of nutrients present in the bulk aqueous phase flowing by.

Bacteria and other particles present in the source water are generally removed before the water is introduced to the DI water system. Because it is a closed system, the biofilm, which

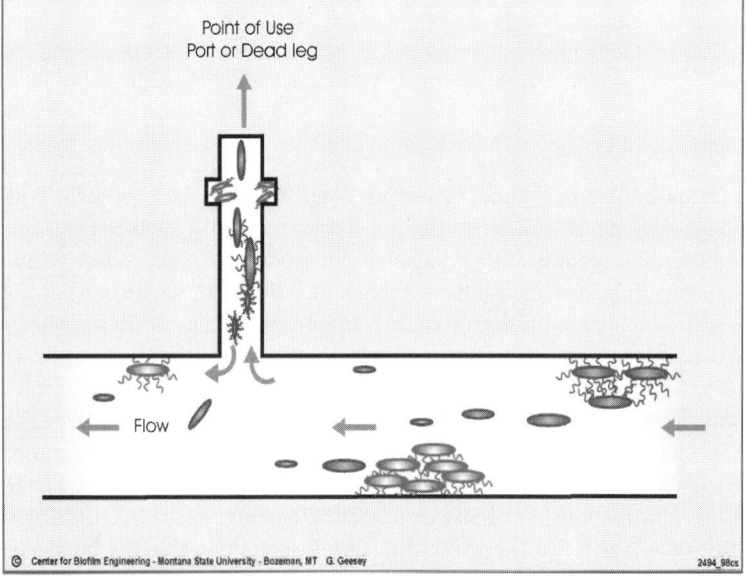

Figure 8.7 Schematic diagram of microbial contamination of a purified water distribution system via erosion of bacterial cells from a surface-associated biofilm.

develops on the surface of system components, represents the only significant source of particle contamination in the system (Patterson et al., 1991). Once deposited on the product surface, cell viability is not an issue. Cell viability is only an issue in terms of the bacteria being able to replicate on the walls of the system and inoculate the water. Controlling this phenomenon to limit system loading can be a daunting challenge. A major device manufacturer has established a limit of 35 bacteria per liter for rinse water for the 4-megabyte (Mb) DRAM chip and 8 bacteria per liter for rinse water for the 16-Mb product. Monitoring such low concentrations of total bacteria has become another challenge to the industry.

The diversity of bacteria recovered from DI water system biofilms is surprisingly broad considering the paucity of available nutrients. Although the vast majority appears to display the gram-negative cell wall structure, numerous gram-positive bacteria have been isolated as well. Using classical microbiological identification methods and phospholipid fatty acid profiles from isolates recovered by culture techniques, 12 broad groups of bacteria have been identified including *Pseudomonas, Moraxella, Morganella, Flexibacter, Caulobacter* and other stalked forms, gram-positive pleomorphic rods, and *Sphaerotilus*-like organisms (Patterson et al., 1991). Many of the isolates, however, do not match any of the known organisms in the existing libraries. Since the detection techniques are biased toward those bacteria that are cultivable on the enrichment and isolation media employed, much of the microbial diversity responsible for biofilm behavior remains to be determined in these water distribution systems.

8.6 BIOFOULING AND HUMAN HEALTH

8.6.1 Biofouling of Food Preparation Surfaces

Initially, food contamination was thought to be the result of entrapment of microorganisms in the food, which proper rinsing could control. However, a portion of bacteria attached to meat and poultry carcasses cannot be removed by spray washing. Today, microbial adhesion to and biofilm development on product or product contact surfaces in food processing plants is considered to be another potential process leading to food contamination. The important problem that arises from surface biofouling is the resultant repeated contamination of food during processing (Notermans, 1994). There is a growing awareness within the food, beverage, and dairy industries that biofilms on food preparation and other contact surfaces cause serious hygiene problems or economic losses due to organoleptic alterations (spoilage) of the product. The reader is referred to Chapter 10 in this volume for details on this topic.

Biofouling of surfaces in the food industry is of particular concern when pathogenic microorganisms are associated with the biofilms. Pathogenic microorganisms such as *Staphylococcus aureus* and *Salmonella* spp. are common inhabitants of poultry processing lines, *Listeria monocytogenes* is common in both meat and dairy plants, and *Bacillus cereus* is common in milking equipment on dairy farms. Of the environmental contamination routes in food processing, surface biofilms are probably the most important route of infection both directly via product contact surfaces or indirectly as reservoirs of infection for product contamination by way of other routes (Timperley et al., 1992). The principal factors controlling contamination of food during processing are believed to be: (1) the total number of microbial cells on the surfaces of contact, and (2) the proportion of the total population associated with the surface that become detached and is picked up by the product (Dunsmore et al., 1981). The formation of biofilms and the consequences in the general food processing environment have been the subjects of several excellent reviews (Notermans et al., 1991; Mattila-Sandholm and Wirtanen, 1992; Carpentier and Cerf, 1993; Zottola and Sashara, 1994).

In the United States, the National Sanitation Foundation has developed a number of guidelines for the dairy industry and food service equipment. These include (1) materials for product contact, (2) surface finish, (3) joints, (4) fasteners, (5) drainage, (6) internal angles and corners, (7) dead spaces, (8) bearings and shaft seals, and (9) instrumentation (Timperley et al., 1992). In the case of fasteners, exposed screw threads must not be used at the product side. Where screwed connections are unavoidable, a metal-backed elastomer gasket should be used (Anonymous, 1993b).

Stainless steel is the material of choice for fabrication of equipment used to process dairy products as well as other food and beverage products because it is durable, corrosion resistant, and easily cleaned. The most commonly used material in the New Zealand dairy industry is unpolished 304L stainless steel with a 2B finish (Flint et al., 1997). In the United States, a No. 4 surface finish provides a surface roughness (R_a) of 1 μm or less. Such a finish is recommended based on the understanding that smoothness is directly related to resistance to bacterial colonization and cleanability. For this reason, welded seams must currently be ground and polished. Although surface roughness is likely to exert some influence on these factors, its significance at circa the 1 μm range is questionable (Tide et al., 1999).

In addition to surface roughness, a number of other environmental factors influence biofilm formation on food preparation surfaces. Bulk liquid phase ionic strength, pH, nutrient content, and temperature all exert an effect on surface biofilm development. Unlike many habitats, food-processing surfaces are repeatedly exposed to food products, which are nutritionally rich and support rapid, balanced microbial growth. Food-derived compounds also adsorb to food processing surfaces during contact to create a conditioning film that influences the development of bacterial biofilms.

The process of bioadhesion on stainless steel surfaces in the food processing industry has been reviewed by Boulange-Petermann (1996) and is presented in Chapter 10 of this volume. Predominant microorganisms that colonize stainless steel surfaces in meat processing plants are *Pseudomonas* spp. and *Klebsiella* spp. (Hood and Zotolla, 1992). Because *Listeria* spp. are ubiquitous in the food preparation environment, have been found in biofilms, and are considered a human pathogen, they have received considerable attention as a contaminant of foods and dairy products. In some cases, adhesion may involve specific interactions between bacterial cell surface proteins and the food. The adhesion of *Salmonella* serotypes to chicken muscle fascia was correlated with the ability to be agglutinated by hyaluronan, a structural polysaccharide component of tissue (Sanderson et al., 1991).

8.6.1.1 Dairy Industry. Biofouling in the dairy industry has been reviewed by Criado et al. (1994) and Flint et al. (1997). Biofouling in the dairy industry can cause serious obstruction, corrosion, and health problems. In spite of the technological advances in the design of milking rooms and machinery, the increased effectiveness of detergents and disinfectants, and the introduction of on-site refrigeration, biofilms still persist in these environments. Biofilms develop rapidly in the food-processing environment, achieving 10^6 bacteria/cm^2 within 12 h under refrigeration conditions (Bouman et al., 1982). However, a period of 48 h is usually necessary for detection of a biofilm on product processing equipment, although this depends on the sensitivity of the detection method (Wirtanen and Mattila-Sandholm, 1992).

8.6 BIOFOULING AND HUMAN HEALTH

A single species often dominates process biofilms in dairy manufacturing plants (Flint et al., 1997). Lewis and Gilmour (1987) found that rubber and stainless steel surfaces exposed to raw milk directly from the udder were colonized by essentially a monoculture bacterial population, which differed from the populations present in the milk. Heat pasteurization may contribute to monoculture biofilms by killing sensitive gram-negative species, allowing thermoduric, gram-positive species such as *Streptococcus thermophilus* to dominate other biofilm populations. In some cases, psychrotrophic microflora have simply replaced mesotrophic bacteria as important agents of contamination. *L. monocytogenes* reportedly displays some degree of heat-resistance when associated with dairy products (Lee and Frank, 1990), so this species may not be as susceptible to the steam rinses applied during surface cleaning as other biofilm populations.

The dairy industry has invested heavily into research on not only the nutritional status of microorganisms that contaminate dairy products but also on the factors that control their adhesion to equipment surfaces where the products become contaminated. When grown as a monoculture, cells of *Listeria monocytogenes* do not attach well to glass and stainless steel surfaces (Sasahara and Zottola, 1993). However, *L. monocytogenes* is often found with other bacteria such as staphylococci and lactobacilli when present in biofilms (Frank and Koffi, 1990). Under some conditions, the presence of other bacteria on the surface does not appear to enhance surface attachment of *L. monocytogenes* (Frank and Koffi, 1990). However, when grown in co-culture with the extracellular polymer-producing bacterium, *Pseudomonas fragi*, *L. monocytogenes* readily colonizes these surfaces (Sasahara and Zottola, 1993). *P. fragi* appears to be the primary surface colonizer, forming an exopolymeric matrix that traps *L. monocytogenes* and other debris in the biofouling layer. The production of exopolymer appears to be more important than hydrophobicity, surface charge, or motility in the attachment of *L. monocytogenes* to inert surface in flowing systems. The success of bacteria such as *L. monocytogenes* in colonizing surfaces is ultimately a function of their competitiveness against other species present on the surface.

Some dairy products reportedly inhibit bacterial attachment to surfaces. Components of whole milk inhibit short-term colonization of surfaces by some bacteria. Adsorbed milk proteins significantly reduce adhesion of *Salmonella typhimurium* and *L. monocytogenes* to steel compared to surfaces without proteins (Helke et al., 1993).

Some spoilage organisms are thought to be transferred to dairy products as a result of their contact with surfaces containing biofilms. Contamination attributed to product contact with equipment surface biofilms has been reported in the manufacturing of cheese, whey, milk powder, and in general milk processing. The psychrotrophs *Acinetobacter*, *Pseudomonas*, and *Flavobacteria*, when found in pasteurized milk, are believed to have been introduced after pasteurization as a result of contact with improperly cleaned surfaces. Table 8.1 lists the different types of microorganisms that form biofilms in various dairy manufacturing processes. A summary of the types of microorganisms that reside on different dairy product surfaces has been prepared by Mucchetti (1995).

A number of questions related to biofouling in the food industry have been raised in a scientific status summary of the Institute of Food Technologists' Expert Panel on Food Safety and Nutrition (Zotolla, 1994). These include:

- Are the biofilms in nonfood systems different from those in food processing systems?
- How long does it take for attached microbes to form biofilms in food systems?
- Does biofilm formation, even after removal from a surface, promote subsequent surface biofouling?

- Does renewal of a food contact surface by regular cleaning and sanitizing encourage or discourage biofilm formation?

The answers to these questions depend on the specific conditions being considered or compared. Our severely limited understanding of biofilms in the food, beverage, and dairy industries precludes drawing many general conclusions except that biofilms are likely to be as unique as the environments in which they are found. For more details on this subject, the reader is directed to Chapter 10.

8.6.2 Biofouling of Drinking Water Systems

Public health standards consider water safe for human consumption when it contains a maximum of 500 CFU/ml, when it is free of coliforms, and when its nephelometric turbidity is less than 2 (Geldreich, 1986). Drinking water standards are established to eliminate frank pathogens–not all microorganisms from potable water. The cost of treatment to eliminate all potential pathogens from public potable water supplies would be unaffordable to most customers.

Microorganisms accumulate on the surfaces of water distribution mains and suspended particular matter in drinking water (Ridgway and Olson, 1981). The community of microorganisms attached to the wall of the pipeline represents the major fraction of biomass in a drinking water distribution system. These biofilm populations contribute to the continuous contamination of the moving water phase through surface growth and subsequent displacement of cells from the pipe surface as a result of fluid shear (Fig. 8.8). Bacteria are not the only microbes present in biological fouling layers on the pipe wall surface. Fungi and protozoa are also present. *Hartmannela vermiformis*, *Vannella mira*, *Cochliopodium minutum*, and *Naegleria* sp. have been identified in an industrial pilot-scale drinking water system operated with treated surface waters at temperatures of 24°C (Block et al., 1993).

A quasi-steady state biofilm is typically achieved within a few weeks of putting a system on-line. Several parameters govern biofilm accumulation including hydraulic regime, flux of nutrients and organisms, and substratum properties, and in the case of drinking water distribution systems, disinfectant concentration and protozoan grazing of attached bacteria. Assimilable organic carbon (AOC) is a term often used to describe the readily degraded fraction of dissolved organic carbon that is measured by the number of organisms that grow in the water. Biodegradable dissolved organic carbon (BDOC) is that quantity of the total organic carbon that has been consumed after supporting all the microbial growth

TABLE 8.1. Biofilms Associated with Dairy Product Processing Equipment

Environment	Organism	Reference
Ultrafiltration membrane, whey processing	*Bacillus subtilis*	Flint and Hartley, 1993
Stainless steel surfaces, milk powder processing	*Bacillus stearothermophilus*	Standhouders et al., 1982
Milk processing equipment	*Thermus thermophilus*	Klijn et al., 1992
Lines handling raw milk	Gram-negative psychrotrophic bacteria	Lewis and Gilmour, 1987

8.6 BIOFOULING AND HUMAN HEALTH

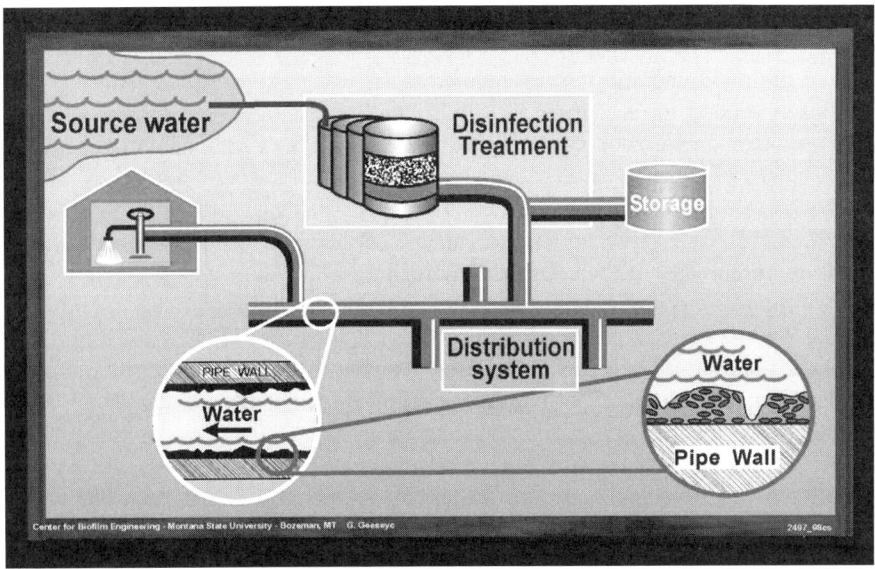

Figure 8.8 Schematic diagram of a public drinking water treatment and distribution system. Expanded views reveal microbial biofilm on pipe wall surface.

in the system (Camper, 1996). Regardless of how it is defined, it represents the nutrients that control biofilm growth in drinking water distribution systems. Although it consists of a complex mixture of compounds, it provides a useful indication of the potential for biofouling in many systems. In a constant hydrodynamic regime, nutrient flux and disinfectant concentration control microbial growth in a biofilm, which in turn controls protozoan grazing on attached bacterial cells and flux of cells into the water phase. The relationship between AOC flux and disinfectant concentration on biofilm growth is complex. Corrosion product generation and accumulation on the pipe wall increases the surface area for biofilm accumulation. Furthermore, corrosion products, along with AOC and biofilm biomass, consume disinfectant.

The mild steel surfaces of water mains provide an excellent habitat for bacterial colonization, including coliforms. Mild steel supports 10 times more heterotrophs and coliforms than more inert surfaces such as polycarbonate (Camper, 1996). Mild steel surfaces also exhibit a higher chlorine demand than more inert surfaces. The high cell densities per unit area and high chlorine demand are believed to be related to an increase in surface area and chlorine reaction afforded by steel corrosion products that accumulate on the pipe surface.

The growth rate of cells on a cement substratum in a pilot-scale drinking water system without chlorine has been reported to be $\mu = 0.0017 \text{ h}^{-1}$ yielding a cell generation time of 17 d (Block et al., 1993). Such high surface-associated growth rates can result in a doubling of bacteria in the water phase along the length of pipe due to sloughing of cells from the biofilm. Whereas approximately 1% of the cultivable bacteria in the water phase are viable, as much as 6% of the bacteria in the biofilm are viable, suggesting that the most active population in these nutrient-depleted systems is the biofilm population.

In the presence of chlorine, the doubling time of attached cells is significantly increased more than 100 d). This slow growth rate represents a very stable community, which is relatively resistant to further inactivation by disinfectant (Paquin et al., 1992). Whereas water quality just downstream from the water treatment plant reflects the efficiency of the plant operation, water quality at the extremity of the distribution system more closely reflects the activity of the biofouling layer on the upstream pipe wall.

The dominant biofilm populations in a drinking water system of one household form a separate cluster of closely related bacteria within the beta 1 subclass of the *Proteobacteria* (Kalmbach et al., 1997). They display less than 97.7% sequence similarity to their closest known relatives. An *Aquabacterium* sp., capable of anaerobic growth on nitrate, has been found to represent 75% of the bacteria in a biofilm formed during daytime flows. A population of *Acinetobacter* sp. in the gamma subdivision was favored in the biofilm at night when the water remained stagnant. The species composition of drinking water biofilm populations also varies from season to season at any one location in the system. It remains to be determined whether the population changes in the system are due primarily to regrowth in the system or to population changes in the raw water supply or in the treatment process.

In situ probing with highly-specific 16S rRNA oligonucleotides suggests that the dominant bacterial species were cultivable on R2A medium and that two-thirds of the autochthonous drinking water population are in a viable but nonculturable state (VBNC) (Kalmbach et al., 1997). VBNC has been demonstrated for a wide variety of bacterial pathogens of concern to the drinking water industry including: *E. coli, Vibrio cholerae, Vibrio vulnificus, Salmonella enteritidis, Shigella sonnei, Shigella flexneri*, and *Campylobacter jejuni* (Barer et al., 1993; Byrd et al., 1991).

Numerous studies have been conducted to determine the influence of biofilms on coliform survival and growth in drinking water distribution systems (Lechevallier, 1990). Both coliform and heterotroph levels in biofilms and the water tend to correlate positively with AOC concentration. However, coliform, but not heterotroph densities increase in biofilms when both AOC and chlorine residual are maintained (Camper, 1996).

Regrowth of coliforms in a drinking water system occurs as a result of their detachment from biofilms and transport into the bulk water. Regrowth is more likely to occur from slow-growing populations introduced from low nutrient sources than from fast-growing populations originating from fecal material (Camper, 1996). Coliforms transported to the bulk water from biofilms can recolonize surfaces downstream in the presence of a free chlorine residual. Coliforms derived from biofilm populations appear to be less susceptible to chlorine disinfection. Thus, biofilms on walls of the distribution system piping appear to negate efforts directed toward their removal from the water at the drinking water treatment plant.

Biofilms have been implicated as a reservoir for pathogenic bacteria in drinking water distribution pipelines (Rogers et al., 1994; Wireman et al., 1993). A mild steel surface, maintained under conditions that simulated a potable water system, developed a biofilm containing 10^6 total bacteria/cm^2. Such a biofilm supported a surface-associated population of *L. pneumophila* to a density of 10^4 cells/cm^2 and a planktonic *L. pneumophila* population density of 5×10^3 CFU/ml in the bulk aqueous phase. *L. pneumophila* has been detected in biofilms of complex microbial consortia by gas chromatography-mass spectrometric analysis of genus-specific hydroxy fatty acids (Walker et al., 1993).

Campylobacter jejuni, an important agent of human bacteria gastroenteritis, has been found to persist for long periods of time in biofilms of autochthonous drinking water microflora propagated in laboratory model drinking water distribution systems when detection is based on nonculturable methods (Buswell et al., 1998).

Morin et al. (1996) demonstrated that carbon fines from the treatment process transported viable cells of *K. pneumoniae* to mixed-population biofilms in drinking water systems. In this habitat, a portion of the *K. pneumoniae* population established in the biofilm was able to resist inactivation by chlorine disinfection.

Mycobacterium spp. have also been recovered from biofilms formed in warm potable water systems (Schulze-Robbecke and Fischeder, 1989). Large numbers of *Mycobacterium kansasii* and *M. flavescens* are distributed as dense aggregations of acid-fast cells within these biofilms. Reporter genes have also been used to monitor the biofilm-induced persistence of *Mycobacterium* in drinking water (Arrage and White, 1997). The resistance of *Mycobacterium* spp. to disinfectants makes them a particular threat to immunocompromised individuals who use the water from *Mycobacterium*-contaminated distribution systems.

Many bacterial pathogens have difficulty competing with other biofilm populations under the oligotrophic conditions characteristic of drinking water distribution systems. Unless protected by other means, the less compettitve pathogens will eventually be displaced from the biofilm. One form of protection that has been demonstrated involves protozoan grazers. Bacteria are capable of intracellular survival and replication within protozoa. Ingestion of *Salmonella typhimurium*, *Yersinia enterocolitica*, *Shigella sonnei*, *Legionella gormanii*, and *Campylobacter jejuni* by the protozoans *Acanthamoeba castellanni* and *Tetrahymena pyriformis* not only protects these bacteria from competitors, but also increases their resistance to chlorine (King et al., 1988). Thus, biofilms offer various forms of protection to pathogens that find their way into drinking water distribution systems.

8.6.3 Biofouling of Purified Water Systems

8.6.3.1 Pharmaceutical Water. Most bacteria in purified water systems are gram-negative, nonfermentative bacilli. These bacteria contain a fever-causing substance in their cell wall, often referred to as endotoxin or pyrogen. The presence of these bacteria and their associated endotoxins in pharmaceutical waters poses the single greatest threat to product quality (Mittelman, 1995). The large surface area created by the various components of a purified water system provides space for surface bacterial colonization. Although purified water systems typically contain very low levels of dissolved solids, sufficient quantities of growth-promoting inorganic and organic substances exist in the flowing bulk aqueous phase or as constituents of the "inert" substratum to promote surface bacterial growth and biofilm development. Despite the dearth of information on the physiology and ecology of microorganisms inhabiting purified water systems, studies in other nutrient-poor environments have repeatedly demonstrated the tendency of "starved" cells to attach to surfaces (Mittelman et al., 1987). The EPS excreted by bacteria attached to surfaces possess a structure and chemistry that enhances nutrient scavenging from the bulk water. EPS can constitute a significant portion of the total surface biomass (bioburden). As a result of its reactivity with oxidizing biocides, EPS can consume a significant portion of a biocide added to inactivate the cells buried in the EPS matrix of a biofouling layer on an industrial surface (Characklis, 1980). The inherent nature of the system therefore selects for EPS-producing, surface biofilm populations. It is not surprising then that most microorganisms in pharmaceutical water systems are associated with surface biofilms.

Sloughing or erosion of biofilm microorganisms contributes the majority of microbes in the bulk water of the fluid handling systems. McFeters et al. (1993) detected 10^2–10^3 CFU/mL in a model laboratory water purifier containing many of the components of a pharmaceutical grade water system. These cell concentrations likely represent a minimum, as many of the cells released from biofilms exist in aggregates containing many cells held together by EPS, which give rise to a single colony-forming unit (CFU). These aggregates confer an infectious dose in a very small volume of water.

Because sloughing is episodic and unpredictable, these aggregates often go undetected during periodic sampling and analysis of the bulk water. Significantly, it is the EPS that affords biofilms protection from chemical treatments and possibly steam/hot water treatment implemented to eradicate bacteria from purified water systems. While both flowing steam and hot water (80°C) appear to be effective at inactivating biofilm bacteria in pharmaceutical waters, neither treatment effectively removes biomass from surfaces, including EPS. Surface residues of this nature facilitate subsequent surface recolonization and biofilm regrowth.

Currently there are no biofilm samplers commercially available for the pharmaceutical industry. Side-stream pipe sections containing a modified Robbins device have been used for recovery of biofilm bacteria in purified water systems, but this type of sampler is cumbersome to manipulate and the analyses are time consuming. A number of on-line biofilm detection systems are being developed, including Fourier transform infrared spectroscopy, fluorimetry, and quartz crystal microbalance.

8.6.3.2 Hemodialysis Units.
Improperly maintained purified water systems have led to a number of endotoxin-related problems in hemodialysis facilities (Laurence and Lapierre, 1995; Murphy et al., 1987). These pyrogenic substances are likely released from the gram-negative bacterial biofilms growing on components of the purified water system or hemodialysis units and transported via the dialysis fluid into the body of the patients receiving treatment (Vincent et al., 1989). Once released and disseminated, endotoxins are extremely difficult to remove from surfaces with which they come in contact.

8.6.3.3 Dental Units.
Every dental unit is equipped with small-bore flexible tubing to deliver water to different handpieces such as the air/water syringe, the ultrasonic descaler, and the high-speed drill. The water supply for dental units may be the potable municipal water plumbed in a building, or distilled or sterile water reservoirs housed in the dental office. The wall of small-bore plastic tubing used to deliver cooling water for these dental handpieces provides an ideal environment for the attachment of bacterial cells and the proliferation of biofilms. The high surface area-to-volume ratio afforded by such tubing ensures a high concentration of bacteria in the water due to sloughing of cells from the biofilm growing on the tubing wall. Since sloughing often involves the release of aggregations of cells, large infectious doses of biofilm-derived cells may be delivered to the patient's oral cavity. Contamination levels in dental treatment water may exceed 10^5 CFU/ml, with many of the colonies arising from not one cell but rather a clump of cells which alone can contain 10^5 cells (Williams et al., 1993). Heating the rinse or irrigation water to body temperature for patient comfort selects for populations of biofilm-forming microorganisms that are well adapted to grow in the body.

Biofouling of dental treatment units has been implicated as a source of infectious agents to dental patients (Schulze-Robbecke et al., 1995; Shearer, 1996). Despite the high levels of microorganisms often reported in dental treatment units, no outbreaks of disease and relatively few clinical case reports have been associated with contamination of dental water lines. Nevertheless, potentially pathogenic bacteria such as *Pseudomonas* spp., nontuberculous *Mycobacteria* and *Legionella* spp. have been isolated from dental treatment water. These same bacteria are opportunistic pathogens that cause nosocomial infections and infections in immuno-compromised patients.

Potable water from public water supplies may be a source of opportunistic pathogens associated with dental treatment units. *Sphyngomonas paucimobilis*, *Acinetobacter calcoaceticus*, *Methylobacterium mesophilicum*, and *Pseudomonas aeruginosa* have been reported as dominant isolates from dental unit water lines (Barbeau et al., 1996). Interestingly, *P. aeruginosa* was never isolated in tap water remote from or near the contaminated dental unit water lines. The source of this organism is likely the biofilm that accumulates in the water lines.

While many opportunistic pathogens are killed by the disinfection treatments applied at the water treatment plant, they are not necessarily eliminated from the source water for the dental treatment units. The providers of potable water are only required to guarantee quality to the point at which the customers connect to the mains—not to the tap or in-house equipment. The building owner, consumer or business operator is responsible for the quality of their water once it leaves the mains of a public drinking water system. Use of sterile deionized water, a costly alternative, offers a source of contamination-free water. However, one must recognize that once this pure source of water comes in contact with any nonsterile component of the dental unit system, it can become contaminated by microorganisms residing on that component's surface as a biofilm. Conscientious use of a separate water reservoir system with periodic or continuous chemical treatment can improve water quality to the extent that it meets the American Dental Association's recommendation of no more than 200 CFU/ml for acceptable water quality.

Retraction of dental treatment water from the mouth of an infected patient into the water lines and subsequent discharge into another patient's mouth has been proposed as a means of the spread of disease. Many dental units today flush the water lines for 20–30 sec after each use and are equipped with check valves or antiretraction valves, which are designed to minimize water line contamination. However, every additional safeguard of this nature offers yet another surface for colonization and biofilm development. Flushing is ineffective due to the adherent nature of the bacteria. Santiago et al. (1994) found that the numbers of bacteria in water of dental units were not reduced by flushing. Nothing short of sterilizing the entire dental treatment unit, which is virtually impossible with today's units, will guarantee a biofilm-free system.

Control of biofouling, presented in Chapter 9, involves measures specific to the type of equipment under consideration. In the case of dental treatment equipment, control depends on technical factors, effective training of personnel using the equipment, and establishing and following validated operating procedures. Judicious compliance with cleaning and treatment protocols appears to be critically important for long-term success in avoiding disease transmission via dental treatment units.

8.6.4 Biofouling of Implanted Prostheses

Implanted devices such as arterial grafts, hip joints, and dental implants have been designed and approved for use in the human body on the basis of host tissue biocompatibility and functional characteristics. Criteria for materials used in the fabrication of the total artificial heart are based on their blood compatibility and structural properties. Lack of tissue integration and associated infections are two of the major factors leading to premature failure of devices implanted in soft tissue and transcutaneous devices, including catheters. Biofouling of implant surfaces by invasive microorganisms is the single most important deterrent to expansion of implant applications and a primary reason for failure of those in use. Infection rates are as high as 100% for certain urinary catheters.

Biofilm formation on devices implanted in the human body can lead to an intractable infection of the patient by pathogenic microbes and/or malfunction of the implanted device (Gristina, 1987). Implanted materials, which have become the locus for bacterial infection, if not replaced, lead to chronic illness (Dickinson and Bisno, 1989). *Staphylococcus epidermidis* and other coagulase-negative staphylococci, and gram-negative bacteria are frequently encountered biofilm-forming microbes on implanted prostheses.

Biofouling of an implanted device provides a source of infectious cells that can be disseminated to other parts of the body via the circulatory system. When an implanted device is surgically removed and replaced due to colonization by microbes, further opportunity for infection occurs. Such biofilms are often unresponsive to antibiotic therapy. When left unchecked, the biofouling can interfere with prosthesis function and ultimately lead to failure. Biofouling of artificial hearts can be life-threatening to patients (Gristina et al., 1988).

The sources of bacteria responsible for the vast majority of implant-associated infections are: (1) perioperative contamination, (2) exit site contamination for percutaneous devices, or (3) hematogenous spread from locations distal to the implant area (Mittelman, 1996). Each of these sources defines a different species diversity, severity of infection, and predisposing host condition. Although there have been numerous studies demonstrating the influence of material properties on bacterial colonization, no clear link to frequency of infection has been established. Circumstances related to device application appear to be the single most important criterion influencing infection.

It has been proposed that most vascular grafts, prosthetic heart valves, and orthopedic implants are colonized by bacteria via hematogenous dissemination. Percutaneous access devices are the major routes of entry of bacteria responsible for early-onset prosthetic valve endocarditis, which is the result of biofilm formation on the surface of heart valves (Keys, 1993). Oral microflora are the most common etiological agents of bacterial endocarditis. Dental hygiene practices can introduce organisms such as *Actinobacillus* spp. into the bloodstream.

Indwelling urethral catheters are tubular, latex, or silicone instruments used to drain urine from the bladder. Prevalence surveys have revealed the large-scale use of indwelling urethral catheters in modern medicine in both hospital and community care. This device provides easy access for bacteria from a contaminated environment into a vulnerable body cavity. Urinary tract infection is a common occurrence in catheterized patients. Organisms colonizing the periurethral skin migrate along the epithelial surface of the urethra and the external surface of the catheter. Bacterial multiplication in the urine allows high numbers of cells to accumulate in the urine drainage bags. These bacteria migrate back up the drainage tube and catheter and into the bladder. The various routes of infection are summarized by Stickler (1996).

In the first week of catheterization, infection is usually by a single species of bacteria, such as *Staphylococcus epidermidis*, *Enterococcus faecalis*, *Escherichia coli*, or *Proteus mirabilis*. As the duration of catheterization increases, the more complex the community of microorganisms becomes. Bacteria such as *Providencia stuartii*, *Pseudomonas aeruginosa*, *Proteus mirabilis*, and *Klebsiella pneumoniae* are common gram-negative, nosocomial species that persist in this type of device-related infection.

Both the external and luminal surfaces of the catheter develop bacterial biofilms. Fibrinogen, along with a number of other blood plasma proteins deposited by the body at the site of tissue injury caused by the catheter, may promote bacterial adhesion and biofilm development on these surfaces. Biofilms range from patchy, discontinuous films across the device surfaces to layers several hundred cells in thickness that cover large areas of surface. The biofilms form encrustations of inorganic minerals on the catheterization device that block fluid flow (Stickler, 1996).

While our bodies have evolved a sophisticated defense system to counter infection, implantation procedures compromise some of the important barriers. By subjecting our bodies to invasive surgery or indwelling catheters, we lose the barrier provided by our skin. Peritoneal dialysis catheters do not allow the skin epithelial cells to form a tight seal at the point of perforation. Instead, necrotic epidermal cells and keratin line the sinus tract, creating a habitat conducive for microbial colonization. Peritonitis is the most common complication associated with peritoneal dialysis and a major cause of morbidity.

By using immune-repressing drugs, we compromise the ability of our immune system to seek and destroy foreign substances that enter the body. Device-related infections possess an intrinsic resistance to topically or systemically applied antimicrobials. Although not yet fully understood, microbial biofilms, which have accumulated on surfaces, resist antibiotic doses that irreversibly inactivate individual, unattached bacterial cells (Nickel and Costerton, 1992; Vrany et al., 1997).

The matrix biopolymers excreted by surface-associated bacteria modulate host immune responses that are usually effective in clearing individual, free-living bacteria. The matrix protects bacterial cells embedded deep in the biofilm from antibodies and other antigen-seeking and neutralizing molecules of our immune system. The matrix also enables the device-associated bacterial populations to remain attached to the surface and to each other, thereby resisting engulfment by macrophages and other cell-mediated processes. Until a weak link in the defense system of biofilm microorganisms is discovered and exploited to control their growth, the potential for prosthetic devices to replace natural human body parts will always come with a certain risk of infection.

8.7 ENGINEERING CONSIDERATIONS OF BIOFILM FORMATION ON INDUSTRIAL SURFACES

8.7.1 How Surface Biofouling Influences Mass Transport

Until recently, most mathematical models treated biofilms as a layer of uniform thickness and cell density. Previously, when biofilms were treated as uniformly porous structures, molecular diffusion was the only mass transport process considered. Although difficult to demonstrate, substrate uptake by bacterial cells attached to a surface is at least as fast as or faster than mixing can replace the substrate removed from the bulk aqueous phase at the liquid–surface interface (Gjaltema et al., 1994). This leads to the formation of substrate concentration gradients within a biofilm at the interface. Localized substrate depletion occurs within the biofilm as the thickness of the biofilm increases. Substrate depletion can lead to biofilm sloughing. In a continuously stirred tank reactor of the Roto-Torque design, sloughing occurs most prominently from the lower surfaces where the biofilm is thicker (Gjaltema et al., 1994).

8.7.1.1 How Biofilm Heterogeneity Affects Mass Transfer. Biofilm communities exhibit specialized biofilm specific activities that are a manifestation of the combined effects of species diversity, surface attachment, and physicochemical gradients. Bacterial growth in biofilms is strongly influenced by local environmental pressures; pH, oxygen, and redox local concentrations are key variables that may affect the colonization structure of a biofilm. The physical microstructure of a biofilm may influence the formation of pH and Eh gradients through its impact on diffusive and convective fluid flow characteristics. These transport phenomena influence the access of substrates for bacterial metabolism and, in turn, the removal of metabolic end-products. Cell:cell communication is dependent upon the exchange of minute levels of signal molecules that must travel through the microstructure of the biofilm. The ability of antimicrobial agents to penetrate to their intended targets depends upon diffusive and convective transport processes, which are directly influenced by the microstructure of the biofilm.

8.7.1.1.1 Surface Area Considerations. The heterogeneity displayed by some biofilms affects mass transport into and within the biofilm. Biofilm heterogeneity can increase contact area between the bulk liquid and biofilm. The larger contact area between the bulk liquid phase and a biofilm of varying thickness promotes mass transport between these phases. Rough surfaces increase external mass transfer by increasing eddy diffusion. The formation of filamentous strands of biofilm extending from a base film into the bulk liquid is estimated to increase the biofilm surface area at the bulk liquid interface from 1.5 to 11 times that of the surface area of the underlying substratum, depending on the density and dimensions of the filaments (Gjaltema et al., 1994). The undulating movement of the filaments in a flow field has been estimated to increase mass transport in and out of the biofilm. A patchy, discontinuous biofilm, on the other hand, offers a smaller biofilm-bulk liquid interface than a homogeneous layer containing the same amount of biomass. The structural heterogeneity of many biofilms makes it extremely difficult to estimate the interfacial area of the biofilm-bulk liquid boundary because direct measurements either give an average over the sampled areas or provide unrelated local values (Gjaltema et al., 1994).

8.7.1.1.2 Volume Considerations. Biofilm heterogeneity extends beneath the biofilm surface-bulk liquid interface. Cell densities vary greatly within a multilayer biofilm. Although it has been known for some time that within trickling filter biofilms, bacteria develop as microcolonies separated by voids or water channels of low cell density (Mack et al., 1975; DiSalvo and Daniels, 1975), quantitative differences have not been established until recently. By cutting biofilms into 10–20 µm thick sections and determining total suspended solids, phospholipid concentrations, and dye absorption for each section, the densities and porosities have been determined using a theoretical model (Zhang and Bishop, 1994a). Biofilm densities were found to be 5–10 times higher at the bottom of the biofilm than at the top. Correspondingly, porosity decreased from 84–93% in the top layers to 58–67% in the bottom layers, but the mean pore radius of the biofilm decreased from the top to the bottom of the biofilm. Both tortuosities and effective diffusivities depend on the density and porosity of the biofilm. The ratio of effective diffusivity within the biofilm to that in the bulk reportedly varies from 68–81% in the top layer to 38–41% in the bottom layer of a biofilm (Zhang and Bishop, 1994b).

Recent evidence indicates that there is convective fluid flow not only at the biofilm-bulk liquid interface but also at the biofilm-substratum interface in biofilms that contain areas of low cell density near the substratum (Lewandowski et al., 1995). Using oxygen microelectrodes and confocal scanning laser microscopy (CLSM), de Beer et al. (1994) found that the flux of oxygen between areas of low cell density and areas of high cell density, within an 80 μm-thick biofilm, was similar to the flux of oxygen between the bulk liquid and the surface of the biofilm. The flux within a biofilm has both a horizontal and vertical component, suggesting the need for a three-dimensional approach to modeling mass transport processes in biofilms. Thus, new models of mass transport within biofilms must take into account the varying effective diffusivity and flux at different planes within a biofilm.

8.7.1.1.3 Advective Mass Transport of Particles into and Within Biofilms.
Biofilms promote the entrainment of particles suspended in the bulk fluid. Drury et al. (1993) showed that latex spheres the diameter of bacterial cells were entrained by bacterial biofilms. The eddies formed as a result of variable biofilm thickness and filament formation entrain particles that would tend to remain in suspension if transported across a biofilm-free surface. The heterogeneity of biofilms allows advective mass transport of particles through voids deep within the biofilm (Lewandowski et al., 1995). Neutrally buoyant, 0.3 μm diameter latex spheres have been tracked at velocities ranging from 10–20 μm/sec in a mixed species, laboratory-cultivated biofilm (Stoodley et al., 1994). Particle velocities within a biofilm vary depending on the biofilm-to-void ratio. Where voids make up a large portion of the volume of a biofilm, the advection term in the mass transport equation increases in importance relative to the diffusion terms (see Eq. (11.6) in Gujer and Wanner, 1990).

The structure of a bacterial biofilm is not the uniform gel-like matrix, entrapping cells, as implied by numerous mathematical models and experimental studies. Rather, a biofilm can be highly heterogeneous in its structural architecture, with very dense cell clusters separated by crevasses or water channels. These water channels are of various sizes and dimensions and render the biofilm gel matrix a labyrinth of tortuous cavities through which fluid, dissolved solutes, and suspended particles can migrate.

Consequently, biofilm properties can no longer be estimated as overall biofilm lumped parameters but rather must be estimated locally. Estimates by the half-cell diffusion chamber method were inadequate to provide an accurate measure of true mass transfer coefficients. Bryers and Drummond (1999) report experiments where pure culture *P. putida* biofilms were cultivated under controlled conditions to a desired overall biofilm thickness, then employed within classical half-cell diffusion chambers to estimate, from transient solute concentrations, the effective diffusion coefficient for several macromolecules of increasing molecular weight and molecular complexity.

The basic idea of a half-cell diffusion system is quite simple. Two compartments are separated by a known thickness of permeable material, the half-cells filled with a solvent, allowed to hydrostatically equilibrate, and at time equal zero, the solute is introduced into one cell and the concentrations of solute measured in both cells as a function in time. Concentrations of solute in both half-cells are then determined as a function of time; diffusion coefficients can then be calculated from Eq. (8.2),

$$ln\left[\frac{C_I(t) - C_{II}(t)}{C_I(t = t_1) - C_{II}(t = t_1)}\right] = \frac{A}{K}\left(\frac{1}{V_I} + \frac{1}{V_{II}}\right)(t - t_1) \qquad (8.2)$$

where V_I = volume of half-cell I (L^3); V_{II} = volume of half-cell II (L^3); $C_I(t)$ = concentration of solute at time t in half-cell I, (M L^{-3}); $C_{II}(t)$ = concentration of solute at time t in half-cell *II*, (M L^{-3}); $C_{I,II}(t = t_1)$ = concentration of solute in half-cell *I* or *II*, respectively, at the onset of pseudo-steady state, time $t = t_1$, (M L^{-3}); t = time (t); A = surface area of transport connecting the two half cells, (L^2); and K is the overall composite mass transfer coefficient, (L t^{-1}). K can be defined for this work by Eq. (8.3),

$$K^{-1} = \left[\frac{L_M}{D_{\text{eff}-m}} + \frac{L_{\text{biofilm}}}{D_{\text{eff}-\text{biofilm}}} \right] \quad (8.3)$$

where L_M = thickness of permeable membrane separating the two half-cells and employed as a support for biofilm accumulation, (L); L_{biofilm} = thickness of the biofilm, (L); $D_{\text{eff}-m}$ = molecular diffusivity of solute through support membrane alone, (L^2 t^{-1}); and $D_{\text{eff}-\text{biofilm}}$ = molecular diffusivity of solute through biofilm, (L^2 t^{-1}).

Diffusivity estimates were based on calculations using Eqs. (8.3) and (8.4) and data considered valid during the steady state portion of each diffusion trial (i.e., after a time period t_1 that varied with solute and biofilm thickness). Results of the effective diffusivity calculations based on Eqs. (8.2) and (8.3) are summarized in Table 8.2. Also reported in Table 8.2 is the ratio of the estimated solute diffusivity in biofilm relative the calculated value for the solute in pure water at the same temperature as estimated from Eq. (8.4),

$$D_{AB} = kT/6\mu_B \pi R \quad (8.4)$$

where k is the Boltzmann constant, T is temperature (°K), μ_B is the solvent viscosity, and R is the solute Stokes radius. Results in Table 8.2 indicate that regardless of solute molecular weight or size (radius of hydration), the diffusivities of solutes considered were all reduced in the biofilm by about the same degree relative to pure water, which is erroneous.

Water channels or perforations in the biofilm structure were observed that were between 10 and 30 μm wide at the outer edge of the biofilm and penetrated deep into the biofilm,

TABLE 8.2. Diffusion Coefficients for Fluorescently Labeled Molecules in a Bacterial Biofilm of 180 μm Using Diffusion Half-Cells

		Diffusion Coefficients (cm^2/sec $\times$ 10^{+7})		
Diffusing Molecule, i	Molecular Weight	In Pure Water[a]	In Biofilm	$D_{\text{biofilm}}/D_{i-\text{water}}$
Fluorescein	332.	55.0	54.0	0.98
Dextran	10,000.	2.2	1.96 (0.1)	0.89
Dextran	70,000.	1.7	1.44 (0.15)	0.85
BSA[b]	68,000.	6.9	5.9 (0.3)	0.86
IgG[b]	146,000.	3.8	NA	NA
HK[b]	102,000.	5.9	4.1 (0.23)	0.69
Catalase	225,000.	4.1	2.78 (0.15)	0.67
DNA	3.2 $\times$ 10^6	0.008	0.005 (0.001)	0.63

[a] Calculated from Einstein's Equation, see Table 8.3 for definition.
[b] Abbreviation definitions found in Table 8.3.
Note: Numerical values in parenthesis are standard deviations.

8.7 ENGINEERING CONSIDERATIONS OF BIOFILM FORMATION

often times completely to the base substratum. The outer edge of the biofilm also exhibited an undulating topography, with the distance between the top and bottom of a "wave" averaging 10 μm. Presence of these perforations was documented by variation in the levels ("heights") of latex particles deposited onto a biofilm sample. Further, the continuous microscopic observation of a latex bead in a water channel as it settled due to gravity allowed estimates of the bead's terminal settling velocity. An average 1.3 μm/sec was observed over several experiments, which agrees well with the value of 1.6 μm/sec calculated from Stokes Law for a pure water system.

These qualitative observations suggest the half-cell diffusion trials just reported may not be able to accurately determine solute transport properties, which apparently vary locally within the biofilm. The presence of perforations in the biofilm coupled with the added surface area for transport afforded by the sides of these water channels, makes it likely that transport of the solutes in question from one half-cell to another was not impeded by the biofilm itself. It is more likely that the solutes were able to diffuse through the water channels at a rate similar to that in pure water. Eventually, solute may diffuse laterally into the biofilm matrix itself.

Thus, an alternative analytical technique was refined by Bryers and Drummond (1999) to determine the local diffusion coefficients on a microscale in order to avoid the errors created by the biofilm architectural irregularities. This technique is based on the fluorescence return after photobleaching (FRAP), which allows image analysis observation of the transport of fluorescently labeled macromolecules as they migrate into a microscale photobleached zone.

In concept FRAP is a simple technique (Axelrod et al., 1976). A small region of a surface or volume containing mobile fluorescent molecules (or fluorescently labeled molecules) is exposed to a brief intense pulse of light, thereby causing irreversible (assumed but often times not true) photochemical bleaching of the fluorophore in that region. Transport properties of the fluorescently labeled molecule are determined by measuring the rate of fluorescence recovery that occurs due to transport of unbleached molecules migrating into the bleached area from the unirradiated surroundings. Overall fluorescence is achieved using a separate source of light with a much attenuated intensity. Because the initial bleaching beam can be focused to bleach a region of very small dimensions, it is possible to acquire several FRAP samples over a sample plane very rapidly.

FRAP measurements provide an estimate of the "intradiffusion" coefficient for a component in a multicomponent system. Intradiffusion is used here rather than the more ambiguous terms of "self" or "tracer" diffusion (Albright and Mills, 1965; Westrin, 1991). Consider a homogeneous sample of a multicomponent system is at a uniform temperature. At time zero, all the molecules of a single component in a small subsection of the entire sample becomes an isotopically labeled form of that component. The labeled solute's concentration can now be uniquely measured as different from that of its unlabeled counterpart. Further, there are no chemical differences between the labeled and unlabeled components with regard to their transport properties. Equivalent and opposite concentration gradients are established for the component. The mutual diffusion between the two chemically equivalent components is defined as intradiffusion. Intradiffusion coefficients are obtained by treating the system as a two-component system and using Fick's Law. Intradiffusion coefficients are numerically equivalent to tracer diffusion coefficients but the terms have two significantly different meanings. In tracer diffusion the tracer molecule may or may not have a chemically equivalent unlabeled counterpart in the system and the tracer is applied in very low concentrations. In intradiffusion, the resultant diffusion coefficient is independent of the ratios of the concentrations of the two components and depends only on the sum of their concentrations and the concentration of other components in the system. Self-diffusion is a special case of intradiffusion that deals with a binary system of only the two diffusing components, labeled and unlabeled.

For pure molecular diffusion, fluorescence recovery is as given by Axelrod et al. (1976) as Eq. (8.5),

$$f(t) = \sum_{n=0}^{\infty} \frac{(-\kappa)^n}{n!} \frac{1}{[1 + n\{1 + 2t/\tau_d\}]} \quad (8.5)$$

where κ is the sample bleach constant and τ_d, is the two-dimensional (2-D) characteristic diffusion time, defined as,

$$\tau_d = \omega^2/4D \quad (8.6)$$

where ω is the e^{-2} laser beam radius for both the bleaching and monitoring phases of the FRAP experiment and D is the sample lateral diffusion coefficient.

Estimates of the local diffusion coefficient are made by correlation between the theoretical fluorescence recovery curve and experimental observation, with the only adjustable parameter being the diffusivity. As the FRAP tests are repeated across a plane in the biofilm, areas of locally specific D_{eff} can be determined for all solutes. As depth is changed and another horizontal scan is carried out, it became evident, within the water channels, that $D_{eff} \sim D_w$. A summary of the various solute (D_{eff}/D_w) ratios taken within biofilm polymer-cell clusters is provided in Table 8.3.

8.7.2 How Biofilms Influence Momentum Transport

Biofouling of a surface in contact with a flowing fluid alters the velocity of that fluid at the fluid–surface interface. The presence of a biofilm may increase or decrease the fluid frictional resistance of the surface depending on the degree of biofouling.

Characklis and co-workers (Zelver, 1979; Picologlou et al., 1980) provided in 1980 what remains today to be the benchmark paper on the effects of biofilm accumulation on momentum transfer (i.e., friction factor) for fluid flow in a tube with an hydraulically smooth surface. With water velocity maintained constant in the tube, then pressure drop will increase as biofilm accumulates. Increasing pressure drop increases pumping costs. If the biofilm forms on a ship hull, power consumption for propulsion at constant velocity will increase. As biofilm accumulates on a tube or pipe surface in turbulent flow, the friction factor increases regardless of whether velocity or pressure drop are maintained constant since it is a dimensionless parameter that takes pressure drop, velocity, and pipe geometry into consideration.

Potential mechanisms, postulated by Zelver (1979) and Zelver et al. (1985), governing the significant losses in momentum due to biofilm accumulation were the following:

- Pipe diameter constriction
- Increase in pipe wall rigid roughness

8.7 ENGINEERING CONSIDERATIONS OF BIOFILM FORMATION

- Biofilm creep or biofilm being pushed along the wall in the axial direction
- Biofilm compliancy
- Biofilm viscoelasticity
- Biofilm filament oscillation
- Increase in fluid viscosity due to soluble biofilm products
- Rigid pipe wall roughness imparted by the biofilm

All but biofilm compliancy, viscoelasticity, and oscillations of the biofilm irregular, filamentous surface structures were proved to be insignificant contributions to the overall dissipation of kinetic energy by a biofilm. In Bryers (1980) experiments indicate that biofilm filamentous nature to some extent is set by the fluid dynamic regime, which in turn dictates the degree to which biofilm influence momentum. At lower Reynolds number (Re), biofilm filaments are longer and for the same overall thickness create larger pressure drops while at higher Re, filaments are short and their impact on frictional resistance is relatively less. Viscoelasticity measurements by Kirkpatrick et al. (1979) indicated that biofilm exhibited both a viscous and an elastic modulus, meaning that fluid flow through a conduit with a compliant viscoelastic wall structure would potentially dissipate significant fluid kinetic energy. These facts were corroborated by Stoodley et al. (1998) and again in Stoodley et al. (1999); a study which for the first time also verified that under certain conditions, biofilm can exhibit a degree of creep (e.g., sliding of biofilm en masse down the tube wall) in material in the axial direction).

Nuclear magnetic resonance imaging studies show that in a rectangular cross-section flow channel the presence of a biofilm can affect the velocity distribution of flow

TABLE 8.3. Diffusion Coefficients for Fluorescently Labeled Molecules in Water and in a Bacterial Biofilm using the FRP Method

		Diffusion Coefficients (cm^2/sec) $\times 10^{+7}$			
Diffusing Molecule, i	Molecular Weight	In Pure Water[a]	Biofilm Location Water Channel	$D_{i-biofilm}$ Cell Cluster	$D_{i-water}$
Fluorescein	332.	55.0	54.0	50. 0.91	
Dextran	10,000.	2.2	2.0	0.62 (0.01)	0.28
Dextran	70,000.	1.7	1.7	0.26 (0.02)	0.15
BSA	68,000.	6.9	6.8	3.73 (0.02)	0.54
IgG[b]	146,000.	3.8	NA	NA	NA
HK[b]	102,000.	5.9	5.0	2.89 (0.01)	0.49
Catalase	225,000.	4.1	4.0	1.27 (0.02)	0.31
DNA	3.2×10^6	0.008	0.007	0.0015 (0.0003)	0.19

Note: Biofilm thickness was 232 μm in overall thickness; results in this table were averaged from FRP scans taken at an overall depth of 120 μm.

[a] Calculated from Einstein's Equation: $D_{i-water} = k_b T/6\pi\mu r_s$ where k_b = Boltzman's constant; T = temperature, °K; μ = viscosity of solvent:solute mixture at temperature, T; and r_s = Stokes molecular radius.

[b] Results for IgG suspect due to binding of the immunoglobulin to biofilm matrix and cell wall. I Ratio reported are for values taken in biofilm cell clusters relative to pure water values.

BSA = bovine serum albumen; IgG = G class of human immunoglobulins, antibodies; HK = hexokinase. NA = not applicable.

(Lewandowski et al., 1995). Whereas a jet is formed at average flow velocities of 4.0 cm/s or greater in the reactor without biofilm, no jet is formed in reactors containing biofilms at these same average fluid velocities (Fig. 8.9). It has been proposed that the presence of a biofilm dramatically increases the wetted perimeter since the biofilm has a well-developed surface. This biofilm-induced increase in wetted perimeter decreases the Reynolds number according to the equation

$$\mathrm{Re} = V4A/vP$$

where v is the kinematic viscosity, A is the cross-sectional area, and P is the wetted perimeter (Lewandowski et al., 1995). The average flow velocity profile is not precisely parabolic near the walls. At flow rates of 4.0 cm/s the flow velocity reaches zero at the reactor surface, not at the biofilm surface.

Fluid frictional resistance or shear increases with increased contact area between biofilm and a flowing fluid. This may promote either cell attachment or biofilm biomass detachment or both. Biofilm desquamation, sloughing, and erosion are three terms used to describe the detachment or loss of surface-associated biomass. Sloughing and desquamation typically refer to a catastrophic event involving the detachment of many cells and associated EPS, and other entrained particles. Sloughing may result in the removal of biomass clear down to the base of the biofilm, exposing the bare substratum. Erosion often refers to the independent detachment of single cells from the surface of the biofilm adjacent to the bulk liquid phase. These processes are very difficult to predict using mathematical models.

Resistance to sloughing is dependent on the limiting nutrient in the case of biofilms of the bacterium, *P. putida*, in a turbulent flow system (Applegate and Bryers, 1991). Oxygen-limited biofilms achieve (1) higher steady state biofilm organic carbon levels, (2) higher extracellular carbon/cellular carbon ratios, and (3) higher biofilm-bound calcium than carbon-limited biofilms. Oxygen-limited biofilms exhibit shear removal rates that are 20–40% of those observed for carbon-limited biofilms.

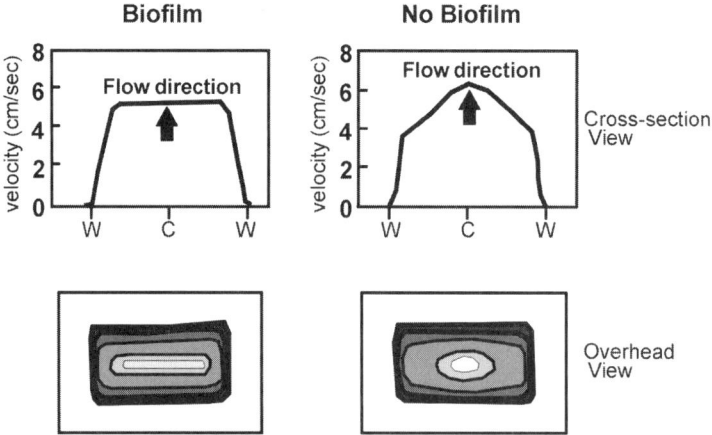

Figure 8.9 Cross-section and overhead views of fluid velocity profiles through a rectangular flow channel in the presence and absence of a biofilm. C is center of flow channel, W is the wall of the flow channel. (Adapted from Lewandowski, 1998.)

8.7.3 How Biofilms Influence Energy Transport

Heat is only one form of energy and, according to the First Law of Thermodynamics, energy and not heat is conserved. Nevertheless, in biofilm systems, the other forms of energy are generally negligible. Heat transport occurs frequently in nature and in process plants (e.g., the cooling of water as it exits a hot spring in Yellowstone Park or the heating of river water as it passes through a power plant condenser). Two mechanisms for heat transport are relevant to biofilm systems: (1) conduction, which is the transport of heat from a high temperature to a low temperature within a phase (intraphase) such as a solid or fluid by the motion of molecules or electrons and (2) advection, which is heat transport that results from bulk fluid motion usually relevant to interphase transport. Despite the tremendous effects of biofilm on heat transfer efficiency and intense research efforts investigating these effects in the late 1970s and early 1980s, ironically very little new research has since been initiated on the fundamental effects of biofilm accumulation on heat transfer.

Conductive heat transfer resistance results from the insulating layer formed by the biofilm and generally increases as the biofilm accumulates. Conductive heat transfer is the difference between overall and advective heat transfer resistance and can be calculated *a priori* if the biofilm thickness and thermal conductivity of the deposit is known. The conductive heat transfer resistance is dependent on thermal conductivity and thickness of the deposit. *Advective* heat transfer resistance results from fluid motion or turbulence and generally decreases as biofilm accumulates, since the roughness of the biofilm increases turbulence in the interfacial region. Advective heat transfer resistance can be calculated from the friction factor and the properties of water. Thus, the advective heat transfer resistance depends on the roughness characteristics of the biofilm and the shear stress at the heat transfer surface.

The overall heat transfer resistance determines the influence of biofilm on heat transfer efficiency. However, advective and conductive heat transfer resistance may be important because differences in deposit properties (apparent roughness and thermal conductivity) can result in significant differences in the relative contributions of conductive and advective processes to overall heat transfer resistance. Measurements related to biofilm accumulation and general fouling processes include monitoring overall heat transfer resistance and characterizing the biofilm or fouling deposit. A variety of different monitoring devices have been developed to obtain accurate fouling data under various experimental conditions. A detailed description of some of the devices can be found elsewhere (Knudsen, 1981). One type of instrument is the thick-walled heat exchanger (Characklis et al., 1981; Turakhia and Characklis, 1984; Zelver et al., 1985). Constant heat is supplied to the thick-walled heat exchanger and temperature at two points in the thick wall are measured, allowing one to calculate the heat flux in the system.

As biofilm accumulates, the overall heat transfer resistance for a tube changes due to increased biofilm conductive resistance and decreased advective heat transfer resistance resulting from increasing biofilm roughness. Characklis et al. (1981) have determined the relative changes in conductive and advective heat transfer resistance in a tube during biofilm accumulation. At the beginning of the experiment, conductive resistance due to the biofilm is zero and only advective resistance exists. As biofilm accumulates, conductive resistance increases proportional to biofilm thickness. Advective resistance decreases as a result of increased roughness reflected by an increase in friction factor. The Colburn analogy was used to calculate the advective resistance from friction factor measurements and therefore represents an approximation to the actual advective resistance. The biofilm thermal conductivity was measured in separate experiments.

8.8 MONITORING BIOFOULING

Monitoring is the key to biofouling control. The key to successful monitoring is (1) knowing exactly what it is your monitor is measuring, (2) deciding what it is you really need to monitor, and (3) relating that which you are monitoring to a performance criteria for a particular process of interest. By passing a process fluid over a substratum, organisms in the process fluid deposit on the substratum and form a biofilm. Selection of the appropriate engineering or biological parameters to monitor depends on the nature of the process that is influenced by the biofouling phenomenon. The most common biological parameters that have been used to monitor biofouling include surface-associated biomass, desquamation, and metabolic activities that impact the substratum or the bulk, aqueous phase (White et al., 1996). Common engineering parameters used to monitor biofouling include heat transfer resistance and fluid frictional resistance (velocity, pressure drop, or increase in torque). Details on methods of monitoring biofouling in various engineered systems may be found in several recent publications (Bryers, 1993; Flemming and Griebe, in press).

8.8.1 Microscopy

Early biofouling monitors consisted of transparent glass slides suspended in the aqueous phase (Henrici, 1933; Zobell and Allen, 1935). Once a biofilm accumulates on the slide, the slide is retrieved and examined by transmitted light microscopy. Following the development of epifluorescence microscopy and scanning electron microscopy, opaque surfaces were substituted for glass slides to examine the influence of different substratum properties on biofouling. These microscopic techniques also allowed better resolution of microbial cells in the presence of other material that was adsorbed to the surface. The recent availability of optical equipment to perform differential interference contrast (DIC) and CSLM has created new opportunities to evaluate the development of biofouling layers nondestructively in real time.

Much has been made of CSLM in recent years as a means of monitoring structural properties of biofilms. CSLM displays objects within a plane of focus (0.3 μm) while eliminating out-of-focus objects in other planes using a pinhole barrier. Images collected in different planes can be merged using computer-controlled image analysis to produce a 3-D reconstruction of a multilayer biofilm, showing depth-dependent relationship of light-absorbing structures (Lawrence et al., 1991; Caldwell et al., 1992). CSLM has been used in conjunction with different types of flow cell reactors, enabling biofilm development to be evaluated in 4 dimensions (x,y,z,t). Surman et al. (1996) have compared the types of biofilm information that can be obtained by the different microscopic techniques.

8.8.2 Sampling Devices

A variety of biofilm samplers have been developed, including the widely used Robbins Device, which can be fabricated of the same material as the fouled industrial surface and mounts flush with the surface of a conduit (McCoy et al., 1981). Coupons remain the most common sampling device, primarily because of their simplicity and versatility. Coupons of any composition and size can be suspended in either a static or dynamic aqueous medium and later sampled by a variety of methods.

8.8.3 Biofouling Reactors

8.8.3.1 Annular Reactors. There are nearly as many variations in the design of reactors to establish biofouling layers on surfaces as there are questions to answer regarding biofouling. The continuous flow stirred tank reactor (CFSTR), described by Characklis (1990), continues to be a popular system to use for development of biofouling layers on coupon surfaces. The Roto Torque, is a CFSTR that offers the opportunity to independently control shear and residence time of the liquid phase (Characklis, 1990). The Roto Torque contains multiple sampling slides that are flush-mounted on the reactor wall. Although wall shear can be controlled by the operator, there is considerable variation in the shear values obtained at different locations in the reactor (Gjaltema et al., 1994). The sampling slides contain a biomass/unit area that is not representative of the entire reactor surface area due to these local differences in shear. Since it is not possible to obtain representative biofilm samples from the Roto Torque, this reactor is not suitable for use in quantitative physiological or kinetic biofilm studies where it is essential that the total amount of active biomass in the system can be measured. The reactor can still be used for morphological and "black box" studies, however.

Substrate gradients and inhomogeneous shear fields are also difficult to avoid in other types of biofilm reactors, including flat plate channel-flow reactors and tubular reactors. Furthermore, because these sampling methods required "off-line" analysis, they offered only a "snapshot" in time of a very dynamic biological phenomenon, thus limiting our ability to detect desquamation or obtain accurate estimates of biofilm production.

8.8.3.2 Radial Flow Reactor. A radial-flow cell or Fowler Cell Adhesion Module, described by Fowler and McKay (1980), provides a continuous shear gradient across a substratum. Mittelman et al. (1990) have utilized a Fowler Cell Adhesion Module to reproducibly demonstrate the changes in cellular concentration and composition of a culture of *P. atlantica* on stainless steel as a function of applied shear force. The apparatus has also been used to observe microscopically through replaceable windows a developing biofilm as a bulk aqueous phase flows over the surface (Mittelman et al., 1992). Flow-through reactors have been developed that accommodate different flush-mounted substrata and surface coatings, offering the opportunity to assess substratum effects on biofilm accumulation under controlled laminar flow (Arrage et al., 1995).

A radial-flow chamber was used by Dickinson and Cooper (1995) to quantitatively describe attachment and detachment of cells of *S. aureus* as a function of shear stress on three different polyurethane surfaces conditioned with various blood fractions. An automated videomicroscope and image analysis system was used to achieve rapid counts of attached bacteria across the surfaces of the treated disc in the radial flow chamber over time. Shear-dependent rate constants were estimated for attachment and detachment by fitting the solutions of phenomenological models to the experimental data. The authors demonstrated the usefulness of the approach to resolve an influence of the various surface-conditioning molecules on cell attachment and detachment.

Dickinson et al. (1995) used the approach described previously to determine the probability of attachment and detachment of cells capable of producing a cell-surface clumping factor, coagulase, or both compared to mutants lacking these cell surface molecules on plasma-, fibrinogen-, or albumin-coated polyurethane surfaces. The result suggested that clumping factor played a pivotal role in enhancing adhesion to surfaces with adsorbed fibrinogen.

8.8.4 Nondestructive Monitoring of Biofilm Biomass

On-line biofilm monitoring techniques have been developed in recent years that offer real promise as practical monitoring tools to industry. Biofilm formation, succession, and stability can now be monitored using in-line, nondestructive techniques in flow-through cells. Biofouling has been quantified on the basis of biofilm microbial biomass determined by on-line monitoring of nicotinamide adenine dinucleotide (NADH), tryptophan, and chlorophyll fluorescence (Angell et al., 1993). The relative tryptophan fluorescence per cell has been used as a nondestructive measure of sublethal toxicity in biofilms exposed to antimicrobial compounds introduced to the bulk aqueous phase or impregnated into the substratum (Arrage et al., 1995). The intensity of the amide protein bands in the infrared has been used as a measure of microbial biofilm biomass (Geesey and Bremer, 1990; Suci et al., 1997). Using attenuated total reflectance Fourier transform infrared (ATR/FT-IR) spectroscopy, Nivens et al. (1993a) were able to detect as few as 5×10^5 cells of *Caulobacter crescentus* per cm^2 surface on a germanium internal reflection element (IRE) in a flow cell based on absorbance in the amide II region of the spectrum.

The quartz crystal microbalance (QCM) has been used to monitor initial fouling and subsequent biofilm development (Nivens et al., 1991). The QCM employs an AT-cut (refers to thickness-shear mode of cut relative to crystallographic axis) quartz crystal, and oscillator circuit, and a frequency counter to obtain a mass measurement. When bacteria in an aqueous suspension adsorb to a submerged quartz crystal containing a thin metal film, the frequency of oscillation decreases in proportion to the mass of adsorbed material. The QCM thus provides a direct measure of biomass. As few as 10^4 cells of *C. crescentus* were detected per square centimeter of a gold-coated quartz crystal (Nivens et al., 1991). A linear relationship was observed between frequency shift and adsorbed biomass over the range of 10^4–10^6 cells/cm^{-1}. One of the main limitations of this monitoring technique is the sensitivity of oscillation frequency to temperature. Furthermore, different crystals respond differently to temperature, making comparison of results among different crystals difficult to interpret.

On-line monitoring of bacterial behavior in biofilms based on bioluminescence has been achieved using a fiber optic probe and photomultiplier detector. Mittelman et al. (1993) monitored accumulation of the bioluminescent marine bacterium *Vibrio harveyi* on antifouling coatings in a flow-through cell containing quartz windows. This approach can be applied to other bioluminescent organisms, including those that have been genetically engineered to express the *lux* or *gfp* gene.

A biofouling monitor just recently introduced uses fiber optic sensors to capture reflected light, which is influenced by the accumulation of adsorbed material at the tip of the fiber (Tamachkiarowa and Flemming, 1996). The sensor can be flush-mounted on the surface undergoing fouling. The intensity of the back-scattered light is proportional to the intensity of the incident light and the projected area of the fouling layer on the sensor tip. However, the sensor is rather insensitive to the nature of the fouling layer in that it does not discriminate between biological and nonbiological material. Selectivity could be achieved through the use of specific wavelengths of energy that are absorbed by specific biological molecules.

8.8.5 Monitoring Biofilm Processes

ATR/FT-IR has also been used to monitor metabolic processes in biofilm populations nondestructively in real time. Nivens et al. (1993b) observed shifts in physiological status, based on an increase in the C=O stretch from the production of polyhydroxy alkanoic acids, in *C. subvibrioides* cells attached to a germanium substratum following ammonium ion depletion. Bremer and Geesey (1991) used the same analytical approach to detect the production of polysaccharide in a bacterial biofilm population growing on a copper film. With this approach, polysaccharide production was related to the deterioration of the underlying copper film.

8.8.5.1 Molecular Biology Tools.
Gene expression has also been monitored in microbial biofilms during surface fouling. Use of heterologous gene expression to characterize biofilm ecology is detailed extensively in Chapter 4. In studies using reporter genes, a *Pseudomonas* strain was constructed with the *lux* gene cassette under control of the promoter for algD, a gene involved in the commitment to synthesize alginic acid (Wallace et al., 1994). The effects of salt concentration and nitrogen balance on alginic acid synthesis by biofilm populations was determined on the basis of bioluminescence emitted from the biofilm (Rice et al., 1995). The brightness of the bioluminescence response in the engineered bacteria is influenced by factors other than promoter activity, however. Factors such as ionic strength, pH, trace metal concentration, carbon source, and plasmid copy number affect the light-emitting reaction (Heitzner et al., 1994).

A *lacZ* reporter gene cassette has been used to monitor the level of expression of *algC*, a "housekeeping" gene that is involved in the production of the extracellular polysaccharide alginic acid in *P. aeruginosa* (Davies et al., 1993). Using the fluorogenic substrate methylumbelliferyl β-D-galactopyranoside (MUG), sufficient quantities of the substrate were converted to fluorescent product by the reporter gene product β-galactosidase, that up-expression could be detected in individual bacterial cells within the biofilm. Three important observations were made in biofilms containing cells carrying these reporter genes (Davies and Geesey, 1995). First, not all members of an isogenic population behave the same in terms of gene expression during residence on a surface. Second, association with a surface promotes expression of genes not normally expressed when cells exist individually in suspension. Third, gene expression in cells on a surface can be dynamic and transient in nature. While sufficient quantities of fluorescent product remained associated with cells to microscopically resolve gene expression in individual cells in the studies cited above, the product often diffuses out of other types of cells as rapidly as it is produced, thus limiting its use in monitoring gene expression in individual cells. This limitation may be overcome by the use of reporter genes whose autofluorescent products remain inside the cells (e.g., green fluorescent protein) or enzyme substrates that not only fluoresce but also precipitate at the site of conversion (e.g., ELF™ substrates; Larison et al., 1995).

The limitations associated with the reporter systems described above can be circumvented by the use of the green fluorescent protein (GFP) reporter system. This system utilizes the *gfp* gene whose product exhibits inherent fluorescence that is less sensitive to conditions in the cell than the *lux* reporter system, and does not require a diffusable substrate like the *lacZ* reporter system. Fluorescence of GFP at 509 nm can be easily detected after excitation at 395 nm. A suicide plasmid containing a promoterless *gfp* gene that recombines with the bacterial chromosome has been constructed to establish *gfp* fusions in a wide variety of gram-negative bacteria (Kalogeraki and Winans, 1997). Alternatively, Tn5- and Tn10-based transposons carrying either a promoterless *gfp* gene or a *gfp* gene expressed from a broad-host-range promoter has been created to tag the chromosome of diverse bacterial species (Matthysse et al., 1996; Stretton et al., 1998). A new *gfp* reporter system has recently been described that results in a fluorescent GFP product that degrades rapidly following synthesis, allowing transient gene expression to be followed nondestructively in real time (Andersen et al., 1998).

GFP fluorescence but not synthesis involves an oxidation reaction requiring molecular oxygen (Heim et al., 1994). This would suggest that GFP may only be useful in anaerobic biofilms. However, recent studies indicate that dissolved oxygen concentrations of 0.73 mg l^{-1} or less is sufficient to promote detectable fluorescence in individual cells of facultatively anaerobic dissimilatory iron reducing bacteria cultured under anaerobic conditions (unpublished results). Furthermore, cultures of the dissimilatory sulfate reducing bacteria, *Desulfovibrio desulfuricans*, carrying a constitutively expressed *gfp* gene, display GFP fluorescence in individual cells under conditions of cell replication (unpublished results). Thus, these genetic tools offer unique opportunities to study the surface behavior of individual cells or populations of a wide variety of surface-fouling bacteria.

Introduction of fluorescent probes that report cell identity, viability, and activity to a flow cell reactor containing mixed population communities offers the opportunity to resolve through CSLM associations and interactions between individual cells and aggregations of cells (Korber et al., 1997; Moller et al., 1997). CSLM has been used to follow expression of chitinase genes in marine bacteria carrying the *gfp* gene under the control of promoters of chitin-degrading genes in cells that have formed a biofilm on squid pen surfaces (Stretton et al., 1998).

8.8.5.2 Microsensors.
Microsensors offer yet another way to evaluate microbial processes in biofilms. Microsensors for N_2O, NH_4^+, NO_2^-, NO_3^-, and O_2 have been used to describe microscale chemical gradients from microbial activities in sediments, fluidized bed bioreactors, and biofilms. Microsensors have also been developed to determine diffusion coefficients and directly measure mass transport into and within biofouling layers on surfaces (Lewandowski, 1993; Rasmussen and Lewandowski, 1998).

Lee and de Beer (1995) have used pH and O_2 microelectrodes to probe respiratory activity in a biofilm on corroding steel. In this instance, the position of the electrodes above the substratum was controlled by a computer and stepper motor. Data was collected at different depths within the biofilm, providing profiles of pH and O_2. Through this approach, it was possible to demonstrate the consumption of O_2 by the biofilm above a tubercle and establish the pH at anodic and cathodic sites.

A highly selective liquid nitrite microsensor has been used to evaluate denitrifying, nitrifying biofilms from wastewater treatment plants (de Beer et al., 1997). Using these microsensors, local areas of elevated nitrite have been located in narrow zones of less than 1 mm. Such sensors have the potential to elucidate which factors control the nitrite concentrations in biofilms.

8.8.6 Concatenation of Reactor Design, Molecular Biology, Microscopy, and Microsensors to Obtain Multidimensional Information on Biofouling of Surfaces

One of the next steps to take in monitoring biofouling is to combine the techniques described previously to characterize surface biofouling processes. This will enable us to obtain a better understanding of the relationships between structure and function in multispecies surface fouling layers. Combining microscopy and the scanning vibrating electrode, Angell et al. (1994) established a relationship between the locations and activities of bacterial microcolonies and the formation of anodic electrochemical activity on a corroding metal coupon. Such small-scale processes must be studied with multiple, nondestructive, on-line techniques that offer high spatial resolution.

Bacterial aggregates from a nitrifying fluidized bed have been investigated with microsensors and rRNA-based oligonucleotide probes (Schramm et al., 1998). Microprofiles of O_2, NH_4^+, NO_2^-, and NO_3^- revealed the occurrence of nitrification in the outer 125 μm of the aggregates. Fluorescent probes identified members of the Nitrosospira group as the primary ammonia oxidizers in this region of the aggregate. Another important step that needs to be taken is the modification of techniques just described to make them useful in the field. While the on-line biofilm monitoring techniques described previously have, so far, been limited to laboratory bench experimentation, theoretically, they should be adaptable to industrial scale biofilm monitoring.

8.9 SUMMARY AND CONCLUSIONS

Biocontamination of engineered materials and systems is difficult to quantify, predict, and assess. Surface biofouling is not governed by easily modeled, predictable processes. Robust three-dimensional models are likely to be required to predict the behavior of complex, spatially dependent populations of surface colonizing species of living organisms under the influence of hydrodynamics and surface chemical phenomena. Indications that such modeling is feasible and often necessary has recently been demonstrated (Picioreanu et al., 1998a; 1998b).

In spite of the heroic financial investments by industry and government to better understand biofilm development and biofouling processes, our ability to manipulate these processes in reproducible, predictable ways has, except in rare instances, been painfully inadequate. Hopefully, the next generation of biofilm researchers will possess the creativity and motivation to develop and effectively utilize the advanced technologies needed to achieve better control over biofilm processes. The development of sensitive devices to monitor biofilm parameters that we consider important is essential before we can demonstrate, at the most rudimentary level, "reproducible" biofilms. Without this capability, progress in understanding fundamental biofilm processes that we seek to control will continue to be slow, while the damage biofilms inflict on engineered materials and processes will continue to occur and cost more to abate. Without a higher level of understanding of biofilm processes, control of biofouling will likely continue to be defined by what amount of damage we can afford to accept in a particular system rather than by a priori knowledge of controllable behavior of the biofilm population in that system.

ACKNOWLEDGMENT

This work was supported by Grant No. N00014-97-1-1062 from the Office of Naval Research DEPSCoR program, National Science Foundation Grants OCE9720151 and EEC8907034, and Grant No. ER62719 from the U.S. Department of Energy.

REFERENCES

Alberte, R. S., S. Snyder, B. J. Zahuranec, and M. Whetstone. 1992. Biofouling research needs for the United States Navy: Program history and goals. *Biofouling* 6: 91–95.

Albright, J. G., and R. Mills. 1965. A study of diffusion in the ternary system, labeled urea-urea-water, at 25 °C by measurements of the intra-diffusion coefficients of urea. *J. Phys. Chem.* 69: 3120–3126.

Andersen, J. B., C. Sternberg, L. K. Poulsen, S. Petersen Bjorn, M. Givskov, and S. Molin. 1998. New unstable variants of green fluorescent protein for studies of transient gene expression in bacteria. *Appl. Environ. Microbiol.* 64: 2240–2246.

Angell, P., A. Arrage, M. W. Mittelman, and D. C. White. 1993. On-line, non-destructive biomass determination of bacterial biofilms by fluorometry. *J. Microbiol. Methods* 18: 317–327.

Angell, P., J. S. Lou, and D. C. White. 1994. *Use of 2-D Vibrating Electrode technique in MIC Studies*. Corrosion/94, paper #266, National Association of Corrosion Engineers, Houston, TX.

Anonymous. 1993a. Welding stainless steel to meet hygienic requirements. *Trends Food Sci. Tech.* 4: 306–310.

Anonymous. 1993b. Hygienic design of closed equipment for the processing of liquid food. *Trends Food Sci. Tech.* 4: 375–379.

Anson, D. 1977. *Availability of Fossil-Fired Steam Power Plants*. Electrical Power Research Institute Report FP-422-SR, Palo Alto, CA.

Applegate, D. H., and J. D. Bryers. 1991. Effects of carbon and oxygen limitation and calcium concentrations on biofilm removal processes. *Biotech. Bioeng.* 37: 17–25.

Arrage, A. A., and D. C. White. 1997. Monitoring biofilm-induced persistence of Mycobacterium in drinking water systems using GFP fluorescence. In *Bioluminescence and Chemiluminescence*, J. W. Hastings, L J. Kricka, and P. E. Stanley, eds. New York: John Wiley, pp. 383–386.

Arrage, A. A., N. Vasishtha, D. Sunberg, G. Baush, H. L. Vincent, and D. C. White. 1995. On-line monitoring of biofilm biomass and activity on antifouling and fouling release surface using bioluminescence and fluorescence measurements during laminar flow. *J. Ind. Microbiol.* 15: 277–282.

Axelrod, A., D. E. Koppel, J. Schlessinger, E. Elsen, and W. W. Webb. 1976. Mobility measurement by analysis of fluorescence photobleaching recovery kinetics. *Biophys. J.* 16: 1055–1069.

Barbeau, J., R. Tanguay, E. Faucher, C. Avezard, L. Trudel, L. Cote, and A. P. Prevost. 1996. Multiparametric analysis of waterline contamination in dental units. *Appl. Environ. Microbiol.* 62: 3954–3959.

Barer, M. R., L. T. Gribbon, C. R. Harwood, and C. E. Nwoguh. 1993. The viable but nonculturable hypothesis and medical bacteriology. *Rev. Med. Microbiol.* 4: 183–191.

Block, J. C., K. Haudidier, J. L. Paquin, J. Miazga, and Y. Levi. 1993. Biofilm accumulation in drinking water distribution systems. *Biofouling* 6: 333–343.

Bott, T. R. 1979. Biological fouling of heat-transfer surfaces. *Effluent and Water Treatment J.* 19(9): 453–461.

Bott, T. R., and P. C. Miller. 1983. Mechanisms of biofilm formation on aluminium tubes. *J. Chem. Tech. Biotech.* 33B: 177–184.

Boulange-Petermann, L. 1996. Processes of bioadhesion on stainless steel surfaces and cleanability: A review with special reference to the food industry. *Biofouling* 10: 275–300.

Bouman, S., D. B. Lund, F. M. Driessen, and D. G. Schmidt. 1982. Growth of thermoresistant streptococci and deposition of milk constituents on plates of heat exchangers during long operating times. *J. Food. Prot.* 45: 806–812.

Bremer, P. J., and G. G. Geesey. 1991. Laboratory-based model of microbiologically induced corrosion of copper. *Appl. Environ. Microbiol.* 57: 1956–1962.

Bryers, J. D. 1980. Dynamics of early biofilm formation in a turbulent flow system. Ph.D. Dissertation, Rice University, Houston, TX.

Bryers, J. D. 1993. Bacterial biofilms. *Curr. Opinion Biotechnol.* 4: 197–204.

Bryers, J. D., and F. Drummond. 1999. Local macromolecule diffusion coefficients in structurally non-uniform bacterial biofilms using fluorescence recovery after photobleaching (FRAP). *Biotechnol. Bioeng.* 60: 462–473.

Buswell, C. M., Y. M. Herlihy, L. M. Lawrence, J. T. M. McGuiggan, P. D. Marsh, C. W. Keevil, and S. A. Leach. 1998. Extended survival and persistence of *Campylobacter* spp. in water and aquatic biofilms and their detection by immunofluorescent-antibody and -rRNA staining. *Appl. Environ. Microbiol.* 64: 733–741.

Byrd, J. J., H.-S. Xu, and R. R. Colwell. 1991. Viable but nonculturable bacteria in drinking water. *Appl. Environ. Microbiol.* 57: 875–878.

Caldwell, D. E., D. R. Korber, and J. R. Lawrence. 1992. Confocal laser microscopy in digital image analysis in microbial ecology. *Adv. Microb. Ecol.* 12: 1–67.

Camper, A. K. 1996. Factors limiting microbial growth in distribution systems: Laboratory and pilot-scale experiments. Denver, CO: American Water Works Association Research Foundation, 121p.

Carpentier, B., and O. Cerf. 1993. Biofilms and their consequences with particular reference to hygiene in the food industry. *J. Appl. Bacteriol.* 75: 499–511.

Characklis, W. G. 1980. Biofilm development and distruction. Electrical Power Research Institute Final Report EPRI CS-1554, Palo Alto, CA.

Characklis, W. G. 1990. Laboratory biofilm reactors. In *Biofilms*, W. G. Characklis and K. C. Marshall, eds. New York: Wiley, pp. 55–89.

Characklis, W. G., M. J. Nimmons, and B. F. Picologlou. 1981. Influence of fouling biofilms on heat transfer. *Heat Transfer Eng.* 3(1): 23.

Cooksey, K. E., and B. Wigglesworth-Cooksey. 1992. The design of antifouling surfaces: Background and some approaches. In *Biofilms—Science and Technology*, L. F. Melo, T. R. Bott, M. Fletcher, and B. Capdeville, eds. Boston: Kluwer, pp. 529–549.

Criado, M.-T., B. Suarez, and C. M. Ferreiros. 1994. The importance of bacterial adhesion in the dairy industry. *Food Technol.* 48(2): 123–126.

Davies, D. G., and G. G. Geesey. 1995. Regulation of alginate biosynthesis gene *algC* in *Pseudomonas aeruginosa* during biofilm development in continuous culture. *Appl. Environ. Microbiol.* 61: 860–867.

Davies, D. G., A. M. Chakrabarty, and G. G. Geesey. 1993. Exopolysaccharide production in biofilms: Substratum activation of alginate gene expression by *Pseudomonas aeruginosa*. *Appl. Environ. Microbiol.* 59: 1181–1186.

de Beer, D., P. Stoodley, F. Roe, and Z. Lewandowski. 1994. Effects of biofilm structures on oxygen distribution and mass transport. *Biotech. Bioeng.* 43: 1131–1138.

de Beer, D., A. Schramm, C. M. Santegoeds, and M. Kuhl. 1997. A nitrite microsensor for profiling environmental biofilms. *Appl. Environ. Microbiol.* 63: 973–977.

Dickinson, G. M., and A. L. Bisno. 1989. Infections associated with indwelling devices: Consequences of pathogenesis: Infections associated with intravascular devices. *Antimicrob. Agents Chemother.* 33: 597–601.

Dickinson, R. B., and S. L. Cooper. 1995. Analysis of shear-dependent bacterial adhesion kinetics to biomaterial surfaces. *AIChE J.* 41: 2160–2174.

Dickinson, R. B., J. A. Nagel, D. McDevitt, T. J. Foster, R. A. Proctor, and S. L. Cooper. 1995. Quantitative comparison of clumping factor- and coagulase-mediated *Staphylococcus aureus* adhesion to surface-bound fibrinogen under flow. *Infect. Immun.* 63: 3143–3150.

DiSalvo, L. H., and G. W. Daniels. 1975. Observations on estuarine microfouling using the scanning electron microscope. *Microb. Ecol.* 2: 234–240.

Drury, W. J., P. S. Stewart, and W. G. Characklis. 1993. Transport of 1-μm latex particles in *Pseudomonas aeruginosa* biofilms. *Biotech. Bioengineer.* 42: 111–117.

Dunsmore, D. G., A. Twomey, W. G. Wittlestone, and H. G. Morgan. 1981. Design and performance of systems for cleaning product-contact surfaces of food equipment: A review. *J. Food Protect.* 44: 220–240.

Field, B. 1981. Marine biology and its control: History and state of the art. *Oceans 81*, IEEE publication no. 81CH1685-7.

Flemming, H.-C., and T. Griebe. In press. *On-line Monitoring of Biofilms*. Berlin: Springer.

Flemming, H.-C., G. Schaule, R. McDonogh, and H. F. Ridgway. 1994. Effects and extent of biofilm accumulation in membrane systems. In *Biofouling and Biocorrosion in Industrial Water Systems*, G. G. Geesey, Z. Lewandowski, and H.-C. Flemming, eds. Boca Raton, FL: Lewis Publishers, pp. 63–89.

Flint, S., and N. J. Hartley. 1993. *An Investigation to Reduce the Contamination of Whey Protein Concentrate with Thermophilic Bacteria*. New Zealand Dairy Institute Report No. SM93CO3.

Flint, S. H., P. J. Bremer, and J. D. Brooks. 1997. Biofilms in dairy manufacturing plant—Description, current concerns, and methods of control. *Biofouling* 11: 81–97.

Fowler, H. W., and A. J. McCay. 1980. The measurement of microbial adhesion. In *Microbial Adhesion to Surfaces*, R. C. W. Berkeley, J. M. Lynch, J. Melling, P. R. Rutter, and B. Vincent, eds. Chichester, UK: Harwood, pp. 143–161.

Frank, J. F., and R. A. Koffi. 1990. Surface-adherent growth of *Listeria monocytogenes* is associated with increased resistance to surfactant sanitizers and heat. *J. Food Protect.* 53: 550–554.

Geesey, G. G., and P. J. Bremer. 1990. Application of Fourier transform infrared spectrometry to studies of copper corrosion under bacterial biofilms. *Mar. Technol. Soc. J.* 24(3): 36–43.

Geesey, G. G., M. W. Mittelman, and V. T. Lieu. 1987. Evaluation of slime-producing bacteria in oil field core flood experiments. *Appl. Environ. Microbiol.* 53: 278–283.

Geldreich, E. E. 1986. Potable water: New directions in microbial regulations. *ASM News* 52: 530–534.

Gjaltema A., P. A. M. Arts, M. C. M. van Loosdrecht, J. G. Keunen, and J. J. Heijnen. 1994. Heterogeneity of biofilms in rotating annular reactors: Occurrence, structure, and consequences. *Biotechnol. Bioeng.* 44: 194–204.

Gristina, A. G. 1987. Biomaterials-centered infection: Microbial adhesion versus tissue integration. *Science* 237: 1588–1595.

Gristina, A. G., J. J. Dobbins, B. Giammara, J. C. Lewis, and W. C. De Vries. 1988. Biomaterial-centered sepsis and the total artificial heart. *JAMA* 259: 870–874.

Gujer, W., and O. Wanner. 1990. Modeling mixed population biofilms. In *Biofilms*, W. G. Characklis and K. C. Marshall, eds. New York: Wiley, pp. 397–443.

Harty, D. W. S., and T. R. Bott. 1981. Deposition and growth of microorganisms on simulated heat exchanger surfaces. In *Fouling of Heat Transfer Equipment*, E. F. C. Somerscales and J. G. Knudsen, eds. Washington, DC: Hemisphere Publishers, pp. 335–344.

Heim, R., D. C. Prasher, and R. Y. Tsien. 1994. Wavelength mutations and posttranslational autooxidation of green fluorescent protein. *Proc. Natl. Acad. Sci.* 91: 12501–12504.

Heitzer, A., K. Malachowksy, J. E. Tonnard, P. R. Bienkowski, D. C. White, and G. S. Sayler. 1994. Optical biosensor for environmental on-line monitoring of naphthalene and salicylate bioavailability using an immobilized bioluminescent catabolite reporter bacterium. *Appl. Environ. Microbiol.* 60: 1487–1494.

Helke, D. M., E. B. Somers, and A. C. L. Wong. 1993. Attachment of *Listeria monocytogenes* and *Salmonella typhimurium* to stainless steel and Buna-N in the presence of milk and individual milk components. *J. Food Prot.* 56: 479–484.

Henrici, A. T. 1993. Studies of freshwater bacteria I. A direct microscopic technique. *J. Bacteriol.* 25: 277–286.

Hood, S. K., and E. A. Zotolla. 1992. Biofilms in meat processing. *Proc. Ann. Mtg. Inst. Food Technol.*, New Orleans, LA.

Kalmbach, S., W. Manz, and U. Szewzyk. 1997. Isolation of new bacterial species from drinking water biofilms and proof of their *in situ* dominance with highly specific 16S rRNA probes. *Appl. Environ. Microbiol.* 63: 4146–4170.

Kalogeraki, V. S., and S. C. Winans. 1997. Suicide plasmids containing promoterless reporter genes can simultaneously disrupt and create gene fusions to target genes of diverse bacteria. *Gene* 188: 69–75.

Keys, T. 1993. Early-onset prosthetic valve endocarditis. *Cleve. Clin. J. Med.* 60: 455–459.

King, C. H., E. B. Shotts, Jr., R. E. Wooley, and K. G. Porter. 1988. Survival of coliforms and bacterial pathogens within protozoa during chlorination. *Appl. Environ. Microbiol.* 54: 3023–3033.

Kirkpatrick, J. P., L. V. McIntire, and W. G. Characklis. 1980. Mass and heat transfer in a circular tube with biofouling. *Water Research* 14: 117–127.

Klijn, N., A. H. Weerkamp, and W. M. deVos. 1992. Detection and identification of bacteria by means of DNA-probes and PCR-amplification. *Voedingsmiddelentechnologie* 25: 10–13.

Knudsen, J. G. 1981. Apparatus and techniques for measurement of fouling of heat transfer surfaces. In *Fouling of Heat Transfer Equipment*, E. F. C. Somerscales and J. G. Knudsen, eds. Washington, DC: Hemisphere Publishing, pp. 57–81.

Korber, D. R., A. Choi, G. M. Wolfaardt, S. C. Ingham, and D. E. Caldwell. 1997. Substratum topography influences susceptibility of *Salmonella entereditis* biofilms to trisodium phosphate. *Appl. Environ. Microbiol.* 63: 3352–3358.

Kutz, S. M., D. L. Bently, N. A. Sinclair, and L. M. Kelley. 1986. Morphological diversity of a bacterium resembling *Selibera*: An electron microscopic evaluation of nutritional effects. *Abstr. Annu. Meet. Am. Soc. Microbiol.*, Washington, DC, p. 189.

Larison, K. D., R. BreMiller, S. K. Wells, I. Clements, and R. P. Haugland. 1995. Use of a new fluorogenic phosphatase substrate in immunohistochemical applications. *J. Histochem. Cytochem.* 43: 77–83.

Laurence, R. A., and S. T. Lapierre. 1995. Quality of hemodialysis water: A 7-year multi-center study. *Am. J. Kidney Dis.* 25: 738–750.

Lawrence, J. R., D. R. Korber, and D. E. Caldwell. 1991. Optical sectioning of microbial biofilms. *J. Bacteriol.* 173: 6558–6567.

Lechevallier, M. W. 1990. Coliform regrowth in drinking water: A review. *Res. Technol. J. Am. Water Works Assoc.* 82: 74–86.

Lee, W., and D. de Beer. 1995. Oxygen and pH microprofiles above corroding mild steel covered with a biofilm. *Biofouling* 8: 273–280.

Lee, S. H., and J. F. Frank. 1990. Resistance of *Listeria monocytogenes* biofilms to hypochlorite and heat. *Proc. XXIII Int. Dairy Congr.*, Montreal, p. 153.

Lewandowski, Z. 1993. Dissolved oxygen gradients near microbially colonized surfaces. In *Biofouling and Biocorrosion in Industrial Water Systems*, G. G. Geesey, Z. Lewandowski, and H.-C. Flemming, eds. Boca Raton, FL: Lewis Publishers, pp. 175–188.

Lewandowski, Z. 1998. *Structure and Function of Bacterial Biofilms*, Paper No. 296, Corrosion 98, NACE International, Houston, TX.

Lewandowski, Z., P. Stoodley, and S. Altobelli. 1995. Experimental and conceptual studies on mass transport in biofilms. *Wat. Sci. Tech.* 31(1): 153–162.

Lewis, S. J., and A. Gilmour. 1987. Microflora associated with the internal surfaces of rubber and stainless steel milk transfer pipeline. *J. Appl. Microbiol.* 62: 327–333.

Mack, W. N., J. P. Mack, and A. O. Ackerson. 1975. Microbial film development in a trickling filter. *Microb. Ecol.* 2: 215–226.

Marshall, K. C. 1992. Biofilms: An overview of bacterial adhesion, activity, and control at surfaces. *ASM News* 58: 202–207.

Matthysse, A. G., S. Stretton, C. Dandie, N. C. McClure, and A. E. Goodman. 1996. Construction of GFP vectors for use in gram-negative bacteria other than *Escherichia coli*. *FEMS Microbiol. Lett.* 145: 87–94.

Mattila-Sandholm, T., and G. Wirtanen. 1992. Biofilm formation in the industry: A review. *Food Rev. Int.* 8: 573–603.

McCoy, W. F., J. D. Bryers, J. Robbins, and J. W. Costerton. 1981. Observations of fouling biofilm formation. *Can. J. Microbiol.* 27: 910–917.

McDonogh, R., G. Schaule, and H.-C. Flemming. 1994. The permeability of biofouling layers on membranes. *J. Membrane. Sci.* 87: 199–217.

McFeters, G. A., S. C. Broadaway, B. H. Pyle, and Y. Egozy. 1993. Distribution of bacteria within operating laboratory water purification systems. *Appl. Environ. Microbiol.* 59: 1410–1415.

Mittelman, M. W. 1995. Biofilm development in purified water systems. In *Microbial Biofilms*, H. L. Lappin-Scott and J. W. Costerton, eds. London: Cambridge Univ. Press, pp. 133–147.

Mittelman, M. W. 1996. Adhesion to biomaterials. In *Bacterial Adhesion: Molecular and Ecological Diversity*. New York: Wiley-Liss, pp. 89–127.

Mittelman, M. W., R. Islander, and R. M. Platt. 1987. Biofilm formation in a closed-loop purified water system. *Med. Device Diagn. Ind.* 10: 50–55.

Mittelman, M. W., D. E. Nivens, C. Low, and D. C. White. 1990. Differential adhesion, activity, and carbohydrate:protein ratios of *Pseudomonas atlantica* monocultures attaching to stainless steel in a linear shear gradient. *Microb. Ecol.* 19: 269–278.

Mittelman, M. W., L. L. Kohring, and D. C. White. 1992. Multi-purpose laminar-flow adhesion cells for the study of bacterial colonization and biofilm formation. *Biofouling* 6: 39–51.

Mittelman, M. W., J. Packard, A. A. Arrage, S. L. Bean, P. Angell, and D. C. White. 1993. Test system for determining bioluminescence and fluorescence in a laminar-flow environment. *J. Microbiol. Methods* 18: 51–60.

Moller, S., D. R. Korber, G. M. Wolfaardt, S. Molin, and D. E. Caldwell. 1997. Impact of nutrient composition on a degradative biofilm community. *Appl. Environ. Microbiol.* 63: 2432–2438.

Morin, P., A. Camper, W. Jones, D. Gatel, and J. C. Goldman. 1996. Colonization and disinfection of biofilms hosting coliform-colonized carbon fines. *Appl. Environ. Microbiol.* 62: 4428–4432.

Mucchetti, G. 1995. Biological fouling and biofilm formation on membranes. In *Fouling and Cleaning in Pressure Driven Membrane Processes*, Chap. 7, IDF General Secretariat, ed. Belgium: International Dairy Federation Group B47, pp. 118–124.

Murphy, J. J., L. A. Bland, B. J. Davies, R. W. Maxey, A. Light, M. S. Favero, and S. L. Solomon. 1987. Pyrogenic reactions with high-flux hemodialysis. *Proceedings ICAAC*. Washington, DC: American Society of Microbiology.

Nakasono, S., J. G. Burgess, K. Takahashi, M. Koike, C. Murayama, S. Nakamura, and T. Matsunaga. 1993. Electrochemical prevention of marine biofouling with a carbon-chloroprene sheet. *Appl. Environ. Microbiol.* 59: 3757–3762.

Nickel, J. C., and J. W. Costerton. 1992. Bacterial biofilms and catheters: A key to understanding bacterial strategies in catheter-associated urinary tract infections. *Can. J. Infect. Dis.* 3: 261–267.

Nivens, D. E., J. Q. Chambers, and D. C. White. 1991. Non-destructive monitoring of microbial biofilms at solid-liquid interface using on-line devices. In *Microbially Influenced Corrosion and Biodeterioration*, N. J. Dowling, M. W. Mittelman, and J. C. Danko, eds. Knoxville, TN, pp. 5-47-5-56.

Nivens, D. E., J. Q. Chambers, T. R. Anderson, A. Tunlid, J. Smit, and D. C. White. 1993a. Monitoring microbial adhesion and biofilm formation by attenuated total reflection/Fourier transform infrared spectroscopy. *J. Microbiol. Methods* 17: 199–213.

Nivens, D. E., J. Schmit, J. Sniatecki, T. Anderson, J. Q. Chambers, and D. C. White. 1993b. Multichannel ATR/FT-IR spectrometer for on-line examination of microbial biofilm. *Appl. Spectrosc.* 47: 688–671.

Notermans, S. 1994. The significance of biofouling to the food industry. *Food Technol.* 48(7): 13–14.

Notermans, S., J. A. M. A. Dormans, and G. C. Mead. 1991. Contribution of surface attachment to the establishment of micro-organisms in food processing plants: A review. *Biofouling* 5: 1–16.

Paquin, J. L., J. C. Block, K. Haudidier, P. Hartemann, F. Colin, J. Miazga, and Y. Levi. 1992. Effet du chlore sur la colonisation bacterienne d'un reseau experimental de distribution d'eau. *Sciences de l'Eau* 5(3): 399–414.

Patterson, M. K., G. R. Husted, A. Rutkowski, and D. C. Mayette. 1991. Bacteria. Isolation, identification and microscopic properties of biofilms in high-purity water distribution systems. *Ultrapure Water* 8(4): 18–24.

Picioreanu, C., M. C. M. van Loosdrecht, and J. J. Heijnen. 1998a. A new combined differential-discrete cellular automaton approach for biofilm modeling: Application for growth in gel beads. *Biotechnol. Bioeng.* 57: 718–731.

Picioreanu, C., M. C. M. van Loosdrecht, and J. J. Heijnen. 1998b. Mathematical modeling of biofilm structure with a hybrid differential-discrete cellular automaton approach. *Biotechnol. Bioeng.* 58: 101–116.

Picologlou, B. F., N. Zelver, and W. G. Characklis. 1980. Biofilm growth and hydraulic performance. *J. Hyd. Div.*, ASCE 106(HY5): 733.

Rasmussen, K., and Z. Lewandowski. 1998. Microelectrode measurements of local mass transport rates in heterogeneous biofilms. *Biotechnol. Bioeng.* 59: 302–309.

Rice, J. F., R. F. Fowler, A. A. Arrage, D. C. White, and G. S. Sayler. 1995. Effects of external stimuli on environmental bacterial strains harboring an *algD-lux* bioluminescent reporter plasmid for studies of corrosive biofilms. *J. Ind. Microbiol.* 15: 318–328.

Ridgway, H. F., and B. H. Olson. 1981. Scanning electron microscope evidence for bacterial colonization of a drinking-water distribution system. *Appl. Environ. Microbiol.* 41: 274–287.

Rogers, J., A. B. Dowsett, P. J. Dennis, J. V. Lee, and C. W. Keevil. 1994. Influence of plumbing materials on biofilm formation and growth of *Legionella pneumophila* in potable water systems. *Appl. Environ. Microbiol.* 60: 1842–1851.

Sanderson, K., C. J. Thomas, and T. A. McMeekin. 1991. Molecular basis of the adhesion of *Salmonella* serotypes to chicken muscle fascia. *Biofouling* 5: 89–101.

Santiago, J. I., M. K. Huntington, A. M. Johnston, and R. S. Quinn. 1994. Microbial contamination of dental unit waterlines: Short- and long-term effects of flushing. *Gen. Dent.* 42(6): 528–535.

Sasahara, K., and E. A. Zotolla. 1993. Biofilm formation by *Listeria monocytogenes* utilizes a primary colonizing microorganism in flowing systems. *J. Food Protect.* 56: 1022–1028.

Schramm, D. deBeer, M. Wagner, and R. Amann. 1998. Identification and activities of *Nitrosospira* and *Nitrospira* spp. as dominant populations in a nitrifying fluidized bed reactor. *Appl. Environ. Microbiol.* 64: 3480–3485.

Schulze-Robbecke, R., C. Feldmann, R. Fischeder, B. Janning, M. Exner, and G. Wahl. 1995. Dental units: An environmental study of sources of potentially pathogenic mycobacteria. *Tubercle Lung Dis.* 76: 318–323.

Schulze-Robbecke, R., and R. Fischeder. 1989. Mycobacteria in biofilms. *Zbl. Hyg.* 188: 385–390.

Shearer, B. G. 1996. Biofilm and the dental office. *J. Am. Dent. Assoc.* 127: 181–189.

Sinclair, N. A. 1982. Microbial degradation of reverse osmosis desalting membranes. *Operation and Maintenance of the Yuma Desalting Test Facility*, Vol. IV. Yuma, AZ: U.S. Department of Interior, Bureau of Reclamation.

Stadhouders, J., G. Hup, and F. Hassing. 1982. The conceptions index and indicator organisms discussed on the basis of the bacteriology of spray-dried milk powder. *Neth. Milk Dairy J.* 36: 231–260.

Stickler, D. J. 1996. Bacterial biofilms and their encrustation of urethral catheters. *Biofouling* 9: 293–305.

Stoodley, P., D. deBeer, and Z. Lewandowski. 1994. Liquid flow in biofilm systems. *Appl. Environ. Microbiol.* 60: 2711–2716.

Stoodley, P., Z. Lewandowsky, J. D. Boyle, and H. M. Lappin-Scott. 1998. Oscillation characteristics of biofilm streamers in turbulent flowing water as related to drag and pressure drop. *Biotechnol. Bioeng.* 57: 536–544.

Stoodley, P., Z. Lewandowsky, J. D. Boyle, and H. M. Lappin-Scott. 1999. Structural deformation of bacterial biofilms caused by short term fluctuations in fluid shear: An in situ investigation of biofilm rheology. *Biotechnol. Bioeng.* 65: 83–92.

Stretton, S., S. Techkarnjanaruk, A. M. McLennan, and A. E. Goodman. 1998. Use of green fluorescent protein to tag and investigate gene expression in marine bacteria. *Appl. Environ. Microbiol.* 64: 2554–2559.

Suci, P. A., K. J. Siedlecki, R. J. Palmer, D. C. White, and G. G. Geesey. 1997. Combined light microscopy and attenuated total reflection Fourier transform infrared spectroscopy for integration of biofilm structure, distribution and chemistry at solid-liquid interfaces. *Appl. Environ. Microbiol.* 63: 4600–4603.

Surman, S. B., J. T. Walker, D. T. Goddard, L. H. G. Morton, C. W. Keevil, W. Weaver, A. Skinner, K. Hanson, D. Caldwell, and J. Kurtz. 1996. Comparison of microscopic techniques for the examination of biofilms. *J. Microbiol. Methods* 25: 57–70.

Tamachkiarowa, A., and H.-C. Flemming. 1996. Glass fiber sensor for biofouling monitoring. In *Proceedings of the 10th International Diodeterioration and Biodegradation Symposium*, W. Sand, H. Brill, E. Heitz, and H.-C. Flemming, eds. DECHEMA Monographs, Vol. 133. Hamburg: VCH Verlagsgesellschaft, pp. 31–36.

Tide, C., S. R. Harkin, G. G. Geesey, P. J. Bremer, and W. Scholz. 1999. The influence of welding procedures on bacterial colonization of stainless steel weldments. *J. Food Engineer.* 42: 85–96.

Timperley, D. A., R. H. Thorpe, and J. T. Holah. 1992. Implications of engineering design in food industry hygiene. In *Biofilms—Science and Technology*, L. F. Melo, T. R. Bott, M. Fletcher, and B. Capdeville, eds. Boston, MA: Kluwer Academic Publishers, pp. 379–393.

Turakhia, M. H., and W. G. Characklis. 1984. Heat transfer effects of biofilms in turbulent flow. *Heat Transf. Eng.* 5: 93–101.

Videla, H. A., and W. G. Characklis. 1992. Biofouling and microbially influenced corrosion. *Internat. Biodeter. Biodegrad.* 29: 195–212.

Vieira, M. J., L. F. Melo, and M. M. Pinheiro. 1993. Biofilm formation: Hydrodynamic effects on internal diffusion and structure. *Biofouling* 7: 67–80.

Vincent, F. D., A. R. Tibi, and J. C. Darbord. 1989. A bacterial biofilm in a hemodialysis system. Assessment of disinfection and crossing of endotoxin. *Asaio Trans.* 35: 310–313.

Vrany, J. D., P. S. Stewart, and P. A. Suci. 1997. Comparison of recalcitrance to ciprofloxacin and levofloxacin exhibited by *Pseudomonas aeruginosa* biofilms displaying rapid-transport characteristics. *Antimicrob. Agents Chemother.* 41: 1352–1358.

Walker, J. T., A. Sonesson, C. W. Keevil, and D. C. White. 1993. Detection of *Legionella pneumophila* in biofilms containing a complex microbial consortium by gas chromatography-mass spectrometric analysis of genus-specific hydroxy fatty acids. *FEMS Microb. Lett.* 113: 139–144.

Wallace, W. H., D. C. White, and G. S. Sayler. 1994. An *algD*-bioluminescent reporter plasmid to monitor alginate production in biofilms. *Microb. Ecol.* 27: 225–239.

Westrin, B. A. 1991. Diffusion measurements in gels. Doctoral diss. Lund University. Lutkdh(TKKA-1003), Lund, Sweden.

White, D. C., A. A. Arrage, D. E. Nivens, R. J. Palmer, J. F. Rice, and G. S. Sayler. 1996. Biofilm ecology: On-line methods bring new insights into MIC and microbial biofouling. *Biofouling* 10(1–3): 3–16.

Williams, C. J., and R. G. Edyvean. 1995. The biological fouling of cartridge filters in aqueous applications. *Biofouling* 9: 115–127.

Williams, J. F., A. M. Johnston, B. Johnson, M. K. Huntington, and C. D. MacKenzie. 1993. Microbial contamination of dental unit waterlines. *J. Am. Dent. Assoc.* 124: 59–65.

Wireman, J. W., A. Schmidt, C. R. Scavo, and D. T. Hutchins. 1993. Biofilm formation by *Legionella pneumophila* in a model domestic hot water system. In *Legionella: Current Status and Emerging Perspectives*, J. Barace, ed. Washington, DC: American Society of Microbiology, pp. 231–234.

Wirtanen, G., and T. Mattila-Sandholm. 1992. Effect of growth phase on foodborne biofilms on their resistance to chlorine sanitizer Part II. *Lebens. m-Wiss. Technol.* 25: 50–54.

Zelver, N. 1979. Biofilm development and associated energy losses in water conduits. M. Sc. thesis, Rice University, Houston, TX.

Zelver, N., F. L. Roe, and W. G. Characklis. 1985. Potential for monitoring fouling in the food industry. In *Fouling and Cleaning in the Food Industry*, D. Lund, E. Plett, and C. Sandu, eds. Madison, WI: University of Wisconsin, Department of Food Science.

Zhang, T. C., and P. L. Bishop. 1994a. Density, porosity and pore structure of biofilms. *Wat. Res.* 28: 2267–2277.

Zhang, T. C., and P. L. Bishop. 1994b. Evaluation of tortuosity factors and effective diffusivities in biofilms. *Wat. Res.* 11: 2279–2287.

Zobell, C. E., and E. C. Allen. 1935. The significance of marine bacteria in the fouling of submerged surfaces. *J. Bacteriol.* 29: 239–251.

Zotolla, E. A. 1994. Microbial attachment and biofilm formation: A new problem for the food industry? *Food Technol.* 48(7): 107–115.

Zotolla, E. A., and K. C. Sashara. 1994. Microbial biofilms in the food processing industry—Should they be a concern? *Int. J. Food. Micro.* 23: 125–148.

9

BIOCORROSION

GILL G. GEESEY
Department of Microbiology and Center for Biofilm Engineering, Montana State University, Bozeman, Montana

IWONA BEECH
School of Pharmacy, Biomedical and Physical Sciences, Portsmouth University, Portsmouth, United Kingdom

PHILIP J. BREMER
Crop and Food Research, University of Otago, Dunedin, New Zealand

BARBARA J. WEBSTER
Industrial Research Ltd., Lower Hutt, New Zealand

D. BRET WELLS
Industrial Research Ltd., Auckland, New Zealand

9.1 INTRODUCTION

9.1.1 What is Biocorrosion?

Corrosion is the naturally occurring process by which materials fabricated of pure metals and/or their mixtures (alloys) undergo chemical oxidation from ground state to an ionized species. The metals used in the fabrication of materials and various types of equipment are in a thermodynamically unfavorable state and upon oxidation achieve a more stable configuration with an accompanying release of energy to the surrounding system. In most instances, the oxidation reaction slows to a low rate after a period of time because the oxidation products adhere to the metal surface and form an oxide/hydroxide layer that serves as a diffusion barrier to other reactants. These "corrosion products" form a protective barrier to further oxidation of the underlying metal. Since environmental conditions can influence the equilibrium concentrations and diffusion rates of reactants and products of the oxidation reaction, metals used for equipment fabrication are matched to the physical/chemical

Biofilms II: Process Analysis and Applications, Edited by James D. Bryers.
ISBN 0-471-29656-2 Copyright © 2000 Wiley-Liss, Inc.

conditions under which the equipment is intended to operate. Changes in the environmental conditions can affect the stability of the protective metal oxide/hydroxide film and, hence, the susceptibility of the metal to corrosion.

Corrosion is an electrochemical process consisting of two half-reactions: an anodic reaction involving the ionization (oxidation) of the metal and a cathodic reaction involving the reduction of a chemical species in contact with the metal surface. Although the anodic and cathodic reactions can occur at different locations on the metal, typically they are adjacent to each other. The sites of the anodic and cathodic reactions are referred to as the anode and cathode, respectively. Because the half-cells are physically separated, all corrosion reactions promote localized attack at the anode. When the half-cells are closely spaced across the surface of the metal, a "generalized corrosion" phenomenon occurs. The formation of a protective oxide film on the metal surface is an example of a generalized corrosion reaction that eventually slows to a rate that causes minimal structural damage to the fabricated material. When the corrosion cells are irregularly distributed across the surface, the resulting corrosion process is referred to as *localized attack*. Crevice corrosion and underdeposit pitting corrosion are examples of such an attack. Under certain conditions, generalized and localized attack cause significant structural damage or deterioration of material performance. Several corrosion textbooks provide excellent overviews of corrosion as well as detailed explanations of the different mechanisms of corrosion (NACE, 1970; Scully, 1990).

When organisms, living in association with a metal surface, influence either one or both of the half-reactions, this is referred to as biocorrosion. Since biocorrosion usually involves surface-associated microbial growth and accumulation, the term *microbiologically influenced corrosion* (MIC) is frequently used in reference to corrosion reactions under a biological influence. Microorganisms, through their mere presence on the surface and their ability to carry out specific biochemical reactions, can alter the physical/chemical conditions at the metal surface. When this results in an increase in the corrosion rate over that observed in their absence, MIC is said to contribute to the overall corrosion process. To our knowledge, MIC does not invoke any new mechanisms of corrosion. Rather, it is the result of a microbiologically influenced change in surface conditions that promotes the establishment or maintenance of cathodic and/or anodic reactions not normally favored under otherwise similar conditions in the absence of the microbes. Case histories demonstrate that altering the chemical characteristics of the system can produce the same topographic changes of electrochemically corroded metals and alloys as observed in the presence of microbes (Stoecker, 1995). A number of mechanisms of biocorrosion, which reflect the variety of physiological activities carried out by different types of microorganisms have been identified. Demonstrating the presence of microorganisms at a corroding site, even those known to produce metabolic by-products aggressive toward metals, is not sufficient evidence for MIC (Ghassem and Adibi, 1995). Some of the myths surrounding MIC have been reviewed by Little and Wagner (1997).

Biocorrosion is difficult to demonstrate and assess in the field because of the problems determining corrosion reactions and rates in the absence of microorganisms. Microbes accumulate on all unprotected surfaces, therefore, their influence on a corrosion reaction in the field is usually measured by changes in the rate of reaction following application of antimicrobial compounds, i.e., biocides (Billman, 1997). However, the action of biocides is not necessarily limited to the control of the biological components of the corrosion process, as the presence of antimicrobials can also influence the electrochemistry at the surface/electrolyte interface. Buchanan et al. (1997) have proposed the use of a MIC factor, defined as the ratio of the biotic corrosion rate in the field over the abiotic corrosion rates measured in laboratory control test, presumably using sterile surfaces.

A role of MIC is often ignored if an abiotic mechanism can be invoked to explain the observed corrosion. MIC, as a significant phenomenon, has thus been viewed with skepticism, particularly among engineers and chemists with little appreciation for the behavior of microorganisms. The slow progress in establishing the importance of MIC in equipment damage is also the result of the paucity of analytical techniques to identify, localize, and control corrosion reactions on metal surfaces with surface-associated microbiological processes. Nevertheless, some progress has been made since this topic was reviewed in the first edition of *Biofilms* (Little et al., 1990a). Much of the new information was recently published in a manual on biocorrosion (Videla, 1996).

9.1.2 Economic Impact of MIC

Assessment of MIC is complicated by a number of factors that have been recently discussed by Flemming (1996). It is difficult to estimate the costs associated with MIC because of the problems industry has distinguishing MIC from abiotic corrosion. A breakdown of the actual cost of MIC on an industry-by-industry basis is not in the public domain. However, some insight on the cost of MIC can be gained from estimates of overall corrosion costs to the military and private industry, and from information that has been made available by individual companies or sectors of industry.

The economic impact of biocorrosion may be estimated from determinations of the losses resulting from all forms of corrosion. The cost of all corrosion experienced by U.S. industry has been estimated to be on the order of several percent of the gross domestic product, or hundreds of billions of dollars annually (Newman and Sieradzki, 1994). United States industry spends $100 billion per year to retard or prevent corrosion on pipes, tanks, pumps, and other equipment in manufacturing plants (Anon., 1991).

The direct cost of all forms of corrosion in U.S. military aircraft exceeds $700 million per year (Roberge et al., 1996). The U.S. Army estimates that the total cost for Department of Defense corrosion-related problems is $10 billion per year (U.S. Department of Defense, 1998).

Blackwood et al. (1994) recently reviewed the cost of MIC to various industries in different parts of the world. Escom, the national power utility of South Africa that provides 90% of that country's power needs, identified MIC of carbon steel used in cooling water systems in virtually all of their power plants. The costs associated with repairs and down time were on the order of millions of dollars annually (Bibb, 1986). Underdeposit pitting corrosion of heat exchanger tubing in nuclear power generating plants operated by Ontario Hydro of Canada has been estimated to cost the corporation $300,000 per unit per day in replacement energy costs (Brennenstuhl and Doherty, 1990). Thirteen million dollars were spent to replace 13 bundles of damaged heat exchanger tubing. The total cost associated with this type of corrosion in the heat exchangers between 1982 and 1990 was estimated at $55 million. Corrosion problems have cost the nuclear power utilities billions of dollars in replacement costs (Jones, 1996).

MIC is believed to contribute significantly to the corrosion failures encountered by the oil and gas industry. Allred et al. (1959) proposed that bacteria were responsible for 77% of the corrosion observed in production wells on one lease operated by one oil company near Griffin, Indiana. Seventy percent of the corrosion costs associated with gas transmission pipelines is heavily influenced by bacterial effects. It has been estimated that 34% of the corrosion damage experienced by one oil company was related to MIC (T. R. Jack, pers. communic.). MIC is believed to play a significant role in corrosion associated with offshore oil production activities, but no estimates of the costs to replace or repair affected equipment is available on an industrywide basis.

Evaluation in the 1950s of MIC-related costs of repair and replacement of piping material used in difference types of service in the United States were on the order of $0.5–2 billion per annum. Booth (1964) suggested that 50% of corrosion failures for pipelines involved MIC. The U.S. Army spent $335 million on the maintenance of their 12,000 miles of water distribution piping, 4000 miles of natural gas distribution piping, and 3000 miles of heat distribution piping in 1996. Replacement costs for gas mains damaged by MIC in the United Kingdom were recently reported to be £250 million per annum. Often, financial losses from damage of equipment from biocorrosion are combined with those resulting from biofouling. Although the two phenomena may be related, they do not necessarily cause the same type of damage. The costs associated with MIC usually include the costs of prevention of both MIC and biofouling. Flemming (1996) proposed that approximately 20% of all corrosion damage to materials fabricated with metal is microbially influenced or enhanced. If Flemming's evaluation is correct, then the annual cost of MIC to U.S. industry alone is on the order of tens of billions of dollars. It should be emphasized that these are approximations based on a limited understanding of MIC, and thus could be underestimated.

9.1.3 Physiological Activities of Microorganisms as the Driving Force for MIC

The microorganisms implicated in biocorrosion of metals such as iron, copper, and aluminum, and their alloys are physiologically diverse and appear to have undergone little adaptive change in order to mediate corrosion of these materials. The bacteria involved in metal corrosion have frequently been grouped by their metabolic demand for different respiratory substrates or electron acceptors. The capability of many microorganisms to substitute alternative oxidizable compounds in place of oxygen as terminal electron acceptors in respiration when oxygen becomes depleted in the environment permits them to be active over a wide range of conditions conducive for corrosion of metals. Examples of bacteria capable of this kind of facultative anaerobic metabolism include those in the genera of *Pseudomonas* or *Vibrio*. The ability to produce a broad spectrum of corrosive metabolic by-products over a wide range of environmental conditions makes bacteria a real threat to the stability of metals that have been engineered for corrosion resistance.

9.2 MIC OF IRON AND FERROUS ALLOYS

The main types of bacteria associated with corrosion failures of cast iron, mild, and stainless steel structures are sulfate-reducing bacteria (Hamilton, 1985), sulfur-oxidizing bacteria (Miller and Tiller, 1971), iron-oxidizing/reducing bacteria (Obuekwe and references therein, 1981), manganese-oxidizing bacteria (Dickinson et al., 1996), and bacteria secreting organic acids and exopolymers or slime (Cragnolino and Tuovinen, 1984; White et al., 1986). These organisms coexist in naturally occurring biofilms often forming synergistic

9.2 MIC OF IRON AND FERROUS ALLOYS

communities able to affect electrochemical processes through cooperative metabolism which individual species have difficulty initiating (Dowling et al., 1991). Recent publications in the area of biocorrosion provide reviews on this subject (Borenstein, 1994; Videla, 1996; Little et al., 1997a).

9.2.1 Sulfate-Reducing Bacteria

Sulfate-reducing bacteria (SRB) are a group of phylogenetically diverse anaerobes that carry out dissimilatory reduction of sulfur compounds such as sulfate, sulfite, thiosulfate, and even sulfur itself to sulfide (Bak and Cypionka, 1987; Lovley and Philips, 1994). Although SRB are strictly anaerobic, some genera tolerate oxygen (Hardy and Hamilton, 1981; Abdollahi and Wimpenny, 1990). It has also been demonstrated that at low dissolved oxygen concentrations certain SRB are able to respire with Fe^{3+} or even oxygen with hydrogen acting as electron donor (Dilling and Cypionka, 1990; Roden and Lovley, 1993). Excellent reviews on SRB are available (Postgate, 1984; Widdel, 1988; Barton, 1995).

The activities of SRB in natural and man-made systems are of great concern to many different industrial operations. In particular, oil, gas, and shipping industries are seriously affected by the sulfides generated by SRB (Hamilton, 1994 and references therein). Biogenic sulfide production also generates health and safety problems, environmental hazards, and severe economic losses due to corrosion of equipment (Odom, 1990; Odom and Singleton, 1992).

The role of SRB has been documented in pitting corrosion of various metals and their alloys in both aquatic and terrestrial environments under anoxic as well as oxygenated conditions. Several models have been proposed to explain the mechanisms by which these bacteria can influence the corrosion process of mild steel (Iverson, 1987; Ford and Mitchell, 1991; Odom and Singleton, 1992). Recent reviews by Lee et al. (1995) and Hamilton (1998) clearly state that one predominant mechanism may not exist and that a number of factors are involved. Parameters relevant to SRB-influenced corrosion phenomenon are discussed in the following sections.

9.2.1.1 Hydrogenase Enzyme as a Depolarizing Agent.
von Wolzogen Kühr and van der Vlugt (1934) proposed what is now referred to as the classical mechanism of anaerobic corrosion of ferrous metal, also known as the cathodic depolarization theory. The essential step in this theory involves the removal of hydrogen (cathodic depolarization) by the activity of the hydrogenase enzyme, catalyzing the reversible oxidation of hydrogen, present in all SRB. The electron removal as a result of hydrogen utilization forces more iron to be dissolved at the anode. The main steps involved in the reaction are listed below and depicted in Figure 9.1

Anodic reaction:	$4Fe \rightarrow 4Fe^{2+} + 8e^-$	(9.1)
Dissociation of water:	$8H_2O \rightarrow 8H^+ + 8OH^-$	(9.2)
Cathodic reaction:	$8e^- + 8H^+ \rightarrow 8H$	(9.3)
Cathodic depolarization:	$SO_4^{2-} + 8H \rightarrow S^{2-} + 4H_2O$	(9.4)
Corrosion products:	$Fe^{2+} + S^{2-} \rightarrow FeS$	(9.5)
Corrosion products:	$3Fe^{2+} + 6OH^- \rightarrow 3Fe(OH)_2$	(9.6)
Overall reaction:	$4Fe + SO_4^{2-} + 4H_2O \rightarrow 3Fe(OH)_2 + FeS + 2OH^-$	(9.7)

Using cathodic polarization measurements, Booth and Tiller (1960; 1962) clearly demonstrated that the SRB with detectable hydrogenase activity, *Desulfovibrio vulgaris*, was able to depolarize the mild steel cathode. No such process was carried out in the cultures of SRB where hydrogenase activity was not detected. Studies by Booth and Wormwell (1962), using batch cultures of several strains of SRB indicated a direct relationship between hydrogenase activity and corrosion rates of steel. Results by Iverson (1966) provided further support for the depolarization theory. He demonstrated the dissolution of iron from a mild steel electrode electrically connected to a similar electrode in contact with a culture of hydrogenase-positive *Desulfovibrio desulfuricans* on an agar surface containing benzyl viologen (BV), acting as electron acceptor. The SRB oxidized cathodic hydrogen and transferred the electrons to the redox dye, BV instead of sulfate. Similar results were obtained by Booth and Tiller (1968).

Costello (1974) criticized the use of benzyl viologen as the electron acceptor, stating that hydrogenase-positive SRB not only catalyzed the uptake of hydrogen but also the liberation of hydrogen by reduced benzyl viologen. The apparent hydrogenase-dependent cathodic depolarization was claimed to be an experimental artifact resulting from the use of BV to "simplify" the system. Since then several studies have correlated the SRB hydrogen metabolism with hydrogenase activity. It has been demonstrated that SRB are able to utilize cathodic H_2 with sulfate as the sole electron acceptor (Hardy, 1983; Cord-Ruwisch and Widdel, 1986; Pankhania et al., 1986). The availability of organic electron donors appeared to be an important factor that influenced the utilization of cathodic H_2 from iron surfaces. Thus in the presence of SRB, anoxic aqueous environments rich in anaerobically degradable organic matter (e.g., interior of sewer pipes) should be more corrosive than environments that are mainly inorganic (e.g., water in geothermal heating plants). These findings led to the re-evaluation of the role of hydrogenase in the corrosion process. Although the influence of sulfide deposited on mild steel electrode surfaces was not negligible, the major mechanism was the oxidation of cathodically produced H_2. The results of the dual cells experiment (one cell was sterile and the other inoculated with hydrogenase-positive or negative SRB) showed that separation of the bacteria from the steel electrode by means of a dialysis membrane pre-

Cathodic depolarization

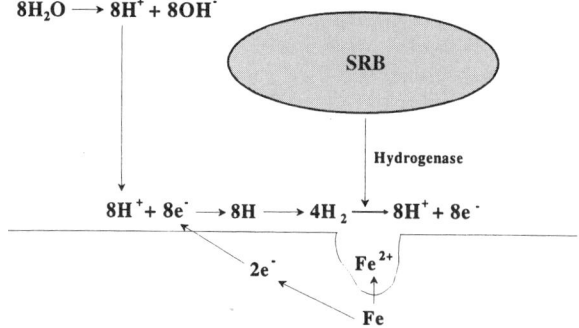

Figure 9.1 Cathodic depolarization of steel due to activity of hydrogenase enzyme.

vented the utilization of cathodic H_2 by the SRB. The current observed in the presence of the hydrogenase-positive SRB did not appear in the presence of hydrogenase-negative SRB (Daumas et al., 1988). It has also been demonstrated that the hydrogenase activity did not depend on the presence of viable cells (Chatelus et al., 1987). The authors emphasized that the results were of particular importance in the understanding of metal biocorrosion by anaerobic bacteria because hydrogenase was still active for months independent of viable cells. Thus bacteria that were still attached to metallic surfaces and that could not be detected by the conventional methods, such as viable counts or production of H_2S, might still retain their depolarizing capabilities, which could be expressed if reducing conditions were created in the environment. A study by Boivin et al. (1990) confirmed that the presence of cell-free extracts of hydrogenase enzyme increased the corrosion of mild steel.

Recently, the importance of hydrogenase activity to corrosion of mild steel was assessed by Bryant et al. (1991). The authors concluded that the susceptibility of a system to biocorrosion depended on the microbial makeup of the mixed population of SRB and the presence of the enzyme hydrogenase.

Another study by Bryant et al. (1993) showed that the hydrogenase of *Desulfovibrio vulgaris* (Hildenborough) was regulated by Fe^{2+} availability. This regulation of enzyme might partly explain why in low nutrient environments such as pipelines, hydrogenase activity was higher than that found in laboratory-maintained cultures. In a field environment, the organisms might have to derepress hydrogenase to consume cathodic hydrogen on the metal surface and thus accelerate the dissolution of Fe^{2+} from the anode to supplement their need for Fe in an iron-depleted environment. It is therefore likely that the bacterial uptake of H_2 governed by the activity of hydrogenase fluctuates depending on the dissolved iron levels. Whether this activity is an important factor in SRB-enhanced corrosion process remains to be determined.

9.2.1.2 Iron Sulfides as a Depolarizer.

In the presence of a semicontinuous or continuous SRB culture in lactate medium containing sulfate, a film of sulfide forms over the iron surface, giving some protection against corrosion (Booth et al., 1965). This film breaks down after some time, leading to an increased corrosion rate.

The quantitative importance of cathode depolarization by solid FeS has been confirmed by the addition of chemically prepared FeS to mild steel test coupons in the presence and absence of *Desulfovibrio desulfuricans* using fumarate as a replacement of sulfate as terminal electron acceptor for bacterial reduction processes (Booth et al., 1968). The extent of corrosion as assayed both by polarization and by weight loss, was proportional to the FeS added and dependent upon direct contact between the sulfide and the metal surface, being greatest when test coupons were mounted horizontally. The authors concluded that although both hydrogenase and FeS could influence the corrosion of steel, the FeS contribution was of major significance.

The foregoing studies were expanded by King and Miller and their colleagues (King and Miller, 1971; King et al, 1973, 1976). They noted that under all conditions, the initial sulfide film formed during corrosion was in the form of a continuous and adherent layer of mackinawite, which was protective. However this film showed a degree of physical disruption in a time-dependent modification to greigite. In the presence of higher soluble iron concentrations there was a similar loss of protective mackinawite film with the conversion to smythite (Fe_3S_4) and pyrrhotite ($Fe_{1-x}S$) rather than greigite. In each case this loss of the uniform protective mackinawite film generated active electrochemical cells between areas of unreacted, and unprotected, steel and deposits of the various ferrous sulfides, with the resultant corrosion occurring by cathodic stimulation. Continued sulfide production from SRB was required to maintain their chemical integrity and electrochemical activity.

The mechanisms of mild steel corrosion by FeS were reviewed by King and Wakerley (1973). The process involved (1) depolarization of cathodic areas by absorption of the polarizing H_2 into the crystal lattice of the FeS or (2) the establishment of a galvanic cell, FeS/Fe, whereby the Fe becomes anodic and FeS acts as the cathode. Comparison of the corrosion of mild steel by chemically prepared FeS and biogenically derived FeS indicated that, in media of high Fe^{2+} concentration, most of the corrosion in a bacterial culture was attributed to the biogenically derived FeS. The corrosion caused by the biogenic FeS appeared identical to that caused by FeS derived by inorganic reactions (Smith and Miller, 1975). By the mid 1970s it was widely believed that corrosion in the presence of SRB proceeded under reduced conditions through cathodic stimulation of electrochemical cells established between areas of unreacted steel (anode) and deposits of various reduced ferrous sulfide corrosion products (cathode) as depicted in Figure 9.2.

9.2.1.3 Iron Sulfide and Bacterial Hydrogenase as Depolarizers.
King and Miller (1971) proposed a new theory of corrosion by the SRB such that the cathodic reaction, H_2 evolution, occurred on the FeS produced by reaction of Fe^{2+} with bacterially produced S^{2-}. They found that the FeS activity diminished with time, possibly as a result of the bonding of atomic hydrogen within the FeS crystals lattice but the activity of the FeS was restored by removal of hydrogen by bacterial hydrogenase activity. They further postulated that the SRB on the surface of FeS continually regenerated or depolarized the FeS by removal of atomic hydrogen as a result of hydrogenase activity and thus explained the continued high rates of corrosion.

9.2.1.4 Hydrogen Sulfate as a Depolarizer.
Costello (1974) carried out a critical analysis of cathodic depolarization. Although demonstrated, it did not arise from the action of hydrogenase in the presence of sulfate. He concluded that the depolarizing agent was a

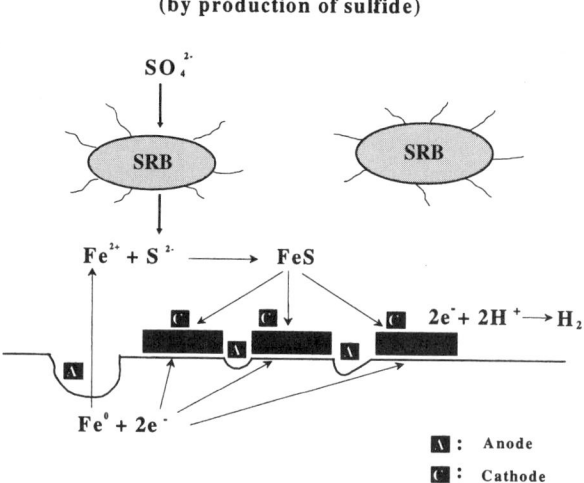

Figure 9.2 Cathodic depolarization of surface by FeS as a result of respiration of sulfate by SRB.

gaseous species, which he assumed to be H_2S, because the addition of H_2S back into the degassed cultures showed the same types of polarization curves as before degassing. He proposed that at neutral pH, H_2S acted as the electron acceptor with hydrogen as the key cathodic product:

$$H_2S + 2e^- \Leftrightarrow 2HS^- + H_2 \tag{9.8}$$

The validity of this mechanism, however, was questioned by Gaylarde and Johnston (1982), because mild steel coupons separated from growing cells by dialysis tubing, which allowed free passage of dissolved gases, did not corrode at rates above control values.

9.2.1.5 Volatile Phosphorus Compound. Iverson and Olson (1983) suggested that anaerobic corrosion in the presence of SRB was due to bacterial production of a highly active volatile phosphorus compound. The corrosion reaction yielded iron phosphide as a corrosion product. Cathodic depolarization experiments on an agar surface in the absence of an electron acceptor resulted in the blackening of the agar at areas in contact with SRB. On extraction and analysis by X-ray diffraction, the black precipitate was found to contain iron phosphide (Fe_2P). It was concluded that the phosphorus compound was released as a result of the activities of SRB and also produced by the action of H_2S on phosphate, phosphite, and hypophosphite. Attempts to repeat Iverson's experiments have been unsuccessful. Recent studies have excluded inorganic phosphate as the source of phosphorus metabolized by the SRB into the corrosive phosphorus product (Iverson, 1998). Organic phosphate in the form of phytic acid, a common metabolite of plants, was proposed as the form of phosphate metabolized by the SRB into the corrosive phosphorus product in the field. To date, the role of a phosphorus compound in SRB-influenced corrosion remains ambiguous.

9.2.1.6 The Influence of Oxygen. The importance of dissolved oxygen on the SRB-influenced corrosion of ferrous metal has been emphasized by several authors (Lee et al., 1993a, 1993b; Nielsen et al., 1993; McKenzie and Hamilton, 1992). Weight loss experiments showed that a corrosion rate as high as 50 mm y^{-1} was obtained when the vessel containing SRB and steel foil was exposed to oxygen on an intermittent basis. This corrosion rate was 50 times higher than that reported under anaerobic conditions (Hardy and Brown, 1984). Nielsen et al. (1993) concluded that under alternating oxic and anoxic conditions, the corrosion mechanism of mild steel was related to the large Fe-S pool that provided surface area and hydrogen to SRB as well as potentially corrosive elemental sulfur. Indeed, under such conditions sulfide may react with oxygen to produce highly corrosive elemental sulfur and polysulfide (Schmitt, 1991). It has been suggested that the reaction sequence might proceed in two stages. First, the iron sulfide on the steel surface may form a protective film of mackinawite, thereby exerting anodic control of metal dissolution by limiting ferrous ion diffusion through the sulfide film. Subsequent breakdown of this protective film would create an electrochemical cell with unreacted steel acting as the anode, iron sulfide acting as the cathode, and sulfur serving as the electron acceptor to promote cathodic depolarization.

9.2.1.7 Anodic Depolarization. Taking a contrary view, other workers concluded that SRB promoted corrosion by a mechanism of anodic depolarization, essentially dependent on sulfide production (Crolet, 1992; Daumas et al., 1993). It has been proposed that an anode was first created by local H^+ production at a focus of SRB metabolic activity, with ensuing metal dissolution. A key element in generating the necessary kinetic conditions was a localized acidification at the anode resulting from the formation of iron sulfide corrosion products:

$$Fe^{2+} + HS^- \rightarrow FeS \downarrow + H^+ \qquad (9.9)$$

This reaction also has the secondary effect of removing HS^-, hence reducing the effectiveness of the H_2S/HS^- buffer system which would resist local acidification. Crolet (1993) has also discussed the effects of the availability of sulfide and free iron on the nature of the corrosion products formed. Where the local sulfide concentration was low, as would be the case with the foregoing reaction in the presence of high soluble iron concentration, the product formed would most likely be mackinawite (FeS_{1-x}) which is considered to be nonprotective. Where sulfide was in excess, the product would be the more protective pyrite (FeS_2). Although pitting corrosion of steel can theoretically be explained by the mechanism of anodic depolarization, the conditions under which locally acidic conditions are maintained remains to be verified.

9.2.1.8 Mineral Signatures of Bacterial Sulfate Reduction.

Many sulfides under near-surface natural environmental conditions may only be produced by enzymatically catalyzed microbiological pathways on specific substrates such as metals. Formation and stability of the sulfide mineral, mackinawite, is dependent on a continuously acting hydrogen sulfide source. While abiotic sources of sulfide such as volcanic venting of reduced gases from subterranean environments may also lead to mackinawite formation in the absence of enzymatically catalyzed biological reactions (Craig, 1991), the most common source of sulfide responsible for the vast majority of industrially related corrosion is that produced by SRB (McNeil and Little, 1990; McNeil et al., 1991a). Thus, even though mackinawite is not favored to exist under standard thermodynamic conditions common to many corrosion scenarios, it is frequently found where SRB have been detected. On continued exposure to SRB, mackinawite alters to greigite (Fe_3S_4) and smythite (Fe_9S_{11}) and finally to pyrrhotite (FeS_{1+x}). Greigite is associated with generalized corrosion, whereas, smythite is associated with pitting corrosion of iron. The presence of mackinawite and greigite among corrosion products of iron is generally proof that SRB participated in the corrosion reaction (McNeil and Little, 1990; McNeil et al., 1991a; Jack et al., 1995). Pyrite is not a common iron corrosion product, but SRB can produce pyrite from mackinawite in contact with elemental sulfur.

Mineral signatures of MIC have been detected using X-ray diffraction (XRD) and energy dispersive X-ray analysis (EDX) as corrosion products on many oil and gas pipeline systems (Jack et al., 1995). Amorphous iron sulfide is also often detected at pipeline corrosion sites employing EDX. Little is known about its subsequent crystallization, although biomineralization around SRB colonies or within biofilms may be a key process.

When anaerobic sites with SRB become oxic (aerobic), new minerals are formed from the original corrosion products. The oxidation process yields lepidocrocite or goethite and rhombic or orthorhomic sulfur. Mixed reducing and oxidizing conditions, such as those commonly found in Alberta, Canada, pipeline systems, often yield the partially oxidized iron oxide magnetite along with lepidocrocite and goethite (Jack et al., 1995). Thus, mineralogical fingerprints of corrosion products provide a window into the conditions leading up to and propagating a corrosion event.

9.2.1.9 Exopolymers as Corrosive Agents.
Both freely suspended and surface-associated SRB are able to synthesize exopolymers (Beech and Gaylarde, 1991; Beech et al., 1991; Zinkevich et al., 1996). Ochynski and Postgate (1963) first reported the chemical properties of extracellular polymeric substances (EPS) produced in SRB batch cultures using paper chromatography. EPS/metal-ion interaction has been proposed as one of the mechanisms of biocorrosion (Geesey and Mittelman, 1985; Geesey et al., 1986) and a number of studies have demonstrated the involvement of exopolymers secreted by different genera of aerobic bacteria in metal deterioration (Geesey and Mittelman, 1985; Geesey et al., 1986; Ford et al. 1987; White et al., 1985; Bremer and Geesey, 1991). Binding of metal ions, in particular Fe and Cr, by SRB exopolymers has been documented (Beech and Cheung, 1995; Beech et al., 1996a). Recent studies confirmed the direct involvement of a high molecular weight, extracellular fraction containing carbohydrate and protein produced by a marine SRB in pitting corrosion of mild steel (Beech et al., 1998a). Thus in addition to metal sulfides, complex organic macromolecules secreted by SRB can influence electrochemical processes at the biofilm/steel interface.

9.2.1.10 Current State of Understanding of SRB-Influenced Corrosion of Iron and Its Alloys.
It is now generally accepted that the number of SRB detected in a system, as either suspended or attached populations, does not necessarily correlate with the extent or rate of corrosion (Little and Wagner, 1994). Rather, the metabolic status of SRB in terms of rates of sulfate reduction and/or hydrogen sulfide production are believed to be more directly related to the corrosion rate (Moosavi et al., 1990). It has been reported that SRB activity based on enzymatic measurements correlated with corrosion rates (Bryant et al., 1993). This simple relationship of cause and effect, however, has not been accepted by all researchers in the field (Lee and Characklis, 1993). It has recently been reported that different types of SRB vary in their ability to influence metal deterioration (Gaylarde, 1992; Beech et al., 1994; Beech and Cheung, 1996). To date, no clear consensus has been reached linking specific bacterial metabolic rates to observed corrosion rates. To properly assess the role of SRB in a corrosion process, it is necessary to evaluate the corrosion reaction in the absence as well as in the presence of these sulfide-producing bacteria. However, it should be emphasized that establishing the mechanism of action, while desirable, is not always necessary in order to achieve control over the corrosion process.

9.2.2 Anaerobic Oxidation of Iron by Methanogens

Although SRB as consumers of cathodically generated hydrogen have received the most attention with respect to cathodic depolarization, other hydrogen-consuming bacteria are also likely to play a similar role in corrosion. Methanogenic bacteria, which grow through anaerobic respiration, according to Eq. (9.10), replicate and produce methane using metallic iron as sole electron source:

$$4H_2 + CO_2 \rightarrow CH_4 + 2H_2O \qquad \Delta G^{0'} = -139 \text{ kJ} \qquad (9.10)$$

The mechanism of Fe^0 oxidation is cathodic depolarization, in which electrons from Fe^0 and H^+ from water produce H_2, which is then used by the methanogens (Daniels et al., 1987). Methanogens can thus be significant contributors to the corrosion of iron-containing materials in anaerobic environments.

9.2.3 Hydrogen-Producing Bacteria

Hydrogen can have a deleterious effect on susceptible alloys when it enters the atomic structure of the metal, causing embrittlement cracking as well as stress corrosion cracking, both of which are major causes of failures of components, structures, pipelines, and pressure vessels in the oil production industry. Hydrogen generated by cathodic protection is known to pose a problem in areas of stress corrosion or corrosion fatigue. Many bacteria release molecular hydrogen as an end product from the fermentation of carbohydrates. It has been demonstrated that environmental cracking of reinforcing steel wire is accelerated in the presence of a biofilm formed by a hydrogen-producing *Clostridium* spp. (Ford and Mitchell, 1991).

Sulfides produced by SRB inhibit the combining of atomic hydrogen into molecular hydrogen, thereby increasing the concentration of atomic hydrogen at the surface and promoting its diffusion into the metal when biofilms are present. Sulfide also disrupts protective surface films of scale and corrosion products that would normally impede diffusion of hydrogen to the bare metal surface. However, the presence of EPS and other organic molecules derived from microbial biofilms may hinder dissolution and dissociation reactions as well as adsorption processes in the cracks, thereby reducing the flux of hydrogen into cracks in the metal. The biological influence from a biofilm consortium containing hydrogen-producing bacteria and SRB on hydrogen permeation through steel and corrosion-fatigue crack growth is complex, and, as yet, not well understood at this time (Edyvean et al., 1998).

9.2.4 Metal-Reducing Bacteria

Microorganisms are known to promote corrosion of iron and its alloys through dissimilatory reduction reactions. The consequence of these reactions is the dissolution of protective oxide/hydroxide films on the metal surface. Thus, passive layers on steel surfaces can be lost or replaced by less stable reduced metal films that allow further corrosion to occur. Obuekwe et al. (1981) demonstrated an increase in the corrosion of mild steel in the presence of an iron-reducing bacterium of the genus *Pseudomonas*. Despite their wide occurrence in nature and likely importance to industrial corrosion, bacterial metal reduction has not been seriously considered in corrosion processes until recently.

Numerous types of bacteria are able to carry out manganese and/or iron oxide reduction (Arnold et al., 1988; Myers and Nealson, 1988; Roden and Zachara, 1996). It has been demonstrated that in cultures of *Shewanella putrefaciens*, iron oxide surface contact was required for bacterial cells to mediate reduction of these metals. The rate of reaction depended on the type of oxide film under attack (Little et al., 1997a). For example after 22 days of exposure, overall integrated rate of iron-reduction for hematite was 50 times slower than rates for geothite and ferrihydrate. The authors proposed that the reduction rate was directly related to surface area, and hence, accessibility of the oxides. Corrosion of carbon steel in the presence of *Shewanella putrefaciens* was monitored for up to 1300 h using current noise measurements. The electrodes examined after this period of time were extensively colonized and pitted. The location of pits coincided with the distribution of bacterial colonies. Subsequent energy dispersive x-ray analysis and X-ray crystallography analysis revealed that mineral replacement reactions had occurred on electrode surfaces.

Interestingly, bacterial reduction of amorphic iron (III) oxyhydroxide has been shown to competitively inhibit sulfate reduction (Lovley and Philips, 1987). Even where sulfate reduction is the predominant terminal electron-accepting process, it can be inhibited by 86–100% by iron reduction. Apparently, Fe(III) reducers can scavenge the electron donors' hydrogen and acetate more efficiently than SRB.

9.2.5 Slime-Producing Bacteria

Virtually any nonsterile surface submerged in an aqueous milieu accumulates an attached microbial population. When conditions are conducive for growth of microorganisms on a surface, a biofilm or slime layer may become visible. Biofilms that develop on metal surfaces typically consist of microbial cells, their extracellular polymeric substances (EPS), which anchor the cells to the surface, and inorganic precipitates derived from the bulk aqueous phase or corrosion products of the metal substratum. EPS consist of a complex mixture of cell-derived polysaccharides, proteins, lipids, and nucleic acids. The chemical composition of EPS varies depending on the microbial species and growth conditions (Sutherland, 1985).

Microorganisms that produce copious quantities of EPS during growth in biofilms have been implicated in localized attack of stainless steels (Pope et al., 1984). The biofilm forms the equivalent of a crevice on the metal surface in regions where mechanical damage or halide attack has breached the protective chromium/iron oxide film. Consumption of oxygen via respiration by microbial cells within the biofilm reduces the oxygen concentration at the metal surface to an extent that the protective oxide film cannot be reformed and the surface not able to repassivate. Biofilm-forming microorganisms that have been recovered from sites of corrosion on stainless steels include, *Clostridium* spp., *Flavobacterium* spp., *Bacillus* spp., *Desulfovibrio* spp., *Desulfotomaculum* spp., and *Pseudomonas* spp.

It has been postulated that EPS are important but not sufficient to induce biocorrosion of stainless steel if their action is not aided by the presence of an as yet ill-defined biocatalyst of oxygen reduction (Scotto et al., 1990). As little as 10 ng cm^{-2} EPS was reported to provoke the onset of MIC of stainless steel in natural seawater (Scotto, 1993). The same study claimed that cathodic protection of the stainless steel, used to prevent MIC, actually increased the amount of EPS in the biofilm. Thus, the role of EPS in MIC of stainless steel remains obscure.

EPS has been suggested to protect metal surfaces from corrosion. A bacterial consortium consisting of a thermophilic *Bacillus* sp. and *Deleya marina* produced metal binding EPS that reduced the rate of corrosion of carbon steel by 94% (Edyvean et al., 1998). Such a mechanism may be responsible for the protection microorganisms afford to mild steel under certain conditions (Soracco et al., 1984).

9.2.6 Acid-Producing Bacteria

Bacteria and fungi can produce copious quantities of either inorganic or organic acids as by-products of metabolism. Acidophilic sulfur-oxidizing bacteria (SOB) such as *Thiobacillus* spp. oxidize reduced forms of sulfur to sulfate with the formation of sulfuric acid. While these microbes can cause severe corrosion damage to equipment used to mine iron and copper ores, their impact extends to streams and surrounding soils where the low pH makes it uninhabitable for other life forms.

Bacteria that ferment organic compounds to low molecular weight organic acids such as acetic acid have also been implicated in the corrosion of iron and its alloys. Organic acid-producing bacteria (APB) have been proposed to be the primary cause of MIC of carbon steel (Soracco et al., 1988). In an electric power station, the APB were the only group of cultivable microorganisms whose abundance was correlated positively with corrosion (Soracco et al., 1988). Acetic, formic, and lactic acids are common metabolic by-products of APB. Little et al. (1992) provide examples of other acids synthesized in the Krebs Cycle, common to most aerobic microorganisms, as contributing to MIC. Little et al. (1988a) demonstrated that an aerobic acetic acid-producing bacterium accelerated the corrosion of cathodically protected stainless steel. Artificially applied acetic acid destabilized or dissolved the calcareous deposits formed during cathodic polarization.

The mechanism of action of acids on corrosion of mild steel is well established in the literature (Shreir, 1963). Unfortunately, the acids produced under the conditions leading to corrosion are rarely identified and their concentrations at the site of corrosion are even less well understood. Pope (1991) reported acetic acid concentrations as high as 708 mg/L producing a pH of 5.5 in the bulk aqueous phase of a recycle test loop reactor containing carbon steel coupons and a mixed population of *Enterobacter* sp., *Clostridium* sp., and *Desulfovibrio* sp. When acids are secreted by slime-producing microorganisms, they are concentrated at the metal surface by the diffusional resistance of the biofilm polymers, making it impossible to accurately assess the acid concentration at the metal surface from measurements made in the overlying bulk aqueous phase. Microsensors have been used to probe the pH gradients within 1 mm thick microbial biofilms growing on corroded mild steel surfaces (Lee and deBeer, 1995). pH values increased from 7.5 at the bulk fluid-biofilm interface to 9.5 at the metal surface at cathodic areas and ranged from 7–5 at the surface of the tubercle in anodic areas.

A group of acid-producing microorganisms whose role in biocorrosion is often overlooked is the slime-producers that excrete acidic extracellular polysaccharides during biofilm formation on metal surfaces. Carboxylic acid functionalities of matrix polysaccharides such as alginic acid, produced by the biofilm-forming bacterium *P. aeruginosa* have been calculated to be on the order of 6 Å apart, and thus, highly concentrated at the metal-biofilm interface (Jang et al., 1989). It is virtually impossible to concentrate dissolved low molecular weight acids to such a high level. As discussed later in Section 9.3.1, binding of metal ions results in the release of protons from these biogenically derived polymers.

9.2.7 Fungi

Fungi are well-known producers of organic acids, and therefore capable of contributing to MIC. Grease-coated wire rope wound on wooden spools stored in a humid environment has been reportedly attacked by *Aspergillus niger* and *Penicillium* spp. (Little et al., 1995). Localized corrosion was detected on wire surfaces where the grease coating was thin or missing. Fungal growth was demonstrated only where the wire rope was in contact with the wood. Both fungal species are known to produce citric acid as a result of hydrocarbon degradation. Citric acid solution at pH 5.0 did penetrate the protective grease and caused corrosion. It was proposed that localized attack of the wire was due either to organic acid production or to initiation of corrosion nucleation sites.

9.2.8 Microbial Consortia

APB are frequently present in the bulk aqueous phase at densities equal to or greater than SRB (Pope et al., 1990). The acids produced by APB serve as nutrients for SRB and methanogens. It has been suggested that SRB occur at sites of corrosion due to the activities of the APB (Soracco et al., 1988). Dowling et al. (1992) compared corrosion of C1020 pipeline steel in the presence and absence of the acetogenic bacterium, *Eubacterium limosum* and SRB populations of a *Desulfovibrio* sp. and *Desulfobacter* spp. *E. limosum* alone had little effect on the corrosion rate compared to sterile controls, but when coinoculated with the *Desulfovibrio* sp., together the consortium induced a significantly higher rate of corrosion than either one alone or sterile controls. It was proposed that by-products of *E. limosum* supported *Desulfovibrio* sp. growth and sulfide production. Corrosion of aluminum alloys in the presence of hydrocarbon-degrading fungi is addressed later in Section 9.4.

9.2.8.1 Metal-depositing Bacteria. Bacteria of the genera *Siderocapsa*, *Gallionella*, *Leptothrix*, *Sphaerotilus*, *Crenothrix*, and *Clonothrix* participate in the biotransformation of oxides of metals such as iron and manganese (Gounot, 1994). Iron-depositing bacteria (e.g., *Gallionella* and *Leptothrix*) oxidize Fe^{2+}, either dissolved in the bulk medium or precipitated on a surface, to Fe^{3+}. The Fe^{3+} precipitates around the organic sheath of these filamentous bacteria as iron hydroxide (Fig. 9.3a, b). Bacteria of genera given above are also capable of oxidizing manganous ions to manganic ions with concomitant deposition of manganese dioxide. Detection of iron and/or manganese-depositing bacteria associated with corroding metals is described in detail by Little et al. (1997b).

A role for metal-depositing bacteria (MDB) in the corrosion of steels has been proposed on the basis of detection of sheathed filamentous bacteria by microscopic techniques in naturally formed corrosion deposits (Kobrin, 1976; Tatnall, 1981; Lutey, 1992). These bacteria have been typically associated with tubercle formation, the creation of an oxygen concentration cell, and consequent underdeposit pitting attack on stainless steel.

The corrosion resistance of stainless steels is due to the formation of a thin passive chromium/iron-rich oxide film. The formation of deposits of organic and inorganic material by MDB on the oxide surface compromises the stability of the protective film. In oxygenated environments, the corrosion cell on stainless steels occurs as a small anodic area surrounded by a relatively large cathodic zone. The corrosion reaction commonly proposed is the reduction of oxygen at the cathode with the formation of hydroxide ions and metal ion hydrolysis by water at the anode with the production of hydroxyl ions. High concentrations of H^+ ions (low pH) may be generated at the anode as a result of the hydrolysis of Cr and Mo, two alloying elements in stainless steels. Aggressive anions such as chloride ions then migrate to the anode to neutralize the charge (Fig. 9.4). Certain bacteria have been shown to help concentrate Cl^- ions in crevices and cracks (Stoecker, 1995). This leads to the formation of $FeCl_3$ and $MnCl_4$, which are highly corrosive to stainless steels. Alloys with high Mo content, originally thought to resist Cl^- attack, appear to be susceptible to pitting corrosion via this mechanism (Scott et al, 1991). Dense accumulations of MDB on the metal surface are thought to promote these reactions via the deposition of cathodically reactive ferric and manganic oxides and the local consumption of oxygen via respiration in the deposit. Difficulty in culturing the sheathed, filamentous MDB has precluded the execution of controlled laboratory simulations of this type of corrosion by these bacteria to verify their involvement in the corrosion of stainless steel in the field.

9.2.8.2 Ennoblement. The principal electrochemical changes associated with a reduction in resistance of stainless steel to corrosion are referred to as ennoblement. These changes include an increase in the open circuit potential (OCP) or corrosion potential (E_{corr})

to noble values approaching the pitting potential (E_{pit}) around +350 mV versus a saturated calomel electrode and a two- to three-order of magnitude increase in the reaction rate at the cathode (Dickinson et al., 1996). Ennoblement is associated with an increased probability of localized corrosion occurring as the OCP approaches the pitting potential E_{pit}. The elevated cathodic reaction rate increases the chances that pit propagation will continue. Sustained ennoblement does not necessarily indicate corrosion, however. At the onset of

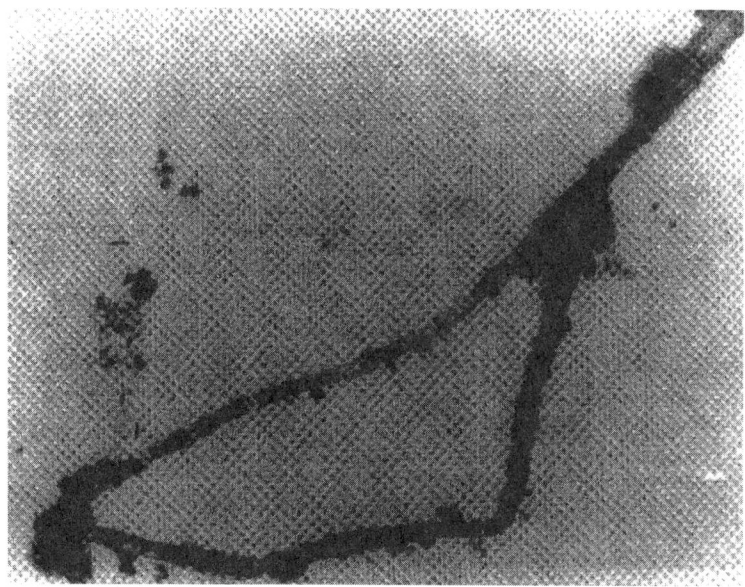

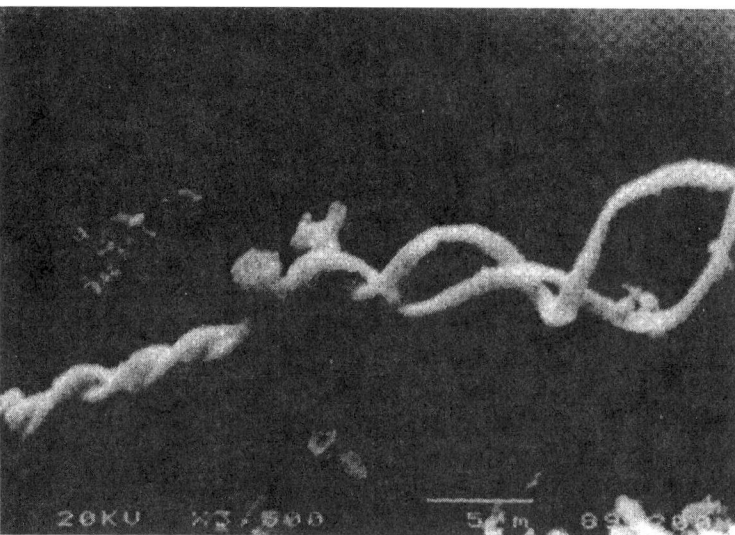

Figure 9.3 *a*. Microphoto of *Leptothrix* sheath showing inclusions of ferric hydroxide (Videla, 1996); *b*. SEM micrograph of a *Gallionella* sheath (Videla, 1996).

9.2 MIC OF IRON AND FERROUS ALLOYS

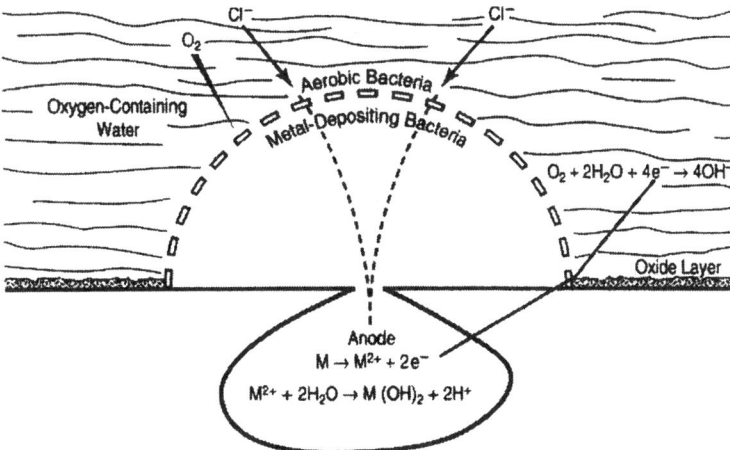

Figure 9.4 Possible reactions under tubercles created by metal-depositing bacteria (Little et al., 1997a).

pitting, E_{corr} shifts from the noble potential to the active region of lower (more negative) potential. Ennoblement has been reported on stainless steels after exposure to natural seawater, brackish water, and fresh water (Scotto et al., 1985, 1986; Dexter and Zhang, 1991; Dexter, 1993; Dickinson et al., 1996; Lewandowski et al., 1997). A number of mechanisms have been proposed by which microorganisms promote ennoblement, including metal deposition at the cathode, oxygen removal, and/or acid production within the deposit (Mansfeld et al., 1992).

Microbial removal of oxygen within a deposit can be achieved by respiration and elaboration of nonrespiratory enzymes that reduce oxygen. Dupont et al. (1998) tested the theory that enzymes catalyze oxygen reduction in biofilms. One such enzyme, glucose oxidase, catalyzed the reduction of oxygen to hydrogen peroxide during the oxidation of glucose. The hydrogen peroxide ennobled the steel.

MDB can fix the redox potential of the Fe^{2+}/Fe^{3+} or Mn^{2+}/Mn^{4+} couples at the metal surface. This, in turn, polarizes the surface potential at the cathode to facilitate the cathodic reaction. In the presence of MDB, MnO_x deposits as a series of concentric rings on stainless steel surfaces to establish anodic and cathodic sites on the metal surface (Dickinson and Lewandowski, 1996; Little et al., 1997c). Ennoblement due to the increased MnO_x-induced acceleration of the cathodic reaction according to the reaction,

$$MnO_2 + H_2O + e^- \rightarrow MnOOH + OH^- \qquad E_{pH\,8.0} = +335\,mV \qquad (9.11)$$

and decreased redox potential via microbial respiration has been proposed to increase the interfacial potential difference within the rings to a value exceeding E_{pit}, which leads to preferential pit nucleation within the rings (Dickinson et al., 1996). Sustained pit growth is favored under conditions of increased probability of pit nucleation. It is proposed that the MnO_x-fixed, cathodic site initially limits surface growth of the anode but eventually, active corrosion currents consume the MnO_x and permit growth of the pit opening at the metal surface. Tubercle formation is proposed to be the result of accumulation of iron corrosion product and biotransformed iron to less soluble products around the pit opening to form the characteristic tubercle (Dickinson and Lewandowski, 1996).

Recently, MDB have been shown to promote the ennoblement stage leading up to pitting corrosion. Formation of a surface biofilm containing the sheath-forming, manganese-depositing bacteria, *Leptothrix discophora*, was shown to be a prerequisite for ennoblement of 316L stainless steel (Dickinson et al., 1997). The biofilm was proposed to be necessary for deposition and electrical contact of cathodically active MnO_x at the metal surface so that electron transfer from the metal to the MnO_x deposit can occur. MnO_x deposition by these bacteria and the resulting ennoblement observed under laboratory conditions mimicked the pattern of ennoblement of stainless steels submerged in natural waters. However, the ennoblement produced in the laboratory study was not accompanied by the characteristic pitting corrosion of the metal it is thought to promote. Thus, while a satisfactory mechanism for ennoblement of stainless steel involving manganese-oxidizing bacteria has been established, their role and that of other MDB in pit initiation and propagation of steels remains elusive.

9.2.8.3 Consortium Interactions.

Consortia of MDB and SRB often exist as biofilms on corroding metal surfaces. The first records reporting isolation of these microorganisms from a corrosion failure dates back to 1910 (Gaines, 1910). It has been proposed that oxygen consumption by MDB creates redox conditions favorable for the growth of SRB (Videla and Characklis, 1992). While the sulfide produced as a result of SRB growth and metabolism itself is not an effective pitting agent at low concentrations, thiosulfate, an oxidation product of sulfide formed in the presence of oxygen, is an aggressive pitting agent on stainless steel (Newman et al., 1989; Moreno et al., 1992). Concerted action of MDB and SRB may promote the breakdown of the passive film on stainless steel as hypothesized by Lewandowski et al. (1997) and depicted in Figure 9.5.

A mixed bacterial species consortium was shown to be necessary for the maintenance of corrosion current of pitted 304L stainless steel in seawater under anaerobic conditions

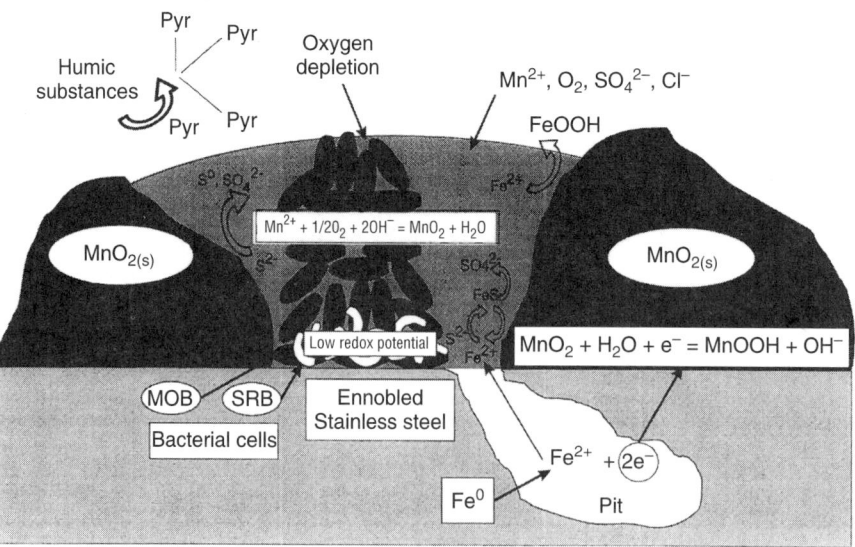

Figure 9.5 Hypothetical model of corrosion processes on stainless steel at sites of biomineralized MnO_2 containing SRB. MOB, metal oxidizing bacteria; Pyr, pyruvate. (Lewandowski et al., 1997).

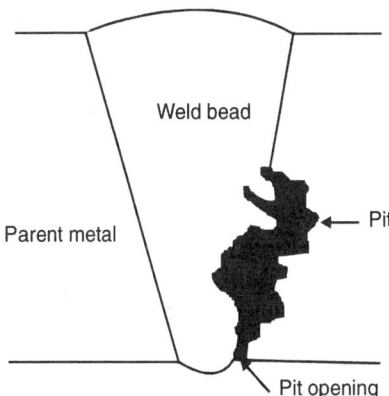

Figure 9.6 Schematic cross-section of a pit at weld seam in stainless steel.

(Angell et al., 1995). Spatial distribution of the bacterial populations was found to influence the corrosion reaction. SRB were necessary on the cathode, leading to a high charge transfer resistance, while a mixed consortium was needed on the anode giving low charge transfer resistance. These results support the involvement of cathodic depolarization as a mechanism for MIC and for another unidentified anodic reaction involving mixed bacterial populations for anaerobic corrosion of stainless steel.

9.2.9 Weldments

For reasons not well understood, welding markedly increases the propensity for MIC (Walsh et al., 1993). The microbial influence is one of significant corrosion acceleration for 304L (UNS S30403), 316L (UNS S31603), and AL-6XN (UNS N08367) stainless steels submerged in estuarine water (Buchanan et al., 1997). The microbial influence was quantified by means of a MIC factor, f_{mic}, which is defined as the ratio of the biotic corrosion rate in the natural water over the abiotic corrosion rate in laboratory control tests. MIC often occurs as pitting attack at or near sensitized structures of weldments of mild steel and austenitic stainless steels or in combination with stress corrosion cracking of these metals (Das and Mishra, 1986). MIC has been shown to occur at welds made with various combinations of base metal and filler material. Pits have been observed at the heat-affected zone (HAZ), at the fusion line, and in the base metal near the weld. Metallographic examination has revealed preferential attack of the stringers or one or both phases of the two-phase weld. Either the austenitic or ferrite phase or both have been attacked (Kobrin, 1976; Borenstein, 1988; Kearns and Borenstein, 1991; Danko and Lundin, 1995). A pit appears as a small hole on the exposed weld surface overlaid by a tubercle composed of a heterogeneous mixture of biofilm and inorganic deposits derived from surface-associated reactions of a corrosive and noncorrosive nature. During pit propagation, the small hole at the surface expands into a cavity in the subsurface region of the weld that may extend into the wrought metal (Borenstein, 1988) (Figure 9.6).

E_{pit} of welded stainless steels is reportedly lower than that of the wrought metal, indicating that welding promotes metal dissolution at anodic sites (Cubicciotti and Licina, 1989). One proposed mechanism for the susceptibility of weldments to MIC is based on the widely accepted theory that an excess phase, rich in chromium, is precipitated along grain boundaries, resulting in a decrease in chemical resistance in those regions that experienced chromium depletion (Das and Mishra, 1986). The weld and HAZ thus contain chromium-depleted regions where preferential anodic dissolution occurs and pitting attack is favored. Similarly, welding-induced segregation of molybdenum in alloys containing this element facilitates pitting attack in Mo-depleted regions. The probability of pitting attack at these compromised areas is increased when the foregoing microbial processes, which influence E_{pit}, also occur on the weld surface (Ibars et al., 1992). Thus, it is likely that a combination of factors of biotic and abiotic origin contribute to the susceptibility of welds and HAZ to pitting corrosion.

Local depletion of Cr has also been demonstrated at grain boundaries of the Cr oxide film of "as received" 316L stainless steel during surface colonization by the bacterium *Citrobacter freundii* (Geesey et al., 1996). The depletion phenomenon was enhanced by the presence of SRB at the oxide grain boundaries. These microbiologically induced surface chemical changes may, like welding, favor the establishment of anodic areas prone to metal dissolution under the oxygen-depleting conditions of a surface experiencing biofilm growth that discourages repassivation of the protective oxide film.

9.3 COPPER

Copper and copper alloys display good resistance to biofouling under most conditions due to the toxicity of the copper corrosion products that are produced when the metal is in contact with water. Resistance to biofouling, however, does not necessarily make copper-based metals immune to biocorrosion. CA 110 (99Cu) has been found to corrode in seawater at three times the rate as CA 706 (90-10 Cu-Ni) and CA 715 (70-30 Cu-Ni) in spite of evidence that all three metals exhibited the same degree of resistance to biofouling (LaQue Report, 1994).

9.3.1 Copper Alloys

Copper alloys are used for a variety of industrial and military piping systems and heat exchangers carrying either seawater or freshwater. Alloys such as 90-10 Cu-Ni (UNS C71590) and 70-30 Cu-Ni (UNS C70600), brasses, aluminum bronzes, and admiralty brass are used in nuclear power plant service in main condensers, feedwater heaters, other heat exchangers, and pumps and valves (Licina, 1989). All of these alloys are susceptible to biocorrosion.

Like other metals, copper alloys experience localized attack as a result of MIC. Mechanisms invoked for the biocorrosion of copper alloys include differential aeration, selective leaching, underdeposit corrosion, and cathodic depolarization. Differential aeration refers to a condition in which the oxygen concentration at a site on the metal surface is lower than at surrounding sites. Surface deposits contribute to reduced oxygen concentrations under the deposit, forming a site that is anodic to surrounding areas of the metal surface. Biofilms containing oxygen-respiring microorganisms can consume oxygen faster than it can diffuse into the deposit (Schiffrin and de Sanchez, 1985), establishing sites for colonization by SRB and other anaerobes on surfaces surrounded by oxygenated water. A number of by-products of microbial metabolism can accelerate localized attack of copper alloys, including ammonia, carbon dioxide, hydrogen sulfide and other sulfides, and organic and inorganic acids. Cathodic depolarization occurs when products of microbial metabolism scavenge electrons or hydrogen from the cathode. Hydrogenase enzymes, elaborated by hydrogen-metabolizing bacteria and some SRB, reportedly promote cathodic depolarization.

In oxygenated seawater a layer of cuprous oxide forms on copper alloy surfaces. Copper ions are formed and deposited as chloride salts such as $Cu_2(OH)_3Cl$, regardless of alloy type (Pollard et al., 1989; Little et al., 1991a; Wagner et al., 1991). Little et al. (1990b) have reviewed the typical cathodic and anodic corrosion reactions that occur on copper alloys in fresh water and seawater. In the presence of SRB, copper alloys form a porous film of cuprous sulfide in place of the cuprous oxide. Copper ions migrate from the bulk alloy to the source of sulfide to form the porous surface film, leaving behind a copper-depleted region enriched in the alloying elements. The corrosion products on 99Cu are nonadherent, meaning that they are easily displaced from the metal surface, promoting further corrosion as a result of periodic exposure of the bare metal to the bulk aqueous phase. In contrast, the corrosion products on 90-10 and 70-30 Cu-Ni are adherent and promote passivation of the underlying metal surface (McNeil et al., 1991b). The mechanical stability of the sulfide films formed in the presence of SRB is poor compared to the oxide films of copper, and offer no corrosion protection.

Dealloying of nickel from 90-10 and 70-30 Cu-Ni has been reported in seawater containing SRB (Little et al., 1990b; Wagner et al., 1992a). Spalling of the nickel-enriched region of the metal occurs during exposure to flowing seawater, exposing fresh metal and causing further dissolution of the alloy. Weldments also exhibit this type of corrosion in the presence of SRB (Little et al., 1988b).

Biofilm processes promote pitting attack of copper alloys through mechanisms other than differential aeration. Although the mechanism is not clearly understood, biofilms have been proposed to concentrate chloride ions, which react readily with copper ions in pits formed under the microbial deposits. $[Cu_2Cl_2]_n^{2-}$ ion scavenging by biofilms has been reported in water with very low Cl^- ion concentrations (Fischer et al., 1988).

9.3.2 Signatures of MIC Associated with Copper and Copper Alloys

Specific copper sulfides are formed during corrosion of copper and its alloys in the presence of SRB. Chalcocite (Cu_2S) and covellite (CuS_{1-x}) are commonly found as SRB-associated corrosion products of copper, and djurleite ($Cu_{31}S_{16}$) is a common SRB-associated corrosion product of copper alloys (McNeil et al., 1991a, 1991b). The formation of thick, nonadherent layers of chalcocite or the formation of hexagonal chalcocite is indicative of SRB-induced corrosion of copper and copper alloys.

9.3.3 Biocorrosion of Copper in Potable Water

Copper is widely accepted as a durable, corrosion resistant conduit for potable water. Tubing constructed of seamless copper alloy UNS C12200, containing a minimum of 99.9% copper is the most commonly used conduit for water systems in the United States (Cohen and Meyers, 1996). More than 7.4 million miles of copper tube have been installed in water service and distribution systems for U.S. buildings (Meyers and Cohen, 1995). This is due to the fact that remarkably few instances of copper tube corrosion failure were reported prior to 1985. On average the Copper Development Association investigates 42 incidents of pitting corrosion of copper water tubes each year. However, more recently, critical assessment of the corrosion performance of copper in potable waters has been triggered by tighter regulatory standards for potable water and wastewater, higher consumer expectations, and the recognition of a possible microbiological role in corrosion failures.

9.3.3.1 Pitting Corrosion.

Several types of pitting corrosion have been identified on the inner surface of copper tube used to distribute potable water in domestic and institutional buildings. Type 1 occurs in cold water lines that deliver water from deep bore holes containing low concentrations of microorganisms and organic carbon but high concentrations of dissolved inorganic ions. Pits contain soft crystalline cuprous oxide under a membrane of cuprous oxide crystals. Microorganisms have not been implicated as a cause for Type 1 pitting corrosion. Type 2 pitting corrosion occurs in systems that circulate hot (> 60°C), soft (bicarbonate/sulfate < 1) water with a pH of less than 7.4. The pits contain hard, crystalline cuprous oxides and basic copper sulfate. Nodules, composed of these corrosion products, form over the top of the pits. No involvement of microorganisms has been demonstrated in Type 2 pitting. Type 3 pitting appears as aggregations of small hemispherical pits under a common basic copper sulfate deposit. This type of pitting occurs in pipes carrying cold water with high pH, low hardness, and low mineral and organic content. Microorganisms are not believed to be involved in this type of corrosion.

At least two unusual types of pitting corrosion, reported in hot and cold water systems of institutional buildings, appear to have microbial origins (Geesey et al., 1994). One type has occurred in hospitals supplied with surface water that contained high concentrations of dissolved organic compounds and suspended particles including microorganisms, a low buffering capacity, and a pH in the range of 7.4 to 9.3. The hot water supply lines, containing water that rarely exceeded 50°C, experienced more severe corrosion than those in which the water was generally maintained above 60°C (Keevil et al., 1989).

The manifestations of this type of pitting corrosion are numerous pits beneath a common basic copper sulfate similar to classical Type 3 pits (see description by Mattson, 1980). Failed tubes contained a film that stained positive with Periodic Schiffs base (PAS) reagent and alcian blue, suggesting the presence of substituted and unsubstituted polysaccharides (Angell et al., 1990). Copious amounts of biofilm detected by scanning electron microscopy were associated with the pitted sites (Keevil et al., 1989), with the most severely corroded tubes containing the most well-developed biofilm (McEvoy and Colbourne, 1988).

The second type of pitting corrosion observed in copper plumbing used in potable water distribution systems, exhibited morphological features of both classical Type 1 (see Campbell, 1950; Lucey, 1967) and Type 2 (see Mattsson and Fredriksson, 1968) pitting. Pits are hemispherical and contain soft crystalline cuprous oxide and varying amounts of cuprous chloride under a cuprous oxide membrane. While the surface between the pits is largely cupric oxide, the mounds above the pits consist principally of basic copper sulfate interspersed with powdery cupric oxide. This type of pitting corrosion has been reported in institutional buildings in Saudi Arabia, southwest England, and Germany, as well as in private houses in Sweden, and is referred to as Type $1\frac{1}{2}$ pitting (Campbell et al., 1993; Fischer et al., 1990; Fischer et al., 1992; Fischer et al., 1994; Geesey et al., 1994; Nuttall, 1993; Shalaby et al., 1989). Type $1\frac{1}{2}$ pitting may also be accompanied by the release of dissolved copper ions and solid copper corrosion products into the water (Fischer et al., 1992). Sections of cold and warm water lines that are subject to stagnation are most prone to Type $1\frac{1}{2}$ pitting corrosion.

Chemical analysis of the adherent copper corrosion products recovered from failed copper tube suggests an interaction between the inorganic products and biologically derived organic molecules. Visible solid, copper corrosion products are located on top of or within a microbial biofilm layer in direct contact with bare metal surface in areas where the pipe is perforated (Fischer et al., 1988; Fischer et al., 1992). The biofilm has been described as linear and/or cross-linked acidic or partly nonionic polysaccharides, oligopeptides, and n-acetylated derivatives of glucose, mannose, and galactose. Corrosion products containing copper complexes of pyruvate, acetate, and histidine have been identified (Paradies et al., 1992). Extended absorption fine structure spectroscopy detected Cu^+ complexes with imidazole residues of histidine in the biofilm (Paradies et al., 1992). Binding of $[Cu_2Cl_2]_n^{2-}$ ions in the biofilm provided a mechanism for Cl^- sequestration into the pits to promote further ionization of the metallic copper (Fischer et al., 1988). Microbiological evaluation of the corrosion deposits determined that while high numbers of bacteria were associated with the pits, the presence of bacteria did not always result in pitting and that the range of cultural bacterial species was quite variable (Wagner et al., 1992b; Wagner et al., 1992c).

Failure of copper pipe displaying Type $1\frac{1}{2}$ pitting corrosion in some institutional buildings has been linked to MIC. In such cases, the pits are almost exclusively associated with a black cupric oxide surface layer. A correlation has been reported between this type of pitting and the presence of certain bacteria (*P. pacimobilis* and *P. solanacearum*) or polysaccharide (Angell et al., 1990; Chamberlain et al., 1988).

Some manifestations of Type $1\frac{1}{2}$ pitting corrosion reported in a German hospital have been reproduced in a test rig installed at that location as well as in a laboratory loop system circulating water taken from the hospital. The release of copper corrosion products into the water was accompanied by a reduction in concentration of microorganisms in the bulk water and an increase in the amount of biofilm constituents on the copper surface. A correlation was reported between the presence of polysaccharide-producing bacterial biofilms and pitting corrosion, but attempts to reproduce the detailed morphological features of pitting corrosion in test rigs and laboratory reactors at the rate observed in the field have failed (Wagner et al., 1992c).

The relationship between pitting propensity and the properties of biofilm polymers has been investigated (Siedlarek et al., 1994). Using cyclic voltammetry, it was shown that artificial biofilms, formed with Xanthan, alginate, and agarose, display cation selectivity (Siedlarek et al., 1994). The artificial biofilms also exert considerable influence on the corrosion reaction(s) of a copper surface in contact with an aqueous phase, particularly at the sites where solid corrosion products precipitate (Wagner et al., 1997). A physicochemical model, incorporating the electrochemical behavior of the artificial biofilms, has been developed to describe the pitting corrosion observed on copper piping of potable water systems. The model takes into account membrane properties and their heterogeneity, and the distribution of exopolymers on the surface of the pipes (Wagner et al., 1997).

9.3.3.2 Copper Corrosion By-product Release.

In addition to pitting attack, copper tube is subject to excessive copper corrosion by-product release ("blue water") under certain conditions where MIC may occur (Anon., 1994a; Arens et al., 1995; Dutkiewicz, 1995; Webster et al., 1996). The corrosion typically results in the formation of shallow cavities with microscopic pitting to a depth of less than 100 μm. The phenomenon governing water characteristics and resultant corrosion morphology has been reviewed elsewhere (Wells et al., 1998).

While blue water corrosion does not necessarily compromise the integrity of the copper tube, the corrosion process does result in the contamination of the water supply and wastewater sediments. The concentration of copper in first-draw (1 L) tape water in piping systems experiencing this type of corrosion may range of 2 to 80 mg L^{-1}, which is well above the United States Environmental Protection Agency 1.3 mg L^{-1} maximum contaminant level for copper (USEPA, 1993). Copper corrosion by-product release is a worldwide phenomenon, as blue water has been reported in New Zealand, Australia, United States, Japan, and Europe (Potter, 1970; Page, 1972; Moss and Potter, 1984; Fischer et al., 1992; Anon., 1993a; Anon., 1993b; Anon., 1994a; Anon., 1994b; Dutkiewicz, 1995; Arens et al., 1995; Anon., 1996; Wells et al., 1998). In Auckland, New Zealand, numerous cases of copper concentrations between 3 and 40 mg L^{-1} have been reported in first-draw water from new copper installations over the past 30 years (Potter, 1970; Page, 1972; Wells et al., 1994; Moss and Potter, 1984; Webster et al., 1996). Copper contamination has been most prevalent in dead-legs or infrequently used cold water services of institutional buildings.

Traditionally, copper corrosion by-product release has been thought to be exclusively controlled by the inorganic chemical properties of the water. Models based strictly on inorganic chemical reactions, however, fail to predict the high concentrations of copper detected in some potable water, particularly that of low alkalinity and elevated pH (7.8 to 9.5). Recent work has demonstrated MIC can potentially contribute excessive copper concentrations in the bulk water even when the water chemistry meets the regulations for potable use (Webster et al., 1998). Studies of excessive by-product release in chlorinated water supplies in New Zealand, Australia, and the United States have shown the problem is restricted to the extremities of the reticulation systems (Page, 1973; Anon., 1994a). At these locations it is difficult to maintain a chlorine residual, so there is no control over microbial processes associated with the walls of the pipe where the reactions associated with copper by-product release occur.

A role for MIC in copper by-product release from copper tubing has been inferred from oxygen consumption rates in corrosion test rigs. Rates were significantly higher in non-sterile tube sections displaying shallow corrosion cavities and pin prick-size pits than in sterile tube sections lacking the manifestations of corrosion (Arens et al., 1995). Furthermore, copper by-product release was correlated with the presence of *Sphingomonas* spp., *P. fluorescens*, and other bacterial species in whose cell wall regions copper is known to accumulate. Additional evidence of microbial involvement was obtained from field tests in which the injection of trace concentrations of Fe_2O_3, pyruvate, and sealing putty (carbon source) stimulated copper corrosion by-product release (Arens et al., 1996). In addition, application of heat or introduction of antibiotics to control microbial growth resulted in a decrease in the copper content of the water (Arens et al., 1995; 1996).

A number of microorganisms isolated from a site in Auckland, New Zealand, experiencing high rates of copper corrosion by-product release all displayed the ability to attach to and grow on copper surfaces in the presence of a simulated potable water medium (Webster et al., 1996). The four most numerically dominant, culturable bacterial species were identified by 16s rRNA gene sequence analysis as *Sphingomonas capsulata* (European Bioinfomatics Institutes (EMBL) Nucleotide sequence database AJ223450), *Staphylococcus warneri* (EMBL AJ223451), *Erythrobacter longus* (EMBL AJ223452) and *Methylobacterium* sp. (EMBL AJ223453). A species of the yeast, *Candida* was also recovered from the copper surface. Biofilms containing these isolates were shown to promote corrosion by-product release on copper surfaces in subsequent laboratory reactor experiments (Webster et al., 1996).

9.3.3.2.1 Mechanisms of Microbiologically Mediated Copper By-product Release.
Davidson et al. (1996) related the production of acidic metabolic products by a biofilm of the bacterium *Acidovorax delafieldii* on a copper surface to an increase in copper concen-

tration in the bulk aqueous phase. The amount of extractable, surface-associated copper was positively correlated with protein and carbohydrate concentrations in the biofilm. Bremer and Geesey (1991) showed a correlation between acidic polysaccharide accumulation in bacterial biofilms on copper films and initiation of copper film dissolution. Thus, organic acid production and concentration at a copper surface during bacterial biofilm development leads to the release of copper species into the surrounding bulk aqueous phase.

As stated in Section 9.1.3, biofilms exert their influence on interfacial chemistry through the metabolic activities of the microorganisms growing on the metal surface. MIC occurs when biofilm activities lead to changes in interfacial chemistry that destabilize the passive film formed initially on the metal surface. Biofilms increase the range of polarization resistance (R_p) of a copper surface (Figure 9.7). Compared to an abiotic surface, the presence of a biofilm shifts the R_p to lower values equivalent to that obtained in the presence of an electrolyte adjusted to pH 6.8. Since corrosion rate is inversely proportional to R_p, biofilms appear to increase the likelihood that interfacial conditions will depolarize the surface. Similar conclusions have been drawn from electrochemical impedance spectroscopy (EIS) studies. EIS spectra and corrosion rates of copper surfaces supporting biofilms in simulated potable water maintained at pH 8 are similar to those observed under abiotic conditions at pH 6.8. EIS data have shown that the oxide structure is more compact and protective at higher pH levels.

Webster et al. (1998) modeled physical interfacial processes extracted from EIS spectra using electrical circuit analogues. Figure 9.8 (a Bode representation of EIS data) displays the breakdown in passivity from spectra for copper surfaces experiencing MIC compared to the passive condition indicated by spectra of a sterile copper surface or one supporting a noncorrosive biofilm. Thus, the biofilm seems to impart characteristics to the copper surface similar to passivity breakdown of other engineering alloys subject to localized corrosion. The increased variability in corrosion rate of the copper surfaces supporting biofilms is likely due to the heterogeneous activities of the associated microbial community, particularly those involving pH changes, and any other processes that contribute to passivity breakdown.

The presence of a biofilm on copper tubing alters the physical properties of the normally protective copper oxide film (Webster et al., 1998). Incorporation of biofilm components into this oxide film increases its porosity and thickness, rendering the underlying copper less protected from corrosion than that conferred by films maintained under sterile conditions (Figure 9.9a,b). Small-spot X-ray photoelectron spectroscopy (XPS) showed

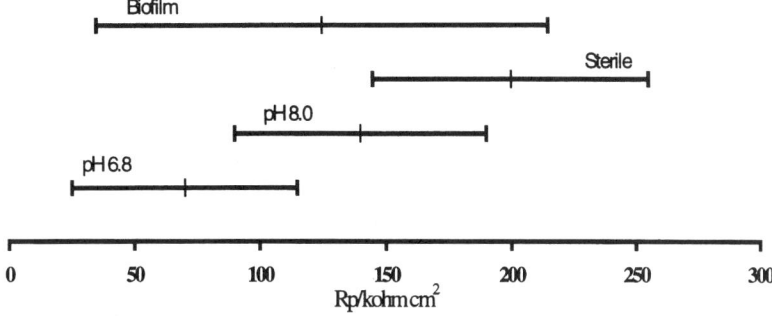

Figure 9.7 Summary of polarization resistance data showing the mean and one standard deviation for copper samples in the presence of a biofilm and under abiotic conditions at pH 8.0 and pH 6.8.

that carbon (presumably originating from a biofilm) was more uniformly dispersed throughout the oxide layer on a copper surface exposed to a culture of microorganisms than in the oxide layer of a copper surface maintained under sterile conditions. Samples of copper tube recovered from water systems experiencing copper by-product release also displayed a surface oxide layer containing organic carbon and nitrogen (Bremer et al., unpublished results). The increased porosity and thickness of the oxide film in the presence of a biofilm contributes to a decrease in film capacitance, which leads to an increase in corrosion rate.

9.3.4 Biocorrosion of Copper in Seawater Medium

A biofilm-forming marine bacterium, *Oceanospirillum* sp. increased the corrosion potential of copper by 145 mV and the corrosion rate fivefold over a 37 day period in seawater medium (Wagner et al., 1991). No pitting was observed, nor were copper corrosion by-products released into the bulk solution. In this case, the copper ions formed by the corrosion reaction became bound to the exopolymers secreted by the adherent bacteria. Copper dissolution by EPS and binding properties of exopolymers of other surface-fouling marine bacteria have been characterized (Jolley et al., 1989; Geesey et al., 1992). Mechanisms for metal deterioration based on metal-binding capabilities of bacterial exopolymers have also been proposed by Ford et al. (1990).

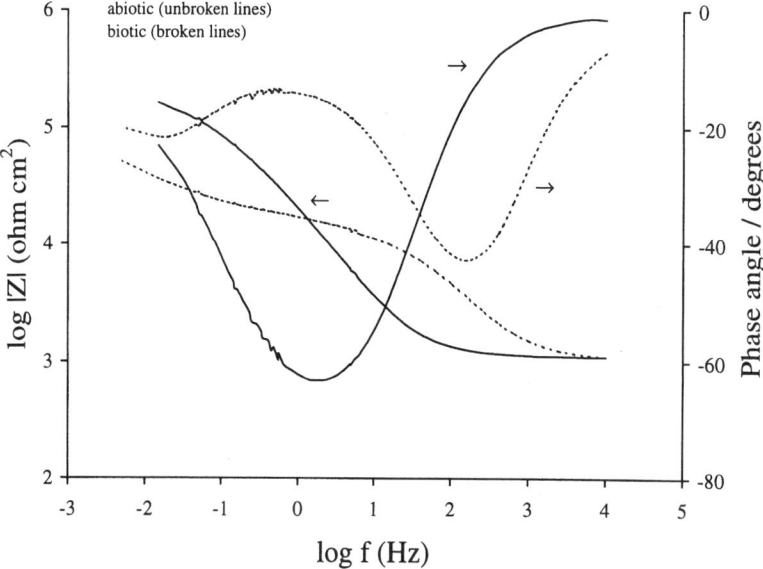

Figure 9.8 EIS data for copper surfaces under abiotic conditions and in the presence of a biofilm (biotic conditions) that promotes MIC presented as a plot of impedance and phase angle versus frequency (Bode plot). The corrosion rate of biotic samples experiencing MIC was higher than that under abiotic conditions; that is, the value of the impedance extrapolation to zero frequency was lower under biotic conditions. The high frequency response (100–1000 Hz) under biotic conditions suggests a decrease in pore resistance and capacitance of the copper oxide film. With the onset of MIC under biotic conditions, the copper oxide film becomes thicker, more porous, and less protective than under abiotic conditions (hence, the higher corrosion rate under biotic conditions).

9.4 ALUMINUM

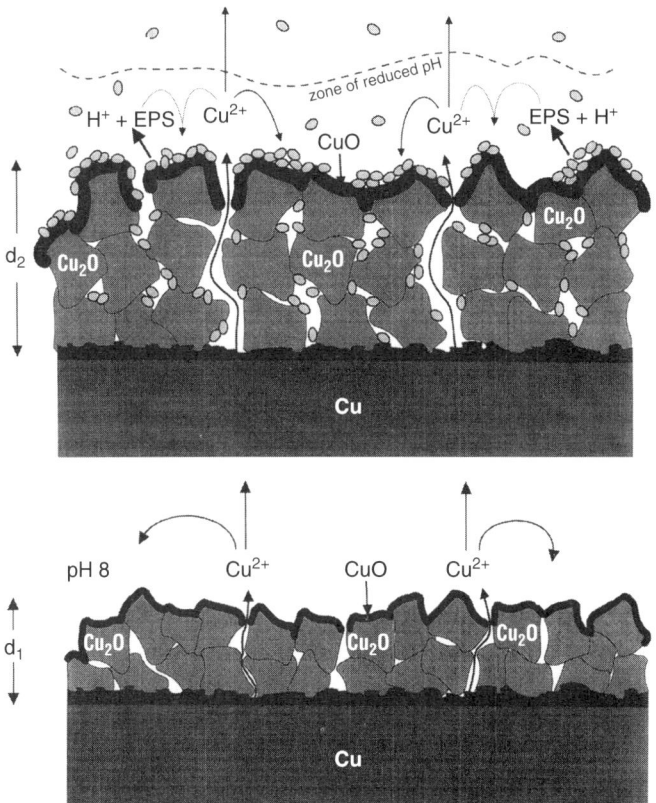

Figure 9.9 Schematic representations of copper experiencing MIC (upper) and copper under abiotic conditions (lower). In instances of MIC, the copper oxide is thicker and more porous than in an equivalent abiotic environment.

In summary, the mechanisms involving surface-associated microbial populations most likely responsible for increased copper by-product release from copper tubing are: (1) the production of acidic metabolites on the metal surface, (2) the binding of copper by microbial cells and their extracellular polymeric products, and (3) alteration of the nature and porosity of the oxide film due to biofilm growth within the film.

9.4 ALUMINUM

Corrosion resistance of aluminum in most environments is due to the thick protective oxide layer that is strongly bonded to the surface. Development of the protective oxide layer upon exposure to water or a humid atmosphere proceeds through three steps at moderate temperatures: beginning with the thickening of an amorphous oxide film, the formation of a Boehmite phase on the amorphous layer, and finally, formation of a Bayerite phase on the Boehmite (Howell, 1995). If this three-layer protective film is breached by mechanical abrasion, a self-healing process occurs under normal conditions to repassivate the surface.

Aluminum alloys are, however, susceptible to different forms of localized corrosion depending on microstructure, heat treatment, grain orientation, and applied stress (Kagwade et al., 1995). One of the most common forms of localized corrosion is pitting. Pitting is the active dissolution of the bare metal sites formed in microfissures and at defect sites in the oxide layer. Al is particularly susceptible to pitting attack in the presence of Cl^- ions, which become incorporated into the oxide film and through chemical reaction, produce a weakness in the protective barrier. The pits can act as fatigue crack initiation sites and harbor acidic conditions that promote further corrosion. The susceptibility of Al and its alloys to localized corrosion makes it particularly vulnerable to microbiological attack. Many of the microbiological processes that influence corrosion of the metals discussed previously also apply to aluminum.

Much of the published work on biocorrosion of aluminum and its alloys has been in association with contamination of jet fuels caused by the fungi *Hormoconis* (*Cladosporium*) *resinae*, *Aspergillus* spp., *Penicillium* spp., and *Fusarium* spp. Surfaces in contact with the aqueous phase of the fuel–water mixtures and sediments are common sites of attack. Exfoliation and granular attack of aluminum alloy occurs under mats of *H. resinae*, leading to skin perforation and fatigue (King and Miller, 1987). The large quantities of organic acid by-products excreted by this fungus selectively dissolve or chelate the copper, zinc, and iron at the grain boundaries of aircraft aluminum alloys, forming pits that do not repassivate under the anaerobic conditions established under the fungal mat. An article by Videla et al. (1993) reviews the topic of Al corrosion in the presence of *H. resinae* and other fungal contaminants of jet fuels.

Fuel rods for some nuclear reactors are clad with Al alloy. During handling of spent fuels, the protective oxide layer of the cladding is occasionally scratched or damaged and then exposed to conditions that prevent repassivation. Upon prolonged storage in cooling water, pitting attack of the cladding has led to the release of fission products into the water causing radionuclide contamination of the storage basin environment (Burke and Howell, 1994). While never implicated in any failures, the potential for MIC is a concern with the arrival of spent fuels from other countries (Santo Domingo et al., 1998).

Controlled laboratory studies have demonstrated that sparsely distributed microbial biofilms formed on galvanically coupled Al coupons exposed to deionized storage basin water led to a significantly higher pit density and maximum pit depth than on coupons exposed to sterilized basin water (Zhang et al., 1999). Although the mechanism was not elucidated, the study showed that MIC of Al occurs in deionized water containing very low concentrations of nutrients sufficient to support the growth of microorganisms. Further studies of biofilm formation on Al and other metals exposed to environments where spent nuclear fuel is stored should yield new insight into the range of extreme conditions under which MIC may occur.

9.5 TECHNIQUES FOR THE STUDY OF MICROBIOLOGICALLY INFLUENCED CORROSION

The forms of corrosion that can be stimulated by the interaction of microorganisms with metals are numerous, ranging from general pitting, crevice attack, and stress corrosion cracking to enhancement of corrosion fatigue, intergranular stress cracking, and hydrogen embrittlement and cracking. Most cases of MIC are associated with some form of localized corrosion. The significant amount of research carried out on MIC has shown this process to be extremely complex. Thus, a variety of techniques are employed to relate the corrosion process to microbial activities at surfaces. Microscopic and culture techniques provide a qualitative or semiquantitative means to establish the presence of microbes in the system experiencing corrosion. However, revealing the presence of microbes, in itself, should not be the sole basis on which the involvement of microorganisms in the corrosion process is made. To verify MIC, one must document a biotic influence that does not occur under strictly abiotic conditions. To establish the mechanism of MIC, specific activities of the microbes at the site where corrosion is occurring should be demonstrated. Microscopic and culture techniques alone rarely provide such evidence.

Chemical spectroscopy at surfaces offers qualitative and semiquantitative information on the nature of the corrosion products that have accumulated on a metal surface. Spatially resolved surface chemistry obtained with spectroscopic instrumentation must then be related to the spatially resolved microbiology at the same location, a procedure that requires image registration and alignment. The latter has only recently been applied to corrosion processes (Geesey et al., 1996; Pendyala et al., 1996; Beech et al., 1998b).

Quantitative measurements of corrosion are commonly obtained with electrochemical methods. Microsensors, which are largely electrochemically based, offer resolution that is needed for the localized corrosion processes influenced by microorganisms. Typically, qualitative and quantitative approaches are used together to implicate microorganisms in a corrosion process.

9.5.1 Qualitative Methods for the Study of Microbiologically Influenced Corrosion

9.5.1.1 Microscopy. Microbial contributions to corrosion have been qualitatively assessed using scanning electron microscopy techniques (SEM) and recently environmental scanning electron microscopy (ESEM) and atomic force microscopy (AFM). Reflected differential interference contrast microscopy (DIC) can also be used where the metal surface provides sufficient reflective properties to resolve attached microorganisms. Microscopy provides information about the morphology of microbial cells and colonies, the distribution of microbial colonies on the surface, presence of EPS, and the nature of corrosion products (crystalline or amorphous). Microscopy also reveals the type of the attack (e.g., pitting or uniform corrosion) by visualizing changes in metal microstructure after removal of the corrosion products (Little et al., 1991b; Wagner and Ray, 1994; Beech et al., 1996b and references therein).

9.5.1.2 Surface Spectroscopy. Surface chemical analysis provides information on the chemical composition of the corrosion products and microbiological deposits, and thus, the opportunity to gain insight to the electrochemical reactions involved in the corrosion process. X-ray diffraction (XRD) and energy dispersive analysis by X-rays (EDS) have been widely used to obtain elemental information on corrosion products on metal surfaces (Marquis, 1988). XRD has limited application to MIC as it does not possess the high spatial resolution to detect localized attack. EDS provides elemental abundance at high spatial resolution over a sampling depth of approximately 1 μm. Depending on the thickness of the corrosion deposit, EDS may provide elemental information on not only corrosion products but also the underlying bulk material. For example, the 2–5 nm-thick oxide film of stainless steel is virtually invisible to EDS.

Auger electron spectroscopy (AES) has been used to investigate biocorrosion of condenser tubes (Chen et al., 1988). The technique provides only elemental abundance in the top few nanometers of the surface at very high spatial resolution. This allows mapping of corrosion products across a metal surface that has experienced localized attack. Surface contamination by organic matter and other material can interfere with detection of the underlying corrosion products when using such a surface-sensitive technique. X-ray photoelectron spectroscopy (XPS) offers information similar to that of AES but has the added capability of resolving the oxidation state of the elements present, facilitating prediction of corrosion product chemistry and, to some extent, chemistry of associated microbial biofilm (Pendyala et al., 1996).

Mössbauer spectroscopy can be applied to detect iron-containing compounds. It has been used to detect "green rust 2" among corrosion products of steel exposed to marine sediments containing SRB (Olowe et al., 1991). Mössbauer spectroscopy was subsequently employed in conjunction with controlled laboratory studies to show that this corrosion product was exclusively associated with SRB induced corrosion (Olowe et al., 1992). Since samples for X-ray diffraction, energy dispersive analysis by X-rays, Auger electron spectroscopy, and X-ray photoelectron spectroscopy must be dehydrated, as they are analyzed in vacuo, some distortion and reorientation of surface-associated material can be anticipated. Care must also be taken to avoid changes in oxidation state of some corrosion products during dehydration.

9.5.2 Quantitative Methods to Evaluate Microbiologically Influence Corrosion

9.5.2.1 Weight Loss. The conventional method for evaluating the corrosion of steel has been to measure the loss of mass of a metal coupon after a period of exposure to a corrosive environment. This measurement can produce useful information on the involvement of microorganisms in corrosion (Gaylard and Johnston, 1986; Dawood and Brozel, 1998). However the method is rather insensitive to corrosion that takes the form of localized attack, because it averages weight loss over the entire coupon surface. It provides no indication of the extent of pitting and whether a through-wall perforation of the metal occurred. Thus, a small weight loss does not indicate the lack of corrosion damage.

9.5.2.2 Electrochemical Measurements. Microorganisms influence corrosion by changing electrochemical conditions at the metal surface, hence electrochemical techniques are often applied in investigations of MIC. Like other approaches to evaluate MIC, electrochemical methods should be combined with other methods to determine the role of microorganisms in the corrosion process. The accumulation of diverse layers of corrosion products, microbial cells, and extracellular polymers may lead to significant changes in the electrochemical behavior of the metal and complicate interpretation of the data. Detailed reviews on the application of various electrochemical approaches to MIC have been published elsewhere (King and Eden, 1989; Mansfield and Little, 1991; Dexter et al., 1991). Thus, only a cursory description of the more commonly used electrochemical methods to evaluate MIC is presented next.

9.5.2.2.1 Potentiodynamic Polarization. This method of corrosion detection is commonly used to assess microbiologically induced pitting and crevice corrosion (Buchanan and Stansbury, 1991). The technique usually involves polarizing the electrode probes with a linear voltage ramp over a wide potential range and monitoring the current

9.5 TECHNIQUES FOR THE STUDY OF MICROBIOLOGICALLY INFLUENCED CORROSION 311

response. The measurements are performed using a high impedance digital voltmeter. The resulting current-potential diagrams (polarization plots) have a variety of shapes, which correspond to different forms of corrosion behavior. A potentiodynamic polarization plot yields information on the ability of a metal to passivate spontaneously in the particular medium, the potential region over which the metal surface remains passive, and the corrosion rate in the active region. The potential at which pitting corrosion initiates, E_{pit}, can also be determined. The use of E_{pit} as an indicator of microbial corrosion of stainless steel in seawater has been described (Dexter et al., 1989). Since potentiodynamic techniques apply unnatural potentials to the system, they may alter the microbiology in ways that may influence the electrochemical phenomena being characterized. Potentiodynamic techniques are rather insensitive to the initial stages of pitting corrosion when microbiological influence may be greatest.

9.5.2.2.2 Electrochemical Impedance Spectroscopy (EIS). Alternating current impedance spectroscopy is finding increasing application in biocorrosion research. Some of the advantages of EIS are the use of very small signals, which do not disturb the electrode properties and associated microbiology (Franklin et al., 1991). EIS also offers the opportunity to characterize the corrosion reaction and measure the corrosion rate in electrolytes with low conductivity. Information obtained by impedance spectroscopy differs from that obtained with more conventional electrochemical techniques such as polarization resistance or corrosion potential measurements. The corrosion process is analyzed at a fixed potential (or current) as a function of frequency of an applied signal. The role of SRB in the localized corrosion of buried pipes has been studied by the electrochemical impedance technique (Dowling et al., 1988). Impedance measurements were used to demonstrate a 20-fold increase in corrosion rates of carbon steel exposed to a three-member bacterial consortium above that of the sterile system (Dowling et al., 1991).

9.5.2.2.3 Electrochemical Noise. Electrochemical noise (ECN) determines the frequency and amplitude of the potential fluctuations as a function of time, and their magnitude is correlated with the magnitude of corrosion rates. ECN offers unique capabilities that should allow quantification of microbial effects on corrosion. The method distinguishes localized and generalized corrosion as well as different types of localized attack (crevice or pitting). The anaerobic corrosion of iron and steel pipelines by SRB has been investigated by ECN in the laboratory and in the field (Iverson et al., 1986; Iverson and Heverly, 1984). Interpretation of ECN spectra can be challenging, however (Mansfeld and Little, 1990). ECN offers promise in identifying MIC and quantifying rates nondestructively in real time (Pope and Dziewulski, 1991).

9.5.2.2.4 Zero-resistance Ammetry (ZRA). This technique is considered to be one of the most suitable methods for evaluating the role of biofilm processes in corrosion (King and Eden, 1989). An electronic ammeter is built into a two-compartment system separated by a semipermeable membrane that isolates abiotic and biotic corrosion processes (Gerchakov et al., 1986). Each compartment contains a working electrode submerged in electrolyte, which is sterile in the abiotic compartment and nonsterile in the biotic compartment. The electrodes are short-circuited together and the current required to maintain electrodes at the same potential is recorded. The current density reflects the microbiological contribution to the total corrosion current. The ZRA method has been used to determine the role of a thermophilic bacterium in corrosion of a failed nickel 201 brazed joint (Little et al., 1986). The technique has also been applied to elucidate the mechanisms of SRB corrosion in austenitic stainless steel (Webster et al., 1991). To date, this technique has only been used to study corrosion under static conditions. Theoretically, it could be operated as a flow-through system to better mimic some field situations, however, this approach would require continuous removal or inactivation of all microorganisms flowing through the abiotic compartment—a difficult challenge for long-term studies.

9.5.2.3 Infrared Spectroscopy.

Attenuated total reflectance Fourier transform infrared spectroscopy (ATR/FT-IR) can be used to quantify the rate of corrosion of a thin metal film deposited on an internal reflectance element exposed to either flowing or stagnant aqueous media. The method is based on the observation that water absorbance in the infrared increases as the thin film decreases in thickness as a result of corrosion. Changes in film thickness corresponding to a few atomic layers can be detected and the measurements can be obtained nondestructively in real time. Quantitative changes in water absorbance are expressed as a corrosion rate. This approach has been used to demonstrate the participation of a microbial biofilm in the localized attack of copper films (Bremer and Geesey, 1991). The work also related the onset of corrosion to the accumulation of polysaccharide during biofilm formation on the copper film. ATR/FT-IR has also been used to demonstrate an influence by the exopolysaccharide of the marine bacterium *Pseudoalteromonas* (*Pseudomonas*) *atlantica* on the corrosion of copper (Jolly et al., 1989).

9.5.2.4 Microsensors.

Microsensors have been used to characterize the chemical gradients within biofilms on corroding metal surfaces. Microsensors were used to show depletion of oxygen within tubercles formed on corroding mild steel surfaces (Lee and deBeer, 1995). The microsensors were also used to show that oxygen was depleted at anodic areas of a corroding mild steel surface containing a 1 mm thick biofilm, but could reach cathodic areas where it can be reduced. This spatially resolved surface chemical approach enabled these investigators to demonstrate the existence of a differential oxygen concentration cell and its role as the driving force for the corrosion reaction.

9.6 SUMMARY AND CONCLUSIONS

Microbial biofilms form on all metal surfaces exposed to environmental conditions conducive for microbial colonization and growth. In most cases, biofilms do not promote corrosion of the underlying metal. In fact, biofilms may protect metal surfaces from corrosion, but this phenomenon needs to be examined more thoroughly before the extent of its importance is appreciated. Only in rare instances does the presence of a biofilm promote or accelerate corrosion of metals. Corrosion under a biological influence is primarily realized as a localized attack, although copper by-product release in certain potable water systems could be considered as a generalized form of corrosion. Localized attack occurs as a result of the activity of physiologically diverse microbial species growing on the metal substratum. These include the consumption of oxygen and production of acids, sulfides, and enzymes that promote the establishment of localized chemical gradients at the metal surface. These chemical gradients lead to the formation of electrochemical cells, which promote anodic and/or cathodic reactions, the ultimate consequence being loss of metal from the discrete locations on the surface. The importance of fluctuating oxic and anoxic conditions in

SRB-influenced corrosion of mild steel, manganese-depositing microorganisms in ennoblement of stainless steels, and consortium activities of mixed population biofilms on the pitting corrosion of copper and other metals are but a few of the examples of the advances that have occurred in recent years in our understanding of MIC. A potentially rich source of future insight to MIC of engineered metals may arise from new discoveries in the biotransformations of natural minerals. Detection of corrosion products through enzymatically catalyzed biological reactions not favored under abiotic conditions offers exciting new ways of documenting MIC.

Detection of biologically formed corrosion products on metal surfaces is now practical with a variety of analytical instrumentation such as XRD and XPS. We are at the threshold of spatially resolving molecular structure or elemental valence state with surface-sensitive techniques such as Raman microscopy, time-of-flight secondary ion mass spectroscopy, and synchroton spectroscopy. Information obtained by these techniques should yield the elusive surface biochemical reactions that give rise to the electrochemistry driving the corrosion reactions. Thus, we are presently at the threshold of distinguishing between MIC and abiotic corrosion, their relative contribution to the overall corrosion reaction rate, and the mechanisms responsible for attack. Only then will an accurate appraisal of the economic consequences of MIC be feasible and the development of strategies for effective, long-term control based on sound scientific knowledge.

REFERENCES

Abdollahi, H., and J. W. T. Wimpenny. 1990. Effects of oxygen on growth of *Desulfovibrio desulfuricans*. *J. Gen. Microbiol.* 136: 1025–1030.

Allred, R. C., J. Sudbury, and D. C. Olson. 1959. Corrosion is controlled by bactericide treatment. *World Oil* 149(6): 111–112.

Angell, P., H. S. Campbell, and A. H. L. Chamberlain. 1990. International Copper Association (ICA), Project No. 405, Interim Report.

Angell, P., J.-S. Luo, and D. C. White. 1995. Microbially sustained pitting corrosion of 304 stainless steel in anaerobic seawater. *Corr. Sci.* 37(7): 1085–1096.

Anon. 1991. Lab Notes: Oysters' protective shell for industrial equipment. *Wall Street Journal*, Apr. 19, p B1.

Anon. 1993a. Copper creating headaches for Florida, Washington utilities. *Waterweek*, March 29, p. 5.

Anon. 1993b. Massachusetts medium systems top EPA's latest copper list. *Waterweek*, June 7, p. 3.

Anon. 1994a. Exposing the East Bay MUD-gate scandal: District's deception put lives at risk. *BIA news* 17: 8–31.

Anon. 1994b. *Investigations on Alleged Copper Contamination in Water*. Final Report No. TPT-0478, Japan Copper Development Association, p. 6.

Anon. 1996. Case Studies in Corrosion Control; Bristol, CT, Copper corrosion/microbially influenced/blue water. National Conference on Integrating Corrosion Control and other Water Quality Goals. Cambridge, MA: New England Water Works Association, May 19–21.

Arens, P., G.-J. Tuschewitzki, M. Wollmann, H. Follner, and H. Jacobi. 1995. Indicators for microbiologically induced corrosion of copper pipes in a cold water plumbing system. *Zbl.Hyg.* 196: 444–454.

Arens, P., G.-J. Tuschewitzki, M. Follner, H. Jacobi, and S. Leuner. 1996. Experiments for stimulating microbiologically induced corrosion of copper pipes in a cold-water plumbing system. *Mat. Corr.* 47: 96–102.

Arnold, R. G., T. DiChristina, and M. Hoffman. 1988. Reductive dissolution of Fe (III) oxides by *Pseudomonas* sp. 200. *Biotechn. Bioeng.* 32: 1081–1096.

Bak, F., and H. Cypionka. 1987. A novel type of energy metabolism involving fermentation of inorganic sulfur compounds. *Nature* 326: 891–892.

Barton, L. L., ed. 1995. *Sulfate-Reducing Bacteria.* New York: Plenum Press.

Beech, I. B., and C. W. S. Cheung. 1995. Interactions of exopolymers produced by suphate-reducing bacteria with metal ions. *Int. Biodet. Biodeg.* 35: 59–72.

Beech, I. B., and C. W. S. Cheung. 1996. The use of biocides to control sulphate-reducing bacteria in biofilms on mild steel surfaces. *Biofouling* 9: 231–249.

Beech, I. B., and C. C. Gaylarde. 1991. Microbial polysaccharides and corrosion. *Int. Biodet.* 27: 95–107.

Beech, I. B., C. C. Gaylard, J. J. Smith, and G. G. Geesey. 1991. Extracellular polysaccharides from *Desulfovibrio desulfuricans* and *Pseudomonas fluorescens* in the presence of mild steel and stainless steels. *Appl. Microbiol. Biotech.* 35: 65–71.

Beech, I. B., C. W. S. Cheung, C. S. P. Chan, M. A. Hill, R. Franco, and A. R. Lino. 1994. Study of parameters implicated in the biodeterioration of mild steel in the presence of different species of sulphate-reducing bacteria. *Int. Biodet. Biodeg.* 34: 289–303.

Beech, I. B., V. Zinkevich, R. Tapper, R. Gubner, and R. Avci. 1996a. The interaction of exopolymers produced by marine sulphate-reducing bacteria with iron. In *Biodeterioration and Biodegradation*, W. Sand, ed., Papers of the 10th International Biodeterioration and Biodegradation Symposium, Hamburg, Dechema Monographs vol. 133, Frankfurt, pp. 333–338.

Beech, I. B., C. W. S. Cheung, D. B. Johnson, and J. R. Smith. 1996b. Comparative studies of bacterial biofilms on steel surfaces using techniques of atomic microscopy and environmental scanning electron microscopy. *Biofouling* 10(1–3): 65–77.

Beech, I. B., V. Zinkevich, R. Tapper, and R. Gubner. 1998a. The direct involvement of extracellular compounds from a marine sulphate-reducing bacterium in deterioration of steel. *Geomicrobiol. J.* 15: 119–132.

Beech, I. B., V. Zinkevich, L. Hanjangsit, and R. Avci. 1998b. Modification of the passive layer of AISI 316 stainless steel in the presence of *Pseudomonas* biofilm. In *Proceedings of the NACE LATINCORR 98*, 3rd Congress of the NACE Latin American Region, National Association of Corrosion Engineers electronic publication, in press.

Bibb, M. 1986. Bacterial corrosion in the South African power industry. In *Biologically Induced Corrosion*, S. C. Dexter, ed. Houston, TX: National Association of Corrosion Engineers, pp. 96–101.

Billman, J. A. 1997. Antibiofoulants: A practical methodology for control of corrosion caused by sulfate-reducing bacteria. *Mat. Perform.* 36(11): 43–48.

Blackwood, D. J., J. L. de Rome, D. Oakley, A. M. Pritchard, and E. R. Walter. 1994. MTS Environmental Degradation Programme: Project 4-Microbially Induced Corrosion. Task 1.2: Critical Review of Microbially Induced Corrosion. AEA-Technology Report AEA-TPD-51, U.K.

Boivin, J., E. J. Laishley, R. D. Bryant, and J. W. Costerton. 1990. The influence of enzyme system on MIC. Paper No. 128. In *Proc. NACE Corrosion '90*. Houston, TX: National Association of Corrosion Engineers.

Booth, G. H. 1964. Sulphur bacteria in relation to corrosion. *J. Appl. Bacteriol.* 27: 174–181.

Booth, G. H., and A. K. Tiller. 1960. Polarization studies of mild steel in cultures of sulphate reducing bacteria. *Trans. Faraday Soc.* 56: 1689–1696.

Booth, G. H., and A. K. Tiller. 1962. Polarization studies of mild steel in cultures of sulphate reducing bacteria, Part 3, halophilic organisms. *Trans. Faraday Soc.* 58: 2510–2516.

Booth, G. H., and A. K. Tiller. 1968. Cathodic characteristic of mild steel in suspensions of sulphate-reducing bacteria. *Corr. Sci.* 8: 583–600.

REFERENCES

Booth, G. H., and F. Wormwell. 1962. Corrosion of mild steel by sulphate reducing bacteria: Effect of different strains of organisms. In *Proceedings of the First International Congress on Metallic Corrosion*, London, pp. 341–344.

Booth, G. H., P. M. Shinn, and D. S. Wakerley. 1965. The influence of various strains of actively growing sulphate reducing bacteria on the anaerobic corrosion of mild steel. *Congr. Met. Corr. Mar.*, Salissures, Cannes, Paris, pp. 542–554.

Booth, G. H., L. Elford, and D. S. Wakerley. 1968. Corrosion of mild steel by sulphate-reducing bacteria: An alternative mechanism. *Br. Corr. J.* 3: 242–245.

Borenstein, S. W. 1991. Microbiologically influenced corrosion of austenitic stainless steel weldments. *Mat. Perform.* 30(1): 51–54.

Borenstein, S. W. 1994. *Microbially Influenced Corrosion Handbook*. Cambridge, England: Woodshead Publishing Limited.

Bremer, P. J., and G. G. Geesey. 1991. Laboratory-based model of microbially induced corrosion of copper. *Appl. Environ. Microbiol.* 57: 1956–1962.

Brennenstuhl, A. M., and P. E. Doherty. 1990. The economic impact of microbiologically influenced corrosion at Ontario Hydro's nuclear plants. In *Microbiologically Influenced Corrosion and Biodeterioration*, N. J. Dowling, M. W. Mittelman, and J. C. Danko, eds. Knoxville, TN: University of Tennessee, pp. 7/5–7/10.

Bryant, R. D., W. Jansen, J. Bovin, E. J. Laishley, and J. W. Costerton. 1991. Effect of hydrogenase and mixed sulfate reducing bacterial populations on the corrosion of steel. *Appl. Environ. Microbiol.* 57: 2804–2809.

Bryant, R. D., F. V. O. Kloeke, and E. J. Laishley. 1993. Regulation of the periplasmic [Fe] hydrogenase by ferrous iron in *Desulfovibrio vulgaris* Hildenborough. *Appl. Environ. Microbiol.* 59(2): 491–495.

Buchanan, R. A., A. L. Kovacs, C. D. Lundin, K. K. Khan, J. C. Danko, P. Angell, and S. C. Dexter. 1997. Microbiologically influenced corrosion of Fe-, Ni-, Cu-, Al-, and Ti-based weldments. *Mat. Perform.* 36(6): 46–55.

Buchanan, R. A., and E. E. Stansbury. 1991. Fundamentals of coupled electrochemical reactions as related to microbially influenced corrosion. In *Microbially Influenced Corrosion and Biodeterioration*, N. J. Dowling, M. W. Mittleman, and J. C. Danko, eds. Knoxville, TN: University of Tennessee, pp. 1/11–1/17.

Burke, S. D., and J. P. Howell. 1994. The impact of prolonged wet storage of DOE reactor irradiated nuclear materials at the Savannah River Site. In *Proc. Topical Mtg. DOE Spent Nuclear Fuel: Challenges and Initiatives*. La Grange Park, IL: The Society, pp. 118–124.

Campbell, H. S. 1950. Pitting corrosion in copper water pipes caused by films of carbonaceous material produced during manufacture. *J. Inst. Metals* 77: 345–356.

Campbell, H. S., A. H. L. Chamberlain, and P. Angell. 1993. An unusual form of microbially influenced corrosion in copper water pipes. In *Corrosion and Related Aspects of Materials for Potable Water Supplies*, P. McIntyre and A. D. Mercer, eds. London: Institute of Materials, pp. 222–231.

Chamberlain, A. H. L., P. Angell, and H. S. Campbell. 1988. Staining procedures for characterising biofilms in corrosion investigations. *Br. Corros. J.* 23: 197–199.

Chatelus, C., P. Carrier, P. Saignes, M. F. Liber, Y. Berlier, P. A. Lespinat, G. Faque, and J. LeGall. 1987. Hydrogenase activity in aged, nonviable *Desulfovibrio vulgaris* cultures and its significance in anaerobic biocorrosion. *Appl. Environ. Microbiol.* 53: 1708–1710.

Chen, J. R., S. D. Chyou, S. I. Lew, C. J. Huang, C. S. Fang, and W. S. Tse. 1988. Investigation of the biological corrosion of condenser tubes by scanning Auger microprobe techniques. *Appl. Surf. Sci.* 33/34: 212–219.

Cohen, A., and J. R. Myers. 1996. Overcoming corrosion concerns in copper tube systems. *Mat. Perform.* 35(9): 53–55.

Cord-Ruwisch, R., and F. Widdel. 1986. Corroding iron as a hydrogen source for sulphate reduction in growing cultures of sulfate-reducing bacteria. *Appl. Microbiol. Biotech.* 25: 169–174.

Costello, J. A. 1974. Cathodic depolarisation by sulfate reducing bacteria. *S. Afr. J. Sci.* 70: 202–204.

Cragnolino, G., and O. H. Tuovinen. 1984. The role of sulfate-reducing and sulfur-oxidising bacteria in the localized corrosion of iron-based alloys, A review. *Int. Biodet.* 20: 9–18.

Craig, B. D. 1991. Discussion of Mackinawite formation during microbial corrosion. *Corrosion* 47(5): 329.

Crolet, J.-L. 1992. From biology and corrosion to biocorrosion. *Oceanologica Acta* 15: 87–94.

Crolet, J.-L. 1993. Mechanism of uniform corrosion under corrosion deposits. *J. Material Sci.* 28: 2589–2606.

Cubicciotti, D., and G. J. Licina. 1989. Electrochemical aspects of MIC. Paper No. 517, *Proc. NACE Corrosion '89*. Houston, TX: National Association of Corrosion Engineers.

Daniels, L., N. Belay, B. S. Rajagopal, and P. J. Weimer. 1987. Bacterial methanogenesis and growth from CO_2 with elemental iron as the sole source of electrons. *Science* 237: 509–511.

Danko, J. C., and C. D. Lundin. 1995. The effect of microstructure on microbially influenced corrosion. In *Proc. Internat. Conf. Microbially Influence Corrosion*, P. Angel, S. W. Borenstein, R. A. Buchanan, S. C. Dexter, N. J. E. Dowling, B. J. Little, C. D. Lundin, M. B. McNeil, D. H. Pope, R. E. Tatnall, D. C. White, and H. G. Ziegenfuss, eds. Houston, TX: NACE International, 14/1–14/12.

Das, C. R., and K. G. Mishra. 1986. Biological corrosion of welded steel due to marine algae. In *Biologically Induced Corrosion*, S. C. Dexter, ed. Houston, TX: National Association of Corrosion Engineers, pp. 114–117.

Daumas, S., Y. Massiani, and J. Crousier. 1988. Microbiological battery induced by sulphate-reducing bacteria. *Corr. Sci.* 28(11): 1041–1050.

Davidson, D., B. Beheshti, and M. W. Mittelman. 1996. Effects of *Arthrobacter* sp., *Acidovorax delafieldii* and *Bacillus megatherium* colonization on copper solvency in a laboratory reactor. *Biofouling* 9: 279–292.

Dawood, Z., and V. S. Brozel. 1998. Corrosion-enhancing potential of *Shewanella putrefaciens* isolated from industrial cooling waters. *J. Appl. Microbiol.* 84: 929–936.

Dexter, S. C. 1993. Role of microfouling organisms in marine corrosion. *Biofouling* 7: 97–127.

Dexter, S. C., and H.-J. Zhang. 1991. Effect of biofilms, sunlight, and salinity on corrosion potential and corrosion initiation. In *Microbially Influenced Corrosion and Biodeterioration*, N. J. Dowling, M. W. Mittelman, and J. C. Danko, eds. Knoxville, TN: University of Tennessee, pp. 8/1–8/3.

Dexter, S. C., O. W. Siebert, D. J. Duquette, and H. A. Videla. 1989. Use and limitation of electrochemical techniques for investigating microbial corrosion. Paper No. 616, *Proc. NACE Corrosion '89*. Houston, TX: National Association of Corrosion Engineers.

Dexter, S. C., D J. Duquette, O. W. Siebert, and H. A. Videla. 1991. Use and limitations of electrochemical techniques for investigating microbiological corrosion. *Corrosion* 47(4): 308–318.

Dickinson, W., and Z. Lewandowski. 1996. Manganese biofouling and corrosion behaviour of stainless steel. *Biofouling* 10: 79–93.

Dickinson, W., F. J. Caccavo, and Z. Lewandowski. 1996. The ennoblement of stainless steel by manganic oxide biofouling. *Corr. Sci.* 38: 1407–1422.

Dickinson, W., F. J. Caccavo, B. Olesen, and Z. Lewandowski. 1997. Ennoblement of stainless steel by the manganese-depositing bacterium *Leptothrix discophora*. *Appl. Environ. Microbiol.* 63: 2502–2506.

Dilling, W., and H. Cypionka. 1990. Aerobic respiration in sulfate-reducing bacteria. *FEMS Microbiol. Lett.* 71: 123–128.

Dowling, N. J., J. Guezennec, M. L. Lemoine, A. Tunlid, and D. C. White. 1988. Analysis of carbon steels affected by bacteria using electrochemical impedance and DC techniques. *Corrosion* 44: 869–874.

Dowling, N. J. E., M. W. Mittelman, and D. C. White. 1991. The role of consortia in microbially influenced corrosion. In *Mixed Cultures in Biotechnology*, J. G. Zeikus, ed. New York: McGraw-Hill, pp. 341–372.

Dowling, N. J. E., S. A. Brooks, T. J. Phelps, and D. C. White. 1992. Effects of selection and fate of substrates supplied to anaerobic bacteria involved in the corrosion of pipe-line steel. *J. Ind. Microbiol.* 10: 207–215.

Daumas, S., M. Magot, and J. L. Crolet. 1993. Measurement of the net production of acidity by a sulphate-reducing bacterium: Experimental checking of theoretical models of microbially influenced corrosion. *Res. Microbiol.* 144: 327–332.

Dupont, I., D. Ferron, and G. Novel. 1998. Effect of glucose oxidase activity on corrosion potential of stainless steels in seawater. *Int. Biodet. Biodeg.* 41: 13–18.

Dutkiewicz, C. M. E. 1995. Cuprosolvency in Adelaide drinking water: Assessment of microbial involvement. Thesis Degree of Bachelor of Science (Honours), Environmental Health, School of Medicine, Flinders University of South Australia, Adelaide.

Edyvean, R. G. J., J. Benson, C. J. Thomas, I. B. Beech, and H. Videla. 1998. Biological influences on hydrogen effects in steel in seawater. *Mat. Perform.* 37(4): 40–44.

Fischer, W. R., I. Hänβel, and H. H. Paradies. 1988. First results of microbial induced corrosion on copper pipes. In *Microbial Corrosion-1*, C. A. C. Sequeira and A. K. Tiller, eds. London: Elsevier Applied Sciences, pp. 300–327.

Fischer, W., H. H. Paradies, I. Hanβel, and D. Wagner. 1990. Copper deterioration in a water distribution system of a county hospital in Germany caused by microbial induced corrosion. In *Microbially Influenced Corrosion and Biodeterioration*, N. J. Dowling, M. W. Mittelman, and J. C. Danko, eds. Knoxville, TN: University of Tennessee Press, 8/47–8/48.

Fischer, W. R., H. H. Paradies, D. Wagner, and I. Hänβel. 1992. Copper deterioration in a water distribution system of a county hospital in Germany caused by microbially induced corrosion. 1. Description of the problem. *Werkst. Korros.* 43: 56–62.

Fischer, W. R., D. Wagner, and H. H. Paradies. 1994. Microbiologically influenced corrosion testing. In J. R. Kearns and B. Little, eds. ASTM STP 1232. Philadelphia: American Society for Testing and Materials, pp. 274–282.

Flemming, H.-C. 1996. Biofouling and microbiologically influenced corrosion (MIC)—An economical and technical overview. In *Microbial Deterioration of Materials*, E. Heitz, W. Sand, and H.-C. Flemming, eds. Heidelberg: Springer, pp. 5–14.

Ford, T., and R. Mitchell. 1991. Microbiological involvement in environmental cracking of high strength steels. In *Microbially Influenced Corrosion and Biodeterioration*, N. J. Dowling, M. W. Mittelman, and J. C. Danko, eds. Knoxville, TN: University of Tennessee, pp. 3/94–3/98.

Ford, T. E., J. S. Maki, and R. Mitchell. 1987. The role of metal-binding bacterial exopolymers in corrosion processes. Paper No. 380, *Proc. NACE Corrosion '87*. Houston, TX: National Association of Corrosion Engineers.

Ford, T., J. P. Black, and R. Mitchell. 1990. Relationship between bacterial exopolymers and corroding metal surfaces. Paper No. 110, *Proc. NACE Corrosion '90*. Houston, TX: National Association of Corrosion Engineers.

Franklin, M. J., D. E. Nivens, J. B. Guckert, and D. C. White. 1991. Effect of electrochemical impedance spectroscopy on microbial biofilm cell numbers, viability, and activity. *Corrosion* 47(7): 519–522.

Gaines, R. H. 1910. Bacterial activity as a corrosive influence in the soil. *J. Eng. Ind. Chem.* 2: 128–130.

Gaylarde, C. C. 1992. Sulphate-reducing bacteria which do not induce accelerated corrosion. *Int. Biodet. Biodeg.* 30(4): 331–338.

Gaylarde, C. C., and J. M. Johnston. 1982. The effect of *Vibrio anguillarum* on the anaerobic corrosion of mild steel by *Desulfovibrio vulgaris*. *Int. Biodeterior. Bull.* 18(4): 111–116.

Gaylarde, C. C., and J. M. Johnston. 1986. Anaerobic metal corrosion in cultures of bacteria from estuarine sediments. In *Biologically Induced Corrosion*, S. C. Dexter, ed. NACE-8, Houston, TX: National Association of Corrosion Engineers, pp. 137–143.

Geesey, G. G., and M. W. Mittelman. 1985. The role of high-affinity, metal-binding exopolymers of adherent bacteria in microbial-enhanced corrosion. Paper No. 85, *Proc. NACE Corrosion '85*. Houston, TX: National Association of Corrosion Engineers.

Geesey, G. G., M. W. Mittelman, T. Iwaoka, and P. R. Griffiths. 1986. Role of bacterial exopolymers in the deterioration of metallic copper surfaces. *Mat. Perform.* 25(2): 37–40.

Geesey, G. G., P. J. Bremer, J. J. Smith, M. Muegge, and L. K. Jang. 1992. Two-phase model for describing the interactions between copper ions and exopolymers from *Alteromonas atlantica*. *Can. J. Microbiol.* 38: 785–793.

Geesey, G. G., P. J. Bremer, W. R. Fischer, D. Wagner, C. W. Keevil, J. Walker, A. H. L. Chamberlain, and P. Angell. 1994. Unusual types of pitting corrosion of copper tubes in potable water systems. In *Biofouling and Biocorrosion in Industrial Water Systems*, G. G. Geesey, Z. Lewandowski, and H.-C. Flemming, eds. Boca Raton, FL: CRC Press, pp. 243–263.

Geesey, G. G., R. J. Gillis, R. Avci, D. Daly, M. Hamilton, P. Shope, and G. Harkin. 1996. Influence of surface features on bacterial colonization and subsequent substratum chemical changes on 316L stainless steel. *Corr. Sci.* 38: 73–95.

Gerchakov, S. M., B. J. Little, and P. Wagner. 1986. Probing microbiologically induced corrosion. *Corrosion* 42(11): 689–692.

Ghassem, H., and N. Adibi. 1995. Bacterial corrosion of reformer heater tubes. *Mat. Perform.* 34(3): 47–48.

Gounot, A. M. 1994. Microbial oxidation and reduction of manganese: Consquences in groundwater and applications. *FEMS Microbiology Reviews* 14: 339–350.

Hamilton, W. A. 1985. Sulphate reducing bacteria and anaerobic corrosion. *Ann. Rev. Microbiol.* 39: 195–217.

Hamilton, W. A. 1994. Metabolic interaction and environmental microniches: Implications for the modeling of biofilm process. In *Biofouling and Biocorrosion in Industrial Water Systems*, 2nd ed., G. G. Geesey, Z. Lewandowski, and H.-C. Flemming, eds. Boca Raton, FL: CRC Press, pp. 27–36.

Hamilton, W. A. 1998. Sulphate-reducing bacteria: Physiology determines their environmental impact. *Geomicrobiol. J.* 15: 19–28.

Hardy, J. A. 1983. Utilization of cathodic hydrogen by sulphate-reducing bacteria. *Br. Corr. J.* 18(4): 190–193.

Hardy, J. A., and J. L. Brown. 1984. The corrosion of mild steel by biogenic sulphide films exposed to air. *Corrosion* 40: 650–654.

Hardy, J. A., and W. A. Hamilton. 1981. The oxygen tolerance of sulfate-reducing bacteria isolated from the North Sea waters. *Curr. Microbiol.* 6: 259–262.

Howell, J. P. 1995. *Corrosion of Aluminum Alloys in Wet Spent Fuel Storage (U)*, Report *WSRC-TR-95-0343* (U). Aiken, SC: Savannah River Technology Center, 49p.

Ibars, J. R., D. A. Moreno, and C. Ranninger. 1992. MIC of stainless steels: A technical review of the influence of microstructure. *Internat. Biodeter. Biodegrad.* 29: 343–355.

Iverson, W. P. 1966. Direct evidence for the cathodic depolarization theory of bacterial corrosion. *Science* 151: 986–988.

Iverson, W. P. 1987. Microbial corrosion of metals. *Adv. Appl. Microbiol.* 32: 1–13.

Iverson, W. P. 1998. Possible source of a phosphorus compound produced by sulfate reducing bacteria that cause anaerobic corrosion of iron. *Mat. Perform.* 37(5): 46–49.

Iverson, W. P., and L. F. Heverly. 1984. Electrochemical noise as an indicator of anaerobic corrosion. *Proceedings of Symposium on Nondestructive Testing and Electrochemical Methods of Monitoring Corrosion in Industrial Plants.* Philadelphia: ASTM.

Iverson, I. P., and G. J. Olson. 1983. Anaerobic corrosion by sulfate-reducing bacteria due to highly reactive volatile phosphorus compound. In *Microbial Corrosion.* London: Metals Society, pp. 46–53.

Iverson, W. P., O. J. Olson, and L. F. Heverly. 1986. The role of phosphorous and hydrogen sulphide in the anaerobic corrosion of iron and the possible detection of this corrosion by an electrochemical noise technique. In *Biologically Induced Corrosion,* S. C. Dexter, ed. Houston, TX: National Association of Corrosion Engineers, pp. 154–161.

Jack, T. R., M. J. Wilmott, and R. L. Sutherby. 1995. Indicator minerals formed during external corrosion of line pipe. *Mat. Perform.* 34(11): 19–22.

Jang, L. K., E. Quintero, G. Gordon, M. Rohricht, and G. G. Geesey. 1989. The osmotic coefficients of the sodium form of some polymers of biological origin. *Biopolymers* 28: 1485–1489.

Jolley, J. G., G. G. Geesey, M. R. Hankins, R. B. Wright, and P. L. Wichlacz. 1989. In situ, real time FR-IR/CIR/ATR study of the biocorrosion of copper by gum arabic, alginic acid, bacterial culture supernatant and *Pseudomonas atlantica* exopolymer. *Appl. Spectrosc.* 43: 1062–1067.

Jones, R. L. 1996. Some critical corrosion issues and mitigation strategies affecting light water reactors. *Mat. Perform.* 35(7): 63–67.

Kagwade, S. V., C. R. Clayton, M. L. Du, and F. P. Chiang. 1995. A surface and optical study of the influence of stress on the corrosion of aluminum alloys. Paper No. 533, *Proc. NACE Corrosion '95.* Houston, TX: National Association of Corrosion Engineers.

Kearns, J. R., and S. W. Borenstein. 1991. Microbiologically influenced corrosion testing of welded stainless alloys for nuclear power plant service water. Paper No. 279, *Proc. NACE Corrosion '91.* Houston, TX: National Association of Corrosion Engineers.

Keevil, C. W., J. T. Walker, J. McEvoy, and J. S. Colbourne. 1989. Detection of biofilms associated with pitting corrosion of copper pipework in Scottish hospitals. In *Biocorrosion.* Proceedings of a Joint Meeting Between the Biodeterioration Society and the French Microbial Corrosion Group, C. C. Gaylarde and L. H. G. Morton, eds., Paris, France, 13–14 September, 1988. Biodeterioration Society Occasional Publication No. 5. Lancashire, UK: Biodeterioration Society, pp. 99–117.

King, R. A., and R. D. Eden. 1989. Evaluation of biofilms by advanced electrochemical monitoring. In *Biocorrosion,* C. C. Gaylarde and L. H. G. Morton, eds. Biodeterioration Society Occasional Publication No. 5, pp. 134–149.

King, R. A., and J. D. A. Miller. 1971. Corrosion by sulphate reducing bacteria. *Nature* 233: 491–492.

King, R. A., and J. D. A. Miller. 1987. An outline of 150 man-years of microbial corrosion studies at UMIST. Paper No. 368, *Proc. NACE Corrosion '87.* Houston, TX: National Association of Corrosion Engineers.

King, R. A., and D. S. Wakerley. 1973. Corrosion of mild steel by ferrous sulphide. *Br. Corr. J.* 8: 41–45.

King, R. A., J. D. A. Miller, and J. S. Smith. 1973. Corrosion of mild steel by iron sulphides. *Br. Corr. J.* 8: 137–141.

King, R. A., C. K. Dittmer, and J. D. A. Miller. 1976. Effect of ferrous iron concentration on the corrosion of iron in semicontinuous cultures of sulphate-reducing bacteria. *Br. Corr. J.* 11: 105–107.

Kobrin, G. 1976. Corrosion by microbiological organisms in natural waters. *Mat. Perform.* 15: 7: 38–43.

LaQue Report. 1994. The interrelation of corrosion and fouling of metals in seawater. 15p.

Lee, W., and W. G. Characklis. 1993. Corrosion of mild steel under anaerobic biofilm. *Corrosion* 49: 186–198.

Lee, W., and D. deBeer. 1995. Oxygen and pH microprofiles above corroding mild steel covered with a biofilm. *Biofouling* 8: 273–280.

Lee, W.-C., Z. Lewandowski, S. Okabe, W. G. Characklis, and R. Avci. 1993a. Corrosion of mild steel underneath aerobic biofilms containing sulfate-reducing bacteria. Part I: At low dissolved oxygen concentration. *Biofouling* 7: 197–216.

Lee, W.-C., Z. Lewandowski, M. Morrison, W. G. Characklis, R. Avci, and P. H. Nielsen. 1993b. Corrosion of mild steel underneath aerobic biofilms containing sulfate-reducing bacteria. Part II: At high dissolved oxygen concentration. *Biofouling* 7: 217–239.

Lee, W., Z. Lewandowski, P. H. Nielsen, and W. A. Hamilton. 1995. Role of sulfate-reducing bacteria in corrosion of mild steel: A review. *Biofouling* 8: 165–194.

Lewandowski, Z., W. Dickinson, and W. Lee. 1997. Electrochemical interactions of biofilms with metal surfaces. *Wat. Sci. Tech.* 36: 295–302.

Licina, G. J. 1989. An overview of microbiologically influenced corrosion in nuclear power plant systems. *Mat. Perform.* 28(10): 55–60.

Little, B., and P. Wagner. 1994. Indicators for sulfate-reducing bacterium microbiologically influenced corrosion. In *Biofouling and Biocorrosion in Industrial Water Systems*, G. G. Geesey, Z. Lewandowski, and H.-C. Flemming, eds. Boca Raton, FL: CRC Press, pp. 213–230.

Little, B., and P. Wagner. 1997. Myths related to microbiologically influenced corrosion. *Mat. Perform.* 36(6): 40–44.

Little, B., P. Wagner, S. M. Gerchakov, M. Walch, and R. Mitchell. 1986. The involvement of a thermophilic bacterium in corrosion processes. *Corrosion* 42(9): 533–536.

Little, B., P. Wagner, and D. Duquette. 1988a. Microbiologically induced increase in corrosion current density of stainless steel under cathodic protection. *Corrosion* 44(5): 270–274.

Little, B., P. Wagner, and J. Jacobus. 1988b. The impact of sulfate-reducing bacteria on welded copper-nickel seawater piping systems. *Mat. Perform.* 27(8): 57–61.

Little, B. J., P. A. Wagner, W. G. Characklis, and W. Lee. 1990a. Microbial corrosion. In *Biofilms*, W. G. Characklis and K. C. Marshall, eds. New York: Wiley, pp. 635–670.

Little, B., P. Wagner, R. Ray, and M. McNeil. 1990b. Microbiologically influenced corrosion in copper and nickel seawater piping systems. *Mar. Technol. Soc. J.* 24(3): 10–17.

Little, B. J., P. Wagner, and F. Mansfeld. 1991a. Microbiologically influenced corrosion of metals and alloys. *Intl. Mat. Rev.* 36: 253–272.

Little, B. J., P. A. Wagner, R. I. Ray, R. Pope, and R. Sheetz. 1991b. Biofilms: An ESEM evaluation of artifacts introduced during SEM preparation. *J. Ind. Microbiol.* 8: 231–222.

Little, B., P. Wagner, and F. Mansfeld. 1992. An overview of microbiologically influenced corrosion. *Electrochim. Acta* 37: 2185–2194.

Little, B., R. Ray, K. Hart, and P. Wagner. 1995. Fungal-induced corrosion of wire rope. *Mat. Perform.* 34(10): 55–58.

Little, B. J., P. Wagner, and F. Mansfeld. 1997a. Microbiological testing. In *Microbiologically Influenced Corrosion. Corrosion Testing Made Easy*, vol. 5. Houston, TX: National Association of Corrosion Engineers International, pp. 29–52.

Little, B. J., P. Wagner, K. Hart, R. Ray, D. Lavoie, K. Nealson, and C. Aguilar. 1997b. The role of metal-reducing bacteria in microbiologically influenced corrosion. Paper No. 215, *Proc. NACE Corrosion '97*. Houston, TX: National Association of Corrosion Engineers International.

Little, B. J., P. A. Wagner, and Z. Lewandowski. 1997c. Spatial relationships between bacteria and mineral surfaces. In *Reviews in Mineralogy*, vol. 35, *Geomicrobiology: Interactions between Microbes and Minerals*, J. F. Banfield and K. H. Nealson, eds. Washington, DC: The Mineralogical Society of America, pp. 123–159.

Lovley, D. R., and E. J. P. Philips. 1987. Competitive mechanisms for inhibition of sulfate reduction and methane production in the zone of ferric iron reduction in sediments. *Appl. Environ. Microbiol.* 53: 2636–2643.

Lovley, D. R., and E. J. P. Philips. 1994. Novel processes for anaerobic sulfate production from elemental sulfur by sulfate-reducing bacteria. *Appl. Environ. Microbiol.* 60: 2394–2399.

Lucey, V. F. 1967. Mechanism of pitting corrosion of copper in supply waters. *Br. Corr. J.* 2: 175–185.

Lutey, R. W. 1992. Identification and detection of microbiologically influenced corrosion. In *Proc. NSF-CONICET Workshop Biocorrosion and Biofouling, Metal/Microbe Interactions*, H. A. Videla, Z. Lewandowski, and R. Lutey, eds. November 1992, Mar del Plata, Argentina. Memphis, TN: Buckman Laboratories International, Inc., pp. 146–158.

Mansfeld, F., and B. Little. 1990. The application of electrochemical techniques for the study of MIC—A critical review. Paper No. 108, *Proc. NACE Corrosion '90*. Houston, TX: National Association of Corrosion Engineers International.

Mansfield, F., and B. J. Little. 1991. A critical review of the application of electrochemical techniques to the study of MIC. In *Microbially Influenced Corrosion and Biodeterioration*, N. J. Dowling, M. W. Mittleman, and J. C. Danko, eds. Knoxville, TN: University of Tennessee, pp. 5/33–5/40.

Mansfeld, F., R. Tsai, H. Shih, B. Little, R. Ray, and P. Wagner. 1992. An electrochemical and surface analytical study of stainless steels and titanium exposed to natural seawater. *Corr. Sci.* 33(3): 445–456.

Marquis, F. D. S. 1988. Strategy of macro and microanalysis in microbial corrosion. In *Microbial Corrosion-1*, C. A. C. Sequeira and A. K. Tiller, eds. London: Elsevier Applied Science, pp. 125–151.

Mattsson, E. 1980. Corrosion of copper and brass: Practical experience related to data. *Br. Corr. J.* 15: 6–13.

Mattsson, E., and A. M. Fredriksson. 1968. Pitting corrosion in copper tubes—Cause of corrosion and counter measures. *Br. Corros. J.* 3: 246–257.

McEvoy, J., and J. S. Colbourne. 1988. Glasgow Hospital Survey Pitting Corrosion of Copper Tube, Report to the International Copper Research Association, New York.

McKenzie, J., and W. A. Hamilton. 1992. The assay of in-situ activities of sulfate reducing bacteria in a laboratory marine corrosion model. *Int. Biodeterior. Biodegrad.* 29: 285–297.

McNeil, M. B., and B. J. Little. 1990. Technical note: Mackinawite formation during microbial corrosion. *Corrosion* 46(7): 599–600.

McNeil, M. B., J. M. Jones, and B. J. Little. 1991a. Mineralogical fingerprinting for corrosion processes induced by sulfate reducing bacteria. Paper No. 580, *Proc. NACE Corrosion '91*. Houston, TX: National Association of Corrosion Engineers.

McNeil, M. B., J. M. Jones, and B. J. Little. 1991b. Production of sulfide minerals by sulfate-reducing bacteria during microbiologically influenced corrosion of copper. *Corrosion* 47(9): 674–677.

Miller, J. D. A., and A. K. Tiller. 1971. Microbial corrosion of buried and immersed metal. In *Microbial Aspects of Metallurgy*, J. D. A. Miller, ed. Aylesbury: Medical and Technical Publication Co., pp. 63–105.

Moosavi, A. N., R. S. Pirie, and W. A. Hamilton. 1990. Effect of sulphate-reducing bacteria activity on performance of sacrificed anodes. In *Proc. Int. Symp. Microbiologically Influenced Corrosion*, N. Dowling, M. M. Mittelman, and J. C. Danko, eds. Knoxville, TN: University of Tennessee, pp. 3/13–3/28.

Moreno, D. A., J. R. Ibars, C. Ranninger, and H. A. Videla. 1992. Use of potentiodynamic polarization to assess pitting corrosion of stainless steels by sulfate-reducing bacteria. *Corrosion* 48: 226–229.

Moss, G., and E. C. Potter. 1984. CSIRO Investigations into the Interactions between Cold Potable Water and Copper Tubes, Concluding Report No. 1534R. Sidney, Australia: CSIRO, 72p.

Myers, C., and K. H. Nealson. 1988. Bacterial manganese reduction and growth with manganese oxide as the sole electron acceptor. *Science* 240: 1319–1321.

Myers, J. R., and A. Cohen. 1995. Pitting corrosion of copper in cold potable water systems. *Mat. Perform.* 34(10): 60–62.

NACE Basic Corrosion Course. 1970. National Association of Corrosion Engineers, Houston, TX.

Newman, R. C., and K. Sieradzki. 1994. Metallic corrosion. *Science* 263: 1708–1709.

Newman, R. C., W. P. Wong, H. Ezuber, and A. Garner. 1989. Pitting of stainless steels by thiosulfate ions. *Corrosion* 45(4): 282–287.

Nielsen, P. H., W.-C. Lee, Z. Lewandowski, M. Morrison, and W. G. Characklis. 1993. Corrosion of mild steel in an alternating oxic and anoxic biofilm system. *Biofouling* 7: 267–284.

Nuttall, J. L. 1993. Copper. In *Corrosion and Related Aspects of Materials for Potable Water Systems*, P. McIntyre and A. D. Mercer, eds. London: Institute of Materials, pp. 65–83.

Obuekwe, C. O., D. W. S. Westlake, F. D. Cook, and J. W. Costerton. 1981. Surface changes in mild steel coupons from the action of corrosion causing bacteria. *Appl. Environ. Microbiol.* 41: 766–774.

Ochynski, F. W., and J. R. Postgate. 1963. Some biological differences between fresh water and salt water strains of sulphate-reducing bacteria. In *Marine Microbiology*, C. H. Openheimer, ed. Springfield, IL: C. C. Thomas Publishers, pp. 426–441.

Odom, J. M. 1990. Industrial and environmental concerns with sulphate-reducing bacteria. *ASM News* 56: 473–476.

Odom, J. M., and R. Singleton, eds. 1992. *The Sulphate-Reducing Bacteria: Contemporary Perspectives*. New York: Springer-Verlag, pp. 1–248.

Olowe, A., J.-M. R. Génin, and J. Guezennec. 1991. Mössbauer effect study of microbially induced corrosion of steel by sulphate reducing bacteria in marine sediments: Role of green rust 2. In *Microbially Influenced Corrosion and Biodeterioration*, N. J. Dowling, M. W. Mittleman, and J. C. Danko, eds. Knoxville, TN: University of Tennessee, pp. 5/65–6/72.

Olowe, A., N. D. Benbouzid-Rollet, J.-M. R. Génin, D. Prieur, M. Confente, and B. Resiak. 1992. La présance simultanée de rouille verte 2 et de bactéries sulfato-réductrices en corrosion perforante de palplanches en zone portuaire. *C.R. Académie des Sciences Paris*, 314, Série II, pp. 1157–1163.

Page, G. G. 1972. Copper corrosion: Discussion on "blue water." *Mat. Perform.* 11(2): 53.

Page, G. G. 1973. Contamination of drinking water by corrosion of copper pipes 1973. *N.Z. J. Sci.* 16: 349–388.

Pankhania, I. P., A. N. Mossavi, and W. A. Hamilton. 1986. Utilization of cathodic hydrogen by *Desulfovibrio vulgaris* (Hildenborough). *J. Gen. Microbiol.* 132: 3357–3365.

Paradies, H. H., W. R. Fischer, I. Haenssel, and D. Wagner. 1992. Characterisation of metal biofilm interactions by extended absorption fine structure spectroscopy. In *Microbial Corrosion-2*, C. A. C. Sequeira and A. K. Tiller, eds. European Federation of Corrosion, Publication No. 8. London: The Institute of Materials, pp. 168–188.

Pendyala, J., R. Avci, G. G. Geesey, P. Stoodley, M. Hamilton, and G. Harkin. 1996. Chemical effects of biofilm colonization on 304 stainless steel. *J. Vac. Sci. Technol.* A 14(3): 1755–1760.

Pollard, A. M., R. G. Thomas, and P. A. Williams. 1989. Synthesis and stabilities of the basic copper (II) chlorides atacamite, paratacamite and botallackite. *Mineralogical Mag.* 53: 557–563.

Pope, D. H. 1991. Mechanisms of microbiologically influenced corrosion of carbon steel: Influence of metallurgical, chemical and biological factors. In *Biodeterioration and Biodegradation 8*, H. W. Rossmore, ed. England: Elsevier Applied Science, pp. 590–592.

Pope, D. H., and D. Dziewulski. 1991. Case histories of microbiologically influenced corrosion in the gas industry: Detection, system analyses and targeted treatment. In *Microbially Influenced Corrosion and Biodeterioration*, N. J. Dowling, M. W. Mittleman, and J. C. Danko, eds. Knoxville, TN: University of Tennessee, pp. 4/11–4/17.

Pope, D. H., D. J. Duquette, A. H. Johannes, and P. C. Wayner. 1984. Microbially influenced corrosion of industrial alloys. *Mat. Perform.* 23(4): 14–18.

Pope, D. H., T. P. Zintel, H. Aldrich, and D. Duquette. 1990. Laboratory and field tests of efficacy of biocides and corrosion inhibitors in the control of microbiologically influenced corrosion. Paper No. 34, *Proc. NACE Corrosion '90*. Houston, TX: National Association of Corrosion Engineers.

Postgate, J. R. 1984. *The Sulphate Reducing Bacteria*, 2nd ed. England: Cambridge University Press.

Potter, E. C. 1970. An investigation of the green-water problem in Auckland, New Zealand and a discussion of possible remedies. *N. Z. Plumbers J.* 22: 35–40.

Roberge, P. R., M. A. A. Tullmin, L. Grenier, and C. Ringas. 1996. Corrosion surveillance for aircraft. *Mat. Perform.* 35(12): 50–54.

Roden, E. E., and D. R. Lovley. 1993. Dissimilatory Fe(III) reduction by marine microorganism *Desulfuromonas acetooxidants*. *Appl. Environ. Microbiol.* 59: 734–742.

Roden, E. E., and J. M. Zachara. 1996. Microbial reduction of crystalline Fe(III) oxides: Influence of oxide surface area and potential for cell growth. *Environ. Sci. Tech.* 30: 1618–1628.

Santo Domingo, J. W., C. J. Berry, M. Summer, and C. B. Fliermans. 1998. Microbiology of spect nuclear fuel storage basins. *Curr. Microbiol.* 37: 387–394.

Schiffrin, D. J., and S. R. de Sanchez. 1985. The effect of pollutants and bacterial microfouling on the corrosion of copper base alloys in seawater. *Corrosion* 41(1): 31–38.

Schmitt, G. 1991. Effect of elemental sulfur on corrosion in sour gas systems. *Corrosion* 47: 285–308.

Scott, P. J. B., J. Goldie, and M. Davies. 1991. Ranking alloys for susceptibility to MIC-A preliminary report on high-Mo alloys. *Mat. Perform.* 30(1): 55–57.

Scotto, V. 1993. *Microbial and Biochemical Factors Affecting the Corrosion Behaviour of Stainless Steels in Seawater*, A working party report on marine corrosion of stainless steels: Chlorination and microbial effects, No. 10. London: Institute of Materials, pp. 21–33.

Scotto, V., R. Di Cintio, and G. Marcenaro. 1985. The influence of marine aerobic microbial film on stainless steel corrosion behavior. *Corr. Sci.* 25(3): 185–194.

Scotto, V., G. Alabiso, and G. Marcenaro. 1986. An example of microbiologically influenced corrosion. The behavior of stainless steel in natural seawater. *Bioelectrochem. Bioeng.* 16: 347–355.

Scotto, V., A. Alabiso, M. Beggiato, G. Marcenaro, and J. Guezennec. 1990. Possible chemical and microbiological factors influencing stainless steel MIC in natural seawater. In *Proceedings of the 5th European Congress on Biotechnology*, C. Christiansen, ed. Copenhagen, Denmark, vol. 2, pp. 866–871.

Scully, J. C. 1990. *Fundamentals of Corrosion*, 3rd ed. Oxford: Pergamon Press.

Shalaby, H. M., F. M. Al-Kharafi, and V. K. Gouda. 1989. A morphological study of pitting corrosion in soft tap water. *Corrosion* 45(7): 536–547.

Shreir, L. L. 1963. The microbiology of corrosion. In *Corrosion*, Vol. 1. New York: J. Wiley, pp. 2.52–2.64.

Siedlarek, H., D. Wagner, W. R. Fischer, and H. H. Paradies. 1994. Microbiologically influenced corrosion of copper: The ionic transport properties of biopolymers. *Corr. Sci.* 36: 1751–1763.

Smith, J. S., and J. D. A. Miller. 1975. Nature of sulphides and their corrosive effect on ferrous metals: A review. *Br. Corros. J.* 10(3): 136–143.

Soracco, R. J., L. R. Berger, J. A. Berger, L. A. Mayack, D. H. Pope, and E. W. Wilde. 1984. Microbiologically mediated reduction in the pitting of mild steel overlaid with plywood. Paper No. 98, *Proc. NACE Corrosion '84*. Houston, TX: National Association of Corrosion Engineers.

Soracco, R. J., D. H. Pope, J. M. Eggers, and T. N. Effinger. 1988. Microbiologically influenced corrosion investigations in electric power generating stations. Paper No. 83, *Proc. NACE Corrosion '88*. Houston, TX: National Association of Corrosion Engineers.

Stoecker II, J. G. 1995. Microbiological and electrochemical types of corrosion: Back to basics. *Mat. Perform.* 34(5): 49–52.

Sutherland, I. W. 1985. Biosynthesis and composition of gram-negative bacterial extracellular and wall polysaccharides. *Ann. Rev. Microbiol.* 39: 243–262.

Tatnall, R. 1981. Fundamentals of bacterial induced corrosion. *Mat. Perform.* 20(9): 32–38.

U.S. Department of Defense. 1998. Program Solicitation 98.2, FY 1998 Small Business Innovation Research (SBIR) Program, August 1998, Army, p. 137.

USEPA. 1993. Maximum contaminant level goals and national primary drinking water regulations for lead and copper; final rule. 40 CFR Parts 141 and 142. Federal Register 56, 110.

Videla, H. A. 1996. *Manual of Biocorrosion*. Boca Raton, FL: Lewis Publishers, 273p.

Videla, H. A., and W. G. Characklis. 1992. Biofouling and microbially influenced corrosion. *International Biodegrad. Biodeter.* 29: 195–207.

Videla, H. A., P. S. Guiamet, S. DoValle, and E. H. Reinoso. 1993. Effects of fungal and bacterial contaminants of kerosene fuels on the corrosion of storage and distribution systems. In *A Practical Manual of Microbiologically Influenced Corrosion*, G. Kobrin, ed. Houston, TX: National Association of Corrosion Engineers International, pp. 125–134.

Von Wolzogen Kuhr, C. A. V., and S. S. Van der Vlugt. 1934. Graphitisation of cast iron as an electrochemical process in anaerobic soils. *Water* (den Haag) 18(16): 147–165.

Wagner, P. A., and R. I. Ray. 1994. Surface analytical techniques for microbiologically influenced corrosion—A review. In *Microbiologically Influenced Corrosion Testing*, ASTM STP 1232, J. R. Kerns and B. J. Little, eds. Philadelphia: American Society for Testing and Materials, pp. 153–169.

Wagner, P. A., B. J. Little, and A. V. Stiffey. 1991. An electrochemical evaluation of copper colonized by a copper-tolerant marine bacterium. Paper No. 109, *Proc. NACE Corrosion '91*. Houston, TX: National Association of Corrosion Engineers.

Wagner, P., B. Little, R. Ray, and J. Jones-Meehan. 1992a. Investigation of microbiologically influenced corrosion using environmental scanning electron microscopy. Paper No. 185, *Proc. NACE Corrosion '92*. Houston, TX: National Association of Corrosion Engineers.

Wagner, D., W. R. Fischer, and H. H. Paradies. 1992b. Copper deterioration in a water distribution system of a county hospital in Germany caused by microbially induced corrosion. 2. Simulation of the corrosion process in two test rigs installed in this hospital. *Werkst. Korros.* 43: 496–504.

Wagner, D., W. R. Fischer, and H. H. Paradies. 1992c. Test methods on microbial induced corrosion in different loops. *Proceedings of the 12th Scandinavian Corrosion Congress and Eurocorr*, Dipoli, Finland, pp. 651–665.

Wagner, D., A. H. L. Chamberlain, W. R. Fischer, J. N. Wardell, and C. A. C. Squeira. 1997. Microbiologically influenced corrosion of copper in potable water installations—A European project review. *Mat. Corr.* 48: 311–321.

Walsh, D., D. Pope, M. Danford, and T. Huff. 1993. The effect of microstructure on microbiologically influenced corrosion. *J. Min. Met. Mat. Soc.* 45(9): 22–30.

Webster, B. J., R. G. Kelly, and R. S. Newman. 1991. The electrochemistry of SRB in austenitic stainless steel. In *Microbially Infulenced Corrosion and Biodeterioration*, N. J. Dowling, M. W. Mittleman, and J. C. Danko, eds. Knoxville, TN: University of Tennessee, pp. 2/9–2/17.

Webster, B. J., D. B. Wells, and P. J. Bremer. 1996. The influence of potable water biofilms on copper corrosion. Paper No. 294, *Proc. NACE Corrosion '96*. Houston, TX: National Association of Corrosion Engineers.

Webster, B. J., S. E. Werner, D. B. Wells, and P. J. Bremer. 1998. Microbiologically influenced corrosion of copper in potable water systems—pH effects. *Corrosion* in press.

REFERENCES

Wells, D. B., B. J. Webster, P. T. Wilson, and P. J. Bremer. 1994. *Blue Water Corrosion in Potable Water*. Australasian Corrosion Association Inc. Conference, Corrosion and Prevention, Adelaide, 27–30 November, 1994, Australasian Corrosion Association, Inc., 10 p.

Wells, D. B., B. J. Webster, P. T. Wilson, and P. J. Bremer. 1998. *Blue Water Corrosion in New Zealand Potable Waters*. Materials Performance Technologies Report 78076.31, Wellington, New Zealand, 26p.

White, D. C., P. D. Nivens, A. T. Nichols, B. D. Kerger, J. M. Henson, G. G. Geesey, and C. K. Clarke. 1985. Corrosion of steel induced by aerobic bacteria and their extracellular polymers. In *Proceedings of International Workshop on Biodeterioration*, University of La Plata, Argentina, March 1985. Sao Paulo: Aquatec Quimica, pp. 73–86.

White, D. C., D. E. Nivens, P. D. Nichols, A. T. Mikell, B. D. Kerger, J. M. Henson, G. G. Geesey, and K. C. Clarke. 1986. Role of aerobic bacteria and their extracellular polymers in the facilitation of corrosion. In *Biologically Induced Corrosion*, NACE-8, S. C. Dexter, ed. Houston, TX: National Association of Corrosion Engineers.

Widdel, F. 1988. Microbiology and ecology of sulfate and sulfur-reducing bacteria. In *Biology of Anaerobic Microorganisms*, A. J. B. Zehnder, ed. New York: Wiley-Liss, pp. 469–586.

Zhang, H.-J., W. J. Dirk, and G. G. Geesey. 1999. Effect of bacterial biofilm on corrosion of galvanically coupled aluminum and stainless steel alloys under conditions simulating wet storage of spent nuclear fuel. Corrosion 55(10): 924–936.

Zinkevich, V., I. Bogdarina, H. Kang, M. A. Hill, R. Tapper, and I. B. Beech. 1996. Characterization of exopolymers produced by different isolates of marine sulphate-reducing bacteria. *Int. Biodet. Biodeg.* 36: 163–172.

10

BIOFILMS—IMPACT ON HYGIENE IN FOOD INDUSTRIES

GUN WIRTANEN
MARIA SAARELA
TIINA MATTILA-SANDHOLM
VTT Biotechnology, Tietotie 2, Espoo, Finland

10.1 FORMATION OF BIOFILM IN THE FOOD INDUSTRY

Biofilm formation causes problems in many areas such as industrial water systems (van der Wende et al., 1989; Flemming, 1991; Mittelman, 1991; Block, 1992; Capdeville and Rols, 1992), medicine (Costerton et al., 1981, 1987; Marrie and Costerton, 1982, 1983; Neu et al., 1992), and in the food processing industry (Pontefract, 1991; Holah et al., 1988; Holah and Kearney, 1992; Mattila-Sandholm and Wirtanen, 1992; Carpentier and Cerf, 1993; Zottola and Sasahara, 1994; Wong and Cerf, 1995). The paper industry has been fighting slime and biofilm formation for years (Harju-Jeanty and Väätäinen, 1984; Nurmiaho-Lassila et al., 1990; Väisänen et al., 1994; Mentu, 1996). Biofilm formation has both positive and negative implications in food-related processes (Wong and Cerf, 2995). Biofilms have been used effectively in wastewater treatment (Janning and Harremoës, 1995) and in the food industry, for example, in the production of vinegar (Zottola and Sasahara, 1994) and beer (Kronlöf, 1994).

The biofilm consists of microbial cell clusters with a network of internal channels or voids (Fig. 10.1) in the extracellular polysaccharide and glycoprotein matrix (Christensen, 1989; Allison, 1992; Carpentier and Cerf, 1993), which allows transport of nutrients and oxygen from the bulk liquid to the cells (Stoodley et al., 1994; Kostyál, 1998). Biofilm can generally be produced by any microbes under suitable conditions, although some microbes naturally have a higher tendency to produce biofilm than others (Fig. 10.2). The most common foodborne biofilm producers belong to the genera *Algaligenes*, *Bacillus*, *Enterobacter*, *Flavobacterium*, *Pseudomonas*, and *Staphylococcus*. Biofilm bound bacteria can accelerate

---

Biofilms II: Process Analysis and Applications, Edited by James D. Bryers.
ISBN 0-471-29656-2 Copyright © 2000 Wiley-Liss, Inc.

corrosion and material deterioration, for example, in sensors and detectors (Costerton et al., 1988; Characklis, 1991; Chamberlain, 1992; Videla and Characklis, 1992; Lee and Characklis, 1993; Giffel et al., 1997; Carpén, 1999). Corrosion is an electrochemical process initiated by a potential difference between two different metals or two zones in the metal forming an electrical circuit with cathodic and anionic areas (Nestor and Cappeline, 1979). Microbial corrosion is also observed in processing equipment, appearing either in combination with or without biofilm formation. Corrosion leads to huge losses in different industrial areas, for example, piping and cooling water systems (Marshall and Blainey, 1991; Chamberlain, 1992; Videla and Characklis, 1992; Morton and Surman, 1994; Carpén, 1999).

Disinfection after removal of biofilms using suitable cleaning procedures is required in food plants where wet surfaces provide favorable conditions for the growth of microbes (Mosteller and Bishop, 1993). In practice, biofilm left on improperly cleaned surfaces is a barrier between microbes and the disinfectants, antibiotics, or biocides used against them (Kinniment and Wimpenny, 1990; Nichols, 1989, 1991; Wirtanen, 1995). The effect of many antibiotics is based on inhibition of active growth. It has been said that most of the bacteria in biofilms are no longer growing actively and their resistance is therefore altered. The resistance of microbes in a biofilm cannot be proved without controlling their growth rate (Gilbert et al., 1990). The permeability barrier of the biofilm considerably affects the efficiency of these agents (Nichols, 1991). Bacterial cells growing in hydrogels can be used to study the effect of various disinfectants on foodborne food spoilage bacteria as well as on pathogens (Wirtanen et al., 1998).

In the food industry, equipment design plays the most important role in combating biofilm formations (Chisti and Moo-Young, 1994; Holah and Timperley, 1999). A system should be designed without edges, crevices, and dead spaces. It should be cleaned frequently to avoid the accumulation of biofilm. The choice of materials and their surface treatments, for example, grinding and polishing, are important factors in inhibiting the for-

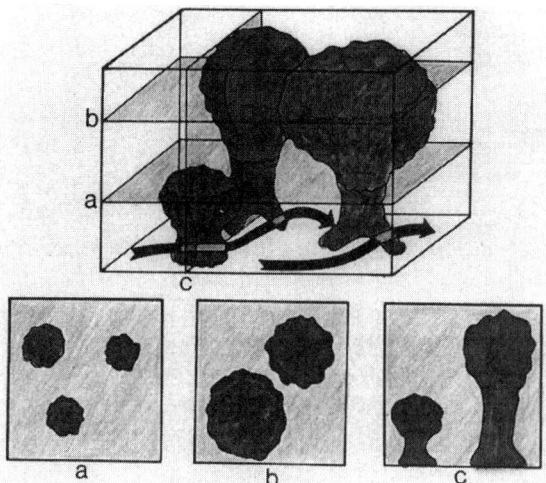

Figure 10.1 Biofilm—a schematic diagram of cell clusters and channel which allow transport of nutrients and oxygen from the bulk solution to cells in lower layers of the biofilm. Horizontal focus planes (*a* and *b*) and Z-scan (*c*) in the biofilm taken with a scanning confocal laser microscope (according to Stoodley et al., 1994, and Kostyál, 1998).

10.1 FORMATION OF BIOFILM IN THE FOOD INDUSTRY

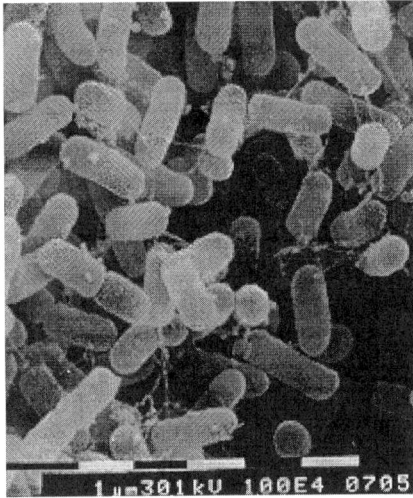 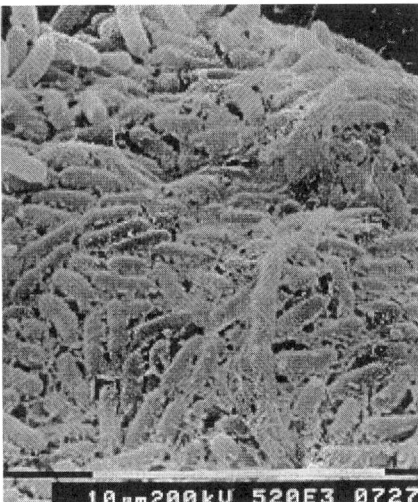

Figure 10.2 Scanning electron microscopy photographs of bacteria without protective biofilm (left) and in a protective matrix called biofilm (right) (according to Mattila-Sandholm and Wirtanen, 1992).

mation of biofilm and in promoting the cleanability of surfaces (Fig. 10.3). Treatments of surface materials to reject biofilms can be performed actively to remove or passively retard biofilm reoccurrence. The cleanliness of surfaces, training of personnel, and good manufacturing and design practices are the most important tools in combating biofilm problems in the food industry (Thorogood, 1999; Holah and Timperley, 1999).

Besides causing problems in cleaning and hygiene (Hood and Zottola, 1995), biofilm can cause energy losses and blockages in condenser tubes, cooling fill materials, water and wastewater circuits, and heat exchangers (Characklis, 1981; Wirtanen, 1995). Biofilm can also cause health risks due to the release of pathogens into drinking water distribution systems. Biofilms may enter a food processing system by causing reduced effectiveness in ion exchange and membrane processes (Whittaker et al., 1984; Flemming et al., 1992; Mucchetti, 1995). In the food processing water supply systems biofilms cause problems in granular activated carbon columns, reverse osmosis membranes, ion exchange systems, degasifiers, water storage tanks, and microporous membrane filters (Mittelman, 1991). The bacterial level of the purified water used and the defects found in microelectronic devices correlate directly. Conductance, electromigration, and corrosion in the oxide layers have been observed in devices manufactured with insufficiently pure water (Mittelman, 1991). Accumulation of mixed population biofilms containing sulphate-reducing bacteria (SRB) causes corrosion in industrial water systems. The mechanisms contributing to the degradation of mild steel are both chemical and biological (Marshall and Blainey, 1991; Videla and Characklis, 1992). SRB-related corrosion is invariably associated with biofilm formation on the metal surface, for example, in closed water systems in paper mills (Lee et al., 1995; Carpén, 1999).

The food industries in which research on biofilms has been carried out produce canned products, meat and poultry products, pastries, biscuits, pizza, fish cakes, cheese, milk products, beer, spices, and vegetable and salad products (Notermans et al., 1991; Austin and Bergeron, 1995; Gibson et al., 1995; Wirtanen, 1995; Flint et al., 1997a, 1997b; Klemetti et al., 1999; Storgårds et al., 1999). According to the literature biofilm has caused problems at various sites: air handling systems (Tarvainen et al., 1994; Gibson et al., 1995; Holah et al., 1995), blancher extractors (Gibson et al., 1995), conveyors (Zottola and Sasahara, 1994; Gibson et al., 1995), cooling systems (Tarvainen et al., 1994; Gibson et al., 1995), floors (Zottola and Sasahara, 1994; Mettler and Carpentier, 1998), floor drains (Zottola and Sasahara, 1994; Gibson et al., 1995), food contact surfaces of stainless steel (Gibson et al., 1995; Wirtanen, 1995; Hood and Zottola, 1995), gaskets (Austin and Bergeron, 1995; Klemetti et al., 1999; Storgårds et al., 1999), heat exchangers (Melo and Pinheiro, 1992; Flint et al., 1997a; Giffel et al., 1997), manufacture line for paper-based packaging material (Carpén, 1999; Raaska, 1999), milk transfer lines (Austin and Bergeron, 1995; Flint et al., 1997a), mixers (Gibson et al., 1995), poultry processing equipment (Lindsay et al., 1996), rubber fingered pluckers (Notermans et al., 1991), slicers (Gibson et al., 1995), packaging machines (Gibson et al., 1995), pasteurizers (Flint et al., 1997a), ultrafiltration and reverse osmosis membranes (Mucchetti, 1995; Flint et al., 1997a), and vegetable lines (Gibson et al., 1995). We can see from the list that problems generating from biofilm can occur anywhere in the food process if the design and maintenance is improper. Biofilm formation in some of the areas important in food production—cooling systems, air handling systems, open and closed food process lines, and manufacture of packaging material—are presented in the sections that follow. A summary of different biofilm effects in various processes is presented in Table 10.1 (Wirtanen, 1995).

10.1.1 Cooling Systems

Biofilm formation has been known for a long time in water systems (Geesey et al., 1977; Lewin, 1984; Freeman et al., 1990; Block, 1992) and the adherance of microbial cells to the surfaces in piping and water distribution systems is also well known (Flemming, 1991; LeChevallier et al., 1987, 1988; Nohata and Taguchi, 1995). The number of free planktonic cells in the water does not necessarily correlate with the amount of biofilm on the pipe surfaces. The numbers of microbial cells adhered to surfaces may be as much as 500 to 50,000 times the numbers of planktonic cells in water (Costerton and Lashen, 1983). In drinking water systems sufficient nutrients are present for planktonic cells to form biofilms, since these microbes require only low levels of nutrients to grow (Block, 1992). Biofilm forma-

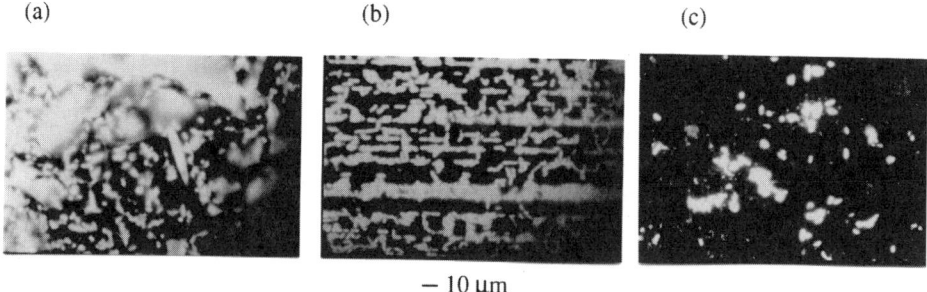

Figure 10.3 Epifluorescent microscopy photographs of 5-d-old *P. fragi* biofilms, which were stained with acridine orange. The biofilms had been grown on various stainless-steel surfaces (AISI 304): (*a*) glass blasted, (*b*) lapped, and (*c*) mechanically polished surfaces. The scale marker is equivalent to 10 μm.

10.1 FORMATION OF BIOFILM IN THE FOOD INDUSTRY

TABLE 10.1. Effect and Relevance of Biofilms in Food Processing Systems and Industrial Processes Producing Material for Food Manufacturer

Biofilm formation in food-related industry can:
* Cause material deterioration in metal condenser tubes, pipelines, and cooling systems
* Reduce the effectiveness on remote sensors, sight glasses, etc.
* Release pathogenic organisms and reduce water quality in cooling systems and in water distribution systems
* Reduce product quality in food production and processes producing goods for food handling, e.g., packaging material
* Disturb specific chemical transformations through contamination of immobilized systems
* Cause energy losses in heating of the process equipment

Source: Wirtanen, 1995.

tion appears in sinks, faucets, valves, and in different joint areas all over the piping systems (Block, 1992; Holah and Thorpe, 1990; Timperley et al., 1992). Mechanical rinsing and shock treatment with chlorine-based agents should be performed when biofilms are observed. The pH value of water often increases when contamination occurs.

It has been found that temperatures below 50°C promote biofilm formation (Miller and Bott, 1982). The average temperature in cooling water systems is 35°C, which is close to the optimum temperature for most microbial growth. From a hygienic and health point of view, biofilm can include organisms that cause infections and diseases. Problems with *Legionella pneumophila*, a dangerous organism that is able to form biofilm, can occur in hot water systems (Schofield and Locci, 1985; Bezanson et al., 1992; Rogers et al., 1994a, 1994b). It has been found that rubber surfaces are the best materials for supporting biofilm formation of *Legionella* and that copper is the most resistant to biofilm formation (Rogers et al., 1994a, 1994b).

Biofilm formation in cooling systems, either open or closed, has been widely investigated (Colturi and Kozelski, 1984; Honneysett et al., 1985; Offringa, 1988; Lewis, 1982; Bott, 1992a; Poulton et al., 1995). In open systems the water is taken directly from natural water sources such as seas and rivers and returned after circulation in the cooling system. Closed systems are, for both functional and hygienic reasons, better than open ones. The level of microbes can be controlled by regular mechanical and chemical cleaning and treatment with biocides (Bott, 1992b). The hygiene in open systems is difficult to control, for example, biocides cannot be used because the water is returned to the water source (Flemming, 1991). The primary colonization of bacteria and molds in cooling water systems is often followed by accumulation of algae and clams. The microbes obtain nutrients from the algal mass, thus initiating a vicious circle. The most common microbes in cooling waters are the slime-forming *Gallionella* and *Pseudomonas*, and the most common algae are *Chloronella*, *Spirogyra*, and *Ulothrix* (Richardson, 1982). The most evident problem caused by biofilm in cooling systems is that heat transfer may be reduced to only 10% of the maximum value. Other problems include health risks, microbial corrosion, increased flow resistance, blockages and accumulation of particles, and decreased efficiency of biocides and corrosion inhibitors (Colbourne and Dennis, 1988). *Pseudomonas aeruginosa* has been isolated from water even with a chlorine concentration of 3–5 mg/kg (Price and Ahearn, 1988).

Investigations simulating real industrial situations have been performed using microbes isolated from water circulation systems. The formation of biofilms takes several days at least, and in low-nutrient water systems even weeks. Biofilm formed in slow-flowing systems are structurally different from those formed in systems with high flow rates (Miller and Bott, 1982; Duddridge et al., 1983). Biofilm formation can be decreased by more than 50% if the oxygen level in the water system is minimized, and by more than 80% when the amounts of both oxygen and nutrients are minimized (Miller and Bott, 1982). Anaerobic conditions favor biofilm formation by anaerobic organisms, which can increase corrosion (Duddridge et al., 1983; Characklis, 1990). Often microbial corrosion is due to biofilm formation and anaerobic conditions in the layers deep in the biofilm (Gaylarde, 1989, 1990; Costerton and Boivin, 1991; Geesey, 1991). The microbes take part in the corrosion directly in the electrochemical reactions, through microbial growth and colony formation that alters the electropotential of the surface or through changes in the metal due to production of metabolic compounds such as organic acids (Sequeira, 1988).

10.1.2 Ventilation and Air Handling Systems

The quality of air in food production facilities is very important for the final product quality. The microbial population in the air channels depends on the environment, filtration membranes, and the sitings of air holes (Shelley, 1990; Tarvainen et al., 1994; Holah et al., 1995; Hugenholtz et al., 1995; Ljungqvist and Reinmüller, 1995). Formation of biofilm in air-conditioning systems does not occur without a water reservoir of some kind. Normally, there is no water in the air-conditioning systems, but it can accumulate unintentionally through condensation (Mattila-Sandholm and Wirtanen, 1992; Tarvainen et al., 1994).

A study carried out in various industrial ventilation ducts has shown that biofilm can accumulate within the ducts. The occurrence of bacterial and fungal biofilms takes place within a few days and weeks (Tarvainen et al., 1994). Biofilms form very easily in humid surroundings where oil has been left on the surfaces. The amount of bacteria was shown not to vary significantly at different times of the year. However, there was less mold growing on the slides placed in the ducts during winter than on those from the autumn sampling period. Biofilms were more easily detected by image analysis connected to epifluorescence microscopy than by conventional cultivation methods. Biofilm growth was prevented in clean-room ducts by using effective filters and appropriate maintenance procedures (Tarvainen et al., 1994). It is therefore very important to monitor clean-room behavior, as most particles in clean rooms originate from the personnel (van Beek, 1994; Ljungqvist and Reinmüller, 1995).

The air filtration equipment must be chosen carefully to suit the processing environment (van Beek, 1994; Lehtimäki, 1995). The membranes in the air-conditioning system and the walls in the air-conditioning channels are places where biofilms start to grow (Shelley, 1990; Hugenholz and Fuerst, 1992; Hugenholz et al., 1995). The dangerous *Legionella pneumophila* has been isolated from water systems that were connected to an air conditioner (Heinzel, 1988; Karsa and Stafford, 1989). When the air-conditioning system is to be cleaned and disinfected, it is very important that the disinfection medium penetrates the biofilm and does not simply flow through the system with the air (Mattila-Sandholm and Wirtanen, 1992).

10.1.3 Open Equipment in Food Processing

Although there is long history of biofilm impacting the food industry, actual biofilm research in this industrial area began in the late 1980s (Holah et al., 1988, 1989). The occurrence of slime-forming microbes is a major problem in the sanitation and disinfection of process equipment. The hygiene of surfaces, instruments, and equipment in the food indus-

10.1 FORMATION OF BIOFILM IN THE FOOD INDUSTRY

try essentially affects the quality of the products processed (Brackett, 1992; Lindsay et al., 1996). Common contaminants on food contact surfaces are enterobacteria, lactic acid bacteria, micrococci, streptococci, and *P. fragi*. It is somewhat alarming that pathogens such as *Listeria monocytogenes*, *Salmonella typhimurium*, and *Yersinia enterocolitica* can readily produce biofilms, causing severe disinfection and cleaning problems on surfaces in the food industry (Mafu et al., 1990; Helke et al., 1993; Ren and Frank, 1993; Ronner and Wong, 1993; Kim and Frank, 1994; Blackman and Frank, 1996; Flint et al., 1997b; Ntsama-Essomba et al., 1997). Inadequate cleaning and sanitation of surfaces coated with biofilm cause contamination, because the biofilm protects the microbes against sanitizers and disinfectants (Czechowski and Banner, 1990; Mosteller and Bishop, 1993). Future studies will certainly focus on enteropathogenic *Escherichia coli*.

Problems with accumulation of particulates and cells occur where cleaning is for any reason inappropriate (Mettler and Carpentier, 1998). Glass surfaces are transparent and hygienic, but glass is unfortunately expensive, has a low pressure tolerance, and breaks easily. The most useful material in processing equipment is steel, which can be treated with mechanical grinding, lapping, and electrolytic or mechanical polishing. In electrolytic polishing, a preground surface is treated in an electrolyte bath to obtain an even surface. The surface structure of stainless steel is very important in avoiding biofilm formation; it has been reported that although the grain boundaries of AISI 316L stainless steel constitute 3–20% of the total surface area, more than 90% of the adherent bacteria were found attached to the grain boundaries (Bryers and Weightman, 1995). Many equipment faults can be avoided by using good design practice guidelines, which are summarized in Table 10.2 (Mattila-Sandholm and Wirtanen, 1992).

The lubricants used in conveyors are a problem, especially in dairies and breweries (Heinzel, 1988; Rossmoore, 1988). Oil-based lubricants containing water are very sensitive

TABLE 10.2. Factors in Good Design Practice

Design Area	Relevant Factors
Equipment layout	Cleanability and cleaning of equipment, servicing of equipment, rationalization of operations, systematic layout
Production	Process parameters, installations, maintenance of hygiene
Equipment	Materials, surface finishing, accessories and joints, checking and cleanability, clearing and cleaning, insulation
Pipelines	Accessories (valves, etc.), materials, surface finishing, clearing and cleaning, consoles, insulation
Automation	Control of production, process reliability
Production control	Working plan, material flow, and disturbance reports
Electrification	Shielding of equipment, lighting
Buildings and Structures	Surface materials, painting, sewerage, tipping, joints, cleanability of surfaces, layout of the connections of different production units, air-conditioning
Organization and Personnel responsibility	Functions, tasks, education, and courses

Source: Mattila-Sandholm and Wirtanen, 1992

to microbial growth. *Acinetobacter* sp., *algaligenes* sp., *Pseudomonas* sp., and sulphate-reducing bacteria have been isolated from lubricants, and the biofilm formed in the lubricant can indirectly promote corrosion (Ortiz et al., 1990; Hamilton, 1991). *Listeria monocytogenes*, an opportunistic pathogen, has been isolated from lubricants in dairies (Rossmoore, 1988). Problems with biofilms are also met in engineering works where lubricants based on vegetable oils are used. Synthetic lubricants containing biocides have been developed, and the problems associated with contamination of lubricants has been decreasing (Henriksson and Haikara, 1990).

10.1.4 Closed Equipment in Food and Drink Processing

Poorly designed sampling valves can destroy an entire process or give rise to incorrect information due to biofilm effects at measuring points. Biofilm is easily formed on measuring probes and the inner parts of equipment because these are usually difficult to clean (Austin and Bergeron, 1995; Flint et al., 1997a). No probe for measuring properties in biofilms can provide an accurate description of the process, when crust has fouled on the probe (Chisti and Moo-Young, 1994). The penetration of disinfectants and heat are hampered by the porous matrix on the surface, which diminishes the effect of sanitation and sterilization procedures. The bacterial slime of one *Bacillus* strain improved the heat resistance of the bacterium, extending the autoclaving time required for successful sterilization to several hours. Biofilm components can also alter the resistance to steam and formalin (Mattila-Sandholm and Wirtanen, 1992).

The choice of surface materials is of great importance in designing and building equipment and processing lines for industrial use (Pirbazari et al., 1990). The process equipment is easy to clean if the surface materials are smooth and in good condition (European Hygienic Equipment Design Group (EHEDG), 1993b). Dead ends, corners, cracks, crevices, gaskets, valves, and joints are vulnerable points for biofilm accumulation (EHEDG, 1993a, 1993b; Chisti and Moo-Young, 1994). Owing to their construction, valves are vulnerable to microbial growth and thus constitute a hygiene risk (Chisti and Moo-Young, 1994; EHEDG, 1994). Hoses, tubes, filters, and such containing polyvinylchloride increase the risk of contamination (Price and Ahearn, 1988). Experiments have been carried out with *L. monocytogenes*, *P. fluorescens*, and *Y. enterocolitica*, which cause contamination of materials used in gaskets (Mosteller and Bishop, 1993). *Pseudomonas* has also been found to attach easily to surfaces of hydrophobic materials such as polystyrene. Rubber- and teflon-based materials used in gaskets are easily contaminated. Some microbes are also able to decompose rubber and use it as a source of energy (Chamberlain and Johal, 1988; Zyska, 1988). Microbial accumulation is known to be a problem in reverse osmosis membranes. The cleaning process in the case of reverse osmosis membranes is more complex because the cellulose acetate membranes can be destroyed if treated with strong chemicals and at high temperatures (Ridgway and Safarik, 1991; Whittaker et al., 1984).

10.1.5 Paper and Cardboard Mills

Biofilm formation causes major problems in paper machines and it affects the production of good quality paper-based packaging material for the food industry (Raaska, 1999; Carpén, 1999). The temperature (20–60°C), acidity (pH 4–10), sedimentation, and crust accumulation, as well as available organic and inorganic nutrients (carbohydrates, starch, nitrogen, phosphorus, etc.) in the process combined with long production runs to provide favorable circumstances for microbes to form biofilm in the processing equipment (Lamot, 1989; Sillanpää, 1998). The nutrients enter the process with water, pulp, and additives. The use of recycled fibers and changes in the acidity of the process to neural or alkaline have increased the total amount of bacteria in the paper-making processes (Sorrelle and Belgard, 1992). The microbes, mostly bacteria and fungi, enter the process with the nutrients as well as in the air (Hughes, 1993). It should be remembered that each paper-making machine has its own microbial flora, which depends on processing parameters such as pH and temperature, the season, and the additives used. The slime-forming microbial flora can be divided into primary, for example, bacteria of the *Bacillus*, *Pseudomonas*, and *Enterobacter* strains and fungi of the *Aspergillus*, *Mucor*, and *Penicillium* strains, and secondary slime-formers, for example, bacteria of the *Alcaligenes*, *Flavobacterium*, *Klebsiella*, *Micrococcus*, and *Staphylococcus* strains and fungi of the *Paecilomyces*, *Tricoderma*, and *Trichosporum* strains (Hughes-van Kregten, 1988). Growth of slime-forming microbes is a problem in the paper mills. The bacterial spores stand the heat treatment during the production and therefore the *Bacillus* strains, which are spore-formers, are the main source of contaminants in final paper and cardboard products (Väätäinen, 1994). The primary slime-formers can form slime without the need for certain additives or symbiosis with other microbes. The secondary slime-formers then cooperate with the primary slime-formers to increase the total slime produced. They are not able to form slime on their own (Harju-Jeanty and Väätäinen, 1984; Hughes-van Kregten, 1988). The microbial growth also causes odor problems, which can destroy both the air in the process facilities and the quality of the end products (Sorrelle and Belgard, 1992).

Practical experiments show that biofilm in pulp and paper processing equipment detaches easily from the surface and the loose slime causes holes in the paper (Nurmiaho-Lassila et al., 1990; Väisänen et al., 1994). The paper machines are connected to water circulation systems, which are prone to rapid biofilm growth (Holt, 1988). The slime in paper machine biofilms contains 10^{11}–10^{12} cfu/g dry mass. The microbial growth causes problems in corrosion, process flow, contamination, and paper quality (Mentu, 1996). Bacteria from manufacturing have been found to be unevenly spread in the end product (Suominen et al., 1997). Corrosion of surfaces in the paper machines is caused by the bacterial biofilms of anaerobic sulphate-reducing bacteria, thermophilic bacteria, and bacteria-producing acids and reducing compounds (Costerton et al., 1987, 1988; Safade, 1988). Excessive concentrations of chlorine have also been found to promote corrosion (Goodman, 1987). More than 50% of all corrosion problems in pipelines in paper-processing machines are due to microbial corrosion. Efforts to combat microbial growth and biofilm formation in paper machiens are important because the annual costs for maintenance of equipment are huge (Walch, 1992; Mentu, 1996; Carpén, 1999).

10.2 ELIMINATION OF BIOFILMS

10.2.1 Chemicals Used in the Elimination of Biofilms

Physical, chemical, and microbiological cleanliness are essential in food plants. Factors governing the selection of detergents and disinfectants in the food industry are that the agent should be efficient, safe, not damage or corrode equipment, be easily rinsable, and not affect the sensory values of the product produced (Dunsmore et al., 1981; Czechowski and Banner, 1990; Troller, 1993). Physical cleanliness means that there is no visible waste, foreign matter, or slime on the equipment surfaces. Chemically clean surfaces are surfaces from which undesirable chemical residues have been removed, whereas microbiologically clean surfaces imply freedom from spoilage microbes and pathogens (Gould, 1994). Attached bacteria or bacteria in biofilms can be a problem in food processing, because they adhere to the surfaces and if the cleaning is insufficient the remaining bacteria start to grow and multiply after the cleaning and contaminate the product (Hood and Zottola, 1995). Once a biofilm is firmly established, cleaning and disinfection becomes much more difficult (Mosteller and Bishop, 1993). Caution should be exercised in selecting appropriate disinfectants for contaminated surfaces (Bourion and Cerf, 1996). Cleaning agents and disinfectants can form residues on surface materials in contact with foods and may contaminate the food (Sjöberg et al., 1995).

Various disinfectants have been developed to destroy microbes (Brown and Gilbert, 1993). Microbes have nevertheless been found in disinfectant solutions. This means that microbial contaminants can be spread on the surface to be cleaned instead of being cleaned. As early as 1967 it was reported that chlorhexidine mixtures were contaminated with *Pseudomonas* sp. *Pseudomonas* sp. have also been found in concentrated iodine solutions (Marrie and Costerton, 1981; Anderson et al., 1990). *Serratia marcescens* was found to be viable even after 27 months in a disinfectant containing 2% chlorhexidine. A concentration of 0.1% chlorhexidine is sufficient to kill the cells of *S. marcescens* if they are freely suspended in liquid (Costerton and Lashen, 1983; Marrie and Costerton, 1981). Microbial contamination has also been found in solutions of aldehydes, quaternary compounds, and amfotensides (Heinzel, 1988). Other microbes that have been isolated from disinfectants include *Alcaligenes faecalis*, *Enterobacter cloacae*, *E. coli*, *Flavobacterium meningosepticum*, and *Pantoea agglomerans*.

Chlorine, a compound in cleaning agents and disinfectants used in the food industry, can also induce biofilm buildup of, for example, *Pseudomonas* sp. (LeChevallier et al., 1988; van der Wende and Characklis, 1990). *Pseudomonas* sp. are relatively resistant to chlorine treatments and can even multiply when chlorine has been used. Free chlorine (1.0 mg/l) showed no effect on coliforms growing in biofilms, and even increased amounts (2.0 mg/l) did not kill *E. coli* grown in biofilm. The material on which microbes form biofilm has a major influence on the effect of the disinfectants used. Capsular *Klebsiella pneumoniae* grown on glass surfaces has been shown to have a 150-fold resistance to chlorine compared with that grown in suspensions. When a low-nutrient liquid is used, the ratio is increased to about 600. On metal and carbon surfaces, the corresponding resistance ratios were about 2400 and 3000 times the value in suspension, respectively (LeChevallier et al., 1988). The effect of surfactant sanitizers against *L. monocytogenes* grown in biofilms was decreased (Frank and Koffi, 1990; Shin-Ho-Lee and Frank, 1991). However, sodium hypochlorite and quaternary ammonium compounds have proved effective against *Listeria* in 24 h old biofilms (Mustapha and Liewen, 1989).

Efficacy of disinfectants and antimicrobial agents are usually determined in free cell suspensions, which do not mimic the growth conditions on surfaces where the agents are required to inactivate the microbes (Anwar et al., 1989, 1990; Frank and Koffi, 1990; Gillatt, 1991). The agent must reduce the microbial populations by 5 log units in suspensions in order to be considered effective. The goal for reduction of surface-attached bacteria with disinfectants is 3 log units (Mosteller and Bishop, 1993). The standard suspension tests have proved sufficiently reliable because the variations of results are within acceptable limits when replication is adequate (Bloomfield et al., 1994). There can, however, be problems

10.2 ELIMINATION OF BIOFILMS

with the repeatability and reproducibility of suspension tests performed with organic load. It is obvious that the surface tests are even more difficult to perform than suspension tests because of the carrier material used and the viability of dried cells on the surfaces (Maris and Fresnel, 1992; Bloomfield et al., 1994). In developing a proposal for testing disinfectants on surfaces to an analytical standard, it is important to identify the major sources of variation in the procedure. Bloomfield et al. (1994) showed that variation is not only due to variability between test laboratories, but also depends on how the surface samples were prepared. Microbes growing or dried on surfaces are not susceptible to disinfectants from all sides as they are in suspensions, and due to the requirement for penetration, disinfectants are used in higher concentrations on surfaces than in suspensions (LeChevallier et al., 1998; Mattila et al., 1990).

10.2.1.1 Cleaning Agents.
Cleaning agents are applied to remove soil, microbes, and biofilms from surfaces (Czechowski and Banner, 1990). The effects of the cleaning agents on biofilms have been thoroughly investigated (Bloomfield, 1988; Anwar et al., 1990; Czechowski and Banner, 1990; Brackett, 1992; Mosteller and Bishop, 1993). Removal of biofilm is important for the maintenance of the equipment, because the debris left on surfaces can act as nutrients for the buildup of new biofilm (Characklis, 1981) or may corrode the equipment material (Videla and Characklis, 1992). In order to minimize biofouling, it is best to clean the equipment at frequent and regular intervals using suitable, efficient cleaning procedures for the process before the biofilm has the opportunity to develop (Dunsmore et al., 1981; de Goederen et al., 1989).

Efficient cleaning of surfaces depends on surface tension, viscosity, chemical reactivity, size of biofilm particles, solubility, and living microorganisms (Busscher and Weerkamp, 1987; Lewis and Gilmour, 1987; Wolfaardt and Cloete, 1992; Troller, 1993). The cleaning agents should be surface active, soluble, nontoxic, rinsable, noncorrosive, and easy to use. They should also emulsify fats and oils, suspend insoluble particulates, and dissolve mineral deposits. Soaps and synthetic detergents clean surfaces in a similar manner, the surfactant molecule having hydrophilic and hydrophobic ends. The hydrophobic end dissolves oily soils (Troller, 1993). The alkaline and acidic detergents are mixtures of several classes of compounds: alkalis, acids, tensides, and surface active agents (Troller, 1993). Chemicals for cleaning are generally formulated for specific cleaning problems based on the following criteria (Gould, 1994):

- Composition and amount of soil
- Nature of the surface to be cleaned (e.g., stainless steel, plastic, or painted)
- Method of cleaning available (e.g., clean-in-place (CIP), high pressure, foam, or hand)
- Physical nature of the cleaning compound
- Quantity and quality of water available
- Time and temperature available

Currently used cleaning agents contain chelators (for example, ethylene diamine tetraacetic acid (EDTA)), surfactants, detergent builders, antideposition agents, enzymes, and disinfectants (Flemming, 1991; Troller, 1993; Johansen et al., 1997). Metal chelators have been used in the breakage of the biofilm layers (Izzat et al., 1981; Turakhia and Characklis, 1989). The chelators are not themselves biocides and should therefore be combined with other antibacterial compounds (Heinzel, 1988). Chelators bind magnesium and calcium ions, thereby destabilizing the outer membranes of microbial cells. The cleaning effect ceases when the surfactant molecules are fully deployed in tying up the soil. An important key in maintaining a good cleaning effect is to use sufficient concentrations of surfactants (Troller, 1993).

Tensides are normally used in chemical compounds to achieve suitable cleaning and foaming effects. Tensides are large molecular compounds with polar and nonpolar groups forming hydrophobic and hydrophilic parts, depending on the group. The tensides are divided into four types: anionic, cationic, nonionic, and amphoteric tensides (Milwidsky and Gabriel, 1982). Sodium alkyl sulphates were the first commercially used anionics in the early 1930s, alkyl benzene sulfonates being developed somewhat later (Troller, 1993). Other anionics include alkyl aryl sulphonates, long-chain alcohol sulphates, sulphonated olefins, and sulphated ethers (Milwidsky and Gabriel, 1982). These compounds are good detergents, but tend to produce foam. They have great interfacial activity and may remove components adhering by hydrophobic groups, such as casein and fat, but are less effective in removing particles adsorbed by electrical forces (Nassauer and Kessler, 1986). Cationic surfactants are composed of various quaternary ammonium compounds (quats). They are expensive and their cleaning effect is only moderate, but they have antimicrobial effects and are therefore used in cleaning agents in food plants (Woollatt, 1985). Nonionic tensides are compounds of alkanolamines, polyethylene glycols, or polyethylene imines and they are moderately good detergents (Wilkinson and Moore, 1982; Troller, 1993) although some of these agents have low-foaming characteristics. The amphoteric tensides contain both anionic and cationic parts and can therefore function both in acidic and alkaline environments or as double ions in the boundary region (Milwidsky and Gabriel, 1982).

10.2.1.2 Disinfectants. Disinfection is required in food plant operations where wet surfaces provide favorable conditions for the growth of microorganisms (Bloomfield, 1988; Bott, 1991; Brackett, 1992; Holah, 1992). The aim of disinfection is to reduce the surface population of viable microorganisms after cleaning by destruction or removal, and to prevent microbial growth on surfaces during the interproduction period (Holah, 1992; Kuo and Smith, 1996). Disinfective agents do not penetrate the polysaccharide and glycoprotein matrix left on the surfaces after an ineffective cleaning procedure very well, and thus do not destroy all the living cells in biofilms (Pontefract, 1991; Carpentier and Cerf, 1993; Stewart et al., 1996). Disinfectants are most effective in the absence of organic material, such as fat- sugar-, and protein-based materials (Czechowski and Banner, 1990). Microbes are less likely to survive disinfection after an effective cleaning (Brackett, 1992). Efficiency of disinfectants is generally controlled by interfering organic substances, pH, temperature, concentration, and contact time (Holah, 1992; Mosteller and Bishop, 1993; Wood et al., 1996). The desired characteristics of disinfectants are the same as for cleaning agents. They must be effective, safe and easy to use, and easily rinsed off surfaces, leaving no toxic residues or residues that affect the sensory values of the product (Troller, 1993).

The use of disinfectants in food plants depends on the material used and the adhering microbes (Troller, 1993). It is therefore recommended that disinfectants used in the sanitation of food processing equipment should be tested on surfaces in circulating systems, for example, in a Robbins device (Green et al., 1987). Disinfectant testing in a circulating system provides many replicates, which makes the evaluation easier and more rapid to perform. The variety in types of soil, cleaners, and cleaning conditions makes it impossible to give an exact, overall statement of required temperatures and concentrations (Gould, 1994). Table 10.3 lists disinfectants approved for use in the food industry, and these contain chlorine-based compounds, iodine compounds, alcohols, quats, oxidants, or surfactants (Holah

10.2 ELIMINATION OF BIOFILMS

TABLE 10.3. Some Disinfectants Used in the Food Processing Industry

Disinfectant	Advantages and Disadvantages in Use
Chlorine-based	Advantages: broad spectrum of activity, active in low concentrations, destroys biofilm matrix, supports detachment, reacts with microbes, easy to use, cheap Disadvantages: toxic by-products, resistance development, residual effect, corrosive, reacts with extracellular polymers, discoloration of product, explosive gas
Peracetic acid	Advantages: very effective in small concentrations, broad spectrum, kills spores, decomposes to acetic acid and water, nontoxic, penetrates biofilms Disadvantages: corrosive, unstable
Hydrogen peroxide	Advantages: decomposes to water and oxygen, relatively nontoxic, easily used in situ; weakens biofilms, supports biofilm detachment Disadvantages: high concentrations ($> 3\%$) necessary, resistance, corrosive
Alcohols	Advantages: nontoxic, easy to use, colorless, effective against vegetative cells, harmless on skin, easily decomposable, most agents soluble in water, volatilizable Disadvantages: microbistatic, ineffective against spores
Iodophor	Advantages: noncorrosive, easy to use, nonirritating, broad spectrum of activity Disadvantages: flavor or odor, form purple compounds with starch, expensive
Glutaraldehyde	Advantages: effective in low concentrations, cheap, nonoxidizing, noncorrosive Disadvantages: does not penetrate biofilms well, degrades to formic acid, increases system dissolve organic carbon (DOC)
Ozone	Advantages: similar effectivity to chlorine, decomposes to oxygen, no residues, weakens biofilm matrix Disadvantages: reacts with organics forming epoxides, corrosive, short half-life, sensitive to water ingredients

Source: Compiled according to Flemming (1991) and Troller (1993).

et al., 1990; Flemming, 1991; Troller, 1993; Norton and LeChevallier, 1997). Choosing a disinfectant for the process is not an easy task. There are mixed cultures in the process environment, which means that some microbes may be resistant to the agent. The concentration of the disinfectant can be altered to provide shock treatments, such as with a 10-fold concentration. It has also been suggested that the disinfectants should be changed continuously (Sequeira et al., 1989). In the maintenance of process hygiene, disinfectants should be chosen according to the process (Sequeira et al., 1989):

- Is the agent effective in the pH range used?
- Is the agent stable when diluted? Does it vaporize?
- Is the agent toxic, safe, or irritating?
- What is the spectrum of the agent?
- How does temperature affect the activity of the agent?

- Is the agent corrosive on the surface?
- Is the agent surface active?
- Is the agent stable when reacting with organic material?
- Is the agent effective, and what are the costs?

Several chlorine or chlorine-based compounds are approved for use in food plants, for example, gaseous chlorine, sodium hypochlorite, calcium hypochlorite, chloramine-T, and chlorine dioxide. The antibacterial active moiety is formed when chlorine gas or a hypochlorite compound is added to water and produces hypochlorous acid. This dissociates further into protons and hypochlorite anions (Troller, 1993; Sanderson and Stewart, 1997). Stabilized hypochlorites are used when disinfection of long duration is required (Troller, 1993). Hypochlorite solutions are more affected by protein-based soil than are chloramines (LeChevallier et al., 1988; Keevil et al., 1990; Troller, 1993; Stewart et al., 1994). The range of microorganisms killed or inhibited by chlorine-based compounds is probably broader than by any other approved sanitizer.

Iodophors are used extensively in the food industry. In the disinfection the iodine compound takes part in the oxidation of essential parts of the microbial cells. Like chlorine-containing products, iodophors are active against gram-positive and gram-negative bacteria, yeasts, and molds (Holah et al., 1990; Holah, 1992; Troller, 1993). Bacterial spores, however, are highly resistant to iodophors. Iodophors cannot be used in food plants where starch-containing products are produced because iodine forms a purple complex with starch (Troller, 1993). Quaternary ammonium compounds are used as sanitizers in dairies and in the food industry because they have good wetting properties and are nonspecifically described as cationic surface active agents in which the cationic part is hydrophobic. The greatest effect of quaternary ammonium compounds is observed against gram-positive bacteria, whereas gram-negative organisms, many of them significant in the contamination of food, may not be affected (Troller, 1993).

Hydrogen peroxide has been found to be effective in removing biofilms from equipment used in hospitals. The effect of hydrogen peroxide is based on the production of free radicals, which affect the polysaccharides and glycoproteins in the biofilm (Christensen, 1989). The microbicidal effect of peracetic acid on microbes in biofilms was shown to vary (Exner et al., 1987; Goroncybermes and Gerresheim, 1996; Pietersen et al., 1996; Kramer, 1997). Aldehydes did not break the biofilm, but rather seemed to improve its stability. The biofilm must be disrupted in some way before chemical agents such as peracetic acid and aldehydes can be used effectively (Exner et al., 1987). The effect of ozone treatment has been found to vary depending on the processing circumstances and the microbes tested, for example, ozonation proved very effective in treatment of cooling water systems (Nakayama et al., 1985; Lin and Yeh, 1993). Physical disinfection can also be performed using ultraviolet irradiation, pulsed laser beams, or steam disinfection (Troller, 1993).

In the NordFood project P93156 "Sanitation in dairies (1994–96)," testing of disinfectants selected by the dairy partners was performed using tests proposed by the European Committee for Standardization CEN (EN 1040 and EN 1276; EN stands for the official standard version written in English). The microbes used were standard test organisms as well as spoilage bacteria, pathogens, and spores of concern in dairies. All disinfectants diluted in distilled water reduced the number of vegetative cells by $\geq$5 log units when tested without skim milk. The EN 1040 did not differentiate between the six disinfectants, whereas high levels of interfering substances such as skimmed milk and hard water used in EN 1276 reduced the efficiency of the majority of the disinfectants by variable degrees. *Staphylococcus aureus* and

10.2 ELIMINATION OF BIOFILMS

P. aeruginosa were the most resistant to all disinfectants tested. *Salmonella infantis* was the most difficult to kill of the strains of industrial concern. Spores of *Bacillus cereus* and *Bacillus thuringiensis* were reduced by less than 1 log unit in the suspension test. The testing of disinfectants using 80% of the lowest recommended concentrations of the agents differentiated well between their efficiency at the described test conditions given in EN 1276. These results showed that alcohol-based agents were the most efficient of those tested (Table 10.4). More stringent tests are, however, clearly needed to simulate the conditions of practical application. Results in surface tests showed that *P. fragi* cells dried on surfaces were more resistant to the disinfectants than were cells in suspension.

Fogging may be defined as "chemical disinfection by means of automatic spraying of disinfectant in a closed room." The aim of disinfection testing on an industrial scale using fogging was to study the efficiency of the disinfection on surfaces at different places in the room. Controlled experiments were carried out at two cheese-producing dairies. Neither of the fogging trials showed clear reduction of the microbial load. Critical points in fogging were the amount of fog used, the disinfectant concentration used for the fog, and thorough rinsing afterward (Wirtanen et al., 1997).

10.2.2 Elimination of Biofilms in Industrial Systems

10.2.2.1 Cleaning in General. The elimination of biofilm is a very difficult and demanding task in which detachment of the biofilm and the microbicidal effect of the agents used should be taken into account (Mafu et al., 1989, 1990; Nichols, 1991; Brown and Gilbert, 1993; Boulange Petermann, 1996). Cleaning agents used in the food industry can be solid or liquid. Alkaline cleaning agents are more commonly used in closed systems. In some cases acidic treatment has been reported to have almost no effect on surface hygiene (Czechowski and Banner, 1990). Depending on the application of the cleaning agent in

TABLE 10.4. Comparison of Efficiency (in log reduction) of Disinfectants Using the Highest Recommended Concentration of Each of the Six Disinfectants

	Hypo-kloran SP (hypo-chlorite)	P3-topax 99 (alkyl-amin-acetat)	P3-oxonia aktiv (peroxide peracetic acid)	P3-triquart (QAC^a, nonionic tenside)	Virkon S (potassium persulfate chelate)	IPA 300 (iso-propanol QAC^a)
Staphylococcus aureus ATCC 6538	2.3	>5	>5	>5	3.6	>5
Pseudomonas aeruginosa ATCC 15442	3.4	>5	>5	3.8	>5	>5
Enterococcus faecium ATCC 10541	>5	>5	>5	>5	>5	>5
Escherichia coli ATCC 10536	1.6	>5	>5	>5	>5	>5
Enterococcus hirae ATCC 8043	>5	>5	>5	>5	>5	>5
Pseudomonas fragi ATCC 4973	3.6	>5	>5	>5	>5	>5
Salmonella infantis (VTT ELI 5)	<1	>5	>5	1.7	>5	>5
Listeria monocytogenes (Valio)	>5	>5	>5	>5	>5	>5

SOURCE: Wirtanen et al., 1997.
^{a}Quaternary ammonium compound.
NOTE: When the testing was carried out in distilled water and absence of skim milk protein, all disinfectants gave ≥5 log kill of all strains. The disinfectants in these tests were diluted in water of standard hardness and 70% (v/v) skim milk.

open or closed processes, the ability to form foam is either desirable or not. The cleaning effect is decreased by the foam in closed processes and in machine washing. In pressure cleaning the agent should form a solid foam or a gel (Troller, 1993). Mechanical treatments in cleaning are the most efficient ways of eliminating biofilms, but frequently the structure of the equipment makes this difficult (Exner et al., 1987; Chisti and Moo-Young, 1994). Multipressure washing systems have limitations because most of them use large amounts of water and the cleaning is thus expensive. Cleaning-in-place (CIP) systems have not been designed to eliminate biofilms but they can prevent biofilm formation if the equipment design and materials are suitable (Schwach and Zottola, 1984; Stone and Zottola, 1985).

Sanitation, that is, cleaning and disinfection, is carried out in food processing plants in order to produce safe products with acceptable shelflife and quality. In the food industry there is a trend toward longer production runs with short intervals for sanitation (Lelieveld, 1985). The sanitation should be performed as cost-effectively and safely as possible, which means as infrequently as possible, in the shortest possible time, with low chemical, energy, and labor costs, producing as little waste as possible, and with no damage to the equipment (Lelieveld, 1985). The mechanical and chemical power, temperature, and contact time in the cleaning regime should be carefully chosen to achieve an adequate cleaning effect (Dhaliwal et al., 1992; Holah, 1992; Sjöberg et al., 1995). Attention should also be paid to the quality of the processing water, steam, and other additives. The process is easily spoilt by using additives of poor quality.

The key to effective cleaning of a food plant is understanding the type and nature of the soil (e.g., sugar, fat, protein, and mineral salts) and of the microbial growth on the surfaces to be removed. The accessibility and type of equipment and accessories to be cleaned and the availability of suitable cleaning agents are also important (Troller, 1993). Intelligent integration and coordination between cleaning programs and manufacturing operation are essential to achieve both successful cleaning and business profit (Troller, 1993). An efficient cleaning procedure consists of a sequence of rinses and detergent and disinfectant applications in various combinations of temperature and concentration (Dunsmore et al., 1981; Exner et al., 1987). This controls accumulation of soil and the development of biofilms on equipment surfaces without corroding the surfaces (Dunsmore et al., 1981; de Goederen et al., 1989). An independent quality control system to monitor the cleaning results for a food plant can be integrated in the program based on Hazard Analysis of Critical Control Points (HACCP). The goal to achieve a clean food plant must be desired by the plant management, which has to invest the necessary time and money to accomplish it. The personnel must be properly trained and responsible to maintain a good level of plant hygiene. The tools and materials used must be properly designed for the plant equipment and the methods used must suit the process (Holah, 1992; Gould, 1994).

Factors affecting the cleaning process are based on mechanical and chemical impact, holding time, and temperature (Czechowski and Banner, 1990; de Goederen et al., 1989; Dunsmore et al., 1981; Mosteller and Bishop, 1993; Troller, 1993). The basic task of detergents is to reduce the interfacial tensions of soils so that the soil becomes miscible in water (Troller, 1993). The effect of the surfactants is increased by the mechanical effect of turbulent flow or water pressure, or of abrasives, for example, salt crystals. A prolonged exposure of the surfaces to the detergent makes the removal efficient. Detergents to be used in the cleaning of open systems are formulated to be effective at room temperature or at slightly elevated temperatures in the range 35–50°C (Troller, 1993). In closed systems the detergents are formulated to be used at temperatures in the range of 55–80°C. Fats are easily removed at temperatures slightly above their melting point. Sugars and other carbohydrates are water soluble at elevated temperatures, but temperatures causing caramelization should

10.2 ELIMINATION OF BIOFILMS

Figure 10.4 The pilot scale equipment at VTT Biotechnology is used for optimization of closed cleaning procedures. It has been built according to European hygienic norms.

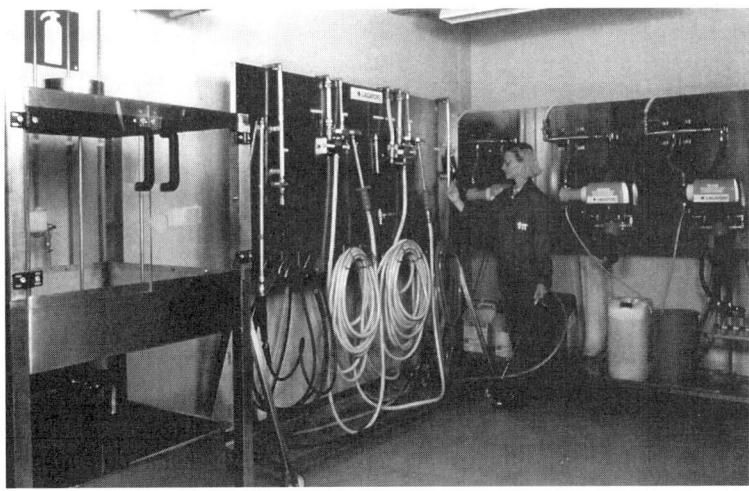

Figure 10.5 Cleaning and disinfection procedures can be optimized with the test rig for open processes. Such a multipressure test rig was designed and built for research purposes at VTT Biotechnology in 1995–96.

be avoided. Proteins are denatured at elevated temperatures and may adhere strongly to surfaces at high temperatures. The use of effective cleaning agents and disinfectants on surface-attached microbes minimizes contamination of the product, enhances shelflife, and reduces the risks of foodborne illness (Czechowski and Banner, 1990). Cleaning and disinfection procedures can be optimized with pilot scale equipment for both closed (Fig. 10.4) and open processes (Fig. 10.5).

10.2.2.2 Elimination of Biofilm in Open Systems.

Gross soil should be removed by dry methods (brushing, scraping, or vacuuming) and visible soil rinsed off with low-pressure water. The cleaning effect is increased by using water of sufficient volume and temperature (Troller, 1993). However, a pure water washing system is not practical due to ineffectiveness and cost limitation. Surfactants, which suspend the adhered particles and microbes from the surfaces in the water, are added to increase the washing effect (Stone and Zottola, 1985; Czechowski and Banner, 1990; Frank and Koffi, 1990; Troller, 1993). Pressure cleaning is used to remove the remainder of the soil and most of the microorganisms present on open surfaces (Table 10.5; Holah, 1992). The elimination of biofilm from open systems, such as surfaces in food processing equipment, has not been widely reported. Holah et al. (1988, 1989) used the biofilm approach in cleaning and disinfection studies for the food industry. After a production run, the equipment should be dismantled and the cleaned utensils should be stored on racks and tables, not on the floor. The cleaning of open process surfaces and surfaces in the processing environment are carried out using either foam or gel cleaning. Foams are most effective in situations where contact with the soil for an extended contact time is necessary. The foam units are constructed to form foam of varying wetness and durability depending on the cleaning to be performed. The application of gels extends the contact time with a soiled surface and can be used with low-pressure systems (Troller, 1993). The cleaning is mostly carried out in combination with a final disinfection, because there are likely to be viable microorganisms on the surfaces that could harm continued production (Exner et al., 1987; Brackett, 1992; Holah, 1992).

The aim of a series of experiments in the cleaning of open systems, parameters given in Table 10.6, was to develop microbial methods for the detection of biofilm and bacterial cells left on surfaces after cleaning and disinfection. Conventional cultivation, combined with impedance measurements (Fig. 10.6) together with image analysis (Fig. 10.7) and cytological (CTC-DAPI) cell staining (Fig. 10.8) gave results that were comparable, complementary, and enabled a total evaluation of both the removal of biofilm and the killing of bacterial cells. Results indicated that the low-pressure application system was not effective in removing all the biofilm unless a foam agent itself was effective. The efficiency of the foam agent was dependent on its ability to remove biofilm from the working surface com-

TABLE 10.5. A Sanitation Sequence for a Production Environment in Food Processing

Sequence	Purpose
Dry cleaning, brushing, vacuuming	Removal of gross soil from production equipment and environmental surfaces
Rinsing	Rinsing of gross soil from the equipment to drain
Cleaning and rinsing of environmental surfaces	Removal of visible soil from environmental surfaces
Cleaning of equipment	Removal of visible soil from production equipment (CIP in closed and foam or gel cleaning in open systems)
Rinsing of equipment	Removal of soil and detergents from equipment surfaces (turbulent flow in closed and multipressure in open systems)
Disinfection	Killing of microorganisms left on surfaces and prevention of microbial growth during interproduction period
Fogging	Disinfection in the production facilities

Source: Compiled according to Holah, 1992.

10.2 ELIMINATION OF BIOFILMS

TABLE 10.6. Parameters in the Experiments for the Cleaning Procedures Performed on Open Surfaces in the NordFood Project P93156 "Sanitation in Dairies (1994–96)"

Variables in Cleaning of Open Surfaces	Parameters
Microbial stains	*P. fragi* VTT E-86249
	L. monocytogenes (Valio)
	⇒ pure and mixed cultures
Organic soil	NordFood soil containing
	⇒ 1% whole milk powder, 1% whey powder, and 0.5% modified starch (E1422) in 20 ml whipping cream and 80 ml water
Surface material	stainless steel AISI 304, 2B (R_a 1.6)
Cleaning agents, treatment time	SU727 Trippel (strong alkaline and chlorine)
	P3-topax 12 (mild alkaline; designed for cleaning of painted surfaces)
	⇒ 15 min at 20°C, rinsing 10 s
Disinfectants, treatment time	P3-Oxonia Aktiv (peroxide and peracetic acid)
	Virkon S (potassium persulfate and chelate)
	⇒ 5 min at 20°C, rinsing 10 s
Detection methods	Cultivation
	Impedance measurement
	Image analysis of acridine orange stained surfaces
	Epifluorescence microscopy using metabolic indicators
	Agar molding
	ATP

Source: Wirtanen et al., 1997

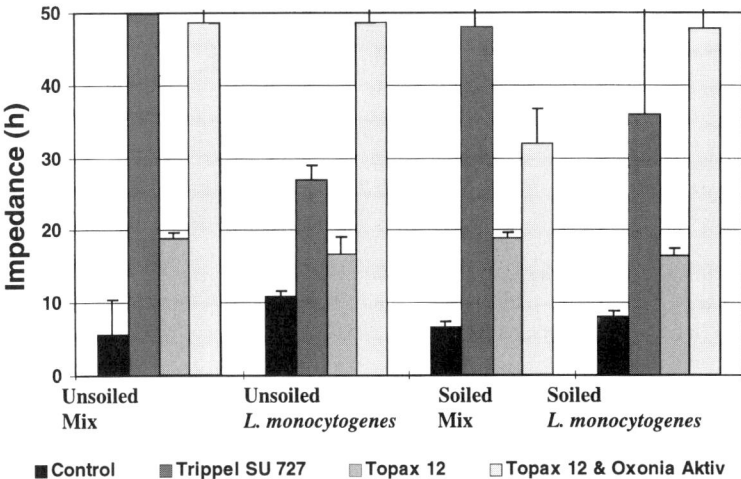

Figure 10.6 The impedance measurements of *L. monocytogenes* and a mixed culture of *P. fragi* and *L. monocytogenes* (= mix) grown as biofilms on unsoiled and soiled stainless steel surfaces showed that chlorine and disinfectant treatments prolonged the reactivation period of the treated cells in the biofilms (Wirtanen et al., 1997).

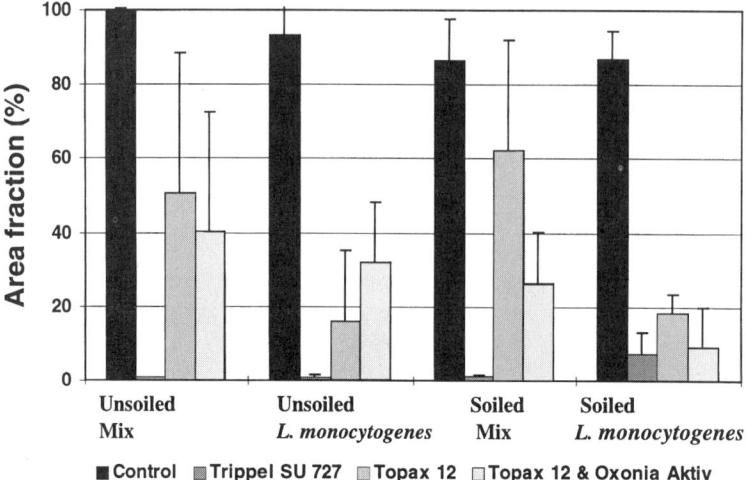

Figure 10.7 The image analysis, after staining with acridine orange, of 4-d-old biofilms of *L. monocytogenes* and a mixed culture of *P. fragi* and *L. monocytogenes* (mix) on unsoiled and solid stainless steel surfaces (AISI 304, 2B) showed that the strong alkaline cleaner Trippel SU 727 was efficient in removing biofilms using low pressure cleaning (Wirtanen et al., 1997).

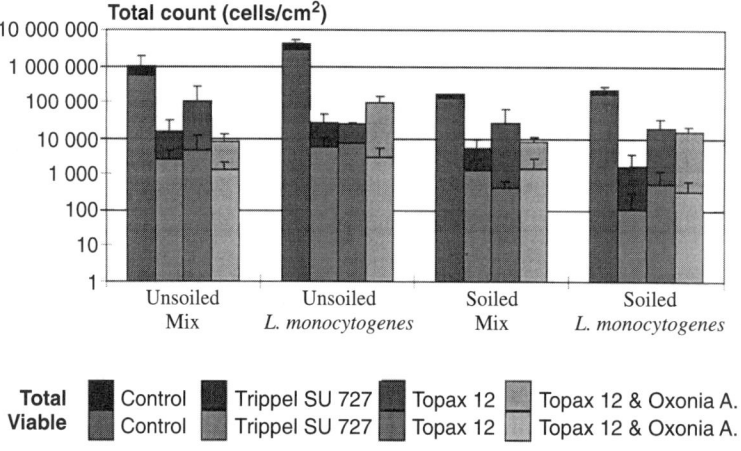

Figure 10.8 Image analysis after CTC-DAPI staining of *L. monocytogenes* and a mixed culture of *P. fragi* and *L. monocytogenes* (= mix) grown as biofilms on unsoiled and presoiled stainless steel surfaces (AISI 304, 2B) showed that most of the bacteria on the surfaces were nonculturable after the cleaning (Wirtanen et al., 1997).

bined with the ability to kill the bacteria present in the biofilm. The foam agent must therefore remain in contact with the surface for a sufficient length of time without drying. The results of the tests using CTC-DAPI staining showed that the effects of all four cleaning procedures were quite similar. Cleaning treatment with Topax 12 combined with disinfectant treatment (results given for Oxonia Aktiv) appeared to be somewhat more effective than the treatment with Topax 12 alone. These results thus indicated that the disinfectants exerted a surprisingly low effect on cleanability. The results obtained by cultivation showed a greater difference between the cleaning methods, which may be partly explained by the stress in the bacterial cells caused by the cleaning and disinfective agents.

10.2.2.3 *Elimination of Biofilm in Closed Systems.* In the cleaning regime applicable to closed processes, prerinsing with cold water is carried out to remove loose soil. The CIP treatment is normally performed using hot cleaning solutions (Chisti and Moo-Young, 1994), but cold solutions can also be used in the processing of fat-free products. The warm alkaline cleaning solution, normally of 1% sodium hydroxide (NaOH), is heated to 75–80°C and the cleaning time is 15–20 min. The equipment is rinsed with cold water before the acid treatment is performed at approximately 60°C for 5 min. The effect of chlorine-based agents can be divided into three phases: loosening of the biofilm from the surface, breakage of the biofilm, and the disinfective effect of the active chlorine (Costerton et al., 1985; Griebe et al., 1994). The cleaning solutions should not be reused in processes aiming at total sterility because the reused cleaning solution can contaminate the equipment (Chisti and Moo-Young, 1994). Single-phase cleaning agents for CIP treatment are nowadays more commonly used because the processing industry wants to save time (Husmark, 1994). In single phase cleaning procedures, the time for one cleaning process, normally the acid treatment, and a rinsing step can be saved.

Studies on CIP cleaning showed that the "cleaning effect" decreased radically when the washing time and the flowing rate were reduced, indicating that the cleaning as a whole was affected by mechanical effects, time, temperature, and chemical parameters (de Goederen et al., 1989; Czechowski and Banner, 1990; Brackett, 1992; Zottola and Sasahara, 1994). Turbulent flow is therefore a decisive factor in combination with chemical compounds in the elimination of biofilms from closed systems. To ensure the removal of soil and to avoid biofilm formation and soil sedimentation, the minimum flow velocity in the CIP treatment must be at least 1.5 m/s, but 2.0 m/s is recommended (Grasshoff, 1992). The flow should be turbulent with a Reynolds number from at least 10,000 to, preferably 30,000, in order to ensure good radial mixing and heat, mass, and momentum transfer (Grasshoff, 1992). Chelating agents in the cleaning solution enhance the removal of biofilms from processing surfaces. The CIP systems used today are based on a combination of alkaline-acid treatment and time-temperature treatment. Problems caused by equipment constructions, valves, and so forth and surface materials cannot be eliminated with CIP, because the CIP treatment was not designed to eliminate biofilms. In normal operation it is impossible to clean pipelines completely by mechanical treatment because there are always bends, corners, pockets, and cracks where biofilm remains (LeChevallier et al., 1988, 1990). The depth of dead zones in the system should be less than one pipe diameter if the dead zone cannot be avoided, in order to ensure adequate cleaning throughout the system using CIP procedures. The maintenance programs should guarantee that excessive biofilm accumulation does not occur. Tanks should be cleaned by applying the cleaning solution through properly installed, removable spray balls or nozzles. The design should ensure that the part of the tank directly above the spray ball is also cleaned (Chisti and Moo-Young, 1994). Drainage, minimization of internal probes, crevices and stagnant areas, arrangement of valves, couplings, and instrument ports, and instrumentation should be planned carefully so that the equipment is easily cleanable (Chisti and Moo-Young, 1994).

European Union legislation on food hygiene and the hygienic design of food machinery, together with the public awareness of product quality and manufacturers' desires to improve product safety, make reliable cleanability testing an important issue. In this type of testing it must be possible to assess the relative cleanability of various equipment components to facilitate the design, testing, and maintenance of hygienic food-processing equipment. The assessment must use standardized test procedures with a sound scientific basis (EHEDG, 1992). The aim of the European Hygienic Equipment Design Group (EHEDG), which is an independent consortium of representatives from research institutes, the food industry, equipment manufacturers, and government organizations, is to develop hygienic equipment on a scientifically and technologically sound basis. The EHEDG has close ties with organizations such as 3-A and the National Sanitation Foundation (NSF) in the USA (Lelieveld, 1999).

There are relatively few international standards for cleanability, and most of these are for the dairy industry. A number of guidelines, for example, the 3-A standards for the dairy industry and NSF standards for food service equipment in the United States, have been developed, but these standards have no benchmarks or test regimes for cleanability (EHEDG, 1992; Lelieveld, 1999). The NSF standards are not applicable to the hygienic design of general food processing equipment. The in-place cleanability of equipment components is currently assessed with a test method based on organic soiling with sour milk containing spores of *Bacillus stearothermophilus* var. *calidolactis* NIZO C953. Areas of poor hygienic design are indicated by growth in SHA agar after a CIP procedure with a mild detergent when the tested piece is compared with reference pipes. This method is intended as a basic screening test for hygienic equipment design and is not indicative of performance in industrial situations (EHEDG, 1992). Evaluation of the microbial cleanability of equipment and process lines already in use are needed. The use of new methods, for example based on the luminescent bacterium *Photobacterium leiognathi*, in cleanability testing of open process lines is needed and must therefore be developed (Wirtanen, 1995).

The aims of the experiment carried out using closed systems, of which experimental details are given in Table 10.7, in the NordFood Project Sanitation in Dairies (1994–96) were to evaluate different CIP procedures and to develop microbial swabbing methods for the de-

TABLE 10.7. Parameters in the Experiments for the Cleaning Procedures Performed on Surfaces in Closed Process in the NordFood Project Sanitation in Dairies (1994–96)

Variables in Cleaning of Closed Surfaces	Parameters
Microbial strains	*P. fragi* VTT E-86249
	B. thuringiensis
	⇒ pure cultures
Organic soil	NordFood soil containing
	⇒ 1% whole milk powder, 1% whey powder, and 0.5% modified starch (E1422) in 20 ml whipping cream and 80 ml water
Surface material	Stainless steel AISI 304, 2B (R_a 1.6)
Cleaning agents, treatment time	Sodium hydroxide
	MIP SP (potassium hydroxide and chelators)
	Sodium hydroxide with EDTA-based P3 Stabicip EA added
	⇒ 7 min at 75°C, cold rinsing 2 min, 1.5 m/s
Detection methods	Cultivation
	Impedance measurement
	Image analysis of acridine orange stained surfaces
	Agar molding

Source: Wirtanen et al., 1997.

10.3 DETECTION OF BIOFILM

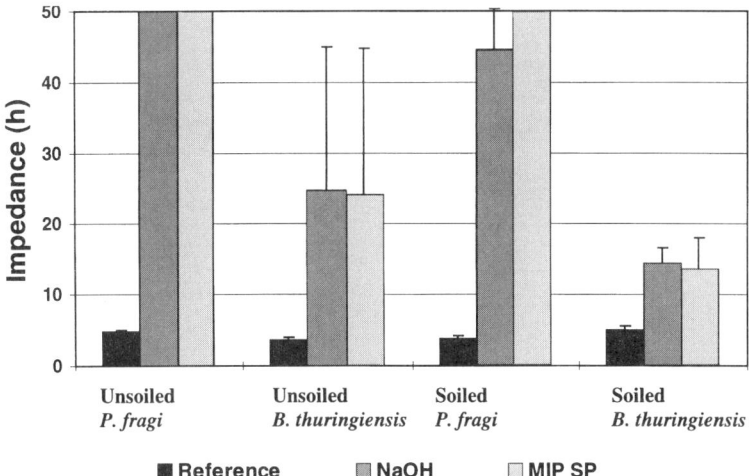

Figure 10.9 The impedance measurements were carried out using coupons in the test cells. The results showed that none of the cleaning agents (sodium hydroxide and potassium peroxide-based MIP SP) could remove *B. thuringiensis* spores. Both cleaning agents proved effective against the *P. fragi* biofilm also on presoiled surfaces (Wirtanen et al., 1997).

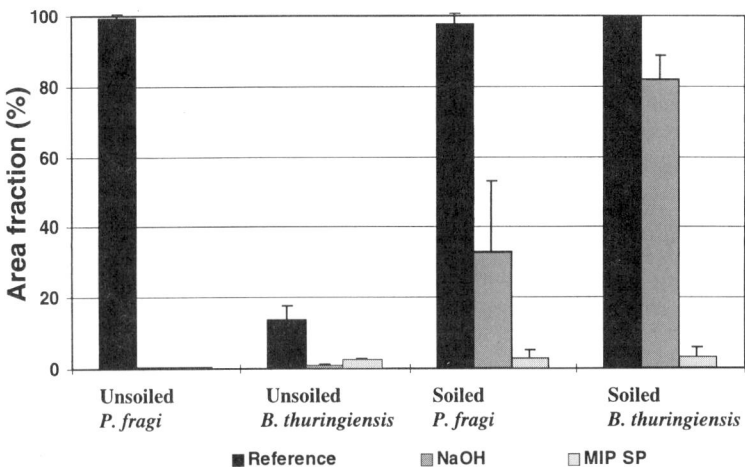

Figure 10.10 Epifluorescence microscopy performed directly on surfaces after staining with acridine orange showed that the cleaning effect was more efficient with potassium hydroxide (KOH)-based MIP SP than with 1% sodium hydroxide (NaOH) (Wirtanen et al., 1997).

tection of biofilm and bacterial cells left on surfaces after CIP treatments. Impedance measurements (Fig. 10.9), image analysis (Fig. 10.10), and conventional cultivation (Fig. 10.11) gave comparable results and complemented each other. The concentration of the cleaning agents used did not destroy all the bacteria growing in biofilms on surfaces. The cleaning effect on unsoiled surfaces was better than on soiled ones. None of the treatments removed all of the biofilm, and low numbers of viable bacteria still remained. The spores of *Bacillus thuringiensis* were more firmly attached to the coupons than the biofilm. Furthermore,

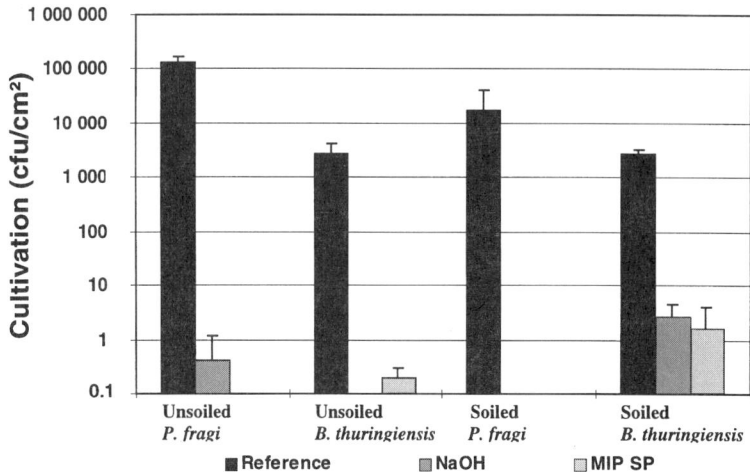

Figure 10.11 The results of swabbed surfaces using conventional cultivation showed that some living cells were present after the treatments with both sodium hydroxide (NaOH) and potassium hydroxide (KOH)-based MIP SP (Wirtanen et al., 1997).

spore-forming foodborne bacteria have high heat and chemical resistance. The results provide interesting information concerning the methodology of hygiene testing and these results showed that the gel matrix on soiled surfaces can give very high area values on surfaces in cases where the cultivable cell number was relatively low (Wirtanen et al., 1997).

The efficiency of the cleaning agent is dependent on its ability to remove biofilm from the working surface, combined with the ability to kill the bacteria present in the biofilm. Chelating agents in the cleaning solution seemed to enhance the removal of biofilms from processing surfaces but frequently leave living bacteria on the surfaces. Treatments with sodium hydroxide containing chelating agents such as MIP SP were more efficient in removing biofilm than sodium hydroxide alone (Wirtanen et al., 1997)

10.3 DETECTION OF BIOFILM

Sessile growth of microbes on equipment and other surfaces involved in food production affects the performance of the processing and the quality of the final product. The organisms, embedded in a slime matrix of extracellular polymer substances, for example, polysaccharides and proteins, are well protected against disinfectants and cleaners. The actual monitoring practice involves sampling of the liquid phase, which does not reflect the location or extent of microbes growing on surfaces. Inadequate cleaning and sanitation of surfaces coated with these films represents a source of contamination within the process (Wirtanen, 1995). The regrowth of microbial contaminants shortens the interval between cleaning cycles and diminishes process efficiency and product quality (Lelieveld, 1985). Practical methods for assessing microbes and organic soil on processing surfaces are needed to establish the optimal cleaning frequency of the equipment.

Reliable monitoring systems that can provide information on-line, directly, and in real time about microbial growth are required within the process. These monitoring techniques

10.3 DETECTION OF BIOFILM

should preferably be adaptable to computer processing. The methods are based on optical and electrochemical measurements, ion mobility, and infrared techniques as well as bioluminescence. The successful transfer of these techniques for on-line monitoring of food and beverage quality and cleanliness of surfaces in food-related processes should be based on microbial reference methods. This means that the threshold values for detected amounts of contaminants should be very low.

10.3.1 Tools in Research and Industrial Use

The formation of biofilm in a laboratory environment is rather difficult because biofilm is not readily produced in laboratory media rich in nutrients. Bacteria usually do not produce biofilm when nutrients are present in excess and the bacteria appear freely suspended (Costerton et al., 1987; Anwar et al., 1990). Many reviews have been published on the formation and detection of biofilms in the laboratory (Duddridge, 1981; Pope and Zintel, 1989; Fletcher, 1990; Ladd and Costerton, 1990). Methods of studying biofilm formation include microbiological, physical, chemical, and microscopical methods (Tables 10.8–10.10; Duddridge, 1981; Pope and Zintel, 1989; Ladd and Costerton, 1990; Mafu et al., 1991;

TABLE 10.8. Methods of Analyzing Biofilms on Food Contact Surfaces

Method	Application and Comments
Contact plate	Used for hygiene testing
	Advantage: identification, selective growth media
	Limitation: sensitivity, time consuming
Swab method (cultivation)	Used for hygiene testing
	Advantage: identification, selective growth media
	Limitation: sensitivity, time consuming
Impedance	Informative in laboratory studies
	Advantage: more rapid than cultivation, selective growth media
	Limitation: sensitivity
ATP	Used for hygiene testing, large scatter when analyzing multilayer biofilms due to different metabolic status of cell layers
	Advantage: on-site ATP-based biofilm detectors
	Limitation: sensitivity
Epifluorescence microscopy	Informative in laboratory studies
	Advantage: DNA stains and metabolic indicators
	Limitation: two-dimensional, research purpose
Electron microscopy (SEM, TEM)	Information in laboratory studies
	Advantage: structural studies of biofilms
	Limitation: not for routine use on-site
Scanning confocal laser microscopy (SCLM)	Information for studies of live biofilms
	Advantage: can be applied to living biofilms for structural studies
	Limitation: research purposes only
Flow cytometry	Informative for studies of microbial densities
	Advantages: accurate, sensitive, and reproducible method; various stains
	Limitation: detachment of cells; at the moment mostly used in research due to expensive equipment

Source: Wirtanen, 1995.

TABLE 10.9. Various Microscopic Techniques for Studying Cell Adhesion and Biofilm Formation on Surfaces

Microscopic Method	Application

Epifluorescence microscopy
  a. DNA stains: AO, DAPI, etc.
  b. Double staining with DNA stains and metabolic indicators: CTC, INT, etc.
  c. Fluorescent immuno and genetic probes
	a. Distribution of microbes; area coverage
	b. Differentiation of respiring cells in the total amount of cells
	c. Identification of microbes; differentiation between types of microbes

Electron microscopy
  d. Immune electron microscopy
  e. Scanning electron microscopy (SEM)
  f. Transmission electron microscopy (TEM)
	d. Identification of microbes
	e. Distribution and morphology of microbes on surfaces and in biofilms
	f. Distribution and morphology of microbes in biofilms

Interference reflection microscopy	Assessment of adhesive interactions between surfaces and microbes
Differential interference contrast (DIC) microscopy	Visualization of individual microbial cells; not for quantitative analysis of biofilm

Scanning confocal laser microscopy (SCLM)
  g. Fluorescent stains and chemical probes
  h. Fluorescent immuno and genetic probes
	g. Distribution of microbes, other objects, and molecules in biofilms; physiological condition in biofilms; biofilm structures
	h. Identification of microbes and their activities in biofilms; ecology of biofilms

Source: Harmsen, 1996; Kostyál, 1998.

TABLE 10.10. Summary of Main Methods Used in Hygiene Research Projects for Assessing the Cleaning Efficiency of Process Surfaces

Detection Method	Application
Conventional cultivation	Assesses living bacteria
Impedance measurement	Assesses the activity of viable and injured bacteria that can recover after reactivation in broth
ATP measurement	Assesses total hygiene
Measurement of protein residues	Assesses the amount of protein-based soil
Image analysis of samples stained with DNA-stains, e.g., AO, DAPI	Assesses biofilm components on the surface (including organic soil, dead and living cells, and slime)
Epifluorescence microscopy of samples stained with metabolic indicators, e.g., CTC-DAPI	Assesses ratio of living and dead cells in biofilms on surfaces and in liquids

Source: Wirtanen et al., 1999.

Nivens et al., 1995; Wirtanen et al., 1999). The biofilm consists of about 85–96% water, which means that only 2–5% of the total biofilm volume is detectable on dry surfaces (Costerton et al., 1981; Duddridge, 1981). In detecting biofilm, the planktonic cell counts of the processing fluids should not be used because they do not represent the sessile organisms

(Cloete et al., 1989). Advances in molecular biology are producing methods with which detection and enumeration of specific organisms on surfaces can be performed (Harmsen, 1996; Kostyál, 1998).

10.3.1.1 Cultivation, Contact Agar, and Sampling Methods.

Organisms from extreme environments are difficult to culture and therefore the standard plate counts do not give accurate estimates. Hygiene testing is currently based on conventional cultivation using swabbing or contact plates (Holah et al., 1988, 1989). These classical evaluation methods suffer from several serious deficiencies (Mittelman, 1991). Conventional cultivation measures only the number of living cells able to grow on the chosen agar. The quantification of cells in the biofilm is hard to perform because they adhere strongly to the surfaces. In the cultivation of microorganisms in the biofilm it is important that the sample is detached and mixed properly (Holah et al., 1988, 1989). Too strong agitation in the detachment of the biofilm from the surface may harm the cells so that they are not able to grow on the agar plates, whereas deficient mixing may result in clumps and inaccurate results (Gilbert and Herbert, 1987; Holah et al., 1988). It has been reported that ultrasonics provide a suitable tool for cleaning biofouled surfaces. The cultivation showed that the use of ultrasonics detached about 10 times the number of cells from the surface compared with swabbing (Zips et al., 1990). The choice of agar and incubation conditions in the cultivation is governed by those characteristics of the microbes considered to be most important (Pope and Zintel, 1989).

The use of various detergent and enzyme solutions in cleansing substances is known per se in the literature. These publications do not, however, report the efficiency of the detergents in removing living microbes. Furthermore, the concentrations and the temperature of the effective substances used are usually so high that the microbes are damaged as the soil is removed (Tuompo et al., 1998). The previously mentioned applications are thus not in themselves suitable for surface hygiene monitoring based on cultivation. At the moment, a method for a reliable analysis of the released microorganisms from process surfaces is still lacking. The use of the various detergent solutions in evaluating the cleaning efficiency of open process surfaces was carried out in both laboratory and industrial environments (Wirtanen et al., 1997). In that work the remaining biofilms were detected by swabbing the surfaces either with swabs moistened in various detergent (alcohol-, surfactant-, and enzyme-based) blends or with dry swabs, the detergents being sprayed on the sampling surface before swabbing. These swabs were placed in inactivation solution to neutralize the chemicals interfering in cultivation on agar plates or Hygicult contact slides. The Hygicult slides (Orion Diagnostica, Finland) were also used in assessing the hygiene of equipment surfaces moistened with detergent blends in advance (Wirtanen et al., 1997; Salo et al., 1999).

The evaluated test kit comprises a composition designed for the pretreatment of process and equipment surfaces before sampling. It provides a sampling method particularly suited to the monitoring of surface hygiene before and after cleaning procedures (Tuompo et al., 1998). The kit contains a sampler as well as a detergent- or alcohol-based solution in spray bottles. The method is suitable for the removal of surface-adherent microorganisms that spoil or poison food, such as *P. fragi*, *L. monocytogenes*, *B. cereus*, *Candida utilis*, and *Aspergillus niger* and other bacteria, yeast, and mold. It is easy to assess the hygiene level with present microbial analysis techniques based on cultivation using either plates or dip slides as well as on ATP and on impedance (Tuompo et al., 1998).

10.3.1.2 ATP Measurement. The chemical methods used in the assessment of biofilm formation are indirect methods based on the utilization or production of specific compounds (organic carbon, oxygen, polysaccharides, and proteins) or on the microbial activity (living cells and ATP (adenosine 5′-triphosphate) content (Characklis et al., 1982)). The ATP measurement is a luminescence method based on the luciferine–luciferase reaction. The ATP content of the biofilm is proportional to the number of living cells in the biofilm and provides information about their metabolic activity (Gilbert and Herbert, 1987; Henriksson and Haikara, 1990). Kinetic data obtained for freely suspended cells should not be used to assess immobilized biomass growth, for example, biofilm. The ATP method is insensitive and therefore not suitable for hygiene measurements in equipment where absolute sterility is needed, because with most of the reagents used today a count of at least 10^3 bacterial cells is needed to obtain a reliable ATP value (Wirtanen, 1995; Lappalainen et al., 1999; Lundin, 1999).

10.3.1.3 Impedance Measurement. In the impedance method, the changes in conductance and capacitance of the samples including biofilm samples on plastic, rubber, and metal surfaces are measured. The changes in impedance depend on the quantity of ions moving in the liquid: cations moving to the negatively charged electrode and anions to the positively charged electrode (Firstenberg-Eden and Eden, 1984). The increase in conductance and capacitance due to the metabolic activity of the microbes leads to a decrease in the impedance. Physical factors affecting the impedance measurement include temperature, composition of the substrate, and electrode type. The temperature must be kept constant so that the impedance measures only the changes due to microbial growth. The measurement of the change in impedance value at suitable time intervals provides an impedance curve and thus the detection time of microbial growth in the sample (Firstenberg-Eden, 1986; Baumgart and Sieker, 1993; Flint et al., 1997c). The detection time depends on the number of microbes in the sample. Handling of the samples, for example, storage and cooling, can affect them, which means that the detection time of microbial growth may be altered. Impedance measurements must therefore be calibrated with cells treated in the same way as the samples. Results are achieved more rapidly with impedance measurements than with cultivation (Johnston and Jones, 1993). Impedance measurement is used in the food industry to control product quality and to assess the effect of cleaning agents and disinfectants (Holah et al., 1990; Mosteller and Bishop, 1993; Wirtanen et al., 1997).

10.3.1.4 Microscopical Methods and Other Research Tools

10.3.1.4.1 Light, Epifluorescence, and Electron Microscopy. Microscopic techniques such as epifluorescence microscopy and, especially, scanning and transmission electron microscopy (SEM and TEM) are very informative tools in biofilm research and hygiene studies (Richards and Turner, 1984; Lewis et al., 1987; Holah et al., 1988, 1989; Ladd and Costerton, 1990; Zottola, 1991). Epifluorescence image analysis of biofilm cells does not correlate with the conventional swabbing method. The drawback of acridine orange staining is that the dye stains both dead and living cells nonspecifically along with noncellular debris (Caldwell et al., 1993; Yu and McFeters, 1994a, 1994b; Yu et al., 1994). One advantage of epifluorescence image analysis is that it measures the adherent cells on the surface, rather than cells that have been detached by some method from the surface. Epifluorescence microscopy of multilayered biofilms can only be counted two-dimensionally (2-D). 2-D imaging techniques have been used to determine the effects of antimicrobial agents and biofilm formation by monitoring attachment rates, detachment rates, growth, motility, viability, morphology, and cell area (Caldwell et al., 1993).

Image analysis is an indispensable tool for obtaining accurate quantitative information about the samples. Video technology and increasingly powerful computers facilitate the application of this technique to a wide range of methodologies, including adhesion, surface hygiene, biomass determination, and biomedical research (Wirtanen, 1995). Electron microscopy has been used to identify biofilm structures. SEM provides accurate information about the surface materials and the position of biofilm cells, but it does not provide quantitative data for statistical analysis (Ladd and Costerton, 1990).

10.3.1.4.2 Confocal Microscopy. Scanning confocal laser microscopy (SCLM) is an optical microscopy technique with significant advantages over conventional light microscopy and SEM (Shotton, 1989). More accurate information is obtained about the chemical and biological relationships between microorganisms and their microenvironment in situ in real time using an SCLM. Attached bacterial cells, microcolonies, and biofilms can be effectively studied with SCLM combined with negative fluorescent staining. Numerous attempts have been made to evaluate the number of living cells in biofilms because it is of interest to count surviving cells after cleaning. Laser microscopy permits optical sectioning of biological materials without optical interference from other focal planes (Caldwell et al., 1992; de Beer et al., 1994; Kostyál, 1998). The use of image analysis and fluorescent probes in combination with SCLM provides a research tool to analyze changes in biofilm structure and chemistry. SCLM also examines the growth and metabolism of living cells in the biofilm; with this technique it is possible to analyze and quantify three-dimensionally (3-D) biofilms and bioaggregates nondestructively, which is not possible with conventional light microscopy (Caldwell et al., 1993).

10.3.1.4.3 Interference Reflection Microscopy. Interference reflection microscopy has been used for investigation of adhesive interactions between the surface and the microorganisms in biofilms. Investigating microbial growth with lasers at frequencies between 190 and 260 nm causes shifts in the wavelength of the scattered photons and can thus be detected as a resonance Raman spectrum. Bacterial biofilm has also been detected on-line with Fourier transformation infrared spectrometry (FTIR), quartz-crystal microbalance, and infrared spectroscopy (Nichols et al., 1985; Marshall et al., 1989; Nivens et al, 1993a, 1993b). FTIR has proved useful for the determination of biomass, but the penetration of the infrared waves work only in biofilms with layers a few cells thick. The quartz-crystal microbalance has proved capable of detecting biomass but it cannot characterize the constituents in the biofilm (Angell et al., 1993). Scanning probe microscopy, including scanning tunneling electron microscopy, atomic force microscopy, scanning ion-conductance microscopy and scanning tunneling microscopy, can provide atomic resolution of the living material. These techniques provide only a view of the surfaces of the objects (Caldwell et al., 1993; Beech, 1996).

10.3.1.4.4 Polymerase Chain Reaction. The polymerase chain reaction (PCR) is a method in which thermostable DNA polymerase is used to exponentially amplify a target DNA sequence defined by two oligonucleotide primers (short segments of synthetic DNA). The amplified DNA fragment can be visualized by agarose gel electrophoresis, which also allows the size determination of the PCR product, or by hybridizing the PCR product with a labelled probe. Combining PCR with a hybridization step improves the sensitivity and specificity of the assay. PCR detection of specific bacteria in a sample material with an abundant microflora (e.g., environmental samples) sets high demands on the specificity of the PCR method applied. In addition, many types of sample materials (e.g., foods) contain factors that can either totally inhibit the amplification reaction or cause partial inhibition leading to a nonexponential amplification of the target DNA (Scheu et al., 1998; Hielm et al., 1999). Inhibition may be avoided or reduced by pre-PCR sample manipulations such as dilution of the sample material, extraction of the DNA from the sample, or by harvesting the bacterial cells from the sample, for example, by using immunomagnetic beads coated with specific monoclonal antibodies. However, even partial inhibition of the PCR reaction inevitably leads to reduced sensitivity and excludes the possibility of performing quantitative PCR. To minimize the risk of obtaining false-negative amplification results, suitable external standards, which are coamplified together with the target DNA in the PCR reaction, should be used (Reischl and Kochanowski, 1995). Sensitivity of the PCR assay can be improved by short enrichment culture treatment of the sample prior to PCR (Agersborg et al., 1997; Wang et al., 1997), but this also precludes attempts to quantitate the number of target organisms in the sample. Thus amplification of target DNA sequences from sample materials containing inhibitory factors for PCR can provide information on the presence, but not on the numbers and usually not on the viability of target organisms. Several variations of the PCR technique have been applied for the detection of food pathogens and contaminants: These include techniques such as multiplex PCR (several targets are amplified in a single PCR reaction, Bhaduri and Cottrell, 1998), nested PCR (same target is amplified using first "outer" and then "inner" primers, Mäntynen et al., 1997), PCR combined to a hybridization step (Deng et al., 1996), reverse transcription PCR (RT-PCR; initial target for PCR amplification in mRNA, which allows the detection of viable bacteria; Klein and Juneja, 1997), and immunomagnetic PCR (IM-PCR; PCR is preceded by immunomagnetic separation of bacterial cells; Docherty et al., 1996). Recently, a technique called in situ PCR, where PCR amplification takes place within bacterial cell, was applied in studies on *Salmonella enterica* serovar Typhimurium gene expression (Fig. 10.12; Tolker-Nielsen et al., 1997, 1998). In situ PCR may in the future prove applicable in biofilm studies.

10.3.1.4.5 Hybridization Techniques. Hybridization techniques can be used in the microbial detection either alone or combined to a preceding PCR step. In the hybridization, the labelled probe, usually a denatured DNA fragment, anneals to a denatured target DNA with sequence homology (e.g., genomic DNA or PCR amplification product). The detection of the hybrids is based on radioactive signal, fluorescence, or color reaction, depending on the type of the label. By determining the intensity of the hybridization signal, the number of

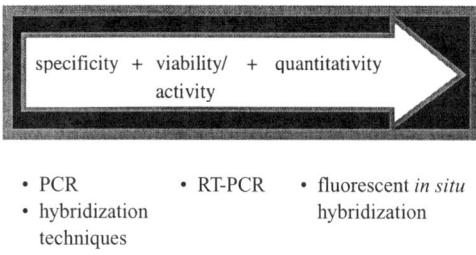

Figure 10.12 Molecular biological detection methods.

10.3 DETECTION OF BIOFILM

target organisms can be estimated. During the past years PCR (alone or combined to a hybridization step) has in large extent replaced the previously applied hybridization techniques such as dot-blot or slot-blot hybridization in the detection of bacteria. However, a technique called in situ hybridization has proved useful in several applications. In this technique the bacterial cell membrane is permeabilized to allow penetration of the labelled oligonucleotide probe and the hybridization then takes place inside the bacterial cell (Amann et al., 1990a, 1990b). Probes for in situ hybridization are usually labelled with fluorescent stains (e.g., fluorescein or rhodamine), which allows the detection of hybrids with techniques such as epifluorescent microscopy, scanning confocal laser microscopy, or flow cytometry (Harmsen, 1996). In situ hybridization has proved an informative tool in detecting uncultureable microorganisms and in identifying bacteria in complex ecosystems such as soil, activated sludge, and especially biofilms (Amann et al., 1992; Wagner et al., 1993; Manz et al., 1995; Neef et al., 1996; Thomas et al., 1997; Santegoeds et al., 1998).

10.3.1.4.6 Flow Cytometry. Flow cytometry has been used to determine the viability of protozoa, fungi, and bacteria. Fluorescent probes revealing the physiological properties of the microbes are useful in measuring microbial populations with flow cytometry, which combines the advantages of microscopy and biochemical analysis of individual cells. Fluorescence of cells stained with rhodamine 123, carboxyfluorescein diacetate, and Chemchrome B in conjunction with forward angle light scatter in the analysis showed that vegetative cells were detected. The advantages of flow cytometry are accuracy, speed, sensitivity, and reproducibility. Flow cytometry measures the viability of a statistically significant number of organisms (5000–25,000 cells per sample) (Kell et al., 1991; Diaper and Edwards, 1994a, 1994b; Pore, 1994; Mason et al., 1995; Lopez-Amoros et al., 1997).

10.3.1.5 Physical Methods. The physical methods used in biofilm assessment are either direct or indirect. Physicochemical methods have been used in studying the surface properties of microbes (Busscher et al., 1990). The direct measurements are based on the mass and thickness of the biofilm and the indirect methods of friction and heat transfer assessments. The thickness of the biofilm can be measured with either dry or wet weight (Characklis et al., 1982; Kristensen and Christensen, 1982). Light microscopy has also been used in thickness measurement (Bakke and Olsson, 1986) as has impedance measurement (Duddridge, 1981). Structural determination measuring the amount of polysaccharides, total organic carbon, total organic nitrogen, oxygen, organic acids, and total and viable cells can also be used for assessing the thickness. It is very important to bear in mind that the biofilm contains 85–96% water (Duddridge, 1981).

10.3.2 Future Prospects for On-line Monitoring of Microbial Deposits

Despite extensive efforts to develop reliable control methods for practical hygiene assessment in the food industry, there are still no such methods available. The conventional swabbing procedure of equipment surfaces should be modified, for instance, by chemical loosening of the remaining biofilm cells from the surface at the moment of assessment (Wirtanen et al., 1997). Practical assessment of microbial films and organic soil on processing surfaces must also be developed for on-line detection, using, for example, FTIR spectrometry. Rapid methods using fluorescent dyes are needed for the detection of living and dead cells left on surfaces after cleaning procedures (Wirtanen et al., 1997). Hopefully new methods will be developed that could provide accurate data for cleaning regimes in practical hygiene in the food industry (Wirtanen, 1995).

The processing equipment, open and closed, should be easily cleanable and cleanability testing methods are therefore needed. Development of practical detection methods for research on biofilm buildup is needed in order to utilize the knowledge of biofilm accumulation in the maintenance of process equipment. The natural luminescence of marine bacteria has been used to assess biofilms in marine systems and for on-line evaluation of biofilm formation as well as antifouling coatings (Mittelman et al., 1993; Duncan et al., 1994). The transformation of luminescence-encoding *lux*-genes to food-spoiling bacteria or opportunistic pathogens makes it possible to use bioluminescence in the detection of these bacteria (Baker et al., 1992; Mittelman et al., 1993; Walker et al., 1993; Duncan et al., 1994; Flemming et al., 1994). Further efforts are needed to optimize the microbial cleanability assay, using, for example, the *Photobacterium leiognathi* procedure, and to adapt the assay to both open and closed systems. The use is now limited to open systems, because the test bacterium is very sensitive to the heat treatments used in closed systems. Efforts should concentrate on optimizing the assay procedure and adapting it to both open and closed systems (Wirtanen, 1995).

On-line monitoring techniques transferred from various industrial fields (glass fiber optics, infrared techniques, ion mobility techniques, bioluminescence, microelectrodes, and heat transfer) could be used for assessing the quality of raw material and products and for control of microbial and chemical contaminants on processing surfaces. The on-line monitoring techniques used should be evaluated using independent microbial reference methods based on cultivation and microscopic techniques in combination with various staining procedures for assessment of cell viability and type of spoilage organisms. To ensure successful transfer of these monitoring techniques to food-processing-related industries, the studies are to be validated on (1) laboratory scale, (2) pilot scale, and (3) full production scale, and with close co-operation between research institutions and industrial partners. The use of measuring techniques in full-scale production must be optimized depending on the threshold values of the method. Future prospects can be divided into the following areas:

- Establishing on-line monitoring techniques in the assessment of the microbial quality of raw material, food, processing water, and equipment surfaces at various stages of production. The techniques will be validated using microbial reference methods.
- Evaluating the cleaning efficacy of equipment surfaces using on-line monitoring techniques. The techniques will be validated using microbial reference methods.
- Evaluating the on-line monitoring techniques using standard methods.
- Adapting the methods to full-scale production.

10.4 CONCLUSIONS

Biofilms can be produced under suitable growth conditions by any microbe. The equipment surfaces in the food industry provide the microbes growing in biofilms with liquids and excess nutrients. This type of phenomena should be combatted using efficient cleaning procedures to improve process hygiene. The cleaning and disinfection procedures represent significant costs in the process industry, and the environmental consequences of these measures are gaining increasing attention. For these reasons, cleaning procedures must be optimized on the basis of accurate data of biofilm formation. The benefits for the industry are reduction in product losses and improvement of quality and consumer safety.

Cleaning in the food industry should be based on systematic planning. Biofilm formation in these systems is a symptom of disturbance in the process (Mattila-Sandholm and Wirtanen,

1992). The knowledge that microbes grow differently on surfaces than in suspensions is the first step in developing advanced regimes in process hygiene (Holah, 1992). Biofilms are less likely to accumulate in well-designed systems that are effectively cleaned. Careful thoughts must be given to the cleaning procedures, including the program, cleaning agents, disinfectants, and cleaning equipment. Results indicate that low pressure cleaning in itself is not effective enough to remove biofilms unless the cleaning agent is effective. The efficiency of cleaning agents is assessed by their ability to remove biofilm from process surfaces together with their ability to kill the bacteria present in the biofilm (Wirtanen et al., 1997). The cleaning effect in open systems can be enhanced using either increased chemical effect through double foaming or added mechanical forces through scrubbing. In closed processes the removal of biofilms from surfaces can be performed using efficient flowing condition in combination with effective cleaning agents (Wirtanen et al., 1997). Strong agents are used to combat microbial deposits on equipment surfaces in problem areas. Satisfactory elimination of biofilms using only disinfectant treatment cannot be achieved, even if the agent is very effective against freely suspended cells (Mosteller and Bishop, 1993).

Methods used for monitoring process hygiene are often based on conventional cultivation using various types of agar plates or ATP. The conventional cultivation requires several days before the result can be obtained and it measures those cells that are able to form colonies on the given agar. ATP is used for measuring the total hygiene (Wirtanen et al., 1997). Detection of deposits built up on equipment surfaces at an early stage enables effective countermeasures and thus results in an improvement of the process hygiene. Successful on-line monitoring of microbiological deterioration in the process industry has many beneficial impacts of economical and environmental value. On-line monitoring saves both expenses and the environment when gentle cleaning methods can be used and unnecessary procedures avoided. Many valuable methodological tools used in biofilm research should be developed for and applied in industrial monitoring.

REFERENCES

Agersborg, A., D. Reidun, and I. Martinez. 1997. Sample preparation and DNA extraction procedures for polymerase chain reaction identification of *Listeria monocytogenes* in seafoods. *Int. J. Food Microbiol.* 35: 275–280.

Allison, D. G. 1992. Polysaccharide interactions in bacterial biofilms. In *Biofilms—Science and Technology*, L. F. Melo, T. R. Bott, M. Fletcher, and B. Capdeville, eds. Dordrecht: Kluwer Academic Publishers, pp. 371–376.

Amann, R. I., L. Krumholz, and D. A. Stahl. 1990a. Fluorescent-oligonucleotide probing of whole cells for determinative, phylogenetic, and environmental studies in microbiology. *J. Bacteriol.* 172: 762–770.

Amann, R. I., B. J. Binder, R. J. Olson, S. W. Chisholm, R. Devereux, and D. A. Stahl. 1990b. Combination of 16S rRNA-targeted oligonucleotide probes with flow cytometry for analyzing mixed microbial populations. *Appl. Environ. Microbiol.* 56: 1919–1925.

Amann, R. I., J. Stromley, R. Devereux, R. Key, and D. A. Stahl. 1992. Molecular and microscopic identification of sulfate-reducing bacteria in multispecies biofilms. *Appl. Environ. Microbiol.* 58: 614–623.

Anderson, R. L., B. W. Holland, J. K. Carr, W. W. Bond, and M. S. Favero. 1990. Effect of disinfectants of pseudomonads colonized on the interior surface of PVC pipes. *Am. J. Publ. Health* 80: 17–21.

Angell, P., A. A. Arrage, M. W. Mittelman, and D. C. White. 1993. On-line, non-destructive biomass determination of bacterial biofilms by fluorometry. *J. Microbiol. Meth.* 18: 317–327.

Anwar, H., T. van Biesen, M. Dasgupta, K. Lam, and J. W. Costerton. 1989. Interaction of biofilm bacteria with antibiotics in a novel in vitro chemostat system. *Antimicrob. Agents Chemother.* 33: 1824–1826.

Anwar, H., M. K. Dasgupta, and J. W. Costerton. 1990. Testing the susceptibility of bacteria in biofilms to antibacterial agents. *Antimicrob. Agents Chemother.* 34: 2043–2046.

Austin, J. W., and G. Bergeron. 1995. Development of bacterial biofilms in dairy processing lines. *J. Dairy Res.* 62: 509–519.

Baker, J. M., M. W. Griffiths, and D. L. Collins-Thompson. 1992. Bacterial bioluminescence: Applications in food microbiology. *J. Food Prot.* 55: 62–70.

Bakke, R., and P. Q. Olsson. 1986. Biofilm thickness measurements by light microscopy. *J. Microbiol. Meth.* 5: 93–98.

Baumgart, J., and S. Sieker. 1993. A comparison between Malthus 2000 and BacTrac 4100 in the measurement of *Listeria* growth with a new medium for impedimetric analysis. *BacTrac 4100 Information 93-3* 1: 66–75, 85–90, and 105–112.

Beech, I. B. 1996. Potential use of atomic force microscopy for studying corrosion of metals in the presence of bacterial biofilms—An overview. *Int. Biodeterior. Biodeg.* 37: 141–149.

Beek, P. van. 1994. Building pharmaceutical cleanrooms according the new EEC-guidelines. *12th International Symposium on Contamination Control*, Yokohama, Japan, 10–14 October, 1994. Tokyo: The International Confederation of Contamination Control Societies, pp. 369–373.

Bezanson, G., S. Burbridge, D. Haldane, and T. Marrie. 1992. *In situ* colonization of polyvinyl chloride, brass, and copper by *Legionella pneumophila*. *Can. J. Microbiol.* 38: 328–330.

Bhaduri, S., and B. Cottrell. 1998. A simplified sample preparation method from various foods for PCR detection of pathogenic *Yersinia enterocolitica*: A possible model for other food pathogens. *Molecul. and Cell. Probes* 12: 79–83.

Blackman, I. C., and J. F. Frank. 1996. Growth of *Listeria monocytogenes* as a biofilm on various food processing surfaces. *J. Food Prot.* 59: 827–831.

Block, J. C. 1992. Biofilms in drinking water distribution systems. In *Biofilms—Science and Technology*, L. F. Melo, T. R. Bott, M. Fletcher, and B. Capdeville, eds. Dordrecht: Kluwer Academic Publishers, pp. 469–485.

Bloomfield, S. F. 1988. Cosmetics and pharmaceuticals: Biodeterioration and disinfectants. In *Biodeterioration 7*, D. R. Houghton, R. N. Smith, and H. O. W. Eggins, eds. Essex: Elsevier Publishers Ltd., pp. 135–145.

Bloomfield, S. F., M. Arthur, B. van Klingeren, W. Pullen, J. T. Holah, and R. Elton. 1994. An evaluation of the repeatability and reproducibility of a surface test for the activity of disinfectants. *J. Appl. Bacteriol.* 76: 86–94.

Bott, T. R. 1991. Ozone as a disinfectant in process plant. *Food Cont.* 2: 44–49.

Bott, T. R. 1992a. Industrial monitoring: Cooling water systems. In *Biofilms—Science and Technology*, L. F. Melo, T. R. Bott, M. Fletcher, and B. Capdeville, eds. Dordrecht: Kluwer Academic Publishers, pp. 661–669.

Bott, T. R. 1992b. The use of biocides in industry. In *Biofilms—Science and Technology*, L. F. Melo, T. R. Bott, M. Fletcher, and B. Capdeville, eds. Dordrecht: Kluwer Academic Publishers, pp. 567–581.

Boulange Petermann, L. 1996. Processes of bioadhesion on stainless steel surfaces and cleanability: A review with special reference to the food industry. *Biofouling* 10: 275–300.

Bourion, F., and O. Cerf. 1996. Disinfection efficacy against pure-culture and mixed-population biofilms of *Listeria innocua* and *Pseudomonas aeruginosa* on stainless-steel, Teflon® and rubber. *Sciences des aliments* 16: 151–166.

Brackett, R. E. 1992. Shelf stability and safety of fresh produce as influenced by sanitation and disinfection. *J. Food Prot.* 55: 808–814.

REFERENCES

Brown, M. R. W., and P. Gilbert. 1993. Sensitivity of biofilms to antimicrobial agents. *J. Appl. Bacteriol.* 74: 87S–97S.

Bryers, J. D., and A. Weightman. 1995. The Centre for Biofilm Engineering: An international resource in managing complex biological systems. *SIM News* 45: 103–111.

Busscher, H. J., and A. H. Weerkamp. 1987. Specific and non-specific interactions in bacterial adhesion to solid substrata. *FEMS Microbiol. Rev.* 46: 165–173.

Busscher, H. J., M.-N. Bellon-Fontaine, N. Mozes, H. C. van der Mei, J. Sjollema, O. Cerf, and P. G. Rouxhet. 1990. An interlaboratory comparison of physicochemical methods for studying the surface properties of microorganisms, Application to *Streptococcus thermophilus* and *Leuconostoc mesenteroides*. *J. Microbiol. Meth.* 12: 101–115.

Caldwell, D. E., D. R. Korber, and J. R. Lawrence. 1992. Confocal laser microscopy and digital image analysis in microbial ecology. In *Advances in Microbial Ecology*, K. C. Marshall, ed. New York: Plenum Press, pp. 1–67.

Caldwell, D. E., D. R. Korber, and J. R. Lawrence. 1993. Analysis of biofilm formation using 2D vs 3D digital imaging. *J. Appl. Bacteriol.* 74: 52S–66S.

Capdeville, B., and J. L. Rols. 1992. Introduction to biofilms in water and wastewater treatment. In *Biofilms—Science and Technology*, L. F. Melo, T. R. Bott, M. Fletcher, and B. Capdeville, eds. Dordrecht: Kluwer Academic Publishers, pp. 13–20.

Carpén, L. 1999. Microbiologically induced corrosion of stainless steels in paper machines. In *30th R³-Nordic Contamination Control Symposium*, G. Wirtanen, S. Salo, and A. Mikkola, eds. Espoo: Libella Painopalvelu Oy, VTT Symposium 193, pp. 153–161.

Carpentier, B., and O. Cerf. 1993. Biofilms and their consequences, with particular reference to hygiene in the food industry. *J. Appl. Bact.* 75: 499–511.

Chamberlain, A. H. L. 1992. Biofilms and corrosion. In *Biofilms—Science and Technology*, L. F. Melo, T. R. Bott, M. Fletcher, and B. Capdeville, eds. Dordrecht: Kluwer Academic Publishers, pp. 207–217.

Chamberlain, A. H. L., and S. Johal. 1988. Biofilms on meat processing surfaces. In *Biodeterioration 7*, D. R. Hougton, R. N. Smith, and H. O. W. Eggins, eds. Essex: Elsevier Publishers Ltd., pp. 57–61.

Characklis, W. G. 1981. Fouling biofilm development: A process analysis. *Biotechnol. Bioeng.* 23: 1923–1960.

Characklis, W. G. 1990. Biofilm processes. In *Biofilms*, W. G. Characklis and K. C. Marshall, eds. New York: John Wiley, pp. 195–231.

Characklis, W. G. 1991. Biofouling: Effects and control. In *Biofouling and Biocorrosion in Industrial Water Systems*, H.-C. Flemming and G. G. Geesey, eds. Berlin Heidelberg: Springer-Verlag, pp. 7–27.

Characklis, W. G., M. G. Trulear, J. D. Bryers, and N. Zelver. 1982. Dynamics of biofilm processes: Methods. *Water Res.* 16: 1207–1216.

Chisti, Y., and M. Moo-Young. 1994. Cleaning-in-place systems for industrial bioreactors: Design, validation and operation. *J. Ind. Microbiol.* 13: 201–207.

Christensen, B. E. 1989. The role of extracellular polysaccharides in biofilms. *J. Biotechnol.* 10: 181–202.

Cloete, T. E., F. Smith, and P. L. Steyn. 1989. The use of planktonic bacterial populations in open and closed recirculating water cooling systems for the evaluation of biocides. *Int. Biodeterior.* 25: 115–122.

Colbourne, J. S., and P. J. Dennis. 1988. Legionella: A biofilm organism in engineered water systems? In *Biodeterioration 7*, D. R. Hougton, R. N. Smith, and H. O. W. Eggins, eds. Essex: Elsevier Publishers Ltd., pp. 36–42.

Colturi, T. F., and K. J. Kozelski. 1984. Corrosion and biofouling control in a cooling tower system. *Mater. Perform.* 238: 43–47.

Costerton, J. W., and J. Boivin. 1991. Biofilms and corrosion. In *Biofouling and Biocorrosion in Industrial Water Systems*, H.-C. Flemming and G. G. Geesey, eds. Berlin, Heidelberg: Springer-Verlag, pp. 195–204.

Costerton, J. W., and E. S. Lashen. 1983. The inherent biocide resistance of corrosion-causing biofilm bacteria. The NACE Annual Conference and Corrosion Show, Corrosion 83, Anaheim, California, USA, 18–22 April, 1983. Houston, TX: National Association of Corrosion Engineers Publication Department, Paper no. 246, 11p.

Costerton, J. W., R. T. Irvin, and K.-J. Cheng. 1981. The bacterial glycocalyx in nature and disease. *Ann. Rev. Microbiol.* 35: 299–324.

Costerton, J. W., G. G. Geesey, and P. A. Jones. 1988. Bacterial biofilms in relation to internal corrosion monitoring and biocide strategies. *Mater. Perform.* 274: 49–53.

Costeron, J. W., T. J. Marrie, and K.-J. Cheng. 1985. Phenomena of bacterial adhesion. In *Bacterial Adhesion*, D. C. Savage and M. Fletcher, eds. New York: Plenum Press, pp. 3–43.

Costeron, J. W., K.-J. Cheng, G. G. Geesey, T. I. Ladd, J. C. Nickel, M. Dasgupta, and T. J. Marrie. 1987. Bacterial biofilms in nature and disease. *Ann. Rev. Microbiol.* 41: 435–464.

Czechowski, M. H., and M. Banner. 1990. Control of biofilms in breweries through cleaning and sanitizing. *Techn. Quart. Mast. Brew. Ass. Am.* 293: 86–88.

de Beer, D., P. Stoodley, F. Roe, and Z. Lewandowski. 1994. Effects of biofilm structures on oxygen distribution and mass transport. *Biotechnol. Bioeng.* 43: 1131–1138.

de Goederen, G., N. J. Pritchard, and A. P. M. Hasting. 1989. Improved cleaning processes for the food industry. In *Fouling and Cleaning in Food Processing*, H. G. Kessler and D. B. Lund, eds. Augsburg: Druckerei Walch, pp. 115–130.

Deng, M. Y., D. O. Cliver, S. P. Day, and P. M. Fratamico. 1996. Enterotoxigenic *Escherichia coli* detected in food by PCR and an enzyme-linked oligonucleotide probe. *Int. J. Food Microbiol.* 30: 217–229.

Dhaliwal, D. S., J. L. Cordier, and L. J. Cox. 1992. Impedimetric evaluation of the efficiency of disinfectants against biofilms. *Lett. Appl. Microbiol.* 15: 217–221.

Diaper, J. P., and C. Edwards. 1994a. Flow cytometric detection of viable bacteria in compost. *FEMS Microbiol. Ecol.* 14: 213–220.

Diaper, J. P., and C. Edwards. 1994b. Survial of *Staphylococcus aureus* in lakewater monitored by flow cytometry. *Microbiology* 140: 35–42.

Docherty, L., M. R. Adams, P. Patel, and J. McFadden. 1996. The magnetic immuno-polymerase chain reaction assay for the detection of *Campylobacter* in milk and poultry. *Lett. Appl. Microbiol.* 22: 288–292.

Duddridge, J. E. 1981. *Some Techniques Commonly Used in the Study of Biological Films*. UKAEA Conference in Progress in the Prevention of Fouling in Industrial Plant. Harwell: UKAEA, pp. 54–67.

Duddridge, J. E., C. A. Kent, P. C. Miller, J. F. Laws, A. M. Pritchard, and T. R. Bott. 1983. *Effects of flow on biofilm development*. Engineering Foundation Conference on Fouling of Heat Exchanger Surfaces, White Haven, 31 October–5 November, 1982. New York: United Engineering Trustees Inc., pp. 717–726.

Duncan, S., L. A. Glover, K. Killham, and J. I. Prosser. 1994. Luminescence-based detection of activity of starved and viable but nonculturable bacteria. *Appl. Environ. Microbiol.* 60: 1308–1316.

Dunsmore, D. G., A. Twomey, W. G. Whittlestone, and H. W. Morgan. 1981. Design and performance of systems for cleaning product-contact surfaces of food equipment: A review. *J. Food Prot.* 44: 220–240.

European Hygienic Equipment Design Group (EHEDG). 1992. A method for assessing the in-line cleanability of food-processing equipment. *Trends Food Sci. Technol.* 3: 325–238.

European Hygienic Equipment Design Group (EHEDG). 1993a. Hygienic design of closed equipment for the processing of liquid food. *Trends Food Sci. Technol.* 4: 375–379.

European Hygienic Equipment Design Group (EHEDG). 1993b. Hygienic equipment design criteria. *Trends Food Sci. Technol.* 4: 225–229.

European Hygienic Equipment Design Group (EHEDG). 1994. Hygienic design of valves for food processing. *Trends Food Sci. Technol.* 5: 169–171.

Exner, M., G.-J. Tuschewitzki, and J. Scharnagel. 1987. Influence of biofilms by chemical disinfectants and mechanical cleaning. *Zbl. Bakteriol, Mikrobiol. Hygiene, Serie B* 183: 549–563.

Firstenberg-Eden, R. 1986. Electrical impedance for determining microbial quality of foods. In *Foodborne Microorganisms and Their Toxins: Developing Methodology*, M. D. Pierson and N. J. Stern, eds. New York: Marcel Dekker Inc., pp. 129–144.

Firstenberg-Eden, R., and G. Eden. 1984. *Impedance Microbiology*. Hertfordshire: Research Studies Press Ltd., pp. 1–121.

Flemming, C. A., H. Lee, and J. T. Trevors. 1994. Bioluminescent most-probable-number method to enumerate *lux*-marked *Pseudomonas aeruginosa* UG2Lr in soil. *Appl. Environ. Microbiol.* 60: 3458–3461.

Flemming, H.-C. 1991. Biofouling in water treatment. In *Biofouling and Biocorrosion in Industrial Water Systems*, H.-C. Flemming and G. G. Geesey, eds. Berlin Heidelberg: Springer-Verlag, pp. 47–80.

Flemming, H.-C., G. Schaule, and R. McDonogh. 1992. Biofouling on membranes—A short review. In *Biofilms—Science and Technology*, L. F. Melo, T. R. Bott, M. Fletcher, and B. Capdeville, eds. Dordrecht: Kluwer Academic Publishers, pp. 487–497.

Fletcher, M. 1990. Methods for studying adhesion and attachment to surfaces. *Meth. Microbiol.* 22: 251–283.

Flint, S. H., P. J. Bremer, and J. D. Brooks. 1997a. Biofilms in dairy manufacturing plant—Description, current concerns and methods of control. *Biofouling* 11: 81–97.

Flint, S. H., J. D. Brooks, and P. J. Bremer. 1997b. The influence of cell surface properties of thermophilic streptococci on attachment to stainless steel. *J. Appl. Microbiol.* 83: 508–517.

Flint, S. H., J. D. Brooks, and P. J. Bremer. 1997c. Use of the Malthus conductance growth analyser to determine numbers of thermophilic streptococci on stainless steel. *J. Appl. Microbiol.* 83: 335–339.

Frank, J. F., and R. A. Koffi. Surface-adherent growth of *Listeria monocytogenes* is associated with increased resistance to surfactant sanitizers and heat. *J. Food Prot.* 53: 550–554.

Freeman, C., M. A. Lock, J. Marxsen, and S. E. Jones. 1990. Inhibitory effects of high molecular weight dissolved organic matter upon metabolic processes in biofilms from contrasting rivers and streams. *Freshwater Biol.* 24: 159–166.

Gaylarde, C. C. 1989. Microbial corrosion of metals. *Environ. Eng.* 24: 30–32.

Gaylarde, C. C. 1990. Advances in detection of microbiologically induced corrosion. *Int. Biodeterior.* 26: 11–22.

Geesey, G. G. 1991. What is biocorrosion? In *Biofouling and Biocorrosion in Industrial Water Systems*, H.-C. Flemming and G. G. Geesey, eds. Berlin Heidelberg: Springer-Verlag, pp. 155–164.

Geesey, G. G., W. T. Richardson, H. G. Yeomans, R. T. Irvin, and J. W. Costerton. 1977. Microscopic examination of natural sessile bacterial populations from an alpine stream. *Can. J. Microbiol.* 23: 1733–1736.

Gibson, H., J. H. Taylor, K. E. Hall, and J. T. Holah. 1995. *Biofilms and Their Detection in the Food Industry*. Chipping Campden: Campden and Chorleywood Food Research Association, 88p.

Giffel, M. C. T., R. R. Beumer, L. P. M. Langeveld, and M. Rombouts. 1997. The role of heat exchangers in the contamination of milk with *Bacillus cereus* in dairy processing plants. *Int. J. Dairy Technol.* 50(2): 43–47.

Gilbert, P. D., and B. N. Herbert. 1987. Monitoring microbial fouling in flowing systems using coupons. In *Industrial Microbiological Testing*, J. W. Hopton and E. C. Hill, eds. Oxford: Blackwell Scientific, pp. 79–98.

Gilbert, P., P. J. Collier, and M. R. W. Brown. 1990. Influence of growth rate on susceptibility to antimicrobial agents: Biofilms, cell cycle, dormancy and stringent response. *Antimicrob. Agents Chemother.* 34: 1865–1868.

Gillatt, J. 1991. Methods for the efficacy testing of industrial biocides 1. Evaluation of wet-state preservatives. *Int. Biodeterior.* 27: 383–394.

Goodman, P. D. 1987. Effect of chlorination on materials for sea water cooling systems: A review of chemical reactions. *Br. Corr. J.* 221: 56–62.

Goroncybermes, P., and S. Gerresheim. 1996. Efficacy of peroxide-containing solutions against microorganisms in biofilms. *Zbl. Hygiene Umweltmedizin* 198: 473–477.

Gould, W. A. 1994. *Current Good Manufacturing Practices: Food Plant Sanitation*, 2nd ed. Baltimore: CTI Publications, pp. 189–215.

Grasshoff, A. 1992. Hygienic design—The basis for computer controlled automation. *Trans. I. Chem. E.* 70(C2): 69–77.

Green, P. N., I. J. Bousfield, and A. Stones. 1987. The laboratory generation of biofilms and their use in biocide evaluation. In *Industrial Microbiological Testing*, J. W. Hopton and E. C. Hills, eds. Oxford: Blackwell Scientific, pp. 99–108.

Griebe, T., C.-I. Chen, R. Srinivasan, and P. S. Stewart. 1994. Analysis of biofilm disinfection by monochloramine and free chlorine. In *Biofouling and Biocorrosion in Industrial Water Systems*, G. G. Geesey, Z. Lewandowski, and H.-C. Flemming, eds. Boca Raton, FL: Lewis, pp. 151–161.

Hamilton, W. A. 1991. Sulphate-reducing bacteria and their role in biocorrosion. In *Biofouling and Biocorrosion in Industrial Water Systems*, H.-C. Flemming and G. G. Geesey, eds. Berlin Heidelberg: Springer-Verlag, pp. 187–193.

Harju-Jeanty, P., and P. Väätäinen. 1984. Detrimental micro-organisms in paper and cardboard mills. *Paperi ja Puu* 66: 245–259.

Harmsen, H. 1996. *Detection, Phylogeny and Population Dynamics of Syntrophic Propionate-oxidizing Bacteria in Anaerobic Granular Sludge*. Wageningen: Landbouwuniversiteit Wageningen, 153p.

Heinzel, M. 1988. The phenomena of resistance to disinfectants and preservatives. In *Industrial Biocides*, K. R. Payne, ed. Chichester: John Wiley, pp. 52–67.

Helke, D. M., E. B. Somers, and A. C. L. Wong. 1993. Attachment of *Listeria monocytogenes* and *Salmonella typhimurium* to stainless steel and Buna-n in the presence of milk and individual milk components. *J. Food Prot.* 56: 479–484.

Henriksson, E., and A. Haikara. 1990. Disinfection of filling halls in breweries. *Mallas ja olut*, pp. 132–139.

Hielm, S., E. Hyytiä, M. Lindström, and H. Korkeala. 1999. DNA-based identification systems for pathogenic bacterial species and strains. In *30th R^3-Nordic Contamination Control Symposium*, VTT Symposium 193, G. Wirtanen, S. Salo, and A. Mikkola, eds. Espoo: Libella Painopalvelu Oy, pp. 97–105.

Holah, J. T. 1992. Industrial monitoring: Hygiene in food processing. In *Biofilms—Science and Technology*, L. F. Melo, T. R. Bott, M. Fletcher, and B. Capdeville, eds. Dordrecht: Kluwer Academic Publishers, pp. 645–659.

Holah, J. T., and L. R. Kearney. 1992. Introduction to biofilms in the food industry. In *Biofilms—Science and Technology*, L. F. Melo, T. R. Bott, M. Fletcher, and B. Capdeville, eds. Dordrecht: Kluwer Academic Publishers, pp. 35–41.

Holah, J., and A. Timperley. 1999. Hygienic design of food processing facilities and equipment. In *30th R^3-Nordic Contamination Control Symposium*, VTT Symposium 193, G. Wirtanen, S. Salo, and A. Mikkola, eds. Espoo: Libella Painopalvelu O, pp. 11–39.

Holah, J. T., and R. H. Thorpe. 1990. Cleanability in relation to bacterial retention on unused and abraded domestic sink materials. *J. Appl. Bact.* 69: 599–608.

Holah, J. T., R. P. Betts, and R. H. Thorpe. 1988. The use of direct epifluoresecent microscopy DEM and the direct epifluorescent filter technique DEFT to assess microbial populations on food contact surfaces. *J. Appl. Bact.* 65: 215–221.

Holah, J. T., R. P. Betts, and R. H. Thorpe. 1989. The use of epifluorescence microscopy to determine surface hygiene. *Int. Biodeterior.* 25: 147–154.

Holah, J. T., C. Higgs, S. Robinson, D. Worthington, and H. Spenceley. 1990. A conductance-based surface disinfection test for food hygiene. *Lett. Appl. Microbiol.* 11: 255–259.

Holah, J. T., K. E. Hall, J. Holder, S. J. Rogers, J. Taylor, and K. L. Brown. 1995. *Airborne Microorganism Levels in Food Processing Environments*. Chipping Campden: Campden and Chorleywood Food Research Association, 22p.

Holt, D. M. 1988. Microbiology of paper and board manufacture. In *Biodeterioration 7*, D. R. Houghton, R. N. Smith, and H. O. W. Eggins, eds. Essex: Elsevier Publishers Ltd., pp. 493–506.

Honneysett, D. G., W. D. van den Bergh, and P. F. O'Brien. 1985. Microbiological corrosion control in a cooling water system. *Mater. Perform.* 2410: 34–39.

Hood, S. K., and E. A. Zottola. 1995. Biofilms in food processing. *Food Control* 6: 9–18.

Hugenholz, P., and J. A. Fuerst. 1992. Heterotrophic bacteria in an air-handling system. *Appl. Environ. Microbiol.* 58: 3914–3920.

Hugenholz, P., M. A. Cunningham, J. K. Hendrikz, and J. A. Fuerst. 1995. Desiccation resistance of bacteria isolated from an air-handling system biofilm determinated using a simple quantitative membrane filter method. *Lett. Appl. Microbiol.* 21: 41–46.

Hughes, M. C. 1993. The effect of some papermaking additives on slime microflora composition. *Appita J.* 46: 194–197.

Hughes-van Kregten, M. C. 1988. Slime flora of New Zealand paper mills. *Appita J.* 41: 470–474.

Husmark, U. 1994. Sanitering i mejeri: Raport 1 "En sammansälling över disk och desinfektionsrutiner i svensk nejeriindustri" [Sanitation in dairies: Report 1 "Cleaning and disinfection procedures used in Swedish dairies"]. Göteborg: SIK, 4p. (in Swedish)

Izzat, I. N., E. O. Bennett, J. E. Gannon, and I. U. Onyekwelu. 1981. Effects of EDTA on the antimicrobial properties of mixtures of cutting fluid preservatives. *Tribol. Int.* 14: 171–173.

Janning, K. F., and P. Harremoës. 1995. Kinetics in a full-scale submerged denitrification filter. *Proceedings of Biofilm Structure, Growth, and Dynamics*, Noordwijkerhout, 30 August–1 September. Delft: Biotechnological Sciences Delft Leiden, pp. 126–133.

Johansen, C., P. Falholt, and L. Gram. 1997. Enzymatic removal and disinfection of bacterial biofilms. *Appl. Environ. Microbiol.* 63: 3724–3728.

Johnston, M. D., and M. V. Jones. 1993. Evaluation of the BacTrac 4100, Sy-Lab Information 93-1. Purkersdorf: Sy-Lab VertriebsgmbH, 24p.

Karsa, D. R., and A. E. Stafford. 1989. BASC; A code of practice; the control of legionellae by the safe and effective operation of cooling systems. Surrey: British Association for Chemical Specialties, 37p.

Keevil, C. W., C. W. Mackerness, and J. S. Colbourne. 1990. Biocide treatment of biofilms. *Int. Biodeterior.* 26: 169–179.

Kell, D. B., H. M. Ryder, A. S. Kaprelyants, and H. V. Westerhoff. 1991. Quantifying heterogeneity: Flow cytometry of bacterial cultures. *Antonie van Leeuwenhoek* 60: 145–158.

Kim, K. Y., and J. F. Frank. 1994. Effect of growth nutrient on attachment of *Listeria monocytogenes* to stainless steel. *J. Food Prot.* 57: 720–726.

Kinniment, S., and J. W. T. Wimpenny. 1990. Biofilms and biocides. *Int. Biodeterior.* 26: 181–194.

Klein, P. G., and V. K. Juneja. 1997. Sensitive detection of viable *Listeria monocytogenes* by reverse transcription-PCR. *Appl. Environ. Microbiol.* 63: 4441–4448.

Klemetti, I., M. Puustinen, and M. Koivisto. 1999. Identification of hazards in the dairy process—A quality control story in Lapinlahti cheese factory. In *30th R^3-Nordic Contamination Control*

Symposium, VTT Symposium 193, G. Wirtanen, S. Salo, and A. Mikkola, eds. Espoo: Libella Painopalvelu Oy, pp. 139–143.

Kostyál, E. 1998. *Removal of Chlorinated Organic Matter from Wastewaters, Chlorinated Ground, and Lake Water by Nitrifying Fluidized-bed Biomass.* Helsinki: Hakapaino Oy, 68p. + 5 app.

Kramer, J. F. 1997. Peracetic acid: A new biocide for industrial water applications. *Mater. Perform.* 36(8): 42–50.

Kristensen, G. H., and F. R. Christensen. 1982. Application of cryo-cut method for measurements of biofilm thickness. *Water Res.* 16: 1619–1621.

Kronlöf, J. 1994. *Immobilized Yeast in Continuous Fermentation of Beer.* VTT Publications 167, Espoo: The Technical Research Centre of Finland, 96p. + app. 47p.

Kuo, J. F., and S. O. Smith. 1996. Disinfection. *Water Environ. Res.* 68: 503–510.

Ladd, T. L., and J. W. Costerton. 1990. Methods for studying biofilm bacteria. *Meth. Microbiol.* 22: 285–307.

Lamot, J. E. 1989. Role of biocides in controlling microbial corrosion. *Microbial Corrosion Conference,* Sintra, 7–9 March, 1988. Essex: Elsevier Applied Science, pp. 224–234.

Lappalainen, J., S. Loikkanen, M. Havana, M. Karp, A.-M. Sjöberg, and G. Wirtanen. 1999. Rapid detection of detergents and disinfectants in food factories. In *30th R^3-Nordic Contamination Control Symposium,* VTT Symposium 193, G. Wirtanen, S. Salo, and A. Mikkola, eds. Espoo: Libella Painopalvelu Oy, pp. 199–205.

LeChevallier, M. W., T. M. Babcock, and R. G. Lee. 1987. Examination and characterization of distribution system biofilms. *Appl. Environ. Microbiol.* 53: 2714–2724.

LeChevallier, M. W., C. D. Cawthon, and R. G. Lee. 1988. Factors promoting survival of bacteria in chlorinated water supplies. *Appl. Environ. Microbiol.* 54: 649–654.

LeChevallier, M. W., C. D. Lowry, and R. G. Lee. 1990. Disinfecting biofilms in a model distribution system. *Am. Water Works Assoc. J.* 827: 87–99.

Lee, W., and W. G. Characklis. 1993. Corrosion of mild steel under anaerobic biofilm. *Corrosion* 49: 186–199.

Lee, W., Z. Lewandowski, P. H. Nielsen, and W. A. Hamilton. 1995. Role of sulfate-reducing bacteria in corrosion of mild steel: A review. *Biofouling* 8: 165–194.

Lehtimäkl, M. 1995. Experiments with electric filters. *26th Nordic R3-Symposium,* Helsinki, 8–10 May, 1995. Helsinki: R3 Nordic, pp. 170–175.

Lelieveld, H. L. M. 1985. Hygienic design and test methods. *J. Soc. Dairy Technol.* 38: 14–16.

Lelieveld, H. L. M. 1999. EHEDG in the 21st century. In *30th R^3-Nordic Contamination Control Symposium,* VTT Symposium 193, G. Wirtanen, S. Salo, and A. Mikkola, eds. Espoo: Libella Painopalvelu Oy, pp. 83–86.

Lewin, R. 1984. Microbial adhesion is a sticky problem. *Science* 224: 375–377.

Lewis, R. O. 1982. The influence of biofouling countermeasures on corrosion of heat exchanger materials in sea water. *Mater. Perform.* 219: 31–38.

Lewis, S. J., and A. Gilmour. 1987. Microflora associated with the internal surfaces of rubber and stainless steel milk transfer pipeline. *J. Appl. Bact.* 62: 327–333.

Lewis, S. J., A. Gilmour, T. W. Fraser, and R. D. McCall. 1987. Scanning electron microscopy of soiled stainless steel inoculated with single bacterial cells. *Int. J. Food Microbiol.* 4: 279–289.

Lin, S. H., and K. L. Yeh. 1993. Cooling water treatment by ozonization. *Chem. Eng. Technol.* 16: 275–278.

Lindsay, D., I. Geornaras, A. von Holy, and A. von Holy. 1996. Biofilms associated with poultry processing equipment. *Microbios* 86: 105–116.

Ljungqvist, B., and B. Reinmüller. 1995. Clean room technologies in aseptic packaging. In *Advances in Aseptic Processing and Packaging Technologies,* T. Ohlsson, ed. Göteborg: Kompendiet, 17p.

REFERENCES

Lopez-Amoros, R., S. Castel, J. Comas-Riu, and J. Vives-Rego. 1997. Assessment of *E. coli* and *Salmonella* viability and starvation by confocal laser microscopy and flow cytometry using rhodamine 123, DiBAC4(3), propidium iodide and CTC. *Cytometry* 29: 298–305.

Lundin, A. 1999. ATP detection of biological contamination. In *30th R^3-Nordic Contamination Control Symposium*, VTT Symposium 193, G. Wirtanen, S. Salo, and A. Mikkola, eds. Espoo: Libella Painopalvelu Oy, pp. 337–352.

Mafu, A. A., D. Roy, L. Savoie, and J. Goulet. 1991. Bioluminescence assay for estimating the hydrophobic properties of bacteria as revealed by hydrophobic interaction chromatography. *Appl. Environ. Microbiol.* 57: 1640–1643.

Mafu, A. A., D. Roy, J. Goulet, L. Savoie, and R. Roy. 1990. Efficiency of sanitizing agents for destroying *Listeria monocytogenes* on contaminated surfaces. *J. Dairy Sci.* 73: 3428–3432.

Mafu, A. A., D. Roy, L. Savoie, R. Roy, and J. Goulet. 1989. Effect of sanitizing agents on *Listeria monocytogenes* attached to milk contact surfaces at different temperatures. *J. Dairy Sci.* 72, suppl. 1: 172.

Mäntynen, V., S. Niemelä, S. Kaijalainen, T. Pirhonen, and K. Lindström. 1997. MPN-PCR-quantification method for staphylcoccal enterotoxin $c1$ gene from fresh cheese. *Int. J. Food Microbiol.* 36: 135–143.

Manz, W., R. Amann, R. Szewzyk, U. Szewzyk, T. A. Stenstrom, P. Hutzler, and K. H. Schleifer. 1995. In situ identification of legionellaceae using 16S ribosomal-RNA-targeted oligonucleotide probes and confocal laser scanning microscopy. *Microbiol.* 141: 29–39.

Maris, P., and R. Fresnel. 1992. Biofilms and disinfection. Development of a microorganism carrier-surface method. *Sci. Aliments* 12: 721–728.

Marrie, T. J., and J. W. Costeron. 1981. Prolonged survival of *Serratia marcescens* in chlorhexidine. *Appl. Environ. Microbiol.* 42: 1093–1102.

Marrie, T. J., and J. W. Costerton. 1982. A scanning and transmission electron microscopic study of an infected endocardial pacemaker lead. *Circulation* 66: 1339–1341.

Marrie, T. J., and J. W. Costerton. 1983. Scanning electron microscopic study of uropathogen adherence to a plastic surface. *Appl. Environ. Microbiol.* 45: 1018–1024.

Marshall, K. C., and B. L. Blainey. 1991. Role of bacterial adhesion in biofilm formation and biocorrosion. In *Biofouling and Biocorrosion in Industrial Water Systems*, in H.-C. Flemming and G. G. Geesey, eds. Berlin Heidelberg: Springer-Verlag, pp. 29–46.

Marshall, P. A., G. I. Loeb, M. M. Cowan, and M. Fletcher. 1989. Response of microbial adhesives and biofilm matrix polymers to chemical treatments as determined by interference reflection microscopy and light section microscopy. *Appl. Environ. Microbiol.* 55: 2827–2831.

Mason, D. J., R. Lopez-Armoros, R. Allman, J. M. Stark, and D. Lloyd. 1995. The ability of membrane potential dyes and calcafluor white to distinguish between viable and non-viable bacteria. *J. Appl. Bacteriol.* 78: 309–315.

Mattila, T., M. Manninen, and A.-L. Kyläsiurola. 1990. Effect of cleaning-in-place disinfectants on wild bacterial strains isolated from a milking line. *J. Dairy Res.* 57: 33–39.

Mattila-Sandholm, T., and G. Wirtanen. 1992. Biofouling in the industry: A review. *Food Rev. Int.* 8: 573–603.

Melo, L. F., and M. M. Pinheiro. 1992. Biofouling in heat exchangers. In *Biofilms—Science and Technology*, L. F. Melo, T. R. Bott, M. Fletcher, and B. Capdeville, eds. Dordrecht: Kluwer Academic Publishers, pp. 499–509.

Mentu, J. 1996. Biofouling in paper machines. In *Future Prospects of Biofouling and Biocides*, VTT Symposium 165, G. Wirtanen, L. Raaska, M. Salkinoja-Salonen, and T. Mattila-Sandholm, eds. Espoo: The Technical Research Centre of Finland, p. 23.

Mettler, E., and B. Carpentier. 1998. Variations over time of microbial load and physicochemical properties of floor materials after cleaning in food industry premises. *J. Food Prot.* 61: 57–65.

Milwidsky, B. M., and D. M. Gabriel. 1982. *Detergent Analysis*. London: George Godwin, pp. 85–104.

Miller, P.C., and T. R. Bott. 1982. Effects of biocide and nutrient availability on microbial contamination of surfaces in cooling-water systems. *J. Chem. Technol. Biotechnol.* 32: 538–546.

Mittelman, M. W. 1991. Bacterial growth and biofouling control in purified water systems. In *Biofouling and Biocorrosion in Industrial Water Systems*, H.-C. Flemming and G. G. Geesey, eds. Berlin Heidelberg: Springer-Verlag, pp. 133–154.

Mittelman, M. W., J. Packard, A. A. Arrage, S. L. Bean, P. Angell, and D. C. White. 1993. Test systems for determining antifouling coating efficacy using on-line detection of bioluminescence and fluorescence in a laminar-flow environment. *J. Microbiol. Meth.* 18: 51–60.

Morton, L. H. G., and S. B. Surman. 1994. Biofilms in biodeterioration—A review. *Int. Biodet. Biodeg.* 34: 203–221.

Mosteller, T. M., and J. R. Bishop. 1993. Sanitizer efficacy against attached bacteria in a milk biofilm. *J. Food Prot.* 56: 34–41.

Mucchetti, G. 1995. Biological fouling and biofilm formation on membranes. *Int. Dairy Fed. Special Issue No. 9504*, 118–124.

Mustapha, A., and M. B. Liewen. 1989. Destruction of *Listeria monocytogenes* by sodium hypochlorite and quaternary ammonium sanitizers. *J. Food Prot.* 52: 306–311.

Nakayama, S., M. Tanaka, S. Yamauchi, and N. Tabata. 1985. Anti-biofouling ozone system for cooling water circuits—An application to seawater. *Ozone Sci. Eng.* 7: 31–46.

Nassauer, J., and H. G. Kessler. 1986. The effect of electrostatic phenomena on the cleaning of surfaces. *Chem. Eng. Proc.* 20: 27–32.

Neef, A., A. Zaglauer, H. Meier, R. Amann, H. Lemmer, and K. H. Schleifer. 1996. Population analysis in a denitrifying sand filter: Conventional and *in situ* identification of *Paracoccus* spp. in methanol-fed biofilms. *Appl. Environ. Microbiol.* 62: 4329–4339.

Nestor, G. J., and G. A. Cappeline. 1979. Water related problems of evaporative cooling systems and control methods. *Ind. Water Eng.* 163: 14–25.

Neu, T. R., H. C. van der Mei, and H. J. Busscher. 1992. Biofilms associated with health. In *Biofilms—Science and Technology*, L. F. Melo, T. R. Bott, M. Fletcher, and B. Capdeville, eds. Dordrecht: Kluwer Academic Publishers, pp. 21–34.

Nichols, P. D., J. M. Henson, J. B. Guckert, D. E. Nivens, and D. C. White. 1985. Fourier transform-infrared spectroscopic methods for microbial ecology: Analysis of bacteria, bacterial polymer mixtures and biofilms. *J. Microbiol. Meth.* 4: 79–94.

Nichols, W. W. 1989. Susceptibility of biofilms to toxic compounds. In *Structure and Function of Biofilms*, W. G. Characklis and P. A. Wilderer, eds. New York: John Wiley, pp. 321–331.

Nichols, W. W. 1991. Biofilms, antibiotics and penetration. *Rev. Med. Microbiol.* 2: 177–181.

Nivens, D. E., J. Q. Chambers, T. R. Anderson, A. Tunlid, J. Smit, and D. C. White. 1993a. Monitoring microbial adhesion and biofilm formation by attenuated total reflection/Fourier transform infrared spectroscopy. *J. Microbiol. Meth.* 17: 199–213.

Nivens, D. E., J. Q. Chambers, T. R. Anderson, and D. C. White. 1993b. Long-term on-line monitoring of microbial biofilms using a quartz crystal microbalance. *Anal. Chem.* 65: 65–69.

Nivens, D. E., R. J. Palmer Jr., and D. C. White. 1995. Continuous nondestructive monitoring of microbial biofilms: A review of analytical techniques. *J. Ind. Microbiol.* 15: 263–276.

Nohata, Y., and H. Taguchi. 1995. Ultrasensitive fouling monitoring system for cooling towers. *Mater. Perform.* 34(3): 43–46.

Norton, C. D., and M. W. LeChevallier. 1997. Chloramination: Its effects on distribution systems water quality. *Am. Water Works Ass.* 89(7): 66–77.

Notermans, S., J. A. M. Dormans, and G. C. Mead. 1991. Contribution of surface attachment to establishment of microorganisms in food processing plants: A review. *Biofouling* 5: 21–36.

Ntsama-Essomba, C., S. Bouttier, M. Ramaldes, F. Dubois-Brissonnet, and J. Fourniat. 1997. Resistance of *Escherichia coli* growing as biofilms to disinfectants. *Vet. Res.* 28: 353–363.

Nurmiaho-Lassila, E.-L., S. A. Lehtinen, S. A. Marmo, and M. S. Salkinoja-Salonen. 1990. Electron microscopical analysis of biological slimes on paper and board machines. *EMAG-MICRO 89*, London, 13–15 September 1989. London: IOP Publishing Ltd., pp. 727–730.

Offringa, G. 1988. The control of microbiological growth in heat exchange systems. *ChemSA* 14: 145–147.

Ortiz, C., P. S. Guiamet, and H. A. Videla. 1990. Relationship between biofilms and corrosion of steel by microbial contaminants of cutting-oil emulsions. *Int. Biodeterior.* 26: 315–326.

Pietersen, B., V. S. Brozel, and T. E. Cloete. 1996. Response of *Pseudomonas aeruginosa* PAO following exposure to hydrogen peroxide. *Water SA* 22: 239–244.

Pirbazari, M., T. C. Voice, and W. J. Weber Jr. 1990. Evaluation of biofilm development on various natural and synthetic media. *Hazard. Waste Hazard. Mater.* 7: 239–250.

Pontefract, R. D. 1991. Bacterial adhesion: Its consequences in food processing. *Can. Inst. Food Sci. Technol. J.* 24: 113–117.

Pope, D. H., and T. P. Zintel. 1989. Methods for investigating underdeposit microbiologically influenced corrosion. *Mater. Perform.* 2811: 46–51.

Pore, R. S. 1994. Antibiotic susceptibility testing by flow cytometry. *J. Antimicrob. Chemother.* 34: 613–627.

Poulton, W. I. J., T. E. Cloete, A. von Holy, and A. von Holy. 1995. Microbiological survey of open recirculating cooling water systems and their raw water supplies at twelve fossil-fired power stations. *Water SA* 21: 357–364.

Price, D., and D. G. Ahearn. 1988. Incidence and persistence of *Pseudomonas aeruginosa* in whirlpools. *J. Clin. Microbiol.* 26: 1650–1654.

Raaska, L. 1999. Process hygiene in manufacture of packaging material. In *30th R^3-Nordic Contamination Control Symposium*, VTT Symposium 193, G. Wirtanen, S. Salo, and A. Mikkola, eds. Espoo: Libella Painopalvelu Oy, pp. 145–152.

Reischl, U., and B. Kochanowski. 1995. Quantitative PCR—A survey of the present technology. *Molecul. Biotechnol.* 3: 55–71.

Ren, T.-J., and J. H. Frank. 1993. Susceptibility of starved planktonic and biofilm *Listeria monocytogenes* to quaternary ammonium sanitizer as determined by direct viable and agar plate counts. *J. Food Prot.* 56: 573–576.

Richards, S. R., and R. J. Turner. 1984. A comparative study of techniques for the examination of biofilms by scanning electron microscopy. *Water Res.* 18: 767–773.

Richardson, D. S. 1982. Cooling—Water system biofouling. *Chem. Eng.* 8925: 103–104.

Ridgway, H. F., and J. Safarik. 1991. Biofouling on reverse osmosis membranes. In *Biofouling and Biocorrosion in Industrial Water Systems*, H.-C. Flemming and G. G. Geesey, eds. Berlin Heidelberg: Springer-Verlag, pp. 81–111.

Rogers, J., A. B. Dowsett, P. J. Dennis, J. V. Lee, and C. W. Keevil. 1994a. Influence of plumbing materials on biofilm formation and growth *Legionella pneumophila* in potable water systems. *Appl. Environ. Microbiol.* 60: 1842–1851.

Rogers, J., A. B. Dowsett, P. J. Dennis, J. V. Lee, and C. W. Keevil. 1994b. Influence of temperature and plumbing material selection on biofilm formation and growth of *Legionella pneumophila* in a model potable water system containing complex microbial flora. *Appl. Environ. Microbiol.* 60: 1585–1592.

Ronner, A. B., and A. C. L. Wong. 1993. Biofilm development and sanitizer inactivation of *Listeria monocytogenes* and *Salmonella typhimurium* on stainless steel and Buna-n rubber. *J. Food Prot.* 56: 750–758.

Rossmoore, K. 1988. The microbial activity of glutaraldehyde in chain conveyor lubricant formulations. In *Biodeterioration 7*, D. R. Houghton, R. N. Smith, and H. O. W. Eggins, eds. Essex: Elsevier Publishers Ltd., pp. 242–247.

Safade, T. L. 1988. Tackling the slime problem in a paper-mill. *Pap. Technol. Ind.* 29: 280–285.

Salo, S., E. Storgårds, and G. Wirtanen. 1999. Alternative methods for sampling from surfaces. In *30th R^3-Nordic Contamination Control Symposium*, G. Wirtanen, S. Salo, and A. Mikkola, eds. Espoo: Libella Painopalvelu Oy, VTT Symposium 193, pp. 187–198.

Sanderson, S. S., and P. S. Stewart. 1997. Evidence of bacterial adaptation to monochloramine in *Pseudomonas aeruginosa* biofilms and evaluation of biocide action model. *Biotech. Bioeng.* 56: 201–209.

Santegoeds, C. M., T. G. Ferdelman, G. Muyzer, and D. de Beer. 1998. Structural and functional dynamics of sulfate-reducing population in bacterial biofilms. *Appl. Environ. Microbiol.* 64: 3731–3739.

Sheu, P. M., K. Berghof, and U. Stahl. 1998. Detection of pathogenic and spoilage micro-organisms in food with the polymerase chain reaction. *Food Microbiol.* 15: 13–31.

Schofield, G. M., and R. Locci. 1985. Colonization of a model hot water system by *Legionalla pneumophila*. *J. Appl. Bacteriol.* 58: 151–162.

Schwach, T. S., and E. A. Zottola. 1984. Scanning electron microscopic study on some effects of sodium hypochlorite on attachment of bacteria to stainless steel. *J. Food Prot.* 47: 756–759.

Sequeira, C. A. C. 1988. Electrochemical techniques for studying microbial corrosion. *Microbial Corrosion Conference*, Sintra, 7–9 March, 1988. Essex: Elsevier Applied Science, pp. 99–118.

Sequeira, C. A. C., P. M. N. A. Carrasquinho, and C. M. Cebola. 1989. Control of microbial corrosion in cooling water systems by the use of biocides. *Microbial Corrosion Conference*, Sintra, 7–9 March, 1988. Essex: Elsevier Applied Science, pp. 240–255.

Shelley, A. 1990. The prevention of microbial contamination in air handling systems. *Austr. Refrig. Air Condit. Heat* 443: 30–36.

Shin-Ho-Lee, and J. F. Frank. 1991. Inactivation of surface-adherent *Listeria monocytogenes*, Hypochlorite and heat. *J. Food Prot.* 54: 4–6.

Shotton, D. M. 1989. Confocal scanning optical microscopy and its applications for biological specimens. *J. Cell Sci.* 94: 175–206.

Sillanpää, J. 1998. *Autocontrol Systems, HACCP and Hygiene in the Paper Industry.* Helsinki: Pikku-Sitomo. Master's thesis at Helsinki University of Technology, 95p. (in Finnish)

Sjöberg, A.-M., G. Wirtanen, and T. Mattila-Sandholm. 1995. Biofilm and residue investigations of detergents on surfaces of food processing equipment. *Trans. Inst. Chem. Eng. C: Food Bioprod. Proc.* 73: 17–21.

Sorrelle, P. H., and W. E. Belgard. 1992. Growth in recycling escalates costs for paper machine biological control. *Pulp & Paper* 66(5): 57–64.

Stewart, P. S., T. Griebe, R. Srinivasan, C.-I. Chen, F. P. Yu, D. de Beer, and G. A. McFeters. 1994. Comparison of respiratory activity and culturability during monochloramine disinfection of binary population biofilms. *Appl. Environ. Microbiol.* 60: 1690–1692.

Stewart, P. S., M. A. Hamilton, B. R. Goldstein, and B. T. Schneider. 1996. Modeling biocide action against biofilms. *Biotech. Bioeng.* 49: 445–455.

Stone, L. S., and E. A. Zottola. 1985. Effect of cleaning and sanitizing on the attachment of *Pseudomonas fragi* to stainless steel. *J. Food Sci.* 50: 951–956.

Stoodley, P., D. de Beer, and Z. Lewandowski. 1994. Liquid flow in biofilm systems. *Appl. Environ. Microbiol.* 60: 2711–2716.

Storgårds, E., H. Simola, A.-M. Sjöberg, and G. Wirtanen. 1999. Hygiene of gasket materials used in food processing equipment. Part 2: Aged materials. *Trans IChemE C* 77: 146–155.

Suominen, I., M.-L. Suihko, and M. Salkinoja-Salonen. 1997. Microscopic study of migration of microbes in food-packaging paper and board. *J. Ind. Microbiol. Biotechnol.* 19: 104–113.

Tarvainen, K., G. Wirtanen, and T. Mattila-Sandholm. 1994. Preventing airborne microbial risks in rooms with special hygiene requirements. *Healthy Buildings '94,* 22–25 August 1994. Budapest: Technical University of Budapest, pp. 217–222.

Thomas, J. C., M. Desrosiers, Y. St-Pierre, P. Lirette, J. G. Bisailloon, R. Beaudet, and R. Villemur. 1997. Quantitative flow cytometric detection of specific micro-organisms in soil samples using rRNA targeted fluorescent probes and ethidium bromide. *Cytometry* 27: 224–232.

Thorogood, D. 1999. Process hygiene—Good manufacturing practices. In *30th R^3-Nordic Contamination Control Symposium*, VTT Symposium 193, G. Wirtanen, S. Salo, and A. Mikkola, eds. Espoo: Libella Painopalvelu Oy, pp. 71–82.

Timperley, D. A., R. H. Thorpe, and J. T. Holah. 1992. Implications of engineering design in food industry hygiene. In *Biofilms—Science and Technology*, L. F. Melo, T. R. Bott, M. Fletcher, and B. Capdeville, eds. Dordrecht: Kluwer Academic Publishers, pp. 379–393.

Tolker-Nielsen, T., K. Holmstrom, and S. Molin. 1997. Visualization of specific gene expression in individual *Salmonella typhimurium* cells by in situ PCR. *Appl. Environ. Microbiol.* 63: 4196–4203.

Tolker-Nielsen, T., K. Holmstrom, L. Boe, and S. Molin. 1998. Non-genetic population heterogeneity studies by in situ polymerase chain reaction. *Molecul. Microbiol.* 27: 1099–1105.

Troller, J. A. 1993. *Sanitation in Food Processing.* San Diego: Academic Press, pp. 30–70 and pp. 263–286.

Tuompo, H., G. Wirtanen, S. Salo, L. Scheinin, A. Båtsman, and S. Levo. 1998. Method and test kit for pretreatment of object surfaces. World Intellectual Property Organization 98/07883, 18 August 1997, 26 February 1998, 33p.

Turakhia, M. H., and W. G. Characklis. 1989. Activity of *Pseudomonas aeruginosa* in biofilms: Effect of calcium. *Biotechnol. Bioeng.* 33: 406–414.

Väätäinen, P. 1994. Microbial control [Microbien torjunta]. *AEL-Insko seminar "Prossesivesijärjestelmät: prosessiveden käsittely ja kierrätys*. Vantaa: AEL-Insko, Part IV. (in Finnish)

Väisänen, O. M., E.-L. Nurmiaho-Lassila, S. A. Marmo, and M. S. Salkinoja-Salonen. 1994. Structure and composition of biological slimes on paper and board machines. *Appl. Environ. Microbiol.* 60: 641–653.

van der Wende, E., W. G. Characklis, and D. B. Smith. 1989. Biofilms and bacterial drinking water quality. *Water Res.* 23: 1313–1322.

van der Wende, E., and W. G. Characklis. 1990. Biofilms in potable water distribution systems. In Brock/Springer Series in Contemporary Bioscience: *Drinking Water Microbiology: Progress and Recent Developments*, G. A. McFeters, ed. New York: Springer-Verlag, pp. 249–268.

Videla, H. A., and W. G. Characklis. 1992. Biofouling and microbially influenced corrosion. *Biodeteriorat. Biodegradat.* 29: 195–212.

Wagner, M., R. Amann, H. Lemmer, and K. H. Schleifer. 1993. Probing activated sludge with oligonucleotides specific for proteobacteria: Inadequacy of culture-dependent methods for describing microbial community structure. *Appl. Environ. Microbiol.* 59: 1520–1525.

Walch, M. 1992. Microbial corrosion. In *Encyclopedia of Microbiology*, Vol. 1, J. Lederberg, ed. San Diego: Academic Press, pp. 585–591.

Walker, A. J., J. T. Holah, S. P. Denyer, and G. S. A. B. Stewart. 1993. The use of bioluminescence to study the behaviour of *Listeria monocytogenes* when attached to surfaces, *Colloids Surfaces A: Physicohem. Eng. Aspects* 77: 225–229.

Wang, R.-F., W.-W. Cao, and C. E. Cerniglia. 1997. A universal protocol for PCR detection of 13 species of foodborne pathogens in foods. *J. Appl. Microbiol.* 83: 727–736.

Whittaker, C., H. Ridgway, and B. H. Olson. 1984. Evaluation of cleaning strategies for removal of biofilms from reverse-osmosis membranes. *Appl. Environ. Microbiol.* 48: 395–403.

Wilkinson, J. B., and R. J. Moore, eds. 1982. *Harry's Cosmeticology*. 7 ed. London: George Godwin.

Wirtanen, G. 1995. *Biofilm Formation and Its Elimination from Food Processing Equipment*. VTT Publication 251, Espoo: The Technical Research Centre of Finland, 106 p. + app. 48p.

Wirtanen, G., S. Salo, D. G. Allison, T. Mattila-Sandholm, and P. Gilbert. 1998. Performance evaluation of disinfectant formulations using poloxamer-hydrogel biofilm-constructs. *J. Appl. Microbiol.* 85: 965–971.

Wirtanen, G., S. Salo, and T. Mattila-Sandholm. 1999. Microbial methods in evaluation of cleaning procedures. In *30th R^3-Nordic Contamination Control Symposium*, VTT Symposium 193, G. Wirtanen, S. Salo, and A. Mikkola, eds. Espoo: Libella Painoplavelu Oy, pp. 207–218.

Wirtanen, G., S. Salo, J. Maukonen, S. Bredholt, and T. Mattila-Sandholm. 1997. *NordFood Sanitation in Dairies*. VTT Publication 309, Espoo: The Technical Research Centre of Finland, 47p. + app. 22p.

Wolfaardt, G. M., and T. E. Cloete. 1992. The effect of some environmental parameters on surface colonization by microorganisms. *Water Res.* 26: 527–537.

Wong, A. C. L., and O. Cerf. 1995. Biofilms: Implications for hygiene monitoring of dairy plant surfaces. *Bull. Int. Dairy Fed.* 302: 40–50.

Wood, P., M. Jones, M. Bhakoo, and P. Gilbert. 1996. A novel strategy for control of microbial biofilms through generation of biocide at the biofilm-surface interface. *Appl. Environ. Microbiol.* 62: 2598–2602.

Woollatt, E. 1985. *The Manufacture of Soaps, Other Detergents and Glyserine*. London: Ellis Horwood Ltd., 473p.

Yu, F. P., and G. A. McFeters. 1994a. Physiological responses of bacteria in biofilms to disinfection. *Appl. Environ. Microbiol.* 60: 2462–2466.

Yu, F. P., and G. A. McFeters. 1994b. Rapid in situ assessment of physiological activities in bacterial biofilms using fluorescent probes. *J. Microbiol. Meth.* 20: 1–10.

Yu, F. P., G. M. Callis, P. S. Stewart, T. Griebe, and G. A. McFeters. 1994. Cryosectioning of biofilms for microscopic examination. *Biofouling* 8: 85–91.

Zips, A., G. Schaule, and H.-C. Flemming. 1990. Ultrasound as a means of detaching biofilms. *Biofouling* 2: 323–333.

Zottola, E. A. 1991. Characterization of the attachment matrix of *Pseudomonas fragi* attached to nonporous surfaces. *Biofouling* 5: 37–55.

Zottola, E. A., and K. C. Sasahara. 1994. Microbial biofilms in the food processing industry—Should they be a concern? *Int. J. Food Microbiol.* 23: 125–148.

Zyska, B. J. 1988. Microbial deterioration of rubber. In *Biodeterioration 7*, D. R. Houghton, R. N. Smith, and H. O. W. Eggins, eds. Essex: Elsevier Publishers Ltd., pp. 535–552.

BIOFILM CONTROL BY ANTIMICROBIAL AGENTS

PHILIP S. STEWART

Center for Biofilm Engineering and Department of Chemical Engineering, Montana State University, Bozeman, Montana

GORDON A. MCFETERS

Center for Biofilm Engineering and Department of Microbiology, Montana State University, Bozeman, Montana

CHING-TSAN HUANG

Department of Agricultural Chemistry, National Taiwan University, Taipei, Taiwan, Republic of China

11.1 INTRODUCTION

One of the hallmarks of the biofilm mode of microbial growth is remarkable resistance to killing by antimicrobial agents. This property frustrates efforts to control detrimental biofouling but it probably also harbors clues to the distinct structure and function of microbial biofilms. In this chapter, quantitative process analysis is applied to several aspects of the treatment of biofilms with antimicrobial agents. This exercise, which is inspired by the approach Bill Characklis established in the first edition of *Biofilms*, is as interesting for the deficiencies in our understanding that it reveals as it is for the areas of insight.

11.1.1 Approaches to Biofilm Control

Mechanical cleaning and antimicrobial chemicals are the most used methods of biofilm control. Mechanical cleaning is effective but can be costly because it often involves equipment downtime or a significant labor expenditure. In many applications, mechanical cleaning is simply not an option because the fouled surface is not physically accessible. Biocides,

---

Biofilms II: Process Analysis and Applications, Edited by James D. Bryers.
ISBN 0-471-29656-2 Copyright © 2000 Wiley-Liss, Inc.

antibiotics, and disinfectants have, therefore, been the principal tools in controlling biofouling. This chapter focuses on the issues involved in controlling biofilms with such antimicrobial agents.

Some alternative biofilm control strategies that are mentioned but not discussed further in this chapter are summarized pictorially in Figure 11.1 The stop growth option refers to the possibility of using a biological pretreatment process to remove the nutrients that support biofouling from the water entering a system. In other words, biofilm accumulation is engineered to occur in a reactor where it can be managed rather than in an operating system where it is difficult to control. Many water treatment systems, from applications ranging from desalination to semiconductor manufacture, might benefit by incorporating a properly engineered biological unit operation. The block attachment option envisions an inherently antifouling material, one to which microorganisms either do not adhere or are so weakly bound that they do not accrete. No such generic material has been discovered. One strategy now being pursued is the incorporation of antimicrobial agents into materials or coatings. In the context of biomaterials used in medical devices, it may be economical to design sophisticated surface chemistries that promote tissue integration while retarding microbial attachment or to develop agents that block specific adhesion pathways. The least investigated of the alternative options illustrated in Figure 11.1 is the promote detachment option. Perhaps if we understood more about what holds a biofilm together and how detachment occurs in nature, methods or chemistries to promote this process could be developed.

11.1.2 Quantitative Literature Survey of Reduced Biofilm Susceptibility

Although the diminished susceptibility of microorganisms growing in biofilms to killing by antimicrobial agents is now widely recognized (Costerton, 1984; Costerton et al., 1987; Nichols, 1989; Gilbert et al., 1990; Hoyle and Costerton, 1991; Brown and Gilbert, 1993; Gilbert and Brown, 1995; Carpentier and Cerf, 1993; Eastwood, 1994; Zottola, 1994; Allison and Gilbert, 1995; Gander, 1996; Wilson, 1996; Gilbert et al., 1997; Morton et al., 1998), there is no generally accepted basis for quantifying the degree of this resistance. We propose two alternative measures of biofilm reduced susceptibility. The first resistance factor (which we denote as RF1) is defined as the log reduction measured in the planktonic state divided by the log reduction measured in the biofilm state in response to the same dose. A second resistance factor (denoted by RF2) could be defined as the concentration required to achieve a particular reduction in viable cell numbers (e.g., 99%) in the biofilm state, divided by the concentration needed to achieve the same reduction in the planktonic state for the same dose duration. A survey of resistance factors calculated in these ways from data in the literature are summarized in Table 11.1. This literature review is not comprehensive.

The biofilm resistance factors presented in Table 11.1 reinforce the familiar statement that biofilm reduced susceptibility is observed consistently across a wide range of antimicrobial agents and microbial species. There is considerable variability in the resistance factor even for the same microbial species and the same antimicrobial agent. Indeed, within a single experimental study, the degree of resistance has been noted to depend on biofilm age, biofilm areal cell density, microbial strain, nutrient medium, and antimicrobial dose. This indicates that the precise way in which the biofilm is grown and treated is important in determining the extent of reduced biofilm susceptibility. We caution against attempting to extrapolate any of the resistance factors reported in Table 11.1 to systems different from the one they were determined in. There is no basis at this point for advising use of one resistance factor over another. The consistent use of at least one of these quantitative measures

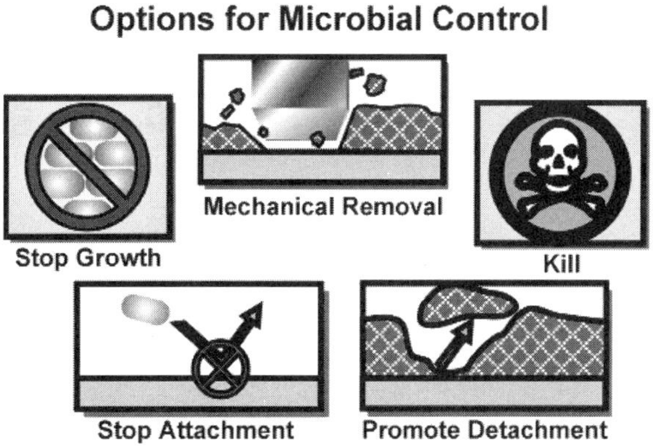

Options for Microbial Control

Figure 11.1 Approaches to biofilm control.

of biofilm reduced susceptibility is recommended to all experimenters as once effects are quantified, it becomes possible to search for correlations.

However, it could be argued that use of a biofilm resistance factor such as previously proposed clings unnecessarily to planktonic microbiological methods. Though our ostensible focus is biofilm susceptibility we remain wed, through this factor, to the planktonic result as a reference state. Planktonic methods are familiar and thereby afford a sense of security. But if their utility in understanding biofilm processes is truly so limited (which it is), then why bother at all with assays based on suspended cultures? In many practical applications, the answer to this question may be that there really is no point in performing planktonic tests. Reliable, user-friendly biofilm testing methods need to be developed, adopted by regulatory agencies and professional groups, and disseminated. One reason to continue to study planktonic microbial responses to antimicrobial agents is to elucidate mechanisms of biofilm resistance to antimicrobials. Differences between planktonic and biofilm susceptibility could provide extremely valuable insights if these differences can be linked to specific physical, chemical, and biological phenomena.

11.2 MECHANISMS OF REDUCED SUSCEPTIBILITY

Biofilms evade antimicrobial challenges by multiple mechanisms. These we have grouped into three broad categories: (1) reduction of the antimicrobial concentration in the bulk fluid surrounding the biofilm; (2) failure of the antimicrobial agent to penetrate the biofilm; and (3) adoption of a resistant physiological state or phenotype by at least a fraction of the cells in a biofilm. In the first scenario, the antimicrobial agent is depleted to ineffectual levels before it gets to the biofilm. In the second scenario, the antimicrobial agent is delivered to the surface of the biofilm, but is not effectively transported into the depth of the biofilm. In the third scenario, the antimicrobial agent permeates the biofilm, but is unable to kill microorganisms because they exist in a phenotypic state that confers reduced susceptibility. These mechanisms of biofilm protection are not mutually exclusive. Indeed, it seems likely that combinations of these three general types of resistance work in concert.

The reduced susceptibility of biofilms has not been attributed to the mechanisms—mutation or acquisition of genetic elements coding for specific resistance genes—that

TABLE 11.1. Selected Biofilm Resistance Factors (defined in text) to Antimicrobial Killing

Antimicrobial Agent	Microorganism	RF1	RF2	Reference
Halogens and Oxidants				
Chlorine	*Listeria monocytogenes*	1.8		Ronner and Wong, 1993
	Salmonella typhimurium	1.4		Ronner and Wong, 1993
	Enterobacter aerogenes	4–20		Stewart et al., 1998
	Escherichia coli		50	Ntsama-Essomba et al., 1997
	Staphylococcus aureus	>1.7		Oie et al., 1996
Iodine	*L. monocytogenes*	1.9		Ronner and Wong, 1993
	S. typhimurium	3		Ronner and Wong, 1993
Povidone-iodine	*Citrobacter diversus*	2–8		Stickler and Hewett, 1991
	Enterococcus faecalis	5–7		Stickler and Hewett, 1991
	Pseudomonas aeruginosa	7–14		Stickler and Hewett, 1991
Hydrogen peroxide	*P. aeruginosa*	1–3		Wood et al., 1996
Peracetic acid	*E. coli*		33	Das et al., 1998
	Staphylococcus epidermidis		2	Das et al., 1988
Potassium monopersulfate	*P. aeruginosa*	1–2		Wood et al., 1996
Biguanides				
Chlorhexidine	*C. diversus*	4–6		Stickler and Hewett, 1991
	E. faecalis	8–9		Stickler and Hewett, 1991
	P. aeruginosa	5		Stickler and Hewett, 1991
	E. coli	1.7		Dusart et al., 1994
	A. actinomycetemcomitans	1.3–3		Thrower et al., 1997
	Staphylococcus aureus	>6		Oie et al., 1996
	Streptococcus sanguis		2–8	Millward and Wilson, 1989
	Streptococcus mutans	2–24		Sutch et al., 1995
Polyhexamethylene biguanide	*E. coli*		3–6	Das et al., 1998
	S. epidermidis		3–10	Das et al., 1998
Quaternary Ammonium Compounds				
Benzalkonium chloride	*E. aerogenes*	2–4		Stewart et al., 1998
	S. aureus	>3		Oie et al., 1996
	E. coli		500	Ntsama-Essomba et al., 1997

11.2 MECHANISMS OF REDUCED SUSCEPTIBILITY

Cetylpyridinium chloride	*Actinobacillus actinomycetemcomitans*	1–4	Thrower et al., 1997
Quat-256	*L. monocytogenes*	1.9	Ronner and Wong, 1993
	S. typhimurium	1.6	Ronner and Wong, 1993
Other Biocides			
Bronopol	*Legionella pneumophila*	3–6	Wright et al., 1991
Glutaraldehyde	*Pseudomonas fluorescens*	>4	Eagar et al., 1988
Glutaraldehyde	*Legionella bozemanii*	1.8–5	Green and Pirrie, 1993
Glutaraldehyde	*E. aerogenes*	2–10	Stewart et al., 1998
Isothiazolone (Kathon)	*L. pneumophila*	3–6	Wright et al., 1991
Isothiazolone (Kathon)	*E. aerogenes*	1–11	Stewart et al., 1998
Phenoxyethanol	*E. coli*	1–2	Das et al., 1998
	S. epidermidis	1–2	Das et al., 1998
Phenolics			
Phenol	*E. coli*	1–2	Ntsama-Essomba et al., 1997
Chloroxylenol	*E. coli*	1	Das et al., 1998
	S. epidermidis	1	Das et al., 1998
Antibiotics			
Ampicillin	*E. coli*	>4	Morck et al., 1994
Amoxicillin	*Lactobacillus acidophilus*	>16	Muli and Struthers, 1998
Ciprofloxacin	*P. aeruginosa*	14	Jass and Lappin-Scott, 1996
	P. aeruginosa	5	Ishida et al., 1998
	P. aeruginosa	1.3	Vrany et al., 1997
	S. epidermidis	1.4	Hamilton-Miller and Shah, 1997
Metronidazole	*Gardnerella vaginalis*	4	Muli and Struthers, 1998
	Porphyromonas gingivalis	160	Wright et al., 1997
Tobramycin	*P. aeruginosa*	1–10	Anwar et al., 1989b
	P. aeruginosa	18–24	Anwar et al., 1989a
	P. aeruginosa	20	Nickel et al., 1985
Vancomycin	*S. aureus*	>14	Williams et al., 1997
	S. epidermidis	>17	Evans and Holmes, 1987
	S. epidermidis	4	Gristina et al., 1989

account for conventional antibiotic resistance. For these mechanisms to explain biofilm resistance, the genetic modifications would have to be present in the biofilm but absent in the planktonic state. This does not appear to be the case in general. Even clearly susceptible microorganisms acquire marked resistance in the biofilm mode of growth (Table 11.1). When dispersed from a biofilm, however, these microorganisms rapidly return to a susceptible state (Nickel et al., 1985; Eagar et al., 1988; Anwar et al., 1989b; Ntsama-Essomba et al., 1997; Williams et al., 1997).

11.2.1 Antimicrobial Depletion in the Bulk Fluid

If a biofilm exerts a chemical demand for the antimicrobial agent with which it is being challenged, then it is possible for these neutralizing reactions to reduce the bulk fluid concentration of the agent. Consider, for example, an antimicrobial susceptibility test performed first against a dilute suspension of planktonic cells and then against a heavily fouled biofilm specimen. It is not difficult to imagine that the antimicrobial concentration could be maintained during the planktonic test but significantly decreased during the course of the biofilm test. One could argue that this phenomenon is not a true resistance mechanism, but simply an unfair comparison. The biofilm is not being subjected to the same antimicrobial concentration as the planktonic reference test. However, it seems that this straightforward mechanism has been often overlooked. It would pay to keep antimicrobial depletion in the bulk fluid in mind as a possible explanation for poor antimicrobial performance against a biofilm. What this mechanism lacks in glamor it may recoup in practical importance.

Bulk fluid antimicrobial depletion can be diagnosed experimentally by measuring antimicrobial residual concentrations during both planktonic and biofilm tests. In some experimental systems, biofilm and planktonic disinfection can be measured in the same fluid, an approach that elegantly eliminates the possibility of unequal treatment concentrations.

11.2.2 Transport Limitation of Antimicrobial Penetration

Failure of an antimicrobial agent to rapidly or completely penetrate a biofilm is perhaps the most intuitively appealing explanation for biofilm resistance. Antoni van Leeuwenhoek invoked this mechanism more than three centuries ago in his seminal studies of dental plaque (which he called "scurf"):

> From whence I conclude, that the Vinegar with which I washt my Teeth, kill'd only those Animals which were on the outside of the scurf, but did not pass thro the whole substance of it.
> —A. van Leeuwenhoek, 1684

We distinguish two versions of the transport limitation resistance mechanism. The first postulates that the biofilm matrix constitutes a barrier to the inherent mobility of antimicrobial agents. According to this hypothesis, the matrix physically excludes antimicrobial compounds from the biofilm. While powerful and generic in its ability to explain antimicrobial resistance, this mechanism poses the paradox of how such a matrix barrier allows nutrients to pass while excluding biocidal molecules of similar size. Measurements of effective diffusion coefficients in biofilms (Stewart, 1998) indicate that, while diffusion is somewhat retarded in biofilms, diffusive transport nevertheless proceeds at rates of the same order of magnitude of those in pure water. Even molecules of the size of oligonucleotide probes, lectins, and others readily penetrate intact biofilms as evidenced by their ability to stain the interior regions of biofilm specimens. The direct experimental demonstration of permeation of certain antimicrobial agents (Dunne et al., 1993; Darouiche et al., 1994; Kumon et al., 1994; Shigeta et al., 1997; Suci et al., 1994; Vrany et al., 1997) through biofilm also argues against a generic barrier to antimicrobial agent access.

The second and more plausible version of antimicrobial transport limitation in biofilms requires an interaction between the antimicrobial agent and the biofilm that neutralizes antimicrobial activity. The barrier to penetration in this case is reactive rather than physical: The rate of deactivation of the antimicrobial exceeds the rate of diffusive penetration. This mechanism is supported by experimental evidence in the case of hypochlorite (de Beer et al., 1994; Chen and Stewart, 1996; Xu et al., 1996), is likely to be important for other highly reactive oxidants, such as ozone and hydrogen peroxide, and may be a factor for some nonoxidizing biocides as well (Stewart et al., 1998). It is also realistic for certain antibiotics, such as the β-lactams, that are subject to rapid enzymatic degradation (Nichols et al., 1989; Nichols, 1991).

11.2.3 Physiological Limitation of Antimicrobial Efficacy

There are many instances in which a biofilm is too thin or is insufficiently reactive with the antimicrobial agent to manifest either of the preceding resistance mechanisms. In these cases, biological explanations for the reduced susceptibility of biofilm microorganisms are called for. We distinguish two general types of biological limitations to biofilm susceptibility. The first type of biological limitation to biofilm susceptibility requires that at least some of the cells in a biofilm experience nutrient limitation and therefore exist in a slow-growing or starved state (Brown et al., 1988; Gilbert and Brown, 1995). Such slow- or non-growing cells are hypothesized (or have been shown experimentally) to be less susceptible to many antimicrobial agents. The second type of biological limitation of biofilm susceptibility invokes the existence of a distinct, and relatively resistant, biofilm phenotype. This phenotype is not the result of a nutrient limitation. The hypothesized biofilm phenotype is adopted by a subset of the microbial population in a biofilm as a result of some other stimulus, for example, contact with a solid surface or attainment of a threshold cell density. Specific biological factors and mechanisms that may be involved in modulating biofilm susceptibility are discussed elsewhere (Gilbert et al., 1990; Gilbert and Brown, 1995).

11.3 PROCESS ANALYSIS

The action of an antimicrobial agent against a biofilm involves the complex interaction of multiple processes. Among the constituent phenomena that must be considered are: (1) bulk fluid flow into and out of the system, (2) consumption of the antimicrobial by reaction with biomass, (3) substrate utilization, (4) diffusive transport within the biofilm, (5) microbial disinfection, and (6) biofilm detachment. Additional processes that might confound the problem include corrosion, accumulation of abiotic particles in the biofilm, and mutation or adaptation. The interactions between these phenomena and the functional outcome of those interactions are not intuitive.

Precisely because of its complexity, the problem of antimicrobial control of biofilm is one that invites quantitative engineering analysis and modeling. Mathematical modeling is one way to organize and integrate phenomena systematically. A model is a means of both formulating a hypothesis and exploring its implications. Ultimately, using a model could be an effective way of performing design and scale-up tasks, such as selecting an appropriate antimicrobial agent for a particular situation or developing optimal antimicrobial dosage protocols. A few examples of mathematical biofilm models that incorporate antimicrobial action can be found in the literature (Nichols et al., 1989; Al-Hoti et al., 1990; Rossman et al., 1994; Stewart, 1994; Lu et al., 1995; Dibdin et al, 1996; Stewart et al., 1996; Sanderson and Stewart, 1997). Rather than present a comprehensive model of this type, we have elected to focus here on the quantitative analysis of the constituent phenomena. These analyses constitute submodels that could be mixed and matched to compose an integrated overall model.

In this section, a series of example calculations are presented to illustrate the concepts presented in the preceding section, their real-world significance, and the application of process analytical approaches to describe these phenomena.

11.3.1 Antimicrobial Reaction

The following examples address the possibility that reaction between an antimicrobial agent and biofilm could reduce the bulk fluid concentration of the agent in all or part of the biofilm system. These analyses do not consider the gradient in antimicrobial concentration that may occur within the biofilm. This aspect is addressed separately in Section 11.3.2.2.

11.3.1.1 Antimicrobial Reaction—Depletion of a Biocide along the Length of a Pipe.
A biocide introduced into a pipe colonized with biofilm can be depleted by reactions with the attached biomass. This reduces the bulk fluid concentration of the biocide as it flows along. Although the reaction may be relatively slow, the cumulative loss of antimicrobial may become significant in the case of a pipe whose length is measured in kilometers. The decay in bulk fluid concentration can be modeled by a differential material balance with the following form:

$$-v\frac{dC}{dy} = \frac{2kC}{R} \quad (11.1)$$

where C denotes biocide concentration (M L^{-3}); k, biocide surface reaction rate coefficient (L T^{-1}); R, pipe radius (L); v, average axial fluid velocity (L T^{-1}); and y, axial distance (L).

This model assumes ideal plug flow and reaction kinetics that are first order with respect to antimicrobial concentration. The surface reaction rate coefficient has units of velocity and can be understood as combining kinetic and mass transfer properties (Rossman et al., 1994). The two terms in the balance represent advection and surface reaction, respectively. The solution to this equation, taking $C = C_o$ at $y = 0$, is simply

$$\frac{C}{C_o} = \exp\left(-\frac{2k}{vR}y\right) \quad (11.2)$$

In other words, the biocide concentration decays exponentially with distance down the pipe. This phenomenon can be anticipated to occur for many biocides when introduced into a fouled pipe.

Example Calculation. Estimate the chlorine residual in a drinking water distribution pipe a distance 3 km (y) downstream from a point where it is measured to be 1 mg L^{-1} (C_o). The pipe diameter ($2R$) is 20 cm, the average velocity (v) is 50 cm s^{-1}, and the surface reaction rate coefficient (k) has been estimated to be 5×10^{-4} cm s^{-1}. The reaction rate coefficient could account for reactions between chlorine and concrete, metallic iron or its corrosion products, as well as with attached biomass. The argument of the exponential is -0.6 and the calculated concentration is 0.55 mg L^{-1}. This represents a significant decrease in the bulk fluid residual which could be a factor in the survival of biofilm microorganisms in the distal regions of a pipe system. Rossman et al. (1994) describe more sophisticated calculations of this type.

11.3.1.2 Antimicrobial Reaction—Depletion of a Biocide in a Well-Mixed, Continuous Flow System.
Here we outline an analysis of the reaction of a biocide introduced into a system whose hydrodynamics approximate those of a continuous stirred tank reactor (CSTR). A material balance of the following form can be derived:

$$V \frac{dC}{dt} = q_f - QC - kAC \qquad (11.3)$$

where the variables and parameters are C, bulk fluid biocide concentration (M L^{-3}); t, time (T); V, system fluid volume (L^3); Q, total fluid flow rate out of system (L^3 T^{-1}); k, surface reaction rate coefficient (L T^{-1}); and q_f, biocide feed rate (M T^{-1}). The terms in Eq. (11.3), from left to right, represent the rate of accumulation of biocide in the circulating water, continuous feed rate of biocide, loss of biocide by fluid flow out of the system as either blowdown (water intentionally purged from the system) and drift (water droplets carried out of the system by wind), and loss of biocide by surface reaction. At steady state, the residual biocide concentration is

$$C = \frac{q_f}{(Q + kA)} \qquad (11.4)$$

Example Calculation. Estimate the residual biocide concentration in a recirculating cooling water system to which the biocide is continuously fed. It is difficult to imagine a better environment for growing biofilms than cooling water towers. In these ubiquitous devices, warm water is circulated over packing material with a high surface area to volume ratio to facilitate evaporative cooling of the water (Fig. 11.2). Cooling water systems are prone to scale formation and biofouling. Consider the illustrative case specified as follows: system fluid volume, V, 200 m^3; combined blowdown and drift flow rate, Q, 2 m^3 h^{-1}; surface area to volume ratio, A/V, 50 m^{-1}; surface reaction rate coefficient, k, 0.01 m h^{-1}; biocide feed rate, q_f, 50 g h^{-1}. Then the steady state residual biocide concentration is calculated to be 0.49 g m^{-3} from Eq. (11.4). What this result shows is that surface associated reaction can easily control the residual biocide concentration that can be maintained in a cooling water system.

11.3.2 Antimicrobial Penetration

Penetration of antimicrobial agents into biofilm is determined by the interaction of diffusion, sorption, and reaction processes. Reaction in this case refers to any deactivating chemical modification or irreversible sequestration of the antimicrobial agent. Sorption refers to any reversible binding of the antimicrobial agent to a biofilm component, either biotic or abiotic. As illustrated by the examples below, when an antimicrobial agent reacts in a biofilm its penetration can be retarded. While sorption can also retard penetration in theory, this aspect is not considered further here because in practice reaction has a far greater potential to limit biofilm penetration than does sorption.

11.3.2.1 *Penetration of a Noninteracting Agent.*
We begin by considering the diffusion of a noninteracting antimicrobial into biofilm. By noninteracting it is meant that neither reaction nor sorption processes act on the antimicrobial agent within the biofilm. In this and other penetration models presented in this section, convective transport within the biofilm is assumed to be negligible. The transient diffusion of an antimicrobial (or any solute for that matter) into a slab biofilm in response to a step increase in the bulk fluid concentration of the solute is given by

$$\frac{\partial C}{\partial t} = D_e \frac{\partial^2 C}{\partial z^2} \qquad (11.5)$$

$$\text{at } z = L_f, C = C_o \text{ for all } t > 0 \qquad (11.6)$$

$$\text{at } z = 0, \frac{\partial C}{\partial z} = 0 \text{ for all } t > 0 \qquad (11.7)$$

$$\text{at } t \leq 0, C = 0 \text{ for all } 0 \leq z \leq L_f \qquad (11.8)$$

where z is a spatial coordinate indicating depth in the biofilm. Equation (11.6) is the diffusion equation, which constitutes an unsteady mass balance on the antimicrobial agent. The term on the left-hand side represents the local accumulation (or, if negative in value, disappearance) of the solute, and the term on the right-hand side represents the net change in concentration due to diffusion. Equations (11.6) and (11.7) are boundary conditions that impose constant concentration at the biofilm-bulk fluid interface and a no-flux condition at the substratum, respectively. In physical terms, the no-flux condition means that the sub-

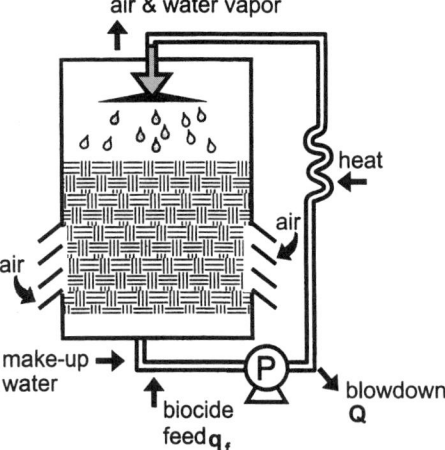

Figure 11.2 Hydraulics of a recirculating cooling water system.

11.3 PROCESS ANALYSIS

TABLE 11.2. Diffusion Coefficients of Antimicrobial Agents in Water at 25°C

Solute	D_{aq} in cm²/s
Alkyl (C_{10}-C_{12}) dimethyl benzyl ammonium chloride	4.3×10^{-6}
Benzylpenicillin	5.0×10^{-6}
Bromine (Br_2)	1.3×10^{-5}
Ceftazidime	3.8×10^{-6}
Chlorhexidine	3.7×10^{-6}
Chlorine (Cl_2)	1.4×10^{-5}
5-Chloro-2-methyl-4-isothiazolin-3-one	9.1×10^{-6}
Ciprofloxacin	5.1×10^{-6}
Gentamicin	4.6×10^{-6}
Glutaraldehyde	9.3×10^{-6}
Hydrogen peroxide[a]	1.4×10^{-5}
Hypochlorous acid[b]	1.9×10^{-5}
Monochloramine	1.9×10^{-5}
Methylenebisisothiocyanate	8.9×10^{-6}
2-Methyl-4-isothiazolin-3-one	1.0×10^{-5}
Ozone	1.9×10^{-5}
Piperacillin	3.9×10^{-6}
Tobramycin	4.2×10^{-6}
Vancomycin	2.1×10^{-6}

[a]Van Stroe-Biezen et al., 1993.
[b]Comparison to ion mobilities of Cl^-, ClO_3^-, and ClO_4^- (Horvath, 1985).
Note: Unless otherwise noted, estimates are based on the Wilke-Chang correlation.

stratum is impermeable. Equation (11.8) is an initial condition that specifies zero concentration throughout the biofilm at time zero.

The parameter D_e is the effective diffusion coefficient in the biofilm. The value of D_e will be reduced compared to the diffusion coefficient in water, D_{aq}, due to the presence in the biofilm of microbial cells, extracellular polymeric substances, abiotic particles, and gas bubbles. Experimental measurements of the ratio D_e/D_{aq} have recently been reviewed (Stewart, 1998) and that article presents guidelines and formulae for estimating D_e/D_{aq}. D_e/D_{aq} depends on the physical-chemical nature of the diffusing solute (its size and whether it is charged) and the biomass density in the biofilm. Estimates and experimental measurements of aqueous diffusion coefficients for a number of antimicrobial agents are given in Table 11.2

The solution to Eqs. (11.5) through (11.8) is

$$\frac{C}{C_o} = 1 - 2 \sum_{n=0}^{\infty} \frac{(-1)^n}{(n + 1/2)\pi} \exp[-(n + 1/2)^2 \pi^2 \alpha] \cos\left[(n + 1/2) \frac{\pi z}{L_f}\right] \quad (11.9)$$

which is reproduced graphically in Figure 11.3. The extent of penetration at any point depends on a single parameter, α, given by

$$\alpha = \frac{tD_e}{L_f^2} \quad (11.10)$$

Two example calculations of diffusive penetration times based on the scenario comprising Eqs. (11.5–11.8) and their solution (Eq. (11.9) Fig. 11.3) are presented next. These calculated times corresponds to the most rapid possible diffusive delivery of antimicrobial into the biofilm as the analysis incorporates neither reaction nor sorption. As mentioned previously, reaction and sorption processes both work to impede or limit penetration of a diffusing solute.

Example Calculation. Estimate the time required for the antibiotic ciprofloxacin to attain, at the base of a 250 μm thick biofilm, one-half the bulk fluid concentration bathing the biofilm. In terms of parameters, we have $C/C_o = 0.5$ at $z = 0$, $L_f = 2.5 \times 10^{-2}$ cm, and $D_{aq} = 6.87 \times 10^{-6}$ cm^2 s^{-1}. From Figure 11.3, this degree of penetration corresponds to $\alpha = 0.38$. In the absence of system-specific information, a reasonable estimate of D_e/D_{aq} is 0.4 (Stewart, 1998). The effective diffusion coefficient of ciprofloxacin within the biofilm is then estimated to be 40% of 6.87×10^{-6} cm^2 s^{-1} or 2.75×10^{-6} cm^2 s^{-1}. The aqueous diffusion coefficient used here is slightly higher than the value quoted in Table 11.2 because it has been estimated at a temperature of 37°C to correspond to the presumed physiological condition. Then, solving Eq. (11.10) for t

$$t = \alpha \frac{L_f^2}{D_e} = 86s$$

That is, the antibiotic should penetrate this biofilm within a minute and a half. This time scale has been confirmed by experimental measurements of rapid penetration of fluoroquinolone antibiotics into *Pseudomonas aeruginosa* biofilms (Suci et al., 1994; Vrany et al., 1997). Certainly when compared to the duration of antimicrobial chemotherapy, which might be 5 or 10 days, this delay in delivery of the agent is insignificant. This calculation suggests that, in the case of an antibiotic that neither sorbs nor reacts in the biofilm, the barrier posed by diffusion alone is not a viable mechanism of biofilm resistance to prolonged antimicrobial chemotherapy.

However, there are many situations in which antimicrobial agents are normally applied for relatively brief durations. Examples could include rinsing with an antibacterial mouthwash, sanitizing a toilet bowl, or disinfecting a hospital counter top. In such cases it is im-

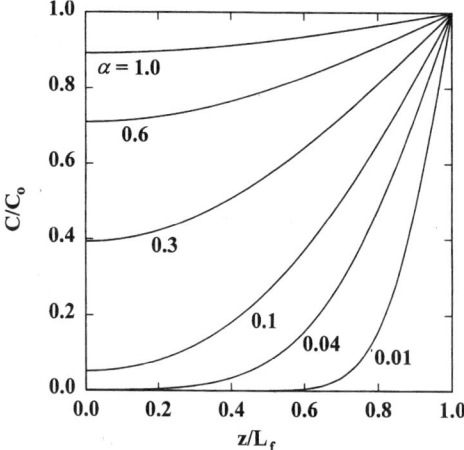

Figure 11.3 Transient diffusion of a noninteracting solute into a slab biofilm.

portant that the dose last sufficiently long to effectively reach the entire depth of the biofilm. The solution to the diffusion equation just presented can be used to estimate the minimum dose duration required.

Example Calculation. Estimate the minimum dose duration necessary to deliver a quaternary ammonium compound based disinfectant, at 90% of the bulk fluid concentration, to the base of a 1.5 mm thick biofilm growing in a crevice on a food-handling surface. The minimum dose duration is calculated by assuming that the disinfectant neither reacts nor is adsorbed within the biofilm; in other words, the result shown in Figure 11.3 applies. Ninety percent penetration at the substratum corresponds approximately to $\alpha = 1$ (Fig. 11.3). Taking 40% of the diffusion coefficient given in Table 11.2, the minimum dose time calculated from Eq. (11.10) is a remarkably long 3.6 h.

11.3.2.2 Penetration of a Reacting Antimicrobial Agent—Catalytic Reaction.

A chemical reaction that deactivates an antimicrobial agent as it diffuses into a biofilm impedes effective penetration into the biofilm. In this section, we consider the case of a catalytic reaction. That is, the deactivating reaction proceeds without consuming any component of the biofilm. Examples of such catalytic deactivation include destruction of β-lactam antibiotics by β-lactamase enzymes and disproportionation of hydrogen peroxide by catalase enzymes. This physical situation is mathematically modeled by deriving a differential material balance that incorporates the processes of diffusion and reaction. The formulation and the solution to this problem depend on the intrinsic kinetics of the deactivating reaction. As a simple illustrative case, consider the simultaneous reaction and diffusion of an antimicrobial agent subject to zero-order reaction kinetics. Zero-order reaction kinetics mean that the reaction rate is independent of the antimicrobial concentration. This approximation can be justified, in the case of an enzyme reaction, when the bulk fluid concentration of antimicrobial agent is much greater than the Michaelis-Menten half saturation coefficient. The material balance takes the form

$$D_e \frac{d^2C}{dz^2} = k_o \qquad (11.11)$$

where the two terms, from left to right, correspond to diffusion and reaction, respectively. The parameter, k_o, is a zero-order reaction rate coefficient with units of concentration per time. There is no time derivative here as there is in Eq. (11.5) because steady state has been assumed. The steady state solution represents the maximal extent of penetration that can be achieved. Boundary conditions on this balance are the same as those given in Eqs. (11.6) and (11.7). In this case of a catalytically reacting antimicrobial, distinct gradients in antimicrobial concentration persist within the biofilm (Fig. 11.4). The solution to this problem is conveniently framed in terms of a single dimensionless parameter known as the Thiele modulus and denoted here by ϕ, where, for the particular case of zero order kinetics,

$$\phi = \left(\frac{k_o L_f^2}{2 C_o D_e} \right)^{1/2} \qquad (11.12)$$

This parameter is actually the ratio of the maximum rate of reaction to the maximum rate of diffusion. When ϕ is small (< 1), diffusion is fast compared to reaction and the solute penetrates well. When ϕ is large (> 1), diffusion is slow compared to reaction and the biofilm is never fully penetrated by the solute. Clearly the situation of $\phi > 1$ would afford an excellent explanation for failure of an antimicrobial agent to kill a biofilm completely: A portion of the biofilm in the deep layers is never exposed to the antimicrobial, no matter how long the exposure period.

The reactions that deactivate the antimicrobial agent might continue long after the microbial cell is dead. For example, neither of the two enzyme reactions mentioned previously are energy-dependent. Provided that the enzyme remains in the cell or trapped in the extracellular matrix, it will continue to catalyze deactivation of the antimicrobial independent of the physiological status of the microorganism from whence it came. In other words, dead cells could contribute to the protection of their viable neighbors because the reactive barrier to penetration is not tied to viability.

Example Calculation. Estimate the penetration depth of 50 mM hydrogen peroxide into a 100 μm thick *P. aeruginosa* biofilm. When zero-order kinetics prevail, the concentration of the penetrating solute will decrease to zero inside the biofilm if it is sufficiently thick. Let the penetration depth, denoted by a, be defined as the depth at which hydrogen peroxide is completely depleted. This distance is given by

$$a = \left(\frac{2C_o D_e}{k_o}\right)^{1/2} \tag{11.13}$$

Take D_e to be 50% of the value given in Table 11.2, or 7.15×10^{-6} cm^2 s^{-1}. Combining a typical biofilm protein density with specific catalase activities in *P. aeruginosa* (Brown et al., 1995), the order of magnitude k_o can be estimated to be 50 mM s^{-1} (Brown et al., 1995). Then the calculated penetration depth is 38 μm. Poor penetration could explain biofilm resistance to disinfection in this case, because 60% of the biofilm will never be exposed to the antimicrobial agent.

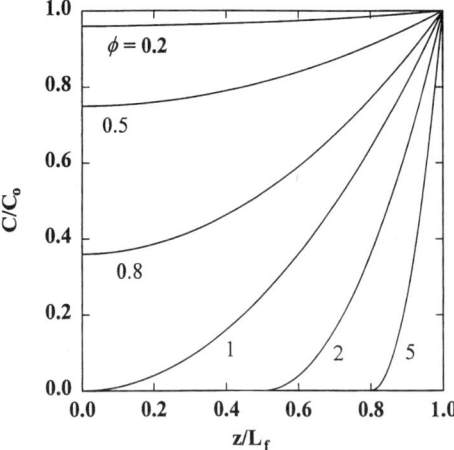

Figure 11.4 Simulated steady state concentration profiles within biofilm for a solute subject to consumption by a zero-order reaction.

Solutions to reaction-diffusion problems of the type outlined can be found in the literature for other kinetic models, such as first order and Michaelis-Menten kinetics. There are also several examples of modeling studies that apply these calculations to the problem of antimicrobial penetration in particular (Nichols et al., 1989; Dibdin et al., 1996).

11.3.2.3 Penetration of a Reacting Antimicrobial Agent—Stoichiometric Reaction.
Here we consider the possibility that the biofilm contains a finite amount of antimicrobial neutralizing capacity that is depleted stoichiometrically as the antimicrobial agent is deactivated. An example of such a stoichiometric reaction might be the neutralization of hypochlorous acid by reaction with nitrogenous organic matter and polysaccharides. In the stoichiometric reaction scenario, the antimicrobial must deplete the neutralizing capacity of the biofilm layer by layer before it can fully penetrate the film. Stewart and coworkers have published full mathematical models of the penetration of a stoichiometrically reacting solute (Stewart and Raquepas, 1995; Stewart, 1997). Presented here is a simplified model that adequately simulates slow penetration. In this conceptual model, the antimicrobial agent diffuses through a continuously expanding zone of the biofilm in which deactivating capacity has been depleted. If the deactivating reaction is very fast, the concentration of the antimicrobial agent will approach zero at the distal side of this depleted zone. With these assumptions, the advance of the antimicrobial front through the biofilm is described by the following differential equation:

$$\frac{X_b}{Y_{xb}} \frac{da}{dt} = D_e \frac{C_o}{a} \tag{11.14}$$

which balances the rate of depletion of neutralizing capacity (left-hand term) with the diffusive flux of antimicrobial through the neutralizing capacity-depleted zone of the biofilm (right-hand term). In Eq. (11.14), a is the penetration depth of the antimicrobial, X_b is the biomass density in the biofilm, and Y_{xb} is a stoichiometric coefficient determining the amount of biomass in which neutralizing capacity is depleted per amount of antimicrobial neutralized. With an initial condition of $a = 0$ at $t = 0$, the solution to this equation can be written in terms of the time required to fully penetrate the biofilm, which is given by

$$t = \frac{X_b L_f^2}{2 D_e C_o Y_{xb}} \tag{11.15}$$

Example Calculation. Estimate the time required for hypochlorite bleach at 10 mg/L chlorine to fully penetrate a 400 μm thick biofilm. Suppose that the biofilm biomass density is 10,000 mg/L and the effective diffusion coefficient of hypochlorite in the biofilm is 1×10^{-5} cm^2 s^{-1}. The yield coefficient for chlorine interacting with bacterial biofilm is approximately 1 mg biomass per mg chlorine (Characklis, 1990; Chen and Stewart, 1996). From Eq. (11.15), the calculated penetration time is 80,000 s or 22 hours. Such long penetration times for chlorine are consistent with direct experimental measurements of profoundly retarded penetration of this antimicrobial agent (de Beer et al., 1994; Chen and Stewart, 1996; Xu et al., 1996).

To be able to use either of the penetration theories outlined previously, one must first determine whether the reaction between antimicrobial and biofilm is catalytic or stoichiometric. Performing a calculation requires an estimate of either a reaction rate coefficient or a stoichiometric coefficient, depending on which type of reaction occurs. For most antimicrobial agents, these coefficients are unknown and data in the literature are inadequate even for the purpose of making rough estimates.

11.3.2.4 Penetration of a Reacting Antimicrobial Agent—Uncharacterized Reaction.
Because the kinetics and stoichiometry of antimicrobial reactions with biofilm are largely uncharacterized, the preceding penetration theories are difficult to apply in most instances. It is possible, nevertheless, to make an assessment of the potential for penetration limitation by comparing the observed overall rate of reaction of the antimicrobial with an estimate of its characteristic diffusion rate. This comparison is made by calculating an observable modulus defined by

$$\Phi = \frac{R_o L_f^2}{D_e C_o} \tag{11.16}$$

where R_o is the observed volumetric reaction rate of antimicrobial agent within the biofilm (Bailey and Ollis, 1986). The units of R_o could be, for example, mg of biocide per cm^3 of biofilm per second. Φ is dimensionless. When its numerical value is clearly greater than one, incomplete penetration is likely. When Φ is clearly less than one, biocide penetrates the biofilm effectively.

Example Calculation. Evaluate the potential for biocide penetration limitation in the hypothetical example considered in Section 11.3.1.2. Assume a biofilm thickness of 1 mm. The measured bulk residual concentration of biocide is 0.5 g m^{-3}. The observed rate of reaction in this case is

$$R_o = \frac{q_f - QC}{AL_f} \tag{11.17}$$

and R_o is calculated to be 4.9 g m^{-3} h^{-1}. The observable modulus, Φ, has a value of 2.7 from Eq. (11.16) indicating probable incomplete penetration of the biocide.

11.3.2.5 Antimicrobial Penetration—Summary.
Biofilms are mostly water and diffusion occurs within a biofilm at rates not too different from that in pure water. Diffusion alone, therefore, does not pose a significant barrier to antimicrobial penetration except in the case of thick biofilms subject to relatively brief challenges. However, if the antimicrobial agent reacts with biofilm constituents in a way that destroys its biocidal properties, then profound penetration failure is possible. A biofilm acts like a porous but reactive sponge that deactivates antimicrobial in the surface layers of the biofilm faster than it can diffuse in. This resistance mechanism is clearly operative in the case of reactive oxidants such as hypochlorite and hydrogen peroxide and contributes to reduced biofilm susceptibility to these agents (de Beer et al., 1994; Chen and Stewart, 1996; Xu et al., 1996; Liu et al., 1998).

11.3.2.6 Antimicrobial Penetration—Biofilm Structural Heterogeneity and External Mass Transfer Resistance.
Two other issues that require comment when discussing transport phenomena in biofilms are biofilm structural heterogeneity and external mass transfer resistance. Some biofilms can exhibit remarkable structural heterogeneity with clusters of microorganisms separated by water channels through which fluid flows

11.3 PROCESS ANALYSIS

(see Chapter 3). This observation appears to invalidate the assumptions common to the preceding penetration models of a uniformly thick slab biofilm in which no convective transport occurs. We contend that these simple models are still useful. Structural heterogeneity alters the geometry of the problem, but does not change the fundamental phenomena at work: diffusion and reaction. Within a cell cluster, diffusion is the dominant transport process (de Beer and Stoodley, 1995). Slab biofilm models correctly simulate penetration phenomena, at least qualitatively if not quantitatively. There are also ways to improve the quantitative accuracy of such models. The obvious way to deal with structural heterogeneity in a penetration model is to assume or specify a more realistic biofilm geometry and then solve the equations for this geometry. The math may be somewhat more complex but in the age of computers this does not represent an obstacle.

Example Calculation. Repeat the calculation of diffusive penetration time of a quaternary ammonium compound present in 11.3.2.1 for a biofilm with the heterogeneous structure illustrated in Figure 11.5. The cell clusters have a radius of $R = 3308$ μm. If the biomass in the heterogenous biofilm were spread out into a uniformly thick biofilm it would be 1500 μm thick, the same depth used in the original calculation. In other words, the biomass in the slab biofilm has simply been redistributed into a heterogenous structure. An exact analytical solution to the problem of unsteady diffusion in spherical coordinates is available (Bird et al., 1960). This solution yields a 90% penetration time of 6.0 h, considerably longer than the 3.6 h calculated for a uniformly thick (1500 μm) biofilm. This comparison suggests that heterogeneity will tend to lengthen diffusive penetration times for noninteracting solutes.

When the antimicrobial reacts in the biofilm and a reaction-diffusion model is called for, different approaches for accounting for biofilm structural heterogeneity are appropriate. One of these is to use the slab biofilm model result but replace biofilm thickness with the biofilm volume to surface area ratio. The surface area is that of the biofilm–fluid interface,

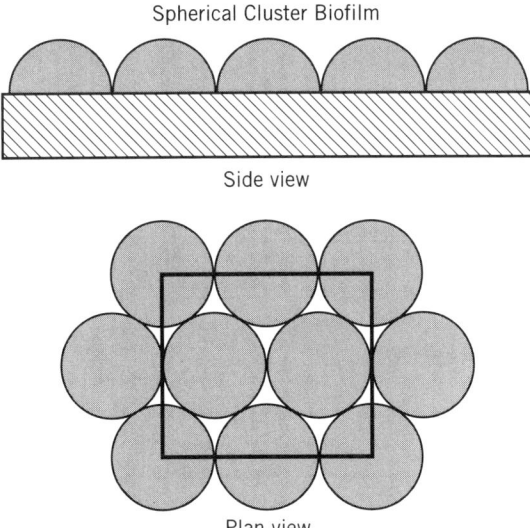

Figure 11.5 Model of a structurally heterogenous biofilm. The biofilm consists of hemispherical cell clusters arranged in a regular array on the substratum.

not the substratum surface area. The theoretical basis for this substitution is well established (Aris, 1957; Bird et al., 1960). In this case, structural heterogeneity enhances access of antimicrobial agent to the biofilm and increases the extent of penetration.

Example Calculation. Repeat the calculation of the penetration depth of hydrogen peroxide presented in 11.3.2.2 for a biofilm with the heterogeneous structure illustrated in Figure 11.5. The cell clusters have a radius of $R = 221$ μm. If the biomass in the heterogeneous biofilm were spread out into a uniformly thick biofilm it would be 100 μm thick, the same depth used in the original calculation. An exact analytical solution to the problem of simultaneous zero-order reaction and diffusion in spherical coordinates can be derived. This solution yields a penetration depth of 40 μm, which corresponds to exposure of 46% of the biofilm volume to the antimicrobial agent. For comparison, recall that 38% of the equivalent slab biofilm was exposed to hydrogen peroxide. Substituting the biofilm volume to surface area ratio, which in the case of a hemisphere is $R/3$, into the slab biofilm model as suggested previously predicts that 51% of the biofilm would be exposed. This correction may be adequate for many purposes. The enhancement of antimicrobial penetration due to heterogenous biofilm structures is relatively modest if this calculation is representative.

External mass transfer resistance refers to the diffusive resistance presented by the slow-moving fluid immediately adjacent to the biofilm. Even in a turbulent flow, flow velocities close to the biofilm are greatly reduced. An antimicrobial agent in the bulk fluid must pass through this quiescent fluid layer before it can move into the biofilm. The additional retardation of antimicrobial penetration posed by external mass transfer resistance was not incorporated in any of the preceding analyses of biofilm penetration. Obviously, external mass transfer resistance, when present, exacerbates the limitation to antimicrobial agent penetration arising from any of the mechanisms discussed previously. External mass transfer resistance can be quantitatively accounted for in any of these penetration models by replacing the boundary condition Eq. (11.6) with a so-called matching flux condition of the form

$$D_e \frac{dC}{dz}\bigg|_{z=L_f} = k(C_o - C_s) \tag{11.18}$$

where C_o is the bulk fluid concentration, C_s is the concentration at the biofilm-bulk fluid interface, and k is a mass transfer coefficient. The mass transfer coefficient can be estimated

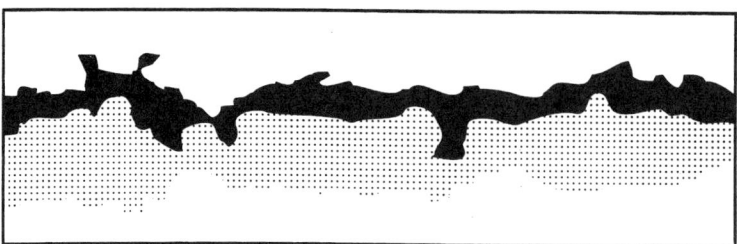

Figure 11.6 Physiological heterogeneity within a *Klebsiella pneumoniae* biofilm cross-section revealed by staining with acridine orange. Dark shading indicates a region where bacteria are growing and light shading indicates a region where bacteria are growing slowly or not at all. The substratum was at the bottom and nutrient medium at the top. After Wentland et al., 1995.

11.3 PROCESS ANALYSIS

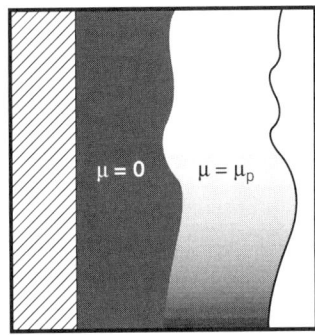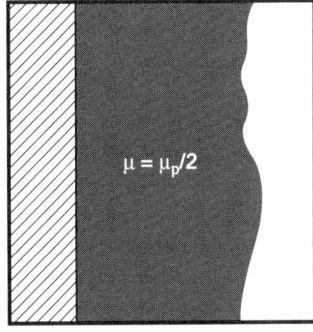

Figure 11.7 Alternative conceptual models for biofilm growth. The planktonic growth rate is denoted by μ_p. The substratum is on the left and the bulk fluid on the right.

by existing correlations or by reference to experimental measurements of this parameter in biofilm systems.

11.3.3 Biofilm Physiology

The physiological status of microorganisms within a biofilm probably plays a central, perhaps dominant, role in governing biofilm susceptibility to antimicrobial agents. Experimental data to date suggest that antimicrobial agent penetration limitation is at best only a partial explanation for reduced biofilm susceptibility (Stewart et al., 1998). Microorganisms are remarkably flexible in the phenotypes they adopt. To complicate matters, we can anticipate microscale spatial heterogeneity in physiological status within biofilms, even within pure culture biofilms (Fig. 11.6). In this section we attempt to provide a quantitative framework for modeling and understanding the potential for physiological limitation of biofilm susceptibility to disinfection.

A basic model for microbial disinfection is

$$r_{dis} = -k_{dis}XC \tag{11.19}$$

where r_{dis} is the rate of disinfection, X is the viable cell density, and k_{dis} is a disinfection rate coefficient. It is important to recognize that the disinfection rate coefficient, in this or in any other model of disinfection, can not be regarded as a constant. Indeed in all likelihood, k_{dis} is hugely variable, not only between organisms, but with genotype and phenotype for a particular species. The disinfection rate coefficient could change depending on physiological status, the implementation of an adaptive stress response, or by mutation. Other models of microbial disinfection are described elsewhere (Wickramanayake and Sproul, 1991).

11.3.3.1 Disinfection of a Physiologically Heterogeneous Biofilm.
The spatial heterogeneity of physiological status within a biofilm may be a crucial issue in determining susceptibility to antimicrobial agents. To illustrate this point, consider the action of

an antimicrobial agent whose killing is growth rate-dependent. The disinfection model is modified in this case to incorporate dependence on the local specific growth rate of the microorganisms, denoted by μ:

$$r_{dis} = -k_{dis} \mu X C \qquad (11.20)$$

Now consider two distinct scenarios regarding microbial growth rate profiles within the biofilm as depicted in Figure 11.7. In the homogeneous scenario, the whole biofilm grows at the same rate, which is reduced by a factor of two from the planktonic growth rate. In the heterogeneous scenario, the top half of the biofilm grows at the planktonic rate while the bottom half of the biofilm is nongrowing. The average growth rate of the biofilm is the same in both cases, namely, one-half the planktonic growth rate. The response to a growth-rate dependent antimicrobial agent is expected to be drastically different in these two scenarios, as shown in Figure 11.8. The growing half of the heterogeneous biofilm will be killed, but the nongrowing (yet still viable) half is impervious to killing. The maximum kill that can be realized in the heterogeneous biofilm is therefore 50%, which corresponds to a log reduction of about 0.3. The homogeneous biofilm can be completely killed, albeit at half the rate of the planktonic cells. This thought experiment underscores the need to go beyond the use of community averages of biofilm activity to a fuller understanding of physiological heterogeneity and its implications.

11.3.3.2 Growth Rate-Dependent Killing in a Nutrient-Limited Biofilm.

Nutrient limitation and growth rate variations within biofilms are well known and these phenomena have been mathematically modeled by civil and environmental engineers for nearly 30 years. Such models are based on an analysis of simultaneous reaction and diffusion exactly like that presented in Eq. (11.11), except that the reaction and diffusion processes pertain to the growth-limiting nutrient rather than to an antimicrobial agent. By coupling such a growth rate model to a model of disinfection, such as Eq. (11.20), one can predict the outcome of biofilm disinfection by a growth rate-dependent antimicrobial agent.

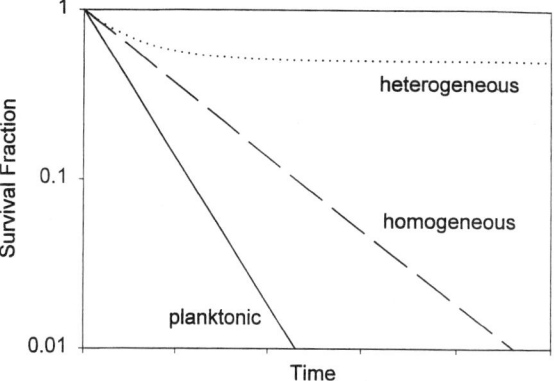

Figure 11.8 Expected killing of planktonic, physiologically homogeneous biofilm, and physiologically heterogeneous biofilm by a growth rate-dependent antimicrobial agent.

11.3 PROCESS ANALYSIS

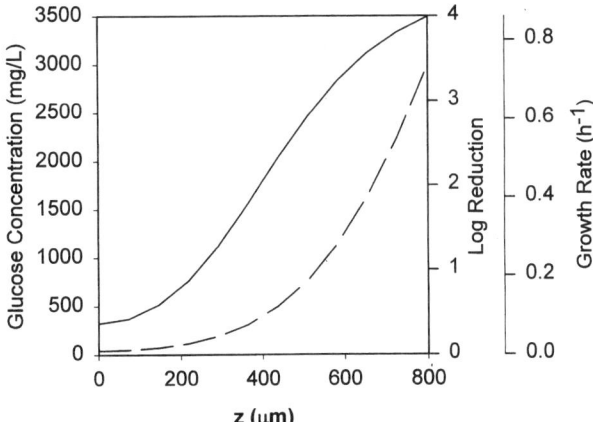

Figure 11.9 Simulated glucose concentration profile (dashed line, left axis), growth rate profile before treatment (solid line, second right axis), and survival profile after treatment (solid line, first right axis) within a biofilm subject to killing by a growth rate-dependent antibiotic.

Example Calculation. Determine the pretreatment growth rate profile and posttreatment survival profile within a sugar-fermenting biofilm subjected to a 4 h dose of ampicillin. The biofilm is 800 μm thick, the biofilm cell density is 20,000 mg L^{-1}, and the limiting nutrient is glucose with a bulk fluid concentration of 3000 mg L^{-1}. Independent measurements with planktonic cells indicate a Monod half saturation coefficient, K_m, of 500 mg L^{-1}, a maximum specific growth rate of 1 h^{-1}, a growth yield, Y_{xs}, of 0.5 mg mg^{-1}. The antibiotic dose results in a 4 log reduction when applied to planktonic cells growing in the bulk fluid. The balance of glucose diffusion and reactive utilization by the microorganisms is expressed by

$$D_e \frac{d^2C}{dz^2} = \frac{\mu_{max} C}{C + K_m} \frac{X_b}{Y_{xs}} \qquad (11.21)$$

Boundary conditions Eqs. (11.6) and (11.7) complete the mathematical statement of this problem. This problem does not have an analytical solution, but can be solved numerically by any number of commercially available software packages. The resulting solution, which takes the form of concentration versus distance information, is substituted into the Monod kinetic expression to solve for the specific growth rate as a function of depth in the biofilm. This growth rate profile is plotted in Figure 11.9 for the case at hand. As expected, the growth rate declines with depth into the biofilm: It is calculated to be 0.08 h^{-1} at the substratum which is less than 10% of the planktonic growth rate of 0.86 h^{-1}.

When the disinfection submodel, Eq. (11.20), is applied to the growth rate profile described previously, the local killing is calculated with the result shown in Figure 11.9. The growing region of the biofilm near the biofilm-bulk fluid interface is readily killed, but the slow growing interior region of the biofilm experiences poor killing. The average log reduction across the entire biofilm is 0.97. Since the same treatment applied to planktonic microorganisms would result in a 4 log reduction, the calculated biofilm resistance factor (see Section 11.1.2) is 4.1.

Specific growth rate is a familiar physiological index that can be quantified and modeled. Certainly for antimicrobial agents for which a clear growth rate dependence has been established, such as for some antibiotics, slow growth in some regions of a biofilm affords a powerful explanation for biofilm reduced susceptibility.

11.3.3.3 The Resistant Biofilm Phenotype.
We explore here the possibility that a subpopulation of the microorganisms in a biofilm adopts a resistant phenotype. Consider a biofilm in which both susceptible and resistant phenotypes are expressed, possibly in a stratified manner as sketched in Figure 11.10. The spatial orientation of these layers is not critical to the following derivations, but is consistent with recent demonstrations of physiological heterogeneity within biofilms (Wentland et al., 1996; Xu et al., 1998). Since we assume complete penetration of the antimicrobial in the following analyses, the exact spatial distribution of the two phenotypes is irrelevant. The two phenotypes each constitute a fraction of the total biofilm population. Let ε_r denote the resistant fraction; $1 - \varepsilon_r$ is, therefore, the susceptible fraction. Suppose that the disinfection rate coefficient of the susceptible population is k_{dis} and the disinfection rate coefficient of the resistant subpopulation is given by pk_{dis}. The parameter p is a fraction less than one that reflects the fact that the disinfection rate of the resistant population is reduced compared to the susceptible population.

When this conceptual model is married to a disinfection model, such as Eq. (11.20), killing curves with a characteristic bilinear shape are obtained as illustrated in Figure 11.11. What is interesting is that clear examples of such bilinear killing of biofilm microorganisms can be found in the literature (Frank and Koffi, 1990; Oie et al., 1996). Of course, biphasic killing is also commonly encountered in planktonic disinfection. The difference is that much higher resistant fractions can be inferred from biofilm disinfection data. While no more than circumstantial evidence, these observations are consistent with the existence of

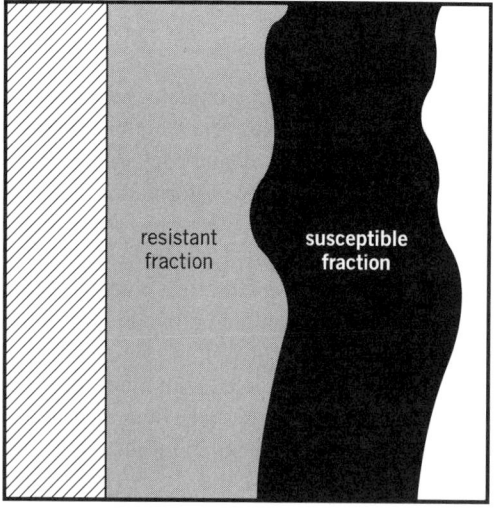

Figure 11.10 Conceptual model of distribution of resistant and susceptible biofilm cells. The substratum is on the left and the bulk fluid on the right.

11.3 PROCESS ANALYSIS

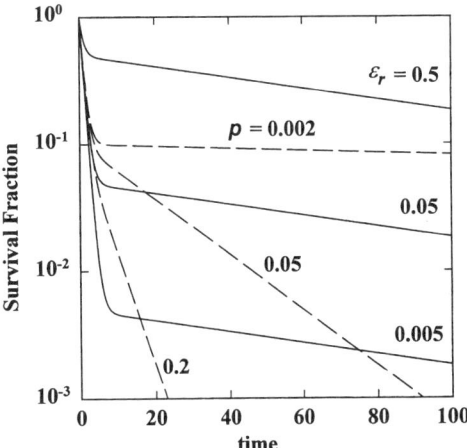

Figure 11.11 Simulated microbial survival in a completely penetrated biofilm expressing susceptible and resistant phenotypes in various proportions. The resistant fraction is denoted by ε_r and the relative susceptibility of the resistant fraction by p. Solid lines show simulations for three different values of ε_r and a fixed value of p (0.01) while the dashed lines show simulations for three different values of p and a fixed value of ε_r (0.1).

a generically resistant biofilm phenotype and should motivate continued investigation of this possibility.

11.3.4 Detachment

Detachment occurs continuously and naturally in every biofilm system. In most cases detachment is the primary process balancing microbial growth and driving the biofilm to a steady state level of accumulation. Despite the central importance of detachment in biofilm development, very little is known about the biological, chemical, and physical mechanisms of detachment. In many biofouling situations, a clean surface is the desired endpoint of a biofilm control program. Certain antimicrobial agents achieve this indirectly by stopping growth and allowing the natural detachment process to slowly remove the biofilm. In this section, detachment subsequent to antimicrobial treatments is discussed.

One of the common mathematical submodels of biofilm detachment, and one for which some phenomenological justification can be argued (Stewart, 1993), states that detachment rate is proportional to the square of biofilm thickness:

$$r_{\text{det}} = k_{\text{det}} L_f^2 \tag{11.22}$$

The order of magnitude of $k_{\text{det}} L_f$ is 0.1 h^{-1} or less, which means that the time scale for detachment is 10 h or more. Detachment is a slow process. It is certainly much slower than disinfection, which occurs over a time scale of minutes or seconds. Consider, for example, a biofilm that has been completely killed by an antimicrobial treatment. The balance that describes this situation is

$$\frac{dL_f}{dt} = -k_{\text{det}} L_f^2 \tag{11.23}$$

and its solution, given the initial condition $L_f = L_{fo}$ at $t = 0$, is

$$\frac{L_f}{L_{fo}} = \frac{1}{1 + k_{det}L_{fo}t} \tag{11.24}$$

This result is plotted in Figure 11.12 for an illustrative case. According to this simulation, treatment of a biofilm with an antimicrobial agent, even if it completely disinfects the biofilm, will not lead to immediate removal of the biofilm. Rather, dead biomass will gradually slough off over a period of many hours. There is some experimental data consistent with this model (Srinivasan et al., 1995; Sanderson and Stewart, 1997).

Stewart (1993) has suggested that part of the detachment process may be growth rate-dependent. If this is the case, then detachment after a killing antimicrobial treatment would be slower than predicted by Eq. (11.24). Other data (Tijhuis et al., 1995) indicate that some portion of biofilm remains attached even after prolonged periods of no microbial growth. This case could be simulated by a detachment model of the form

$$r_{det} = \begin{cases} k_{det}(L_f - L_{f*})^2 & L_f > L_{f*} \\ 0 & L_f \leq L_{f*} \end{cases} \tag{11.25}$$

where L_{f*} is the thickness at which detachment ceases. An illustrative result is shown in Figure 11.12.

However, some antimicrobial treatments might be expected to induce detachment, of at least a part of the biofilm, much more rapidly than predicted by the above models. This possibility could be modeled by assuming that a fraction of the biofilm thickness, denoted by

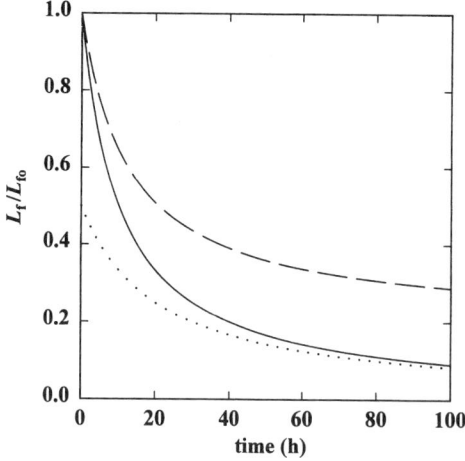

Figure 11.12 Biofilm detachment following a lethal antimicrobial treatment as predicted by three different models. Basic model (Eqs. (11.22) to (11.24), $k_{det}L_{fo} = 0.05$ h^{-1}), solid line; basic model with a fraction of the biofilm that never detaches (Eq. (11.25); $L_{f*} = 0.2 L_{fo}$), dashed line; basic model with instantaneous detachment of a fraction of the biofilm (Eq. 11.26; $A = 0.5$), dotted line.

11.4 PRACTICAL CONCERNS

A, detaches instantaneously upon antimicrobial treatment. This leads to the following equation for a completely killed biofilm

$$\frac{L_f}{L_{fo}} = \frac{1 - A}{1 + k_{\text{det}} L_{fo}(1 - A)t} \tag{11.26}$$

which is also illustrated in Figure 11.12.

These models are mostly speculation. Clearly, experimental data on detachment rates following antimicrobial treatment are needed to advance the ability to model this process.

11.4 PRACTICAL CONCERNS

The preceding examples of process analysis, many of which contain inaccessible parameters and hypothetical mechanisms, might well frustrate practitioners faced with the challenge of controlling biofouling in the real world. This section attempts to link the ideas regarding fundamental biofilm resistance mechanisms presented previously to more practical aspects of biofilm control. In particular, we discuss possible new approaches and technologies for improved biofilm control and conclude with some remarks about the need for further development of biofilm testing methods.

11.4.1 Suggestions for Improving Biofilm Control When Penetration Is Limiting

In this section we consider practical suggestions for improving antimicrobial performance against a biofilm in the particular case of a poorly penetrating biocide. The first suggestion is to treat the biofilm when it is thin. Biofilm penetration times are more sensitive to biofilm thickness than to any other single parameter. In the case of a stoichiometrically reacting biocide, for example, penetration time is proportional to the square of biofilm thickness (Eq. (11.15)). In other words, a doubling of biofilm thickness would necessitate using four times the biocide concentration or four times as long a dose duration to achieve full penetration.

While treating the biofilm when thin may seem to be an obvious suggestion, it led us to a dosing strategy that was not immediately intuitive. This strategy, which emerged from exploratory simulations with a computer model (Stewart et al., 1996), calls for two sequential and appropriately timed biocide doses. The second dose is delivered when the biofilm is at its thinnest point following the first dose. The time required to reach minimum thickness after the first dose is typically considerably longer than the time required to reach the maximum kill in the biofilm. This results because detachment is a much slower process than disinfection. According to this concept, the second dose will have maximal effect if added when the system is least fouled. This contradicts most current practices, which call for dosing when an unacceptable threshold of planktonic or biofilm cell numbers is reached.

We recently tested this double-dosing strategy experimentally and met with failure (Sanderson and Stewart, 1997). A second dose of monochloramine delivered 40 to 60 h after a first dose, rather than being more effective than the first dose, was actually clearly less effective. This result contradicted the model prediction. It was hypothesized that bacteria were able to adapt after exposure to the first dose to become less susceptible to monochloramine.

A second recommendation to improve reactive biocide penetration and performance is to decrease external mass transfer resistance to biocide delivery into the biofilm. In practical terms, this equates to keeping the bulk fluid moving. If slow penetration of a biocide into biofilm is indeed an issue, then there is no surer way to exacerbate this limitation than by applying the antimicrobial agent in a static soak. The delivery of a biocide to a reactive surface (i.e., biofilm) from a stagnant fluid is highly inefficient.

A third suggestion for improving biocide performance against biofilms is to weigh both biocide disinfecting power and reactivity when selecting a biocide. A strong disinfectant, if it is also highly reactive, may fail to penetrate biofilm well and thus be limited to incomplete disinfection of the biofilm. This phenomenon is underscored by an experimental comparison of chlorine and monochloramine made using alginate gel bead artificial biofilms. Equal quantities of gel beads containing *P. aeruginosa* were added to stirred beakers containing equivalent concentrations of chlorine, one in the form of free chlorine (pH 7) and one in the form of monochloramine (pH 8.9). Free chlorine was consumed much more rapidly than was monochloramine (Fig. 11.13A), indicating more rapid reaction of free chlorine than monochloramine in the gel beads. Monochloramine was able to disinfect the gel beads well, while free chlorine, the stronger disinfectant, scarcely caused any drop in cell numbers in the beads (Fig. 11.13B). The explanation for the poor performance of free

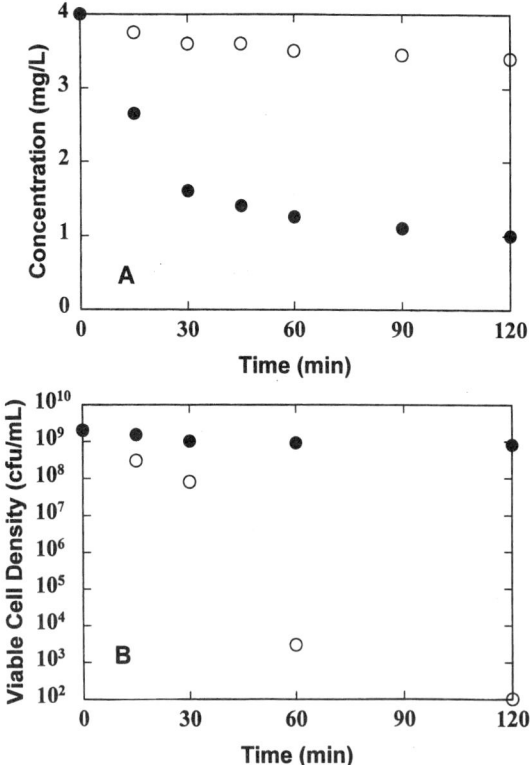

Figure 11.13 Comparison of monochloramine at pH 9 (○) and hypochlorite at pH 7 (●) reaction with (*A*) and disinfection of (*B*) gel bead artificial biofilms. See text for discussion. Unpublished data of Xu and Stewart; see Xu et al., 1996 for methods.

chlorine in this case is that it fails to penetrate the beads because of its high reactivity. The less reactive monochloramine penetrates more effectively and, even though it is a weaker disinfectant than free chlorine, can reach and disinfect the bead interior.

11.4.2 Suggestions for Improving Biofilm Control When Physiology Is Limiting

In this section, we speculate about possible approaches for manipulating or bypassing the physiological status of microorganisms within a biofilm to improve antimicrobial efficacy. The potential to subvert physiological resistance would be greatly increased by knowledge of the specific biological nature of the barrier. For example, if reduced susceptibility stems solely from low specific growth rates in the biofilm, then one could imagine prescribing a pretreatment with nutrients to stimulate growth and place the microorganisms in a more susceptible state. If the resistant biofilm phenotype is based on reduced permeability of the cell envelope, then cotreatment with a permeabilizing agent, such as an organic solvent, might improve antimicrobial performance. Targeted prescriptions such as these await better definition of the particular physiological changes that render biofilms cells less susceptible.

Another approach that can be envisioned for overcoming physiological limitations to biofilm disinfection is to discover alternative antimicrobial agents that are effective against resistant phenotypes. There probably exist antibiotics and biocides that have been dismissed as inferior based on tests against planktonic microorganisms but whose relative efficacy against biofilms would make them superior for this application. As an example, consider the possibility of overcoming the limitations of growth rate-dependent antibiotics by screening for agents based on their efficacy against nongrowing bacteria. Such agents are unlikely to prevail in traditional planktonic or agar plate screening approaches, but might well offer improved performance against slow-growing biofilms. Similarly, we predict the existence of antimicrobial agents or formulations that target resistant biofilm phenotypes and will therefore be more effective than many currently used agents. The challenge is to develop suitable screening protocols centered around biofilm testing methods because existing planktonic methodologies are simply incapable of identifying biofilm-effective antimicrobials.

11.4.3 Novel Control Strategies

A few interesting potential strategies for controlling biofilms have been described in recent years and deserve mention here. Wood et al. (1996) created catalytic surfaces that generate highly reactive hydroxyl radicals when exposed to a suitable biocide precursor such as hydrogen peroxide or potassium monopersulfate. The hydroxyl radicals are generated right at the base of the biofilm where they degrade the matrix and cause detachment of the biofilm. By generating the active agent in-situ, this scheme evades the penetration limitation that would inevitably occur in attempting to deliver a species as reactive as the hydroxyl radical from the bulk fluid.

Costerton and co-workers described the electrical enhancement of antimicrobial efficacy, which they termed the "bioelectric effect" (Blenkinsopp et al., 1992). Subsequent reports have reproduced this effect (Jass et al., 1995; Jass and Lappin-Scott, 1996; Wellman et al., 1996). For example, the efficacy of killing of *P. aeruginosa* biofilm by the antibiotic tobramycin can be enhanced as much as 4 to nearly 8 log reductions by the application of direct current (Wellman et al., 1996). The bioelectric effect is interesting not only as an alternative control strategy but also for the light it may shed on the underlying mechanisms of reduced biofilm susceptibility. In the case of *P. aeruginosa* and tobramycin, unpublished data from our lab suggest that the mechanism of enhancement is either alleviation of growth limitation by provision of the limiting nutrient (oxygen) or potentiation of antibiotic efficacy by induction of oxidative stress in the cell.

The discovery of a role for cell-to-cell communication in the development of microbial biofilm (Davies et al., 1998) raises the possibility that signaling molecules may mediate biofilm resistance to antimicrobial agents. Davies showed, for example, that a mutant *P. aeruginosa* biofilm unable to synthesize an acyl-homoserine lactone signaling molecule was more susceptible to removal by sodiuim dodecyl sulfate. If signaling does play a role in determining biofilm susceptibility to antimicrobial agents, then it might be possible to find analogues of the natural signaling molecules that would interfere with the communication and thereby render the biofilm more susceptible to disinfection or removal.

11.4.4 Biofilm Testing Methods

At this juncture in the evolution of the biofilm concept it is still impossible to predict the efficacy of an antimicrobial agent against a biofilm based on experiments performed with planktonic microorganisms. The performance of antimicrobial agents against biofilm is modulated by such phenomena as expression of biofilm phenotypes, reactive neutralization of antimicrobial agents, detachment, and diffusion that are manifested distinctly in the biofilm mode of growth. These essential phenomena are not captured by tests that use microorganisms in suspension. To put it very bluntly, planktonic disinfection measurements do not even remotely simulate the biofilm state and should not be used as a basis for designing biofilm control strategies.

Biofilm control will only be advanced by the deliberate application of biofilm-based testing methods. The current lack of recognized standard biofilm testing methodologies poses a real barrier to the discovery, understanding, and development of practical biofilm control technologies. Many biofilm methods dwell in the technical literature, but a new generation of methods is now needed. These methods should be developed with an awareness of the mechanisms of biofilm resistance discussed in this chapter, must be validated by comparison with field or in-vivo systems, and, crucially, must win the attention and acceptance of regulatory agencies.

ACKNOWLEDGMENT

This work was supported by Cooperative Agreement EEC-8907039 between the National Science Foundation and the Center for Biofilm Engineering and by National Science Foundation grant BES-9623233.

REFERENCES

Al-Hoti, B., T. Waite, and W. Chow. 1990. Development and calibration of a model for predicting optimum chlorination scenarios for biofouling control. In *Water Chlorination: Chemistry, Environmental Impact and Health Effects*, 6th ed., R. L. Jolley, L. W. Condie, J. D. Johnson,

REFERENCES

S. Katz, R. A. Minear, J. S. Mattice, and V. A. Jabobs, eds. Chelsea, MI: Lewis Publishers, pp. 521–534.

Allison, D. G., and P. Gilbert. 1995. Modification by surface association of antimicrobial susceptibility of bacterial populations. *J. Indust. Microbiol.* 15: 311–317.

Anwar, H., M. Dasgupta, K. Lam, and J. W. Costerton. 1989a. Tobramycin resistance of mucoid *Pseudomonas aeruginosa* biofilm grown under iron limiation. *J. Antimicrob. Chemother.* 24: 647–655.

Anwar, H., T. van Biesen, M. Dasgupta, K. Lam, and J. W. Costerton. 1989b. Interaction of biofilm bacteria with antibiotics in a novel *in vitro* chemostat system. *Antimicrob. Agents Chemother.* 33: 1824–1826.

Aris, R. 1957. On shape factors for irregular particles—I. The steady state problem. Diffusion and reaction. *Chem. Eng. Sci.* 6: 262–268.

Bailey, J. E., and D. F. Ollis. 1986. *Biochemical Engineering Fundamentals*. New York: McGraw-Hill.

Bird, R. B., W. E. Stewart, and E. N. Lightfoot. 1960. *Transport Phenomena*. New York: McGraw-Hill.

Blenkinsopp, S. A., A. E. Khoury, and J. W. Costerton. 1992. Electrical enhancement of biocide efficacy against *Pseudomonas aeruginosa* biofilms. *Appl. Environ. Microbiol.* 58: 3770–3773.

Brown, M. R. W., and P. Gilbert. 1993. Sensitivity of biofilms to antimicrobial agents. *J. Appl. Bacteriol.* 74: 87S–97S.

Brown, M. R. W., D. G. Allison, and P. Gilbert. 1988. Resistance of bacterial biofilms to antibiotics: A growth-rate related effect? *J. Antimicrob. Chemother.* 22: 777–783.

Brown, S. M., M. L. Howell, M. L. Vasil, A. J. Anderson, and D. J. Hassett. 1995. Cloning and characterization of the *kat*B gene of *Pseudomonas aeruginosa* encoding a hydrogen peroxide-inducible catalase: Purification of KatB, cellular localization, and demonstration that it is essential for optimal resistance to hydrogen peroxide. *J. Bacteriol.* 177: 6536–6544.

Carpentier, B., and O. Cerf. 1993. Biofilms and their consequences, with particular reference to hygiene in the food industry. *J. Appl. Bacteriol.* 75: 499–511.

Characklis, W. G. 1990. Microbial biofouling control. In *Biofilms*, 1st ed., W. G. Characklis and K. C. Marshall, eds. New York: John Wiley.

Chen, X., and P. S. Stewart. 1996. Chlorine penetration into artificial biofilm is limited by a reaction-diffusion interaction. *Environ. Sci. Technol.* 30: 2078–2083.

Costerton, J. W. 1984. The formation of biocide-resistant biofilms in industrial, natural, and medical systems. *Dev. Ind. Microbiol.* 25: 363–372.

Costerton, J. W., K.-J. Cheng, G. G. Geesey, T. I. Ladd, J. C. Nickel, M. Dasgupta, and T. J. Marrie. 1987. Bacterial biofilms in nature and disease. *Ann. Rev. Microbiol.* 41: 435–464.

Darouiche, R. O., A. Dhir, A. J. Miller, G. C. Landon, I. I. Raad, and D. M. Musher. 1994. Vancomycin penetration into biofilm covering infected prostheses and effect on bacteria. *J. Infect. Dis.* 170: 720–723.

Das, J. R., M. Bhakoo, M. V. Jones, and P. Gilbert. 1998. Changes in the biocide susceptibility of *Staphylococcus epidermidis* and *Escherichia coli* cells associated with rapid attachment to plastic surfaces. *J. Appl. Microbiol.* 84: 852–858.

Davies, D. G., M. R. Parsek, J. P. Pearson, B. H. Iglewski, J. W. Costerton, and E. P. Greenberg. 1998. The involvement of cell-to-cell signals in the development of a bacterial biofilm. *Science* 280: 295–298.

de Beer, D., and P. Stoodley. 1995. Relation between the structure of an aerobic biofilm and transport phenomena. *Wat. Sci. Tech.* 32: 11–18.

de Beer, D., R. Srinivasan, and P. S. Stewart. 1994. Direct measurement of chlorine penetration into biofilms during disinfection. *Appl. Environ. Microbiol.* 60: 4339–4344.

Dibdin, G. H., S. J. Assinder, W. W. Nichols, and P. A. Lambert. 1996. Mathematical model of beta-lactam penetration into a biofilm of *Pseudomonas aeruginosa* while undergoing simultaneous inactivation by released beta-lactamases. *J. Antimicrob. Chemother.* 38: 757–769.

Dunne, W. M., Jr., E. O. Mason, Jr., and S. L. Kaplan. 1993. Diffusion of rifampin and vancomycin through a *Staphylococcus epidermidis* biofilm. *Antimicrob. Agents Chemother.* 37: 2522–2526.

Dusart, G., M. Zuccarelli, J. L. Jeannot, and M. Simeon de Buochberg. 1994. Activity of chlorhexidine against *Escherichia coli* growing as biofilms on inert solid phase. *Coll. Surf. Int.* B 2: 83–88.

Eagar, R. G., J. Leder, J. P. Stanley, and A. B. Theis. 1988. The use of glutaraldehyde for microbiological control of waterflood systems. *Matl. Perf.* 27: 40–45.

Eastwood, I. M. 1994. Problems with biocides and biofilms. In *Bacterial Biofilms and Their Control in Medicine and Industry*, J. Wimpenny, W. Nichols, D. Stickler, and H. Lappin-Scott, eds. Cardiff, UK: BioLine, pp. 169–172.

Evans, R. C., and C. J. Holmes. 1987. Effect of vancomycin hydrochloride on *Staphylococcus epidermidis* biofilm associated with silicone elastomer. *Antimicrob. Agents Chemother.* 31: 889–894.

Frank, J. F., and R. A. Koffi. 1990. Surface-adherent growth of *Listeria monocytogenes* is associated with increased resistance to surfactant sanitizers and heat. *J. Food Prot.* 53: 550–554.

Gander, S. 1996. Bacterial biofilms: Resistance to antimicrobial agents. *J. Antimicrob. Chemother.* 37: 1047–1050.

Gilbert, P., and M. R. W. Brown. 1995. Mechanisms of the protection of bacterial biofilms from antimicrobial agents. In *Microbial Biofilms*, H. M. Lappin-Scott and J. W. Costerton, eds. Cambridge: Cambridge University Press, pp. 118–130.

Gilbert, P., P. J. Collier, and M. R. W. Brown. 1990. Influence of growth rate on susceptibility to antimicrobial agents: Biofilms, cell cycle, dormancy, and stringent response. *Antimicrob. Agents Chemother.* 34: 1865–1868.

Gilbert, P., J. Das, and I. Foley. 1997. Biofilm susceptibility to antibiotics. *Adv. Dent. Res.* 11: 160–167.

Green, P. N., and R. S. Pirrie. 1993. A laboratory apparatus for the generation and biocide efficacy testing of *Legionella* biofilms. *J. Appl. Bacteriol.* 74: 388–393.

Gristina, A. G., R. A. Jennings, P. T. Naylor, Q. N. Myrvik, and L. X. Webb. 1989. Comparative in vitro antibiotic resistance of surface-colonizing coagulase-negative staphylococci. *Antimicrob. Agents Chemother.* 33: 813–816.

Hamilton-Miller, J. M. T., and S. Shah. 1997. Activity of quinupristin/dalfopristin against *Staphylococcus epidermidis* in biofilms: A comparison with ciprofloxacin. *J. Antimicrob. Chemother.* 39A: 103–108.

Horvath, A. L. 1985. *Handbook of Aqueous Electrolyte Solutions*. New York: John Wiley.

Hoyle, B. D., and J. W. Costerton. 1991. Bacterial resistance to antibiotics: The role of biofilms. *Prog. Drug Res.* 37: 91–105.

Ishida, H., Y. Ishida, Y. Kurosaka, T. Otani, K. Sato, and H. Kobayashi. 1998. In vitro and in vivo activities of levofloxacin against biofilm-producing *Pseudomonas aerugionosa*. *Antimicrob. Agents Chemother.* 42: 1641–1645.

Jass, J., and H. M. Lappin-Scott. 1996. The efficacy of antibiotics enhanced by electrical currents against *Pseudomonas aeruginosa* biofilms. *J. Antimicrob. Chemother.* 38: 987–1000.

Jass, J., J. W. Costerton, and H. M. Lappin-Scott. 1995. The effect of electrical currents and tobramycin on *Pseudomonas aeruginosa* biofilms. *J. Indust. Microbiol.* 15: 234–242.

Kumon, H., K. Tomochika, T. Matunaga, M. Ogawa, and H. Ohmori. 1994. A sandwich cup method for the penetration assay of antimicrobial agents through *Pseudomonas* exopolysaccharides. *Microbiol. Immunol.* 38: 615–619.

Liu, X., F. Roe, A. Jesaitis, and Z. Lewandowski. 1998. Resistance of biofilms to the catalase inhibitor 3-amino-1,2,4-triazole. *Biotechnol. Bioeng.* 59: 156–162.

Lu, C., P. Biswas, and R. M. Clark. 1995. Simultaneous transport of substrates, disinfectants and microorganisms in water pipes. *Wat. Res.* 29: 881–894.

Millward, T. A., and M. Wilson. 1989. The effect of chlorhexidine on *Streptococcus sanguis* biofilms. *Microbios* 58: 155–164.

Morck, D. W., K. Lam, S. G. McKay, M. E. Olson, B. Prosser, B. D. Ellis, R. Cleeland, and J. W. Costerton. 1994. Comparative evaluation of fleroxacin, ampicillin, trimethoprimsulfamethaoxazole, and gentamicin as treatments of catheter-associated urinary tract infection in a rabbit model. *J. Antimicrob. Agents* Suppl. 2: S21–S27.

Morton, L. H. G., D. L. A. Greenway, C. C. Gaylarde, and S. B. Surman. 1998. Consideration of some implications of the resistance of biofilms to biocides. *Intl. Biodet. Biodeg.* 41: 247–259.

Muli, F., and J. K. Struthers. 1998. Use of a continuous-culture biofilm system to study the antimicrobial susceptibilities of *Gardnerella vaginalis* and *Lactobacillus acidophilus*. *Antimicrob. Agents Chemother.* 42: 1428–1432.

Nichols, W. W. 1989. Susceptibility of biofilms to toxic compounds. In *Structure and Function of Biofilms*, W. G. Characklis and P. A. Wilderer, eds. New York: John Wiley, pp. 321–331.

Nichols, W. W. 1991. Biofilms, antibiotics, and penetration. *Rev. Med. Microbiol.* 2: 177–181.

Nichols, W. W., M. J. Evans, M. P. E. Slack, and H. L. Walmsley. 1989. The penetration of antibiotics into aggregates of mucoid and non-mucoid *Pseudomonas aeruginosa*. *J. Gen. Microbiol.* 135: 1291–1303.

Nickel, J. C., J. B. Wright, I. Ruseska, T. J. Marrie, C. Whitfield, and J. W. Costerton. 1985. Antibiotic resistance of *Pseudomonas aeruginosa* colonizing a urinary catheter in vitro. *Eur. J. Clin. Microbiol.* 4: 213–218.

Ntsama-Essomba, C., S. Bouttier, M. Ramaldes, F. Dubois-Brissonnet, and J. Fourniat. 1997. Resistance of *Echerichia coli* growing as biofilms to disinfectants. *Vet. Res.* 28: 353–363.

Oie, S., Y. Huang, A. Kamiya, H. Konishi, and T. Nakazawa. 1996. Efficacy of disinfectants against biofilm cells of methicillin-resistant *Staphylococcus aureus*. *Microbios* 85: 223–230.

Ronner, A. B., and A. C. L. Wong. 1993. Biofilm development and sanitizer inactivation of *Listeria monocytogenes* and *Salmonella typhimurium* on stainless steel and Buna-n rubber. *J. Food Prot.* 56: 750–758.

Rossmann, L. A., R. M. Clark, and W. M. Grayman. 1994. Modeling chlorine residuals in drinking-water distribution systems. *J. Environ. Eng.* 120: 803–820.

Sanderson, S. S., and P. S. Stewart. 1997. Evidence of bacterial adaptation to monochloramine in *Pseudomonas aeruginosa* biofilms and evaluation of biocide action model. *Biotechnol. Bioeng.* 56: 201–209.

Shigeta, M., G. Tanaka, H. Komatsuzawa, M. Sugai, H. Suginaka, and T. Usui. 1997. Permeation of antimicrobial agents through *Pseudomonas aeruginosa* biofilms: A simple method. *Chemotherapy* 43: 340–345.

Srinivasan, R., P. S. Stewart, T. Griebe, C.-I. Chen, and X. Xu. 1995. Biofilm parameters influencing biocide efficacy. *Biotechnol. Bioeng.* 46: 553–560.

Stewart, P. S. 1993. A model of biofilm detachment. *Biotechnol. Bioeng.* 41: 111–117.

Stewart, P. S. 1994. Biofilm accumulation model that predicts antibiotic resistance of *Pseudomonas aeruginosa* biofilms. *Antimicrob. Agents Chemother.* 38: 1052–1058.

Stewart, P. S. 1997. Theoretical aspects of antibiotic diffusion into microbial biofilms. *Antimicrob. Agents Chemother.* 40: 2517–2522.

Stewart, P. S. 1998. A review of experimental measurements of effective diffusive permeabilities and effective diffusion coefficients in biofilms. *Biotechnol. Bioeng.* 59: 261–272.

Stewart, P. S., and J. B. Raquepas. 1995. Implications of reaction-diffusion theory for the disinfection of microbial biofilms by reactive antimicrobial agents. *Chem. Eng. Sci.* 50: 3099–3104.

Stewart, P. S., M. A. Hamilton, B. R. Goldstein, and B. T. Schneider. 1996. Modeling biocide action against biofilms. *Biotechnol. Bioeng.* 49: 445–455.

Stewart, P. S., L. Grab, and J. A. Diemer. 1998. Analysis of biocide transport limitation in an artificial biofilm system. *J. Appl. Microbiol.* 85: 495–500.

Stickler, D., and P. Hewett. 1991. Activity of antiseptics against biofilms of mixed bacterial species growing on silicone surfaces. *Eur. J. Clin. Microbiol. Infect. Dis.* 10: 416–421.

Suci, P., M. W. Mittelman, F. P. Yu, and G. G. Geesey. 1994. Investigation of ciprofloxacin penetration into *Pseudomonas aeruginosa* biofilms. *Antimicrob. Agents Chemother.* 38: 2125–2133.

Sutch, J. C. D., R. J. Pinney, and M. Wilson. 1995. Activity of chlorhexidine and cetylpyridinium on biofilms of *Streptococcus mutans*. *J. Pharm. Pharmacol.* 47: 1094.

Thrower, Y., R. J. Pinney, and M. Wilson. 1997. Susceptibilities of *Actinobacillus actinomycetemcomitans* biofilms to oral antiseptics. *J. Med. Microbiol.* 46: 425–429.

Tijhuis, L., M. C. M. van Loosdrecht, and J. J. Heijnen. 1995. Dynamics of biofilm detachment in biofilm airlift suspension reactors. *Biotechnol. Bioeng.* 45: 481–487.

van Stroe-Biezen, S. A. M., F. M. Everaerts, L. J. J. Janssen, and R. A. Tacken. 1993. Diffusion coefficients of oxygen, hydrogen peroxide, and glucose in a hydrogel. *Anal. Chimica Acta* 273: 553–560.

Vrany, J. D., P. S. Stewart, and P. A. Suci. 1997. Comparison of recalcitrance to ciprofloxacin and levofloxacin exhibited by *Pseudomonas aeruginosa* biofilms displaying rapid-transport characteristics. *Antimicrob. Agents Chemother.* 41: 1352–1358.

Wellman, N., S. M. Fortun, and B. R. McLeod. 1996. Bacterial biofilms and the bioelectric effect. *Antimicrob. Agents Chemother.* 40: 2012–2014.

Wentland, E., P. S. Stewart, C.-T. Huang, and G. A. McFeters. 1996. Spatial variations in growth rate within *Klebsiella pneumoniae* colonies and biofilm. *Biotechnol. Prog.* 12: 316–321.

Wickramanayake, G. B., and O. J. Sproul. 1991. Kinetics of the inactivation of microorganisms. In *Disinfection, Sterilization, and Preservation*, 4th ed., S. S. Block, ed. Philadelphia: Lea & Febiger, pp. 72–84.

Williams, I., W. A. Venables, D. Lloyd, F. Paul, and I. Critchley. 1997. The effects of adherence to silicone surfaces on antibiotic susceptibility in *Staphlyococcus aureus*. *Microbiology* 143: 2407–2413.

Wilson, M. 1996. Susceptibility of oral bacterial biofilms to antimicrobial agents. *J. Med. Microbiol.* 44: 79–87.

Wood, P., M. Jones, M. Bhakoo, and P. Gilbert. 1996. A novel strategy for control of microbial biofilms through generation of biocide at the biofilm-surface interface. *Appl. Environ. Microbiol.* 62: 2598–2602.

Wright, J. B., I. Ruseska, and J. W. Costerton. 1991. Decreased biocide susceptibility of adherent *Legionella pneumophila*. *J. Appl. Bacteriol.* 71: 531–538.

Wright, T. L., R. P. Ellen, J.-M. Lacroix, S. Sinnadurai, and M. W. Mittelman. 1997. Effects of metronidazole on *Porphyromonas gingivalis* biofilms. *J. Periodon. Res.* 32: 473–477.

Xu, K. D., P. S. Stewart, F. Xia, C.-T. Huang, and G. A. McFeters. 1998. Spatial physiological heterogeneity in *Pseudomonas aeruginosa* biofilm is determined by oxygen availability. *Appl. Environ. Microbiol.* 64: 4035–4039.

Xu, X., P. S. Stewart, and X. Chen. 1996. Transport limitation of chlorine disinfection of *Pseudomonas aeruginosa* entrapped in alginate beads. *Biotechnol. Bioeng.* 49: 93–100.

Zottola, E. A. 1994. Microbial attachment and biofilm formation: A new problem for the food industry? *Food Technol.* 48: 107–114.

INDEX

Accumulation mechanisms, biofilms in porous media, 135–141
  hydrodynamics, 136–137
    Darcy's law, 136–137
    permeability, 137
  influences on permeability and hydrodynamics, 139–141
  initial deposition influence, 138–139
Acid-producing bacteria
  microbiologically influenced corrosion (MIC):
    copper alloys, 304–306
    iron and ferrous alloys, 293–294
    microbial consortia, 294–299
Acinetobacter sp., drinking water systems, engineered biofouling bioremediation, 252–253
Adhesion efficiency
  bacterial interactions on porous media, 128–129
  drinking water systems, engineered biofouling bioremediation, 251–253
  engineered biofouling bioremediation, food preparation surfaces, 247–250
Adhesion rates
  bacterial interactions on porous media, 128–129
  bacterial specific adhesions, receptor-ligand interactions, mathematical models, 61–63
  bacterial transport in porous media, 130–135
  biofilm formation and performance, surface topography, 50–52
Adhesion strength per bond (F_{bond}), bacterial specific adhesions, receptor-ligand interactions, 60
Adsorption mechanism
  conditioning film, bacterial adhesion studies, 53–54
  hazardous waste treatment, adsorbing substratum interactions, 222–229
Advective (convective) transport
  bacterial accumulation statistics, 125–126
  industrial surfaces, mass transfer, engineered biofouling bioremediation, 259–262
Advective heat transfer, industrial surfaces, engineered biofouling bioremediation, 265
Aerobic biofilm processes
  fluidized bed reactors (FBR), 171–172
  industrial wastewater treatment, 195–196
Aerobic environment, electron acceptors in, 20–21
Aerosol contaminants, engineered biofouling bioremediation, evaporative heat exchangers, 241–242
Aggregate accumulation
  biofilm accumulation in porous media, 141
  hazardous waste treatment, biofilms as microbial aggregates, 208–209
Air conditioning units
  engineered biofouling bioremediation, evaporative heat exchangers, 241–242
  food industry biofilm formation, 332
Air-lift reactors
  biofilm wastewater treatment technologies, moving bed processes, 175–177
  industrial wastewater treatment
    aerobic biofilm processes, 196
    anaerobic biofilm processes, 194
  municipal wastewater treatment, total nitrogen removal, 190
algC gene
  cell:cell signal control, biofilm reactivity, 63–64

407

algC gene *(continued)*
  reactivity dynamics of biofilms, molecular control of detachment, 72–73
Alkaline cleaning agents
  food industry biofilm, elimination in closed equipment, 345–348
  food industry biofilm elimination, 341–348
Aluminum alloys, microbiologically influenced corrosion (MIC), 307–308
Anaerobic biofilm processes
  carbon removal, two-phase liquid-solid fluidized bed technology, 168–172
  hazardous waste treatment, inhibition protection mechanisms, 212–213
  industrial wastewater treatment, 194–195
Anaerobic methanogen oxidation, microbiologically influenced corrosion (MIC), iron and ferrous alloys, 291
Anaflux fluidized bed reactors, industrial wastewater treatment, anaerobic biofilm processes, 194
Annular biofouling reactors, engineered biofouling bioremediation monitoring, 267
Anodic depolarization, microbiologically influenced corrosion (MIC), iron and ferrous alloys, sulfate-reducing bacteria, 289–290
Antibiotic resistance, molecular ecology of biofilms, plasmid introduction in microbial communities, 112
Antimicrobial agents, biofilm control
  detachment mechanisms, 395–398
  penetration mechanisms
    catalytic reactions, 386–388
    improvement of, 398–400
    noninteracting agents, 384–386
    stoichiometric reactions, 388–389
    structural heterogeneity and mass transfer resistance, 390–392
    uncharacterized reaction, 389–390
  physiological characteristics, 392–395
    growth rate-dependent killing, nutrient limitations, 393–394
    heterogeneity, 393
    improvement of, 400
    resistant phenotypes, 395
  process analysis, 379–398
  reaction mechanisms
    biocide depletion on pipe lengths, 380–381
    continuous flow systems, biocide depletion, 381
    oxidant depletion from immune response, 381–383
  reduced susceptibility mechanisms, 375–379
    bulk fluid depletion, 378
    physiological limits of efficacy, 379
    transport limitation, 378–379
  research background, 373–375
  technology trends in, 401
  testing methods, 401–402
Aquabacterium sp., drinking water systems, engineered biofouling bioremediation, 252–253
AQUASIM program, biofilm accumulation in porous media, 140–141
Aquifers, bacterial accumulation statistics, 125–127
Assimilable organic carbon (AOC), drinking water systems, engineered biofouling bioremediation, 250–253
Atomic force microscopy (AFM), microbiologically influenced corrosion (MIC), 309
ATP measurement, food industry biofilm detection, 352
Attenuated total reflectance Fourier transform infrared (ATR/FT-IR) spectroscopy:
  biofilm biomass monitoring, 268
  biofilm metabolism, 268–270
  microbiologically influenced corrosion (MIC), 312
Auger electron spectroscopy (AES), microbiologically influenced corrosion (MIC), 309–310
Autoinducer molecules, cell:cell signal control, biofilm reactivity, gram-negative bacteria, 63–64
Autoinduction, gene expression, 71–72
Autotrophic denitrification, drinking water treatment, 185
Average density, three-dimensional biofilm architecture, 77–79

Bacterial adhesion studies
  batch reactors, limits of, 26
  membrane system biofouling, engineered bioremediation, 243
  molecular aspects
    nonspecific adhesion processes, 55–58
    receptor-ligand mediated specific adhesion, 58–63
    substrata preconditioning, 52–55

Bacterial deposition mechanisms, biofilm accumulation in porous media, 138–139
Bacterial growth
  mathematics of, 67–68
  molecular ecology of biofilms, 97–109
    environmental shifts, community reactions to, 108–109
    miscellaneous cellular reporters, 106–107
    mRNA growth indicator experssion in single cells, 104–106
    plasmid transfer and, 114–116
    ribosomal RNA synthesis rates I, 101–103
    ribosomal RNA synthesis rates II, 103–104
    in situ rRNA hybridization, 98–101
  organic contaminants, metabolism of, subsurface contamination remediation, 142–143
  replication and, 66–67
Bacterial metabolic rates, microbiologically influenced corrosion (MIC), iron and ferrous alloys, sulfate-reducing bacteria, 291
Bacterial swarming, molecular ecology of biofilms, communication, 96–97
Bacterial transport, in porous media, 129–135
Bacteriophages, molecular ecology of biofilms, gene transfer and transduction by, 110
B2A filters, biofilm wastewater treatment, 165–167
Balance equation, microbial growth, 67–68
Barrier mechanisms, molecular ecology of biofilms, growth rate assessment, 100–101
Batch reactors, biological transformation kinetics, 25–26
Bead experiments, biofilm formation and performance, surface topography, 51–52
Benzyl viologen (BV), microbiologically influenced corrosion (MIC), iron and ferrous alloys, hydrogenase depolarization, 286–287
Binary fission, microbial growth, 66–67
Bioactivated zones, subsurface contamination remediation, biofilms in porous media, 146–150
Bioavailability, subsurface contamination remediation, biofilms in porous media, 146–147

Biobarriers, subsurface contamination remediation, biofilms in porous media, 146–150
Biobead system, biofilm wastewater treatment, submerged biofilters, 167
Biocarbone process
  biofilm wastewater treatment, 165–166
  municipal wastewater treatment
    carbon removal, 187
    total nitrogen removal, 188–189
Biocorrosion. *See also* Microbiologically influenced corrosion (MIC)
  research background, 281–283
Biodegradable dissolved organic carbon (BDOC), drinking water systems, engineered biofouling bioremediation, 250–253
Biodegradable inhibition, hazardous waste treatment, protection mechanisms, 212–213
Biodegradation rate, subsurface contamination, biofilms in porous media
  bioavailability and, 146–147
  intrinsic bioremediation, 144–145
Biodenit process, drinking water treatment, heterotrophic denitrification, 184
Bioelectric effect, antimicrobial agents for biofilm control, 401
Biofilm Air-lift Suspension Reactor (BAS), moving bed reactors, air-lift bioreactors, 176–177
Biofilm bacteria, oil and gas platforms, engineered biofouling bioremediation, 238
Biofilm carriers, design parameters, 179
Biofilm front, three-dimensional biofilm architecture, 76–79
Biofilm on activated carbon (BFAC), hazardous waste treatment, adsorbing substratum interactions, 225–229
Biofilm on an inert surface (BFIS), hazardous waste treatment, adsorbing substratum interactions, 225–229
Biofilm reactor design
  heterogeneous reactions, 34–37
  homogeneous reactions, 24–25
  ideal reactors
    batch reactors, 25–26
    continuously fed well-mixed reactor, 26–30
    plug flow reactor, 30–33

Biofilm reactor design *(continued)*
  innovative design for wastewater treatment, 178–183
    biofilm carriers, 179
    denitrification performance parameters, 183
    loading rates, 179–181
    nitrification performance parameters, 181–182
  process analysis, 13–14
Biofilms
  architecture and global reactivity:
    heterogeneity, 74
    three-dimensional modeling, 74–79
  beneficial biofilms, 4–5
  defined, 3
  detrimental biofilms, 5
  formation and performance, 45–52
    accumulation as function of time, 45–46
    laminar flow, 48
    quiescent conditions, 47–48
    surface topography, 49–52
    turbulent flow, 48–49
  impact of, 4–5
  kinetics studies, batch reactors, limits of, 26
  layering mechanisms, 4
  reactivity dynamics
    cell replication, maintenance and endogenous decay, 66–69
    detachment processes, 72–73
    gene retention and transfer, 70–72
    metabolic responses to adhesion, 63–66
  research background, 6–9
Biofilm study system
  continuously stirred tank reactor (CSTR) behavior, 37–40
  laboratory reactors, 37–38
  plug flow reactor behavior, 39–42
Biofilm surface enlargement, three-dimensional biofilm architecture, 77–79
Biofor reactor
  biofilm wastewater treatment, 163–166
  industrial wastewater treatment, aerobic biofilm processes, 195–196
  municipal wastewater treatment
    carbon removal, 187
    tertiary nitrification, 191–192
    tertiary postdenitrification, methanol addition, 192–193
    total nitrogen removal, 188–189
Biofouling
  detrimental biofilms and, 5
  engineered bioremediation
    electronics-grade water systems, 246–247
    health effects
      drinking water systems, 250–253
      food preparation surfaces, 247–250
      implanted prostheses, 255–257
      purified water systems, 253–255
    heat exchangers, 239–242
      evaporative exchangers, 242
      tubular structures, 240–242
    industrial surfaces
      energy transport, 265
      mass transport mechanisms, 257–262
      momentum transport, 262–264
    marine environment
      oil and gas platforms, 238
      ship hulls, 239
    membrane system biofouling and biocontamination, 242–245
      cartridge filters, 245
      microporous membranes, 245
      reverse osmosis (RO) membranes, 243–245
    monitoring procedures, 265–271
      biofilm process monitoring, 268–270
      microscopy, 266
      multidimensional analysis, 270–271
      nondestructive monitoring, 268
      reactor design, 267
      sampling devices, 266
    research background, 237
  reactor design, engineered biofouling bioremediation, 267
Biolift process, municipal wastewater treatment, total nitrogen removal, 189
Biolite medium
  biofilm wastewater treatment, heavy granular media, fixed bed processes, 163–166
  drinking water treatment, heterotrophic denitrification, 184
Biological aerated filters (BAF), biofilm wastewater treatment, heavy granular media, fixed bed processes, 163–166
Biomass growth rate
  engineered biofouling bioremediation, ship hulls, 239
  hazardous waste treatment, inhibitory substrate protection, mathematical model, 219–222
  molecular ecology of biofilms, environmental shifts and response to, 108–109

three-dimensional biofilm architecture, 78–79
Bioremediation, subsurface contamination, biofilms in porous media, 142–150
  bioactivated zones and biobarriers, 146–150
  bioavailability, 146–147
  intrinsic bioremediation, 144–145
  microbial metabolism of organic contaminants, 142–143
  in situ bioremediation engineering, 145–146
Biostyr reactor
  biofilm wastewater treatment, submerged biofilters, 166–167, 169
  municipal wastewater treatment, total nitrogen removal, 189
Block attachment technology, biofilm control, 374–375
Blocking factor, bacterial interactions on porous media, 133–135
Blue water corrosion, microbiologically influenced corrosion (MIC), copper alloys, 303–306
Boltzmann constant, bacterial interactions on porous media, surface-substratum interactions, 127–129
Boundary conditions
  antimicrobial agents for biofilm control, 393–394
  hazardous waste treatment, adsorbing substratum interactions, 224–229
  heterogeneous biofilm reactions, 36–37
  plug flow reactor design, 33
Brownian diffusion, biofilm formation and performance:
  laminar flow, 48
  quiescent conditions, 47–48
Bulk fluid phase
  antimicrobial agents for biofilm control:
    biocide depletion on pipe lengths, 380–381
    depletion in, 378
  biological transformation kinetics, plug flow reactor (PFR), 32–33
Bulk oxygen concentration, biofilm reactors, performance parameters, nitrification performance, 181–182
Bulk transport, biofilm process analysis, 16
N-Butylyl-L-homoserene lactone (BHL), cell:cell signal control, biofilm reactivity, gram-negative bacteria, 64
By-product release mechanisms, microbiologically influenced corrosion (MIC), copper alloys, 303–306

Calcium carbonate scales, heat exchangers, engineered biofouling bioremediation, 230–242
Campylobacter sp., drinking water systems, engineered biofouling bioremediation, 252–253
Captor sponge biofilm process:
  municipal wastewater treatment, carbon removal, 187–188
  wastewater treatment technologies, 174–175
Carbon removal, municipal wastewater treatment, 186–188
Cardboard mills, food industry biofilm formation, 334–335
Cartridge filters, membrane system biofouling, engineered bioremediation, 242, 245
Catalytic reactions, antimicrobial agents for biofilm control, penetration mechanisms, 386–388
Catheterization procedures, implanted prostheses, engineered biofouling bioremediation, 256–257
Cathodic depolarization theory, microbiologically influenced corrosion (MIC), iron and ferrous alloys, hydrogenase depolarization, 285–287
Cathodic protection, microbiologically influenced corrosion (MIC), iron and ferrous alloys, hydrogen-producing bacteria, 292
Cell automata (CA) models
  biofilm process analysis and, 14–15
  three-dimensional biofilm modeling, 74–79
Cell:cell adhesion
  nonspecific adhesion interactions, colloidal techniques, 57–58
  signal control, biofilm reactivity, 63–66
    gram-negative bacteria, 63–64
    gram-positive bacteria, 64–66
Cell replication, reactivity dynamics of biofilms, 66–68
  microbial growth, 66–67
  mathematics of, 67–68
Cell-solid interaction, bacterial interactions on porous media, 132–135
Cellstate program, molecular ecology of biofilms, growth rate assessment, 99–101
Cell-surface adhesion, nonspecific adhesion interactions, colloidal techniques, 57–58

Cell surface receptors, bacterial specific adhesions, receptor-ligand interactions, 59–60
Cell-to-cell communication, antimicrobial agents for biofilm control, 401
Cellular biomass, stoichiometric analysis, 20
Chelators, food industry biofilm elimination, cleansers containing, 337–338
Chemical oxygen demand (COD), soichiometric analysis, 16
Chemostat, biofilm kinetics using, 27–30
Chlorhexidine disinfectant, food industry biofilm elimination, 336
Chlorine
  drinking water systems, engineered biofouling bioremediation, 251–253
  food industry biofilm elimination using, 336–341
Chromium depletion, microbiologically influenced corrosion (MIC), iron and ferrous alloys, weldments, 300
CIP systems, food industry biofilm, elimination in closed equipment, 345–348
Circox process, moving bed reactors, air-lift bioreactors, 176–177
Circular geometry, reactor behavior, in biofilm laboratories, 39–42
Circulating bed reactors
  biofilm wastewater treatment technologies, moving bed processes, 175–177
  municipal wastewater treatment, total nitrogen removal, 190
Cleaning chemicals
  biofilm control, antimicrobial agents, 373–375
  food industry biofilm detection, 351–352
  food industry biofilm elimination, 335–341
Closed equipment design, food industry biofilm
  elimination in, 345–348
  formation
    cooling systems, 331–332
    processing operations, 334
Coliform bacteria, drinking water systems, engineered biofouling bioremediation, 251–253
Collision efficiency factors, bacterial transport in porous media, 131–135
Colloidal fouling, membrane system biofouling, engineered bioremediation, 243
Colloidal method, nonspecific adhesion interactions, 56–58

Colony-forming units (CFU)
  dental units, purified water systems, engineered biofouling bioremediation, 254–255
  pharmaceutical purified water, engineered biofouling bioremediation, 254
Column type fluidized bed reactors (FBR), three-phase gas-liquid-solid technology, 173–174
Cometabolism, hazardous waste treatment, 208
Concentration polarization, membrane system biofouling, engineered bioremediation, reverse osmosis (RO) membranes, 243–244
Conditioning film, bacterial adhesion studies
  adsorption mechanism, 53–54
  composition, 54–55
  substratum properties, influence on, 55
Conductive heat transfer, industrial surfaces, engineered biofouling bioremediation, 265
Confocal scanning laser microscopy (CSLM)
  engineered biofouling bioremediation monitoring, 266
  food industry biofilm detection, 354–355
  industrial surfaces, mass transfer, engineered biofouling bioremediation, 258–259
  molecular ecology of biofilms, structural organization, 93–94
Conjugation, biofilm reactivity dynamics, gene retention and transfer, 70–72
Consortium interactions, microbiologically influenced corrosion (MIC), iron and ferrous alloys, MDB/SRB interactions, 298–299
Contact agar, food industry biofilm detection, 350–352
Continuous backwashing reactors, biofilm wastewater treatment, fixed bed reactors, 166–168
Continuous film formation, biofilm accumulation in porous media, 141
Continuous flow stirred reactor (CFSTR)
  antimicrobial agents for biofilm control, biocide depletion, 381
  engineered biofouling bioremediation monitoring, 267
Continuously stirred tank reactor (CSTR)
  antimicrobial agents for biofilm control, biocide depletion, 381
  behavior, in biofilm laboratories, 37–41
  biofilm kinetics studies, 26–30

INDEX **413**

Continuum models, biofilm process analysis and, 14–15
Control volume, biofilm process analysis and, 15
Convection-dispersion equation, bacterial transport in porous media, 131–135
Convective flow mechanics, industrial surfaces, mass transfer, engineered biofouling bioremediation, 258–259
Cooling systems, food industry biofilm formation, 330–332
Coordination, molecular ecology of biofilms, 90, 95–97
  environmental shifts and response to, 108–109
Copiotrophic bacterial, growth mechanisms, 67
Copper alloys, microbiologically influenced corrosion (MIC), 300–307
  MIC signatures, 301
  potable water biocorrosion, 301–306
    by-product release, 303–306
    pitting corrosion, 302–303
  seawater, 306–307
Copper metal tubing, tubular heat exchangers, engineered biofouling bioremediation, 239–242
Corrosion potential, microbiologically influenced corrosion (MIC), iron and ferrous alloys, ennoblement process, 295–298
Corrosion products, drinking water systems, engineered biofouling bioremediation, 251–253
Coulomb interaction, nonspecific adhesion interactions, 57
Coupon sampling, engineered biofouling bioremediation monitoring, 266
Crystalline fouling, membrane system biofouling, engineered bioremediation, 243
Cultivation techniques, food industry biofilm detection, 350–352
Cytological (CTC-DAPI) analysis, food industry biofilms, open processing, elimination in, 344–345

Dairy industry
  biofilm elimination, disinfectants for, 340–341
    closed equipment design, 347–348
    biofouling, engineered bioremediation, 248–250

Darcy's law, biofilm accumulation in porous media, 136–137
Deep-shaft process, moving bed reactors, airlift bioreactors, 176–177
Deionized (DI) water lines
  biofouling, engineered bioremediation, 246–247
  dental units, purified water systems, engineered biofouling bioremediation, 255
Denipor, drinking water treatment, heterotrophic denitrification, 184
Denitrification
  biofilm reactor performance, 183
  drinking water treatment:
    autotrophic denitrification, 185
    heterotrophic denitrification, 184–185
  moving bed circulating reactors, 177
  municipal wastewater treatment, tertiary postdenitrification, methanol addition, 192–193
  two-phase liquid-solid fluidized bed technology, 168–172
Denitropur, drinking water treatment, autotrophic denitrification, 185
Dental units, purified water systems, engineered biofouling bioremediation, 254–255
Depletion mechanisms, antimicrobial agents for biofilm control
  biocide depletion on pipe lengths, 380–381
  continuous flow systems, biocide depletion, 381
  oxidant depletion from immune response, 381–383
Derjaguin, Landau, Verwey, and Overbeek (DLVO) theory
  bacterial adhesion, nonspecific adhesion interactions, 55, 56–58
  bacterial interactions on porous media, surface-substratum interactions, 127–129
  biofilm formation and performance, motility, quiescent conditions, 48
Desulfovibrio vulgaris, microbiologically influenced corrosion (MIC), iron and ferrous alloys, hydrogenase depolarization, 286–287
Detachment processes
  antimicrobial agents for biofilm control, 395–398
  reactivity dynamics of biofilms, 72–73
    molecular control of, 72–73

Detachment rate constant, bacterial specific adhesions, receptor-ligand interactions, mathematical models, 62–63
Detergent solutions, food industry biofilm detection, 351–352
Detriments, biofilms as, 5
Differential interference contrast (DIC)
  engineered biofouling bioremediation monitoring, 266
  microbiologically influenced corrosion (MIC), 309
Diffusion coefficients, antimicrobial agents for biofilm control, 384–385
Diffusion-limited aggregation model, three-dimensional biofilm modeling, 74–75
Diffusivity estimates, industrial surfaces, mass transfer, engineered biofouling bioremediation, 260–262
Dilution rate, chemostat analysis, 28–30
Dimensionless parameters, hazardous waste treatment
  adsorbing substratum interactions, 224–229
  inhibitory substrate protection, mathematical model, 218–222
Discrete spreading algorithm, three-dimensional biofilm modeling, 74–75
Disinfectants
  antimicrobial agents for biofilm control, penetration mechanisms, 400
  drinking water systems, concentration in, engineered biofouling bioremediation, 251–253
  food industry biofilm elimination, 336–341
Dissociation rate constant, bacterial specific adhesions, receptor-ligand interactions, 59–60
DNA
  food industry biofilm detection
    hybridization techniques, 356–357
    polymerase chain reaction (PCR), 355–356
  hazardous waste treatment, substrata alteration of microbial aggregates, 210–211
  molecular ecology of biofilms, gene transfer and transformation of, 110
Down-flow filtration
  biofilm wastewater treatment, heavy granular media, fixed bed processes, 163–166
  three-phase gas-liquid-solid technology, 174

Draft-tube fluidized bed reactors (FBR), three-phase gas-liquid-solid technology, 173–174
Drainage forces, biofilm formation and performance, turbulent flow mechanisms, 49
Drinking water treatment:
  biofilm technologies, 183–185
    autotrophic denitrification, 185
    heterotrophic denitrification, 184–185
  engineered biofouling bioremediation, 250–253
  food industry biofilm formation, 330–332
  microbiologically influenced corrosion (MIC), copper and copper alloys, 301–306
Dual substrate kinetics, homogeneous reactions, 25
DynaSand process:
  biofilm wastewater treatment, fixed bed reactors, 166–168
  municipal wastewater treatment, tertiary postdenitrification, methanol addition, 192–193

E. coli, molecular biology of, 90
Economic consequences
  food industry biofilm elimination, 342–343
  microbiologically influenced corrosion (MIC), 283–284
Eddy diffusion
  biofilm formation and performance, turbulent flow mechanisms, 49
  industrial surfaces, mass transfer, engineered biofouling bioremediation, 259–262
Effectiveness factor, heterogeneous biofilm reactions, 35–37
Electrochemical impedance spectroscopy (EIS), microbiologically influenced corrosion (MIC)
  copper alloys, 305–306
  quantitative analysis, 311
Electrochemical measurements, microbiologically influenced corrosion (MIC), 310–312
Electrochemical noise (ECN), microbiologically influenced corrosion (MIC), 311
Electron acceptors, stoichiometric analysis, 20–21
Electron donors, stoichiometric analysis, 21–22

Electron equivalents. *See* Chemical oxygen demand (COD)
Electronics-grade water systems, biofouling, engineered bioremediation, 246–247
Electrostatic forces
   bacterial interactions on porous media, 128–129
   nonspecific adhesion, 52–53
   nonspecific adhesion interactions, 57
Endogenous metabolism, reactivity dynamics of biofilms, 69
Endotoxins
   hemodialysis units, purified water systems, engineered biofouling bioremediation, 254
   pharmaceutical purified water, engineered biofouling bioremediation, 253–254
Energy dispersive X-ray analysis (EDX)
   microbiologically influenced corrosion (MIC), 309
      iron and ferrous alloys, sulfate-reducing bacteria, 290–291
Energy distribution, stoichiometric estimation, 22
Energy transport, industrial surfaces, engineered biofouling bioremediation, 265
Engineered bioremediation
   biofouling mechanisms
      electronics-grade water systems, 246–247
      health effects
         drinking water systems, 250–253
         food preparation surfaces, 247–250
         implanted prostheses, 255–257
         purified water systems, 253–255
      heat exchangers, 239–242
         evaporative exchangers, 242
         tubular structures, 240–242
      industrial surfaces
         energy transport, 265
         mass transport mechanisms, 257–262
         momentum transport, 262–264
      marine environment
         oil and gas platforms, 238
         ship hulls, 239
      membrane system biofouling and biocontamination, 242–245
         cartridge filters, 245
         microporous membranes, 245
         reverse osmosis (RO) membranes, 243–245
      monitoring procedures, 265–271

         biofilm process monitoring, 268–270
         microscopy, 266
         multidimensional analysis, 270–271
         nondestructive monitoring, 268
         reactor design, 267
         sampling devices, 266
   research background, 237
   subsurface contamination, biofilms in porous media, 145–146
      bioavailability and, 147
Ennoblement process, microbiologically influenced corrosion (MIC), iron and ferrous alloys, 295–298
Enterococcus faecalis, cell:cell signal control, biofilm reactivity, gram-positive bacteria, 65–66
Envirex fluidized bed process, municipal wastewater treatment, tertiary postdenitrification, methanol addition, 193
Environmental scanning electron microscopy (ESEM), microbiologically influenced corrosion (MIC), 309
Environmental shifts, molecular ecology of biofilms, 108–109
Epifluorescence microscopy, food industry biofilm detection, 353–354
Equation of state approach, nonspecific adhesion interactions, thermodynamics, 56
Equipment design, food industry biofilm formation, 328–330
Erosion, industrial surfaces, momentum transport, engineered biofouling bioremediation, 264
European Hygienic Equipment Design Group (EHEDG), food industry biofilm, elimination in closed equipment, 347–348
Evaporative heat exchangers, engineered biofouling bioremediation, 242
Excessive loss rate, hazardous waste treatment, protection mechanisms, 213–216
Exopolymers (EPS)
   copper and copper alloys, pitting corrosion and, 306
   food industry biofilms, detection of, 348–349
   membrane system biofouling, engineered bioremediation:
      cartridge filters, 245
      reverse osmosis (RO) membranes, 244

Exopolymers (EPS) *(continued)*
  microbiologically influenced corrosion (MIC)
    copper alloys in seawater, 306–307
    iron and ferrous alloys
      hydrogen-producing bacteria, 292
      slime-producing bacteria, 293
      sulfate-reducing bacteria, 291
  pharmaceutical purified water, engineered biofouling bioremediation, 253–254
  subsurface contamination remediation, biofilms in porous media, 147–150
Extracellular matrices (ECM), bacterial specific adhesion, receptor-ligand mediated adhesion, 58–60
Extracellular polymeric substances. *See* Exopolymers (EPS)

Fiber optic sensors, biofilm biomass monitoring, 268
Fibrinogen, implanted prostheses, engineered biofouling bioremediation, 256–257
Fick's laws
  first law of diffusion
    biofilm formation and performance, laminar flow, 48
    molecular transport, bacterial adhesion studies, 52–53
  second law of diffusion, hazardous waste treatment, inhibitory substrate protection, mathematical model, 217–222
Filamentous growth
  industrial surfaces, momentum transport, engineered biofouling bioremediation, 263–264
  microbial growth as, 66–67
Fixed bed processes
  biofilm wastewater treatment, 161–167
    granular media, submerged biofilter, 162–166
    light and floating media, submerged biofilter, 166–167
  drinking water treatment, heterotrophic denitrification, 184
  industrial wastewater treatment
    aerobic biofilm processes, 195–196
    anaerobic biofilm processes, 194
  municipal wastewater treatment:
    carbon removal, 186–187
    tertiary nitrification, 191–192
    tertiary postdenitrification, methanol addition, 192–193
    total nitrogen removal, 188–189
Floating bed processes, biofilm wastewater treatment, light and floating media, submerged biofilter, 166–167, 169
Flow cytometry, food industry biofilm detection, 357
Flow porosity, membrane system biofouling, engineered bioremediation, reverse osmosis (RO) membranes, 244
Fluid definition, three-dimensional biofilm architecture, 76–79
Fluid density, biofilm formation and performance, turbulent flow mechanisms, 49
Fluid flow velocity
  engineered biofouling bioremediation
    electronics-grade water systems, 246–247
    tubular heat exchangers, 241–242
  industrial surfaces, momentum transport, engineered biofouling bioremediation, 264
  tubular heat exchangers, engineered biofouling bioremediation, 240–242
Fluidized bed reactors (FBR)
  biofilm wastewater treatment
    three-phase gas-liquid-solid technology, 172–174
    two-phase liquid-solid technology, 169–172
  drinking water treatment, heterotrophic denitrification, 185
  industrial wastewater treatment
    aerobic biofilm processes, 196
    anaerobic biofilm processes, 194
  municipal wastewater treatment
    tertiary postdenitrification, methanol addition, 193
    total nitrogen removal
      tertiary nitrification, 192
      three-phase gas-liquid-solid technology, 189
Fluorescence measurement techniques, molecular ecology of biofilms, in situ ribosome counting, 98–101
Fluorescence return after photobleaching (FRAP), industrial surfaces, mass transfer, engineered biofouling bioremediation 261–262
Fogging technique, food industry biofilm elimination, 341

Food industry
   biofilm detection, 348–358
      ATP measurement, 352
      cultivation, contact agar and sampling, 350–351
      impedance measurement, 352–353
      microscopical methods, 353–357
      on-line monitoring techniques, 357–358
      physical detection, 357
      research and industrial tools, 349–350
   biofilm elimination
      chemicals for, 335–341
      cleaning agents, 337–338
      closed system elimination, 345–348
      disinfectants, 338–341
      industrial cleaning systems, 341–343
      open systems elimination, 344–345
   biofilm formation, 327–335
      closed equipment, 334
      cooling systems, 330–332
      open equipment, 332–334
      paper and cardboard mills, 334–335
      ventilation and air handling systems, 332
   food preparation surfaces, biofouling, engineered bioremediation, 247–250
Fourier transform interference reflection microscopy (FTIR), food industry biofilm detection, 355
Fowler Cell Adhesion Molecule, engineered biofouling bioremediation monitoring, 267
Frictional drag forces
   biofilm formation and performance, turbulent flow mechanisms, 49
   industrial surfaces, momentum transport, engineered biofouling bioremediation, 264
Front length, three-dimensional biofilm architecture, 77
Fungi accumulation
   drinking water systems, engineered biofouling bioremediation, 250–253
   microbiologically influenced corrosion (MIC):
      aluminum alloys, 308
      iron and ferrous alloys, 294
Fuzzy layers, hazardous waste treatment, excessive loss protection, 215–216

Gel mediums, cell immobilization, moving bed reactors, 177–178

Gene expression, biofilm metabolism monitoring, 269–270
Gene reaction and transfer, molecular ecology of biofilms, 109–116
   growth activity and plasmid transfer, 114–116
   natural systems, 110–111
   plasmid introduction, 111–112
   plasmid transfer on-line monitoring, 112–114
Gene retention and transfer:
   autoinduction of gene expression, 71–72
   biofilm reactivity dynamics, 70–72
   hazardous waste treatment, substrata alteration of microbial aggregates, 210–211
Geometric mean approach, nonspecific adhesion interactions, thermodynamics, 56
Gibbs free energy, bacterial adhesion, nonspecific adhesion interactions, 55–58
Glucose oxidase, microbiologically influenced corrosion (MIC), iron and ferrous alloys, ennoblement process, 297–298
Glucose reactions, stoichiometric analysis, 18–19
Gram-negative bacteria
   cell:cell signal control, 63–64
   engineered biofouling bioremediation, electronics-grade water systems, 247
   gene retention and transfer, 71
Gram-positive bacteria
   cell:cell signal control, 64–66
   engineered biofouling bioremediation, electronics-grade water systems, 247
   gene retention and transfer, 71
Granular activated carbon (GAC)
   biofilm wastewater treatment technology, fluidized bed reactors (FBR), 171–172
   hazardous waste treatment:
      adsorbing substratum interactions, 223–229
      inhibition protection mechanisms, 211–213
      microbial aggregates, 209
   municipal wastewater treatment, 186–188
Granular media, biofilm wastewater treatment, heavy granular media, fixed bed processes, 162–166

Gravity force, biofilm formation and performance, turbulent flow mechanisms, 49
Green fluorescent protein (GFP) reporter system
  biofilm metabolism monitoring, 269–270
  molecular ecology of biofilms
    interactions, 94–95
    microbial growth and plasmid transfer, 114–116
    mRNA growth indicator in single cells, 104–106
    plasmid transfer monitoring, 112–114
    ribosomal RNA synthesis, 104
Greigite, microbiologically influenced corrosion (MIC), iron and ferrous alloys, sulfate-reducing bacteria, 287–291
groEL gene, molecular ecology of biofilms, mRNA growth indicator in single cells, 106–107
Groove parameters, biofilm formation and performance, surface topography, 50–52
Growth potential
  antimicrobial agents for biofilm control:
    detachment process and, 397–398
    variations in rate-dependent killing, 393–394
  drinking water systems, engineered biofouling bioremediation, 251–253
  engineered biofouling bioremediation, dairy industry contaminants, 249–250
  hazardous waste treatment
    adsorbing substratum interactions, 228–229
    inhibitory substrate protection, mathematical model, 218–222

Haldane relationship, hazardous waste treatment, inhibitory substrate protection, mathematical model, 217–222
Half-cell diffusion, industrial surfaces, mass transfer, engineered biofouling bioremediation, 259–262
Half reactions
  antimicrobial agents for biofilm control, catalytic reactions, 387–388
  stoichiometric analysis, 20–21
Halorespiring anaerobes, hazardous waste treatment, 216
Hazard Analysis of Critical Control Points (HACCP), food industry biofilm elimination, 343

Hazardous waste treatment, biofilm technology
  adsorbing substratum protection, 222–230
  inhibitory substrate protection, steady state biofilm, 217–222
  microbial aggregates, 208–209
  protection mechanisms, 211–216
    excessive loss rate protection, 213–216
    inhibition protection, 211–213
  research background, 207–208
  substrata effects on microbes, 210–211
Health effects
  biofouling, engineered bioremediation
    dairy industry, 248–250
    drinking water systems, 250–253
    food preparation surfaces, 247–250
    implanted prostheses, 255–257
    purified water systems, 253–255
  food industry biofilm formation (*See* Food industry)
Heat-affect zone (HAZ), microbiologically influenced corrosion (MIC), iron and ferrous alloys, weldments, 299–300
Heat exchangers, engineered biofouling bioremediation, 239–242
  evaporative exchangers, 242
  tubular structures, 240–242
Heat shock induction, molecular ecology of biofilms, mRNA growth indicator in single cells, 106–107
Heat transfer, industrial surfaces, engineered biofouling bioremediation, 265
Heat transfer efficiency, tubular heat exchangers, engineered biofouling bioremediation, 240–242
Hemodialysis units, purified water systems, engineered biofouling bioremediation, 254
Heterogeneous reactions
  biofilm architecture and global reactivity, 74
    three-dimensional structural descriptors, 76–79
  biological transformation kinetics, 34–37
  defined, 24
  industrial surfaces, mass transfer, engineered biofouling bioremediation, 257–259
Heterotrophic denitrification, drinking water treatment, 184–185
Homogeneous reactions
  defined, 24
  kinetic transformations, 24–25

Homoserine lactones
  cell:cell signal control, biofilm reactivity, gram-negative bacteria, 63–64
  molecular ecology of biofilms, communication, 96–97
Hydraulic conductivity
  biofilm reactor loading rates, 179–181
  subsurface contamination remediation, biofilms in porous media, biobarrier formation, 148–150
Hydraulic models, fluidized bed reactors (FBR), 169–172
Hydrodynamics
  biofilm accumulation in porous media, 136–137
    Darcy's law, 136–137
    influences on, 139–141
    permeability, 137
  fluidized bed reactors (FBR), hydraulic models, 169–172
  moving bed reactors, air-lift bioreactors, 175–177
  tubular heat exchangers, engineered biofouling bioremediation, 240–242
Hydrogenase depolarization, microbiologically influenced corrosion (MIC), iron and ferrous alloys, sulfate-reducing bacteria (SRB), 285–288
Hydrogen peroxide
  food industry biofilm elimination, disinfectants containing, 340–341
  microbiologically influenced corrosion (MIC), iron and ferrous alloys, ennoblement process, 297–298
Hydrogen-producing bacteria, microbiologically influenced corrosion (MIC), iron and ferrous alloys, 292
Hydrogen sulfate depolarizer, microbiologically influenced corrosion (MIC), iron and ferrous alloys, sulfate-reducing bacteria (SRB), 288–289
Hydrophobicity
  bacterial adhesion studies, conditioning film, influence of, 55
  biofilm formation and performance, surface topography, 51–52
  subsurface contamination, biofilms in porous media, bioavailability and, 146–147
Hydroxyl radicals, antimicrobial agents for biofilm control, 401
Hygicult slides, food industry biofilm detection, 351–352

Ideal reactors, kinetics of biological transformations
  batch reactors, 25–26
  continuously fed well-mixed reactor, 26–30
  plug flow reactor, 30–33
Immobilized cell process
  gel mediums, moving bed reactors, 177–178
  municipal wastewater treatment, total nitrogen removal, 191
  reactor design, 5
Immune response, antimicrobial agents for biofilm control, oxidant depletion, 381–382
Impedance measurement, food industry biofilm detection, 352–353
Implanted prostheses, engineered biofouling bioremediation, 255–257
Industrial surfaces
  engineered biofouling bioremediation:
    energy transport, 265
    mass transport mechanisms, 257–262
    momentum transport, 262–264
  food industry biofilm elimination, 341–348
Industrial wastewater treatment, biofilm technologies, 183–196, 193–196
  aerobic film processes, 195–196
  anaerobic processes, 194–195
  drinking water treatment, 183–185
  industrial wastewater treatment, 193–196
  municipal wastewater treatment, 185–193
Inert media, biofilm wastewater treatment technology, 171–172
Infectious bacteria
  engineered biofouling bioremediation, evaporative heat exchangers, 241–242
  implanted prostheses, engineered biofouling bioremediation, 256–257
Infrared spectroscsopy, microbiologically influenced corrosion (MIC), 312
Inhibitory contaminants
  engineered biofouling bioremediation, dairy industry contaminants, 249–250
  hazardous waste treatment, 208
    adsorbing substratum interactions, 227–229
    protection mechanisms, 211–213
    substrate protection, steady state biofilms, 217–222
Initial adhesion, bacterial interactions on porous media, 133–135

In situ bioremediation
  drinking water systems, engineered biofouling bioremediation, 252–253
  organic contaminants, metabolism of, 142–143
    engineering processes, 145–146
In situ hybridization
  food industry biofilm detection, 356–357
  molecular ecology of biofilms, 91–92
    growth and stationarity mechanisms, 98–101
    interactions, 94–95
    rRNA synthesis rates, 101–105
Interactions, molecular ecology of biofilms, 90, 94–95
Interfacial (interphase) transport, biofilm process analysis, 16
Interference reflection microscopy, food industry biofilm detection, 355
Internal diffusivity, engineered biofouling bioremediation, tubular heat exchangers, 241–242
Intradiffusion coefficient, industrial surfaces, mass transfer, engineered biofouling bioremediation, 261–262
Intraphase transport, biofilm process analysis, 16
Intrinsic bioremediation, subsurface contamination, biofilms in porous media, 144–145
Inverse fluidized bed reactors (FBR), three-phase gas-liquid-solid technology, 173–174
Iodophors, food industry biofilm elimination, disinfectants containing, 340–341
Ionic strength, bacterial interactions on porous media, 128–130, 132–135
Iron and ferrous alloys, microbiologically influenced corrosion (MIC), 284–300
  acid-producing bacteria, 293–294
  anaerobic methanogen oxidation, 291
  fungi, 294
  hydrogen-producing bacteria, 292
  metal-reducing bacteria, 202–203
  microbial consortia, 294–299
    consortium interactions, 298–299
    ennoblement, 295–298
    metal-depositing bacteria, 295
  slime-producing bacteria, 293
  sulfate-reducing bacteria (SRB), 285–291
    anodic depolarization, 289–290
    bacterial hydrogenase depolarizer, 288
    expolymer corrosive agents, 291
    hydrogenase enzyme depolarizing agent, 285–287
    hydrogen sulfate depolarizer, 288–289
    iron sulfide depolarizer, 287–288
    mineral signatures, 290
    oxygen influence, 289
    rate correlations, 291
    volatile phosphorus compound, 289
  weldments, 299–300
Iron sulfide depolarizer, microbiologically influenced corrosion (MIC), iron and ferrous alloys, sulfate-reducing bacteria (SRB), 287–288
Irreversible adhesion, defined, 55
Isopropyl β-D-thiogalactoside (IPTG), reactivity dynamics of biofilms, molecular control of detachment, 73

Kaldness moving bed biofilm reactor (MBBR), wastewater treatment technologies, 175
Kinetics
  bacterial specific adhesions, receptor-ligand interactions, 59–60
    mathematical models, 61–63
  biological transformations, 24–37
    heterogeneous reactions, 34–37
    homogeneous reactions, 24–25
    ideal reactors, 25–33
  defined, 24
  fluidized bed reactors (FBR), hydraulic models, 169–172
Klebsiella sp.
  drinking water systems, engineered biofouling bioremediation, 253
  engineered biofouling bioremediation, food preparation surfaces, 247–250

lacZ reporter gene
  biofilm metabolism monitoring, 269–270
  cell:cell signal control, biofilm reactivity, 63–64
Laminar flow, biofilm formation and performance, 48
Laser microscopy, food industry biofilm detection, 354–355
lasI gene, cell:cell signal control, biofilm reactivity, gram-negative bacteria, 64

Legionella sp.
  dental units, purified water systems, engineered biofouling bioremediation, 254–255
  drinking water systems, engineered biofouling bioremediation, 252–253
  evaporative heat exchangers, 242
  food industry biofilm formation, 331–332
Ligand binding sites, bacterial specific adhesions, receptor-ligand interactions, 58–60
Light media, biofilm wastewater treatment, submerged biofilters, 166–167, 169
Linpor sponge biofilm process
  municipal wastewater treatment, carbon removal, 188
  wastewater treatment technologies, 174–175
Listeria sp., engineered biofouling bioremediation, food preparation surfaces, 247–250
Loading rates, biofilm reactors, 179–181
Loss rates, hazardous waste treatment, protection mechanisms from excessive loss, 213–216
Lubricants, food industry biofilm formation, open processing operations, 333–334
lux gene, biofilm metabolism monitoring, 269–270

Mackinawie film, microbiologically influenced corrosion (MIC), iron and ferrous alloys, sulfate-reducing bacteria, 287–291
Macrofouling, engineered biofouling bioremediation, ship hulls, 239
Maintenance rates, reactivity dynamics of biofilms, 69
Marine environment
  engineered biofouling bioremediation:
    oil and gas platforms, 238
    ship hulls, 239
  microbiologically influenced corrosion (MIC), copper alloys in seawater, 306–307
  molecular ecology of biofilms, plasmid transfer in, 110–111
Mass-basis, stoichiometric analysis, 16
  glucose reactions, 18–19
  multiple reactions, 22–23
Mass conservation principle, continuously stirred tank reactor (CSTR) and, 26–30

Mass density measurements, biofilm formation and performance, turbulent flow mechanisms, 49
Mass transfer
  antimicrobial agents for biofilm control: penetration mechanisms, 390–392
    resistance of, 399–400
  biological transformation kinetics, heterogeneous reactions, 34–37
  industrial surfaces, engineered biofouling bioremediation, 257–262
Mass transport
  industrial surfaces, engineered biofouling bioremediation, 257–262
  three-dimensional biofilm architecture, 78–79
Mathematical models
  bacterial specific adhesions, receptor-ligand interactions, 60–63
  hazardous waste treatment, substrate protection, steady state biofilms, 217–222
  microbial growth, 67–68
Matrix biopolymers, implanted prostheses, engineered biofouling bioremediation, 257
McCarty framework, stoichiometric analysis, biological reactions, 20
Mechanism development, stoichiometric analysis, 17–19
Membrane systems, biofouling, engineered bioremediation, 242–245
  cartridge filters, 245
  microporous membranes, 245
  reverse osmosis (RO) membranes, 243–245
Metal-depositing bacteria (MDB), microbiologically influenced corrosion (MIC), iron and ferrous alloys, 295
  ennoblement process, 297–298
  MDB/SRB interactions, 298–299
Metal-reducing bacteria, microbiologically influenced corrosion (MIC), iron and ferrous alloys, 292–293
Metal surface deposits, heat exchangers, engineered biofouling bioremediation, 239–242
Methanogens
  hazardous waste treatment, excessive loss protection, 216
  microbiologically influenced corrosion (MIC), iron and ferrous alloys, anaerobic oxidation, 291

Methanol addition, municipal wastewater treatment, tertiary postdenitrification, 192–193
Michaelis Menten half saturation coefficient, antimicrobial agents for biofilm control, catalytic reactions, 387–388
Microbial aggregates
  hazardous waste treatment, 208–209
    substrata alteration of, 210–211
  microbiologically influenced corrosion (MIC), iron and ferrous alloys, 294–299
    consortium interactions, 298–299
    ennoblement, 295–298
    metal-depositing bacteria, 295
  molecular ecology of biofilms, plasmid introduction in microbial communities, 111–112
Microbial chemical energy, stoichiometric analysis, 19–22
Microbial surface components recognizing adhesion matrix molecules (MSCRAMMs), bacterial specific adhesions, receptor-ligand interactions, 58–60
Microbiological fouling, membrane system biofouling, engineered bioremediation, 243
Microbiologically influenced corrosion (MIC)
  aluminum alloys, 307–308
  copper alloys, 300–307
    MIC signatures, 301
    potable water biocorrosion, 301–306
      by-product release, 303–306
      pitting corrosion, 302–303
    seawater, 306–307
  defined, 282–283
  economic impact of, 283–284
  food industry:
    research background, 327–330
    water cooling systems, 331–332
  iron and ferrous alloys, 284–300
    acid-producing bacteria, 293–294
    anaerobic methanogen oxidation, 291
    fungi, 294
    hydrogen-producing bacteria, 292
    metal-reducing bacteria, 292–293
    microbial consortia, 294–299
      consortium interactions, 298–299
      ennoblement, 295–298
      metal-depositing bacteria, 295
    slime-producing bacteria, 293
  sulfate-reducing bacteria (SRB), 285–291
    anodic depolarization, 289–290
    bacterial hydrogenase depolarizer, 288
    expolymer corrosive agents, 291
    hydrogenase enzyme depolarizing agent, 285–287
    hydrogen sulfate depolarizer, 288–289
    iron sulfide depolarizer, 287–288
    mineral signatures, 290
    oxygen influence, 289
    rate correlations, 291
    volatile phosphorus compound, 289
  weldments, 299–300
  microorganism physiological activities and, 284
  research techniques for, 308–313
    qualitative analysis, 309–310
    quantitative analysis, 310–312
Microbiologically influenced corrosion (MIC) factor
  defined, 282–283
  iron and ferrous alloys, weldments and, 299–300
Microfouling, engineered biofouling bioremediation, ship hulls, 239
Microporous (MP) membranes, membrane system biofouling, engineered bioremediation, 242, 245
Microscopy
  engineered biofouling bioremediation monitoring, 266
  food industry biofilm detection, 353–357
    confocal microscopy, 354–355
    flow cytometry, 357
    hybridization techniques, 356–357
    interference reflection microscopy, 355
    light, epifluorescence and electron microscopy, 353–354
    polymerase chain reaction, 353–356
  microbiologically influenced corrosion (MIC), 309
Microsensors
  biofouling monitoring, 270
  microbiologically influenced corrosion (MIC), 312
Mild steel surfaces
  drinking water systems, engineered biofouling bioremediation, 251–253
  microbiologically influenced corrosion (MIC)
    acid-producing bacteria, 294
    iron sulfide depolarizer, 288

microbiologically influenced corrosion (MIC), iron and ferrous alloys, oxygen and, 289
Mineral particles
  biofilm carriers, 179
  microbiologically influenced corrosion (MIC), iron and ferrous alloys, sulfate-reducing bacteria, 290–291
Mixazur process, moving bed circulating reactors, 177
Mobil bed processes, municipal wastewater treatment, carbon removal, 187–188
Mobilization, biofilm reactivity dynamics, gene retention and transfer, 70–72
Modeling, biofilm process analysis and, 14
Molar relationship, stoichiometric analysis
  glucose reactions, 18–19, 23
  multiple reactions, 22–23
  single reaction mechanisms, 17–19
Molecular biology tools
  biofilm metabolism monitoring, 269–270
  molecular microbial ecology and, 89–90
Molecular ecology of biofilms
  bacterial growth and stationarity, 97–109
    environmental shifts, community reactions to, 108–109
    miscellaneous cellular reporters, 106–107
    mRNA growth indicator experssion in single cells, 104–106
    ribosomal RNA synthesis rates I, 101–103
    ribosomal RNA synthesis rates II, 103–104
    in situ rRNA hybridization, 98–101
  coordination, 90, 95–97
  gene transfer in communities, 109–116
    growth activity and plasmid transfer, 114–116
    natural systems, 110–111
    plasmid introduction, 111–112
    plasmid transfer on-line monitoring, 112–114
  interactions, 90, 94–95
  organism classification, 90–92
  research background, 89–90
  structural characteristics, 90, 92–94
Molecular microbial ecology, development of, 89–90
Molecular tagging techniques, molecular ecology of biofilms, 92
Molecular transport, bacterial adhesion studies, substrata preconditioning, 52–53

Momentum transport, industrial surfaces
  engineered biofouling bioremediation, 262–264
  mass transfer, engineered biofouling bioremediation, 262–264
Monitoring procedures
  engineered biofouling bioremediation, 265–271
    biofilm process monitoring, 268–270
    microscopy, 266
    multidimensional analysis, 270–271
    nondestructive monitoring, 268
    reactor design, 267
    sampling devices, 266
  food industry biofilm formation, detection of, 348–350
Monod kinetics
  homogeneous reactions, 25
  microbial growth equation, 68
Mössbauer spectroscopy, microbiologically influenced corrosion (MIC), 310
Motility, biofilm formation and performance
  quiescent conditions, 47–48
  surface topography, 50–52
Moving bed processes
  biofilm wastewater treatment technologies, 167–178
    air-lifts and circulating bed reactors, 175–177
    gel mediums, cell immobilization, 177–178
    three-phase gas-liquid-solid fluidized beds, 172–174
    turbulent and moving bed reactors, 174–175
    two-phase liquid-solid fluidized beds, 168–172
  industrial wastewater treatment, aerobic biofilm processes, 196
  municipal wastewater treatment, total nitrogen removal, 189–190
mRNA, molecular ecology of biofilms, growth indicator in single cells, 104–106
Multidimensional analysis, biofouling monitoring using, 270–271
Multiple reactions, stoichiometric analysis, 22–23
Municipal wastewater treatment
  biofilm technologies, 185–193
    carbon removal, 186–188
    tertiary nitrification, 191–192
    tertiary postdenitrification with methanol addition, 192–193
    total nitrogen removal, 188–191

Municipal wastewater treatment *(continued)*
  fixed bed processes, 161–163
  moving bed circulating reactors, 177
Mycobacterium sp.
  dental units, purified water systems, engineered biofouling bioremediation, 254–255
  drinking water systems, engineered biofouling bioremediation, 253

National Sanitation Foundation (NSF), biofilm elimination, disinfectants for, closed equipment design, 347–348
Natural bioremediation, defined, 144
Natural environments, molecular ecology of biofilms, plasmid transfer in, 110–111
Natural gels, moving bed reactors, cell immobilization, 177–178
Net accumulation rate, biofilm process analysis and, 15
Nickel compounds, microbiologically influenced corrosion (MIC), copper alloys, 301
Nitrate removal, two-phase liquid-solid fluidized bed technology, 168–172
Nitrazur process, drinking water treatment, heterotrophic denitrification, 184
Nitrification
  biofilm reactors, performance parameters, 181–182
  biofouling monitoring, multidimensional analysis, 270–271
  municipal wastewater treatment, tertiary nitrification, 191–192
Nitrifiers, hazardous waste treatment, excessive loss protection, 216
Nitrogen loading rates, biofilm reactors, 179–181
Nitrogen removal, municipal wastewater treatment, 188–191
  air-lift and circulating bed processes, 190–191
  fixed bed processes, 188–189
  immobilized cell processes, 191
  moving bed processes, 189–190
  three-phase fluidized bed process, 189
Non-Brownian diffusion, biofilm formation and performance, motility, quiescent conditions, 47–48

Nondestructive monitoring techniques, biofilm biomass monitoring, 268
Noninteracting agents, antimicrobial agents for biofilm control, penetration mechanisms, 384–386
Nonpoint source contamination phenomenon, engineered biofouling bioremediation, electronics-grade water systems, 246–247
Nonspecific adhesion interactions
  bacterial adhesion, 55–63
    colloidal approach, 56–58
    thermodynamic techniques, 56
  defined, 52–53
Nonsteady state models, hazardous waste treatment, adsorbing substratum interactions, 223–229
Normalized density, three-dimensional biofilm architecture, 77–79
Normalizing stoichiometric coefficient, stoichiometric analysis, multiple reactions, 23
Nutrient limitations, antimicrobial agents for biofilm control, 393–394
Nutritional shifts, molecular ecology of biofilms, community reponse to, 108–109

Oil and gas platforms, engineered biofouling bioremediation, 238
Oil/water separation systems, oil and gas platforms, engineered biofouling bioremediation, 238
On-line monitoring
  biofilm biomass monitoring, 268
  food industry biofilm detection, 357–358
  molecular ecology of biofilms, plasmid transfer monitoring, 112–114
Open circuit potential (OCP), microbiologically influenced corrosion (MIC), iron and ferrous alloys, ennoblement process, 295–298
Open equipment design, food industry biofilms
  elimination in, 344–345
  formation
    cooling systems, 331–332
    food processing operations, 332–334
Opportunistic pathogens, dental units, purified water systems, engineered biofouling bioremediation, 255

Organic contaminants
  membrane system biofouling, engineered bioremediation, 243
  microbial metabolism of, subsurface contamination remediation, 142–143
Organic loading rates, biofilm reactors, 179–181
Organism identification, molecular ecology of biofilms, 90–92
Orthogonal collocation, hazardous waste treatment
  adsorbing substratum interactions, 225–229
  inhibitory substrate protection, mathematical model, 219–222
Oxidant depletion, antimicrobial agents for biofilm control, 381–382
Oxidation, microbiologically influenced corrosion (MIC), iron and ferrous alloys, anaerobic oxidation by methanogens, 291
N-(3-Oxododecanoyl)-L-homoserine lactone (OdDhl), cell:cell signal control, biofilm reactivity, gram-negative bacteria, 64
Oxygen, microbiologically influenced corrosion (MIC), iron and ferrous alloys:
  ennoblement process, 297–298
  sulfate-reducing bacteria (SRB), 289
Oxygen flux, industrial surfaces, mass transfer, engineered biofouling bioremediation, 258–259

Paper mills, food industry biofilm formation, 334–335
Parallel plate flow cells, biofilm laboratories, 38
Particle flux
  biofilm formation and performance, turbulent flow, 48–49
  industrial surfaces, mass transfer, engineered biofouling bioremediation, 259–262
Particle fouling, membrane system biofouling, engineered bioremediation, 243
Pasteurization, engineered biofouling bioremediation, dairy industry contaminants, 249–250
Pathogenic bacteria
  dental units, purified water systems, engineered biofouling bioremediation, 255
  food preparation surfaces, biofouling, engineered bioremediation, 247–250
Pegasus moving bed process
  cell immobilization, 178
  municipal wastewater treatment, total nitrogen removal, 191
Penetration mechanisms, antimicrobial agents for biofilm control
  catalytic reactions, 386–388
  improvement of, 398–400
  limitations of, 398–400
  noninteracting agents, 384–386
  stoichiometric reactions, 388–389
  structural heterogeneity and mass transfer resistance, 390–392
  uncharacterized reaction, 389–390
Perforation measurements, industrial surfaces, mass transfer, engineered biofouling bioremediation, 260–262
Permeability, biofilm accumulation in porous media, hydrodynamics, 137–138
Pharmaceutical purified water, engineered biofouling bioremediation, 253–254
Pheromone-induced adhesins, cell:cell signal control, biofilm reactivity, gram-positive bacteria, 65–66
Phosphate compounds, microbiologically influenced corrosion (MIC), iron and ferrous alloys, sulfate-reducing bacteria (SRB), 289
Phosphorus compound, microbiologically influenced corrosion (MIC), iron and ferrous alloys, sulfate-reducing bacteria (SRB), 289
Phosphorus removal
  biofilm wastewater treatment, tertiary nitrification biofilters, 192
  municipal wastewater treatment, tertiary postdenitrification, methanol addition, 192–193
Physicochemical analysis, food industry biofilm detection, 357
Physiological limitations, antimicrobial agents for biofilm control, 379, 392–395
  disinfection in heterogeneous biofilm, 393
  growth rate-dependent killing in nutrient-limited film, 393–395
  improvements of, 400
  resistant phenotypes, 395
Pitting corrosion
  aluminum alloys, 308

Pitting corrosion *(continued)*
    copper and copper alloys, drinking water systems, 302–306
    potentiodynamic polarization, pitting corrosion potential (E_{pit}), 311
Plasmid conjugation
    biofilm reactivity dynamics, gene retention and transfer, 70–72
    cell:cell signal control, biofilm reactivity, gram-positive bacteria, 65–66
    hazardous waste treatment, substrata alteration of microbial aggregates, 210–211
    molecular ecology of biofilms
        bacterial growth activity and, 114–116
        communication, 96–97
        gene transfer and, 110
        natural system gene transfer, 110–111
        on-line monitoring of, 112–114
        plasmid introduction in microbial communities, 111–112
Plastic low-density materials, biofilm carriers, 179
Plug flow reactor (PFR)
    behavior, in biofilm laboratories, 39–42
    biological transformation kinetics, 30–33
Polarization resistance, microbiologically influenced corrosion (MIC), copper alloys, 305–306
Polymerase chain reaction (PCR)
    food industry biofilm detection, 355–356
    molecular ecology of biofilms, rRNA identification, 91–92
Polymer surface adhesions
    cell:cell signal control, biofilm reactivity, gram-positive bacteria, 65–66
    copper and copper alloys, pitting corrosion and, 306
Population models, biofilm process analysis and, 14–15
Porous media
    biofilms in
        accumulation mechanisms, 135–141
        hydrodynamics, 136–137
        initial deposition influence, 138–139
        permeability and hydrodynamics, 139–140
        numbers and distribution, 124–127
        research background, 123–124
        solid surface interactions, 127–129
        subsurface contamination remediation, 142–150
        bioactivated zones and biobarriers, 146–150
        bioavailability, 146–147
        intrinsic bioremediation, 144–145
        microbial metabolism of organic contaminants, 142–143
        in situ bioremediation engineering, 145–146
        transport mechanisms, 129–135
    wastewater treatment technologies, turbulent and moving-bed reactors, 175
Potentiodynamic polarization, microbiologically influenced corrosion (MIC), 310–311
Power-law expression, kinetic transformations, homogeneous reactions, 24–25
Predation losses, hazardous waste treatment, protection mechanisms, 214–216
Pre-RNA hybridization technique, molecular ecology of biofilms, rRNA synthesis rates, 102–103
Process analysis
    biofilm research, 13–16
    defined, 13
Process rate, biofilm process analysis, 16
"Promiscuous" plasmids, biofilm reactivity dynamics, gene retention and transfer, 71
Promote detachment technology, biofilm control, 374–375
Protection mechanisms, hazardous waste treatment, 211–216
    adsorbing substratum interactions, 222–229
    nonsteady state biofilm model, 223–227
    excessive loss rate protection, 213–216
    inhibition protection, 211–213
    inhibitory substrate protection, steady state biofilm, 217–222
Proteobacteria, drinking water systems, engineered biofouling bioremediation, 252–253
Protozoa accumulation, drinking water systems, engineered biofouling bioremediation, 250–253
Protozoan grazers, drinking water systems, engineered biofouling bioremediation, 253
Pseudomonas sp.
    cell:cell signal control, biofilm reactivity, 63–64

dental units, purified water systems, engineered biofouling bioremediation, 254–255
engineered biofouling bioremediation, food preparation surfaces, 247–250
microbiologically influenced corrosion (MIC), iron and ferrous alloys, metal-reduction by, 292–293
reactivity dynamics of biofilms, molecular control of detachment, 72–73
Psychrotrophic organisms, engineered biofouling bioremediation, dairy industry contaminants, 249–250
Purified water systems, engineered biofouling bioremediation, 253–255
  dental units, 254–255
  hemodialysis units, 254
  pharmaceutical water, 253–254
Pyrrhotite, microbiologically influenced corrosion (MIC), iron and ferrous alloys, sulfate-reducing bacteria, 287–291

Quadratic equations, hazardous waste treatment, inhibitory substrate protection, mathematical model, 218–219
Qualitative analysis, microbiologically influenced corrosion (MIC), 309–310
Quantitative analysis
  biofilm formation and performance, surface topography, 51–52
  biofilm susceptibility and control, 374–377
  microbiologically influenced corrosion (MIC), 310–312
  three-dimensional biofilm architecture, 76–79
Quartz crystal microbalance (QCM), biofilm biomass monitoring, 268
Quasi-steady state biofilms, drinking water systems, engineered biofouling bioremediation, 250–253
Quiescent conditions, biofilm formation and performance, 47–48
Quorum sensing
  cell:cell signal control, biofilm reactivity, gram-negative bacteria, 64
  gene expression, autoinduction of, 72
  molecular ecology of biofilms, communication, 95–97

Radial flow devices, 39
  engineered biofouling bioremediation monitoring, 267
  waste gas bioreactor, 39
Random walk modeling
  biofilm formation and performance, motility, quiescent conditions, 47–48
  three-dimensional biofilm modeling, 74–75
Rate concept, process analysis and, 14
Rate equation, microbial growth, 67–68
Rate expressions, kinetic transformations, homogeneous reactions, 24–25
Rate of appearance, stoichiometric analysis, 17–19
Reaction mechanisms, antimicrobial agents for biofilm control:
  biocide depletion on pipe lengths, 380–381
  continuous flow systems, biocide depletion, 381
  oxidant depletion from immune response, 381–383
Reaction stoichometry, equations, 16–17
Reactivity dynamics of biofilms
  architecture and global reactivity heterogeneity, 74
    three-dimensional modeling, 74–79
  cell replication, maintenance and endogenous decay, 66–69
  detachment processes, 72–73
  gene retention and transfer, 70–72
  metabolic responses to adhesion, 63–66
Reactor area measurements, heterogeneous biofilm reactions, 35–37
Receptor-ligand mediated adhesion, 58–63
  bacterial specific adhesion, 58–60
  mathematical model, 60–63
Redox potential
  biofilm nitrification rates, 182–183
  microbiologically influenced corrosion (MIC), iron and ferrous alloys, ennoblement process, 297–298
Reduced susceptibility mechanisms, antimicrobial agents for biofilm control, 375–379
  bulk fluid depletion, 378
  physiological limits of efficacy, 379
  transport limitation, 378–379
Replication
  engineered biofouling bioremediation, electronics-grade water systems, 247
  microbial growth, 66–67

Reporter genes
  biofilm metabolism monitoring, 269–270
  molecular ecology of biofilms
    interactions, 94–95
    miscellaneous cellular reports of physiological states, 106–107
    plasmid transfer monitoring, 112–114
    ribosomal RNA synthesis, 104
Residence time of cells, benificial biofilm immobilization, 4–5
Resistance factors
  antimicrobial agents for biofilm control, phenotypes for, 395
  biofilm susceptibility and control, 374–377
Respiratory burst, antimicrobial agents for biofilm control, oxidant depletion, 382–383
Reverse osmosis (RO) membranes, membrane system biofouling, engineered bioremediation, 242–245
Reversible adhesion, defined, 55
Reynolds number
  biofilm accumulation in porous media, Darcy's law, 136–137
  industrial surfaces, momentum transport, engineered biofouling bioremediation, 263–264
RHII gene, cell:cell signal control, biofilm reactivity, gram-negative bacteria, 64
Ribosomal RNA systhesis, molecular ecology of biofilms, determination of rates, 101–103
Ribosome counting, molecular ecology of biofilms, growth rate assessment, 98–101
Robbins flow devices, biofilm laboratories, 38
Rotating annular reactor, biofilm laboratories, 38
Rotating biological contactors (RBCs), biofilm wastewater treatment, fixed bed processes, 162
Roto Torque CFSTR, engineered biofouling bioremediation monitoring, 267
RP62A bacterial strain, cell:cell signal control, biofilm reactivity, gram-positive bacteria, 65–66
RpoS factor, cell:cell signal control, biofilm reactivity, gram-negative bacteria, 64
rRNA identification, molecular ecology of biofilms, 91–92
  interactions, 94–95

RT-PCR technique, molecular ecology of biofilms, mRNA growth indicator in single cells, 105–107
Rubber surfaces, engineered biofouling bioremediation, dairy industry contaminants, 249–250
Runge-Kutta algorithm, bacterial specific adhesions, receptor-ligand interactions, 63

Sampling devices
  engineered biofouling bioremediation monitoring, 266
  food industry biofilm detection, 350–352
Sanitation equipment, food industry biofilm elimination, 342
Saturation rate equation, microbial growth, 68
Scanning confocal laser microscopy. *See* Confocal scanning laser microscopy
Scanning electron microscopy (SEM)
  food industry biofilm detection, 353–354
  microbiologically influenced corrosion (MIC), 309
Seawater handling system
  microbiologically influenced corrosion (MIC): copper alloys, 306–307
  oil and gas platforms, engineered biofouling bioremediation, 238
Secondary minimum, nonspecific adhesion interactions, colloidal techniques, 57–58
Secondary substrate, hazardous waste treatment, 208
  inhibition protection mechanisms, 212–213
Sedimentation rates, biofilm formation and performance, quiescent conditions, 47–48
Sediments, bacterial accumulation statistics, 125–127
Self-assembled monolayers (SAMs), bacterial adhesion studies, conditioning film composition, 54–55
Self-inhibition kinetics, hazardous waste treatment
  protection mechanisms, 212–213
  substrate protection, 217–222
Self-inhibitory substrates, hazardous waste treatment, future development of, 229–230
Sessile bacterial cell accumulation
  cell:cell signal control, biofilm reactivity, gram-positive bacteria, 65–66

food industry biofilms, detection of, 348–349
oil and gas platforms, engineered biofouling bioremediation, 238
Shewanella putrefaciens, microbiologically influenced corrosion (MIC), iron and ferrous alloys, metal-reduction by, 292–293
Ship hulls, engineered biofouling bioremediation, 239
Signaling molecules
  antimicrobial agents for biofilm control, 401
  molecular ecology of biofilms, communication, 95–97
Silicone-based polymers, engineered biofouling bioremediation, ship hulls, 239
Slime-producing bacteria
  food industry biofilm formation
    closed processing equipment, 334
    detection of, 348–349
    open processing equipment, 332–334
    paper and cardboard mills, 334–335
  microbiologically influenced corrosion (MIC), iron and ferrous alloys, 293
    acid production by, 294
Sloughing
  defined, 4
  industrial surfaces, momentum transport, engineered biofouling bioremediation, 264
  pharmaceutical purified water, engineered biofouling bioremediation, 253–254
Slow-growing species, hazardous waste treatment, excessive loss protection, 216
Soil analysis
  bacterial accumulation statistics, 125–127
  food industry biofilm elimination, 343–345
  subsurface contamination, biofilms in porous media, bioavailability and, 146–147
Solid-liquid fluidization, fluidized bed reactors (FBR), 169–172
Solid surfaces, bacterial interactions on porous media, 127–129
Solute characteristics, bacterial transport in porous media, 131–135
Specific adhesion interactions. *See also* Bacterial adhesion studies
  defined, 52, 54
  receptor-ligand mediated adhesion, 58–63
    bacterial specific adhesion, 58–60
    mathematical model, 60–63
Spectroscopic techniques, biofilm analysis, 8
Stainless steel surfaces
  food industry biofilm formation, open processing operations, 333–334
  food preparation contaminants, biofouling, engineered bioremediation, 248–250
  microbiologically influenced corrosion (MIC), iron and ferrous alloys
    metal-depositing bacteria, 295
    slime-producing bacteria, 293
    weldments, 299–300
Staphyloccocus epidermidis
  cell:cell signal control, biofilm reactivity, gram-positive bacteria, 65–66
  implanted prostheses, engineered biofouling bioremediation, 256–257
Starvation survival
  antimicrobial agents for biofilm control, physiological limitations, 379
  bacterial transport in porous media, 130–135
  microbial growth, 67
Stationarity mechanisms, molecular ecology of biofilms, 97–109
  environmental shifts, community reactions to, 108–109
  miscellaneous cellular reporters, 106–107
  mRNA growth indicator experssion in single cells, 104–106
  ribosomal RNA synthesis rates I, 101–103
  ribosomal RNA synthesis rates II, 103–104
  in situ rRNA hybridization, 98–101
Steady state biofilm models
  hazardous waste treatment
    protection mechanisms, 212–213
    substrate protection, 217–222
  heterogeneous biofilm reactions, 35–37
Steric stabilization
  bacterial interactions on porous media, 128–129
  nonspecific adhesion, 52–53
  colloidal techniques, 58
Stoichiometry
  antimicrobial agents for biofilm control, penetration mechanisms, 388–389
  biological processes, 16–23
    multiple reactions, 22–23
    reaction estimations, 19–22
    single reaction mechanisms, 17–19
Stokes radius, industrial surfaces, mass transfer, engineered biofouling bioremediation, 260–262

"Stop growth" technology, biofilm control, 374–375
Stratification mechanisms, molecular ecology of biofilms, growth rate assessment, 100–101
Strength of inhibition, hazardous waste treatment, inhibitory substrate protection, mathematical model, 218–222
Streptococcus sp., engineered biofouling bioremediation, dairy industry contaminants, 249–250
Structural heterogeneity, antimicrobial agents for biofilm control, penetration mechanisms, 390–392
Structural organization, molecular ecology of biofilms, 90, 92–94
Submerged biofilters, biofilm wastewater treatment
  heavy granular media, fixed bed processes, 162–166
  light and floating media, fixed bed processes, 166–167, 169
Substrata preconditioning, bacterial adhesion studies, 52–55
  conditioning film
    adsorption, 53–54
    composition, 54–55
    influence on substratum properties, 55
  conditioning film adsorption, 53–54
  molecular transport, 52–53
Substrate removal rates, continuously stirred tank reactor (CSTR) and, 28–30
Substratum
  bacterial interactions on porous media, surface-substratum interactions, 127–129
  biofilm formation and performance, surface topography, 50–52
  defined, 3–4
  hazardous waste treatment
    adsorbing substratum interactions, 222–229
    microbial aggregates, 209
    microorganism changes, 210–211
Subsurface contamination remediation, biofilms in porous media, 142–150
  bioactivated zones and biobarriers, 146–150
  bioavailability, 146–147
  intrinsic bioremediation, 144–145
  microbial metabolism of organic contaminants, 142–143

in situ bioremediation engineering, 145–146
Sulfate-reducing bacteria (SRB):
  microbiologically influenced corrosion (MIC)
    copper and copper alloys, 301
    iron and ferrous alloys, 285–291
      amorphic iron (III) oxyhydroxide reduction, 292–293
      anodic depolarization, 289–290
      bacterial hydrogenase depolarizer, 288
      expolymer corrosive agents, 291
      hydrogenase enzyme depolarizing agent, 285–287
      hydrogen sulfate depolarizer, 288–289
      iron sulfide depolarizer, 287–288
      mineral signatures, 290
      oxygen influence, 289
      rate correlations, 291
      volatile phosphorus compound, 289
    Mössbauer spectroscopy, 310
  microbiologically influenced corrosion (MIC), iron and ferrous alloys, MDB/SRB interactions, 298–299
  oil and gas platforms, engineered biofouling bioremediation, 238
Sulfides, microbiologically influenced corrosion (MIC), iron and ferrous alloys, hydrogen-producing bacteria, 292
Sulfur-oxidizing bacteria (SOB), microbiologically influenced corrosion (MIC), iron and ferrous alloys, 293–294
Surface area parameters, industrial surfaces, mass transfer, engineered biofouling bioremediation, 258
Surface biofilms, food preparation surfaces, biofouling, engineered bioremediation, 247–250
Surface enlargement, three-dimensional biofilm architecture, 77
Surface roughness
  engineered biofouling bioremediation
    food preparation surfaces, 248–250
    tubular heat exchangers, 241–242
  industrial surfaces, mass transfer, engineered biofouling bioremediation, 258
  three-dimensional biofilm architecture, 78–79
Surface shear stress, biofilm formation and performance, surface topography, 50–52
Surface spectroscopy, microbiologically influenced corrosion (MIC), 309–310

Surface topography, biofilm formation and performance, 49–52
Surfactants, food industry biofilms, open processing, elimination in, 344–345
Suspension testing, food industry biofilm elimination using, 336–337
Synthetic gels, moving bed reactors, cell immobilization, 177–178
System rates, biofilm process analysis, 15–16

Tensides, food industry biofilm elimination, cleansers containing, 338
Tertiary nitrification, municipal wastewater treatment, 191–192
Tessier kinetics, homogeneous reactions, 25
Thermophoresis, biofilm formation and performance, turbulent flow mechanisms, 49
Thickness variables
    biofilm accumulation in porous media, hydrodynamics and permeability, influence on, 139–141
    engineered biofouling bioremediation, tubular heat exchangers, 241–242
Thiele modulus, heterogeneous biofilm reactions, 36–37
Three-dimensional architecture, biofilm modeling, 74–79
    derivation techniques, 75–76
    quantitative descriptors, 76–79
Three-phase gas-liquid-solid technology, fluidized bed reactors (FBR):
    biofilm wastewater treatment, 172–174
    municipal wastewater treatment, total nitrogen removal, 189
Topax 12 agent, food industry biofilms, open processing, elimination in, 345
Toxic contaminants, hazardous waste treatment, 208
Transcription rates, molecular ecology of biofilms, rRNA synthesis, 101–103
Transport limitation resistance mechanism, antimicrobial agents for biofilm control, penetration limitation, 379–380
Transport rates
    bacterial transport in porous media, 129–135
    biofilm process analysis, 16
Trickling filters, biofilm wastewater treatment, fixed bed processes, 161–163
Tubular heat exchangers, engineered biofouling bioremediation, 240–242

TURBO N reactor
    moving bed circulating reactors, 176–177
    municipal wastewater treatment, total nitrogen removal, 190
    nitrification performance, 181–182
Turbulent bed reactors, biofilm wastewater treatment technologies, 174–175
"Turbulent bursts," biofilm formation and performance, turbulent flow mechanisms, 49
Turbulent flow
    biofilm formation and performance, 48–49
    food industry biofilm, elimination in closed equipment, 346–348
Two-phase liquid-solid fluidized beds, biofilm wastewater treatment technologies, moving bed processes, 168–172
Two-photon confocal microscopy, molecular ecology of biofilms, structural organization, 93–94

Ultrafiltration (UF) membranes, membrane system biofouling, engineered bioremediation, 242, 245
Ultramicrobacteria (UMB), subsurface contamination remediation, biofilms in porous media, 147–150
Ultrasonics, food industry biofilm detection, 350–352
Uncharacterized reations, antimicrobial agents for biofilm control, penetration mechanisms, 389–390
Unit-mass based relationship, stoichiometric analysis
    glucose reactions, 18–19
    single reaction mechanisms, 17–19
Unsteady state models, heterogeneous biofilm reactions, 36–37
Up-flow filtration
    biofilm wastewater treatment, heavy granular media, fixed bed processes, 163–166
    fluidized bed reactors (FBR), three-phase gas-liquid-solid technology, 173–174

Van der Waals forces
    nonspecific adhesion, 52–53
    nonspecific adhesion interactions, 57

Ventilation and air handling systems, food industry biofilm formation, 332
Viable but nonculturable (VBNC) state, drinking water systems, engineered biofouling bioremediation, 252–253
Viscoelasticity measurements, industrial surfaces, momentum transport, engineered biofouling bioremediation, 263–264
Volatile phosphorus, microbiologically influenced corrosion (MIC), iron and ferrous alloys, sulfate-reducing bacteria (SRB), 289
Volume parameters, industrial surfaces, mass transfer, engineered biofouling bioremediation, 258–259

Wastewater treatment
  biofilm reactor configurations and, 5
  biofilm technologies
    advantages of, 197–198
    fixed bed processes, 161–167
      granular media, submerged biofilter, 162–166
      light and floating media, submerged biofilter, 166–167
    industrial applications, 183–196
      drinking water treatment, 183–185
      industrial wastewater treatment, 193–196
      municipal wastewater treatment, 185–193
    innovative design, 178–183
      biofilm carriers, 179
      denitrification performance parameters, 183
      loading rates, 179–181
      nitrification performance parameters, 181–182
    moving bed processes, 167–178
      air-lifts and circulating bed reactors, 175–177
      gel mediums, cell immobilization, 177–178
      three-phase gas-liquid-solid fluidized beds, 172–174
      turbulent and moving bed reactors, 174–175
      two-phase liquid-solid fluidized beds, 168–172
    research background, 159–161
Water circulation systems, food industry biofilm formation, 331–332
Water transport limits, membrane system biofouling, engineered bioremediation, reverse osmosis (RO) membranes, 244
Weight loss measurements, microbiologically influenced corrosion (MIC), 310
Weldments, microbiologically influenced corrosion (MIC), iron and ferrous alloys, 299–300

X-ray diffraction (XRD), microbiologically influenced corrosion (MIC)
  iron and ferrous alloys, sulfate-reducing bacteria, 290–291
  qualitative analysis, 309–310
X-ray photoelectron spectroscopy (XPS), microbiologically influenced corrosion (MIC), copper alloys, 305–306

Zero-order reaction kinetics, antimicrobial agents for biofilm control, catalytic reactions, 386–388
Zero-resistance ammetry (ZRA), microbiologically influenced corrosion (MIC), 311–312
Zygotic induction system, molecular ecology of biofilms, plasmid transfer monitoring, 112–114